Strigolactone in Plants

A Regulator for Plant Growth and Development

Strigolactone in Plants

A Regulator for Plant Growth and Development

Edited by

Muhammad Waseem
School of Breeding and Multiplication (Sanya Institute of Breeding and Multiplication), Hainan University, Sanya, China

Pingwu Liu
School of Breeding and Multiplication (Sanya Institute of Breeding and Multiplication), Hainan University, Sanya, China

Channapatna S. Prakash
College of Arts and Sciences at Tuskegee University, USA

Sajid Fiaz
Institute of Molecular Biology and Biotechnology (IMBB), The University of Lahore, Punjab, Pakistan

CABI

CABI is a trading name of CAB International

CABI
Nosworthy Way
Wallingford
Oxfordshire OX10 8DE
UK
Tel: +44 (0)1491 832111
E-mail: info@cabi.org
Website: www.cabi.org

CABI
200 Portland Street
Boston
MA 02114
USA
T: +1 (617)682-9015
E-mail: cabi-nao@cabi.org

A catalogue record for this book is available from the British Library, London, UK.

Library of Congress Cataloging-in-Publication Data

Names: Waseem, Muhammad, 1988- editor | Liu, Pingwu, 1980- editor | Prakash, Channapatna S. editor | Fiaz, Sajid editor Title: Strigolactone in plants / edited by Muhammad Waseem, Pingwu Li, Channapatna S. Prakash, Sajid Fiaz.
Description: First edition. | Boston : CAB International, 2026. | Includes bibliographical references and index. | Summary: "A comprehensive reference covering the latest advancements in strigolactone research, which provides insights into how these hormones can be leveraged to enhance plant growth and stress tolerance"– Provided by publisher.
Identifiers: LCCN 2026008424 | ISBN 9781836990574 hardcover | ISBN 9781836990581 ebook | ISBN 9781836990598 epub
Subjects: LCSH: Plant hormones | Plants–Effect of stress on
Classification: LCC SB128 .S76 2026
LC record available at https://lccn.loc.gov/2026008424

ISBN-13: 9781836990574 (hardback)
9781836990581 (ePDF)
9781836990598 (ePub)

DOI: 10.1079/9781836990598.0000

Commissioning Editor: Rebecca Stubbs
Editorial Assistant: Theresa Regueira
Production Editor: Helen Elliott

Typeset by Exeter Premedia Services Pvt Ltd, Chennai, India
Printed and bound in the UK by CPI Group (UK) Ltd, Croydon, CR0 4YY

Contents

Contributors

Banzeer Ahsan Abbasi, Department of Botany, Rawalpindi Women University, 6th Road, Satellite Town, Rawalpindi 46300, Pakistan. Email: benazirahsanabbasi786@gmail.com

Zainul Abideen, Dr. Muhammad Ajmal Khan Institute of Sustainable Halophyte Utilization, University of Karachi, Karachi, Pakistan; Sharjah Seed Bank and Herbarium, Environment and Protected Areas Authority, Sharjah, United Arab Emirates. Email: zuabideen@uok.edu.pk

Farooq Ahmad, Department of Botany, University of Agriculture Faisalabad, Faisalabad 38040, Pakistan. Email: farooqbot@yahoo.com

Muhammad Sajid Aqeel Ahmad, Department of Botany, University of Agriculture Faisalabad, Faisalabad 38040, Pakistan. Email: sajidakeel@yahoo.com

Kamran Akbar, Department of Botany, Bacha Khan University, Charsadda, 24420, Khyber Pakhtunkhwa, Pakistan. Email: durrani_kamran0911@yahoo.com

Naila Ali, Institute of Molecular Biology and Biotechnology, University of Lahore, Lahore, Pakistan. Email: naila.ali@imbb.uol.edu.pk

Amina Ameer, College of Geographical Sciences, Faculty of Geographical Science and Engineering, Henan University, Zhengzhou, 450046, China. Email: amnaameer127@gmail.com

Muhammad Anas, Department of Soil Science & Plant Nutrition, Selcuk University Campus, Konya 42079, Turkey; Department of Plant Sciences, Faculty of Biological Sciences, Quaid-I-Azam University, Islamabad 45320, Pakistan. Email: muhammed.anas@selcuk.edu.tr

Aribah, Department of Botany, University of Karachi, Karachi, Pakistan. Email: 22aribah@gmail.com

Shikha Arora, Ranjan Plant Physiology and Biochemistry Laboratory, Department of Botany, University of Allahabad, Prayagraj 211002, India. Email: Shikha.arora111@gmail.com

Muhammad Munawar Arshad, Institute of Molecular Biology and Biotechnology, University of Lahore, Lahore, Pakistan. Email: ranamunawar5151@gmail.com

Ansa Asghar, Department of Botany, University of Agriculture Faisalabad, Faisalabad 38040, Pakistan. Email: ansaasghar1122@gmail.com

Asima Asghar, Department of Zoology, Lahore College for Women University, Lahore, Pakistan. Email: hunjraasma@gmail.com

Ayesha, Institute of Molecular Biology and Biotechnology, University of Lahore, Lahore. Email: ayesha@imbb.uol.edu.pk

K.S. Baig, Vasantrao Naik Marathwada Krishi Vidyapeeth, Parbhani 431402, Maharashtra, India. Email: baig.vnmkv@gmail.com

Sana Basharat, School of Breeding and Multiplication (Sanya Institute of Breeding and Multiplication), School of Tropical Agriculture and Forestry, Hainan University, Sanya 572025, China. Email: sanabasharat143@gmail.com

Shivangi Chaudhary, Ranjan Plant Physiology and Biochemistry Laboratory, Department of Botany, University of Allahabad, Prayagraj 211002, India. Email: shivangichaudhary0013@gmail.com

Noor Fatima, Department of Basic and Applied Chemistry, Faculty of Science and Technology, University of Central Punjab, Lahore 54590, Pakistan. Email: nf9026607@gmail.com

Syeda Noureen Fatima, Department of Botany, Faculty of Natural Sciences, Government College Women University, Sialkot 51310, Pakistan. Email: snfatima678@gmail.com

Sanjay Gairola, Sharjah Seed Bank and Herbarium, Environment and Protected Areas Authority, Sharjah, United Arab Emirates. Email: sanjay.prasa@uodh.ac.ae

Amjad Shahzad Gondal, Department of Plant Pathology, Bahauddin Zakariya University, Multan 60800, Pakistan. Email: amjadshahzad@bzu.edu.pk

Dasari Venu Gopal, Loyola Academy, Degree & P.G. College, Hyderabad, Telangana, India. Email: venu.junnu261@gmail.com

Aneesa Gul, Department of Botany, Bacha Khan University, Charsadda, 24420, Khyber Pakhtunkhwa, Pakistan. Email: aneesagul2789@gmail.com

Erdogan E. Hakki, Department of Soil Science & Plant Nutrition, Selcuk University Campus, Konya 42079, Turkey. Email: eehakki@selcuk.edu.tr

Mansoor Hameed, Department of Botany, University of Agriculture Faisalabad, Faisalabad38040, Pakistan. Email: hameedmansoor@yahoo.com

Sameel Hassan, Department of Agriculture & Forest Science (DAFNE), University of Tuscia, Viterbo 01100, Italy. Email: sameel.hassan@studenti.unitus.it

Umer Hayat, Department of Life Sciences, Khwaja Fareed University of Engineering and Information Technology, Rahim Yar Khan, Pakistan. Email: umerhayatbot@gmail.com

Tajammul Husain, Ranjan Plant Physiology and Biochemistry Laboratory, Department of Botany, University of Allahabad, Prayagraj 211002, India. Email: tajammulhusain198@gmail.com

Waleed Hussain, Department of Botany, University of Narowal, Pakistan. Email: mwaleed@gmail.com

Sajjad Hyder, Department of Botany, Faculty of Natural Sciences, Government College Women University, Sialkot 51310, Pakistan. Email: sajjad.hyder@gcwus.edu.pk

Aliya Ibrar, Department of Chemistry, Faculty of Physical and Applied Sciences, University of Haripur, Haripur, KPK 22620, Pakistan. Email: aliya.ibrar@uoh.edu.pk

Muhammad Ilyas, Institute of Molecular Biology and Biotechnology, University of Lahore, Lahore, Pakistan. Email: ilyas7149@gmail.com

Javed Iqbal, Department of Botany, Bacha Khan University, Charsadda 24420, Khyber Pakhtunkhwa, Pakistan. Email: Email: javed89qau@gmail.com

Mudassir Iqbal, Department of Plant Protection Biology, Swedish University of Agricultural Sciences, Lomma, Sweden. Email: mudassir.iqbal@slu.se

Ummar Iqbal, College of Biological Sciences, China Agricultural University, Beijing 100083, China. Email: ummariqbal@yahoo.com

Hina Jabeen, Department of Biotechnology, Women University Mardan, Mardan 23200, Khyber Pakhtunkhwa, Pakistan. Email: dr.hinajabeen@wumardan.edu.pk

Anum Javaid, Department of Botany, University of Agriculture Faisalabad, Faisalabad 38040, Pakistan. Email: anumjavaid729@gmail.com

Hirakant V. Kalpande, Department of Agricultural Botany, Vasantrao Naik Marathwada Krishi Vidyapeeth, Parbhani 431402, Maharashtra, India. Email: hvkalpande@gmail.com

Razia Kalsoom, Pakistan Council of Scientific and Industrial Research (PCSIR) Labs, Plot No. 16, Sector H-9, Islamabad, Pakistan. Email: raziawasif80@gmail.com

Rizwana Kanwal, Department of Botany, University of Agriculture Faisalabad, Faisalabad 38040, Pakistan. Email: rizwanakanwal653@gmail.com

Aimen Razzaq Khan, Department of Botany, Faculty of Natural Sciences, Government College Women University, Sialkot 51310, Pakistan. Email: aimenkhan34603@gmail.com

Ameer Khan, Department of Botany, Division of Science and Technology, University of Education, Lahore, Pakistan. Email: ameer.khan@ue.edu.pk

Amjid Khan, UNESCO UNISA Africa Chair in Nanoscience and Nanotechnology, College of Graduate Studies, University of South Africa, Pretoria, South Africa. Email: amjidniazi8@gmail.com

Imtiaz Khan, Department of Chemistry, Manchester Institute of Biotechnology, University of Manchester, 131 Princess Street, Manchester M1 7DN, UK. Email: kimtiaz@hotmail.co.uk

Waseem Ahmed Khattak, Department of Plant Sciences, Faculty of Biological Sciences, Quaid-I-Azam University, Islamabad 45320, Pakistan. Email: waseemkhattack@bs.qau.edu.pk

G. Harish Kumar, Department of Genetics and Plant Breeding, School of Agriculture, SR University, Warangal 506371, Telangana, India. Email: harishkumargollangi@gmail.com

Kewat Sanjay Kumar, Ranjan Plant Physiology and Biochemistry Laboratory, Department of Botany, University of Allahabad, Prayagraj 211002, India. Email: kewatsk@allduniv.ac.in

Zunaira Malik, Institute of Molecular Biology and Biotechnology, University of Lahore, Lahore, Pakistan. Email: zunairamalik@gmail.com

Sana Maqsood, Department of Botany, University of Agriculture Faisalabad, Faisalabad 38040, Pakistan. Email: sanauaf98@yahoo.com

Saba Mazhar, Department of Botany, Faculty of Natural Sciences, Government College Women University, Sialkot 51310, Pakistan. Email: sabamazhar0786@gmail.com

Mohammad Mehdizadeh, Department of Agronomy and Plant Breeding, Faculty of Agriculture and Natural Resources, University of Mohaghegh Ardabili, Ardabil, Iran. Email: mehdizade.mohammad@gmail.com

Samavia Mobeen, School of Breeding and Multiplication (Sanya Institute of Breeding and Multiplication), School of Tropical Agriculture and Forestry, Hainan University, Sanya 572025, China. Email: samawia@gmail.com

Mashal Nadeem, Department of Botany, Faculty of Natural Sciences, Government College Women University, Sialkot 51310, Pakistan. Email: mashalrajput786@gmail.com

Vasavi Nagula, Loyola Academy, Degree & P.G. College, Hyderabad, Telangana, India. Email: vasavi1610nagula@gmail.com

Sabavat Raju Naik, International Crops Research Institute for Semi-Arid Tropics, Patancheru, Hyderabad, Telangana 502 324, India; Department of Genetics and Plant Breeding, Vasantrao Naik Marathwada Krishi Vidyapeeth, Parbhani, Maharashtra 431 402, India. Email: rajuplantbreeder@gmail.com

Hummera Nawaz, Department of Botany, Division of Science and Technology, University of Education, Lahore, Pakistan. Email: hummera.nawaz@ue.edu.pk

Salma Noureen, Department of Botany, Bacha Khan University, Charsadda, 24420, Khyber Pakhtunkhwa, Pakistan. Email: salmanoreen738@gmail.com

Emmaculate Akinyi Ogutu, Department of Agricultural Production, College of Agriculture and Environmental Sciences, Makerere University, PO Box 7062, Kampala, Uganda. Email: emmaculateogutu@gmail.com

Fannie Isela Parra-Cota, Campo Experimental Norman E. Borlaug, Instituto Nacional de Investigaciones Forestales, Agrícolas y Pecuarias (INIFAP), Norman E. Borlaug Km. 12, C.P. 85000, Cd. Obregón, Sonora, Mexico. Email: fipc04@gmail.com

Zahida Parveen, Department of Botany, University of Agriculture, Faisalabad 38040, Pakistan. Email: zahidaperveen811@gmail.com

Om Sajeevan Patel, Ranjan Plant Physiology and Biochemistry Laboratory, Department of Botany, University of Allahabad, Prayagraj 211002, India. Email: ompatel8115@gmail.com

Sheo Mohan Prasad, Ranjan Plant Physiology and Biochemistry Laboratory, Department of Botany, University of Allahabad, Prayagraj 211002, India. Email: profsmprasad@gmail.com

Shahzadi Rafique, Department of Botany, Faculty of Natural Sciences, Government College Women University, Sialkot 51310, Pakistan. Email: shahzadirafique0@gmail.com

Ahmed Raza, Department of Botany, University of Agriculture Faisalabad, Faisalabad 38040, Pakistan. Email: ahmadmsm37@gmail.com

Rakhshinda Sadiq, Department of Biotechnology, Women University Mardan, Khyber Pakhtunkhwa, Pakistan. Email: rakhshinda_sadiq@wumardan.edu.pk

Wajid Saeed, School of Breeding and Multiplication (Sanya Institute of Breeding and Multiplication), School of Tropical Agriculture and Forestry, Hainan University, Sanya 572025, China. Email: wajid@hainanu.edu.cn

Saifullah, Department of Botany, University of Agriculture, Faisalabad, Pakistan. Email: saif1600686@gmail.com

Sergio de los Santos-Villalobos, Departamento de Ciencias Agronómicas y Veterinarias, Instituto Tecnológico de Sonora, Ciudad Obregón, Sonora, Mexico. Email: sergio.delossantos@itson.edu.mx

Fouzia Shabbir, Department of Botany, University of Agriculture Faisalabad, Faisalabad 38040, Pakistan. Email: fouziashabbir1996@gmail.com

Jazab Shafqat, Department of Botany, University of Agriculture Faisalabad, Faisalabad 38040, Pakistan. Email: jazabshafqat0308@gmail.com

Syed Mohsan Raza Shah, Department of Plant Pathology, Bahauddin Zakariya University, Multan, 60800, Pakistan. Email: syed.mohsan@ue.edu.pk

Nimra Shahid, Department of Botany, University of Narowal, Pakistan. Email: nimra.shahid@gmail.com

Nadia Sharif, Department of Biotechnology, Women University Mardan, Mardan 23200, Khyber Pakhtunkhwa, Pakistan. Email: dnadiasharif@wumardan.edu.pk

Muhammad Faheem Siddiqui, Department of Botany, University of Karachi, Karachi, Pakistan. Email: mfsiddiqui@uok.edu.pk

Sobia, Department of Botany, Bacha Khan University, Charsadda, 24420, Khyber Pakhtunkhwa, Pakistan. Email: Khaliqsobia7@gmail.com

Muhammad Sufyan, College of Biological Sciences, China Agricultural University, Beijing 100083, China. Email: sufyan11.uaf@yahoo.com

Ashwini Talakayala, International Crops Research Institute for Semi-Arid Tropics, Patancheru, Hyderabad, Telangana 502 324, India. Email: Ashwini.Talakayala@icrisat.org

Usman Tufail, Department of Botany, University of Agriculture Faisalabad, Faisalabad 38040, Pakistan. Email: usman3383ar@gmail.com

Mushfaq Ul-Nisa, Department of Botany, Faculty of Natural Sciences, Government College Women University, Sialkot 51310, Pakistan. Email: mushafaq.tajammal@gmail.com

Shabir H. Wani, National Seed Project Laboratory, Sher-e-Kashmir University of Agricultural Sciences and Technology of Kashmir, Shalimar, Srinagar, Jammu and Kashmir, 190025, India. Email: shabirhwani@skuastkashmir.ac.in

Narkhede Gopal Wasudeo, Department of Genetics and Plant Breeding, School of Agriculture, SR University, Warangal 506371, Telangana, India. Email: n.gopalwasudeo@sru.edu.in

Tabassum Yaseen, Department of Botany, Bacha Khan University, Charsadda, 24420, Khyber Pakhtunkhwa, Pakistan. Email: tabassumyaseen@bkuc.edu.pk

Ayesha Zafar, Department of Botany, University of Agriculture Faisalabad, Faisalabad 38040, Pakistan. Email: Ayesharana4050@gmail.com

Sadia Zafar, Department of Botany, Division of Science and Technology, University of Education, Lahore, 54770 Punjab, Pakistan. Email: sadia.zafar@ue.edu.pk

Sumera Zaib, Department of Basic and Applied Chemistry, Faculty of Science and Technology, University of Central Punjab, Lahore 54590, Pakistan. Email: sumera.zaib@ucp.edu.pk

Zainab Zaib, Department of Dermatology, Combined Military Hospital Abbottabad, 22010, Pakistan. Email: zaibzainab092@gmail.com

Zaib-un-Nisa, Institute of Molecular Biology and Biotechnology, University of Lahore, Lahore, Pakistan. Email: zaib.nisa@imbb.uol.edu.pk

Farishta Zarshan, Department of Botany, Bacha Khan University, Charsadda, 24420, Khyber Pakhtunkhwa, Pakistan. Email: farishta24bkuc@gmail.com

Preface

Strigolactones (SLs) have evolved from obscure root exudates into central regulators of plant development, stress response, and environmental adaptation. This book compendium reflects the maturation of SL biology into a discipline that not only interrogates fundamental signaling mechanisms but also translates molecular insight into agricultural innovation. This book charts the trajectory of SL research from chemical diversity and biosynthetic pathways to the multi-faceted roles of SLs in development, abiotic stress resilience, and immunity.

The first part anchors SLs at the nexus of plant growth and morphogenesis. It opens with foundational chapters on SL discovery, structural diversity, and biosynthetic machinery, delineating canonical (ABC-D) and non-canonical (D-ring-containing) variants and their generation via DWARF27, CCD7/8, MAX1, and related enzymes. Developmental functions are extensively treated, encompassing axillary bud suppression, root system architecture modulation, and SL coordination with arbuscular mycorrhizal symbioses. Redox and nitrogen signaling chapters explore reciprocal regulation between SLs and reactive oxygen/nitrogen species, revealing redox as a crucial code in hormonal cross-talk. Hormonal integration is a pervasive theme, with SLs interacting dynamically with auxin, cytokinins, abscisic acid, ethylene, gibberellins, jasmonates, and salicylic acid. The book culminates in analyses of SL-mediated control over photosynthesis and respiration, thereby bridging molecular regulation and whole-plant metabolism.

The second part expands this foundation into the domain of stress biology and plant–microbe interactions. It compiles case studies and integrative reviews that position SLs as mediators of abiotic stress responses—drought, salinity, cold, and waterlogging—and as modulators of nutrient-use efficiency and pathogen defense. Individual chapters link SL biosynthesis and signaling to antioxidant induction, hormonal reprogramming, and architectural adaptation under environmental duress. Mechanistic insights are augmented by functional genomic approaches, including CRISPR-Cas9, transcriptomics, metabolomics, and physiological phenotyping. Notably, the book explores SLs as agents of immune priming against fungal, bacterial, and viral pathogens, illuminating their roles in balancing growth and defense within diverse ecological and agronomic contexts. Within this book, several unifying threads emerge:

- *Chemical diversity as a determinant of functional specificity.* Structural variation in SLs underpins their tissue selectivity, environmental responsiveness, and species-specific phenotypic outcomes.

- *Network logic over linearity*. SL signaling is deeply embedded within a multi-layered hormonal and redox communication system, wherein outcomes emerge from context-sensitive crosstalk rather than isolated pathways.
- *From mechanistic understanding to agronomic application*. Advances in biosensors, synthetic analogs, and translational phenotyping provide actionable insights into how SLs can inform ideotype breeding, optimize symbiotic associations, and improve stress resilience.

This collection is intended for researchers, advanced students, breeders, and agronomists seeking a comprehensive, mechanistic, and application-oriented understanding of SL biology. The contributors have integrated diverse methodologies and disciplinary lenses, while candidly acknowledging current knowledge gaps, particularly in spatial–temporal signaling logic, ecological ramifications of SL manipulation, and genotype-by-environment interactions.

We express our profound gratitude to the authors for their intellectual rigor, to the reviewers for enhancing clarity and coherence, and to our institutions and families for unwavering support. It is our hope that this book resource will not only consolidate current knowledge but also catalyze the next generation of discoveries, linking molecular mechanisms to sustainable agricultural practices under a changing climate.

Editors
Muhammad Waseem
Pingwu Liu
Channapatna S. Prakash
Sajid Fiaz

1 Strigolactone Discovery and Divergent Biosynthetic Pathways in Plants

Sumera Zaib[1]*, Imtiaz Khan[2], Noor Fatima[1], Zainab Zaib[3] and Aliya Ibrar[4]
[1]Department of Basic and Applied Chemistry, Faculty of Science and Technology, University of Central Punjab, Lahore, Pakistan; [2]Department of Chemistry, Manchester Institute of Biotechnology, University of Manchester, Manchester, UK; [3]Department of Dermatology, Combined Military Hospital Abbottabad, Pakistan; [4]Department of Chemistry, Faculty of Physical and Applied Sciences, University of Haripur, Haripur, Pakistan

Abstract
Strigolactones (SLs) are a unique class of plant hormones with their origin in carotenoids, and were initially found to be germination stimulants for the parasitic plants of the genera *Striga* and *Orobanche*, but subsequently have been found to play more general regulatory functions in plant growth and adaptation to the environment. They take part in fundamental developmental processes including the development of root architecture, repression of shoot branching, and induction of symbiotic relationships with arbuscular mycorrhizal fungi. Non-canonical SLs are ringless but nevertheless possess the necessary D-ring, whereas canonical SLs are identified by their tricyclic ABC-ring structure. Based on species-specific modifications including methylation and hydroxylation, approximately 25 structurally distinct variants have been identified. Analytical methods such as liquid chromatography-tandem mass spectrometry have enabled the precise detection of SLs, and improved our understanding of their biosynthesis with the enzymes DWARF27, carotenoid cleavage dioxygenase 7, and carotenoid cleavage dioxygenase 8. As well as plant growth, SLs help plants response to nutrient deficiency and drought stress, and have other possible agricultural applications, particularly in controlling parasitic weeds by means of synthetic analogs of SLs (e.g. GR24), signifying their evolutionary and ecological significance.

Keywords: arbuscular mycorrhizal fungi, carotenoids, canonical strigolactones, non-canonical strigolactones, parasitic weeds, strigolactones

1.1 Introduction

Strigolactones (SLs) represent a fascinating class of plant hormones that have gained significant scientific attention over the past decade. Initially identified in the 1960s as germination stimulants for parasitic weeds in the genera *Striga* and *Orobanche*, these compounds derived from carotenoids are now recognized as crucial multi-functional signaling molecules with diverse roles in plant development and environmental adaptation (Waters *et al.*, 2017). The chemical structure of canonical SLs consists of

*Corresponding author: sumera.zaib@ucp.edu.pk

DOI: 10.1079/9781836990598.0001

tricyclic lactone (ABC-ring) connected *via* an enol-ether bridge to a butanolide group (D-ring), which serves as the moiety determining their bioactivity (Ito, 2023). The general structure of SLs is shown in Fig. 1.1.

The discovery of SLs began with the isolation of strigol from root exudates of *Gossypium hirsutum* (cotton) as a compound that stimulated the germination of the parasitic weed *Striga lutea* (Cook *et al.*, 1966). However, SLs were first identified as endogenous plant hormones in the early 2000s and were believed to regulate shoot branching inhibition (Gomez-Roldan *et al.*, 2008; Umehara *et al.*, 2008). This discovery marked the beginning of extensive research into the molecular, biochemical, and physiological aspects of SLs.

More than 25 naturally occurring SLs have been discovered and classified into two main structural groups: canonical and non-canonical SLs (Jia *et al.*, 2018). Canonical SLs include strigol-type and orobanchol-type compounds, distinguished by the stereochemistry at the B- and C-ring junction. Non-canonical SLs, such as carlactone (CL) and methyl carlactonoate, lack the typical ABC-ring structure but retain the essential D-ring connected through an enol-ether bridge (Wang *et al.*, 2019). Recent advancements in analytical techniques, particularly liquid chromatography-tandem mass spectrometry (LC-MS/MS), have significantly improved the ability to detect and quantify these compounds, even at femtomolar concentrations, in plant tissues, facilitating more detailed investigations into their distribution and functions (Yoneyama *et al.*, 2013).

The diverse structures of SLs across different plant species reflect their evolutionary adaptation to various ecological niches and functions. This diversity is generated through modifications of core systems specific for species such as hydroxylation, acetylation, and methylation, which influence their stability, mobility, and receptor-binding specificity (Brewer *et al.*, 2013). These modifications are also responsible for the effectiveness of SLs in mediating various physiological processes and ecological interactions. The prime focus of this chapter is to provide an understanding of SLs ranging from their discovery to their importance in the plant kingdom and their biosynthetic routes.

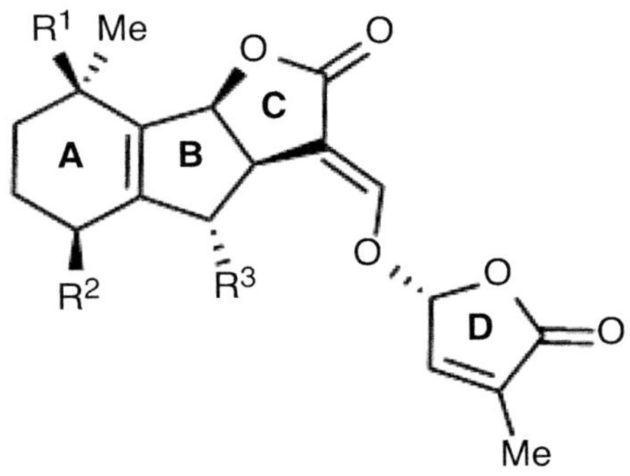

Fig. 1.1. General structure of strigolactones.

1.2 Discovery of Strigolactones

1.2.1 Early studies on germination stimulants

SLs were discovered when it was observed that the seeds of certain parasitic plants would only germinate when they were in close proximity to their host plants. This peculiar behavior made researchers curious enough to hypothesize the existence of chemical signals that could be responsible for germination. Early research in the 1960s and 1970s was focused mainly on parasitic weeds of the genera *Striga* and *Orobanche*, which were causing devastating crop losses across Asia, Africa, and the Mediterranean region (Cook *et al.*, 1966).

The first significant contribution to this discovery happened when Musselman (1980) showed that root exudates from the host plants could be responsible for the germination of *Striga* seeds, even in the absence of the host plant itself. This breakthrough explained the chemical communication between host and parasite. Moreover, the germination-inducing substance was water soluble and could diffuse easily through soil, further explaining how parasitic plants could detect potential hosts from a distance (Brun *et al.*, 2018).

These early studies laid the foundation for more detailed research into the chemical nature of these germination stimulants. However, the isolation and characterization of these compounds proved to be incredibly challenging because of their extremely low concentrations in root exudates and their inherent instability (el Zahar Haichar *et al.*, 2014). The aim of identifying these elusive compounds would continue for several decades before yielding definitive results.

1.2.2 Identification of strigol as a germination stimulant

The very first milestone in SL research was achieved in 1966 when Cook *et al.* (1966) successfully isolated a compound from cotton (*G. hirsutum*) root exudates that could induce germination of *Striga* seeds at extremely low concentrations. The complete chemical structure was named strigol but remained elusive for almost a decade due to its complex nature and the limited amount of material available for analysis.

The structural elucidation was realized when Cook *et al.* (1972) published the complete chemical structure of strigol, identifying it as a lactone with a unique tricyclic ring system. This breakthrough was very significant, as it advanced our understanding of how host plants might communicate with parasitic plants. The structure was validated through chemical synthesis, allowing the creation of adequate quantities of the substance for future biological research (MacAlpine *et al.*, 1976).

Strigol was found to be extraordinary potent, capable of stimulating *Striga* seed germination at concentrations as low as 10^{-11} M. This exceptional potency showed why the compound had been so difficult to isolate and study, as host plants needed to produce only minute amounts to effectively signal to parasitic seeds (Hauck *et al.*, 1992).

1.2.3 Breakthroughs in SL research

The research into SLs experienced a period of comparatively slow progress throughout the 1980s. However, the field was revitalized in the 1990s with the discovery of additional compounds structurally related to strigol in the root exudates of various plant species. A groundbreaking discovery came when a compound named sorgolactone was isolated and identified from sorghum root exudates. This discovery was particularly important because sorghum is a main crop severely affected by *Striga* parasitism in sub-Saharan Africa (Yoneyama *et al.*, 2010). Shortly after, another chemical, alectrol, was identified from cowpea roots, another key crop in parasitic weed-infested areas. These findings substantially improved our understanding of SL diversity and its significance in host–parasite interactions (Ueno *et al.*, 2015).

These discoveries led to the recognition that strigol was not an isolated compound but rather a member of a family of related compounds, later named “strigolactones.” The identification of multiple SL variants raised intriguing questions about their biosynthesis, ecological significance, and potential roles beyond stimulating parasitic seed germination (Xie *et al.*, 2010).

By the late 1990s, advancements in analytical techniques, especially LC-MS, enabled researchers to identify and characterize SLs at the ultra-trace levels at which they naturally occur (Yokota *et al.*, 1998). This technological advancement accelerated the pace of SL research and set the stage for discoveries that would expand our understanding of the biological functions of these compounds.

1.2.4 Chemical nature and first structural elucidation of SLs

SLs are a class of plant hormone derived from carotenoids, and feature a core butenolide (D-ring) structure linked via an enol-ether bond to a variable tricyclic lactone (ABC-ring) configuration. This unique molecular arrangement enables SLs to act both as endogenous regulators of plant growth and as signaling molecules that are released into the soil environment (Mashiguchi *et al.*, 2021). More than 30 naturally occurring SLs have been discovered, each displaying notable structural variation across different plant species. This diversity is thought to be responsible for their particular functions in different physiological processes and ecological interactions (Nomura *et al.*, 2024).

The early structural knowledge of SLs came from bioassays and chemical isolation procedures, which resulted in the identification of strigol, a chemical emitted from cotton roots, as a germination stimulant for parasitic plants (Nelson, 2021). As research progressed, technologies such as nuclear magnetic resonance (NMR) spectroscopy, MS, and X-ray crystallography were instrumental in precisely uncovering the stereochemistry and detailed molecular structure of SLs (Guercio *et al.*, 2023).

Recent research has significantly deepened our knowledge of the biosynthesis and structural diversity of SLs. The central enzymes DWARF27 (D27), carotenoid cleavage dioxygenase 7 (CCD7), and carotenoid cleavage dioxygenase 8 (CCD8) are involved in converting carotenoids to CL, a key SL biosynthesis intermediate. Species-specific chemical modifications of CL produce an enormous diversity of SL compounds (Al-Babili and Bouwmeester, 2015). Synthetic analogs of SLs, such as GR24, are utilized for research due to their ability to replicate the function of natural SLs precisely, providing useful information about their involvement in the plant communication process (Nomura *et al.*, 2024).

1.3 Types of Strigolactones

1.3.1 Canonical SLs

Canonical SLs are distinguished by a three-ring lactone structure (known as the ABC framework), which is linked to a shared butenolide moiety (D-ring) via a conserved enol-ether bond at the 2OR configuration (Wakabayashi *et al.*, 2022a). The ongoing discovery of other SLs accentuates structural variations in the C-ring orientation, and hence a more precise categorization. SLs from the 5-deoxystrigol origin, which have a β-oriented C-ring, are classified under the strigol-type group (Wakabayashi *et al.*, 2022a). In contrast, SLs that come from the 4-deoxyorobanchol and have an α-oriented C-ring are classified as orobanchol-type group members (Al-Babili, 2024). While the C-ring shows structural variability, the AB-rings in all canonical SLs can be functionally modified by reactions such as hydroxylation, methylation, acetylation, ketonization, and epoxidation (Guercio *et al.*, 2023). Some examples of this type are shown in Fig. 1.2.

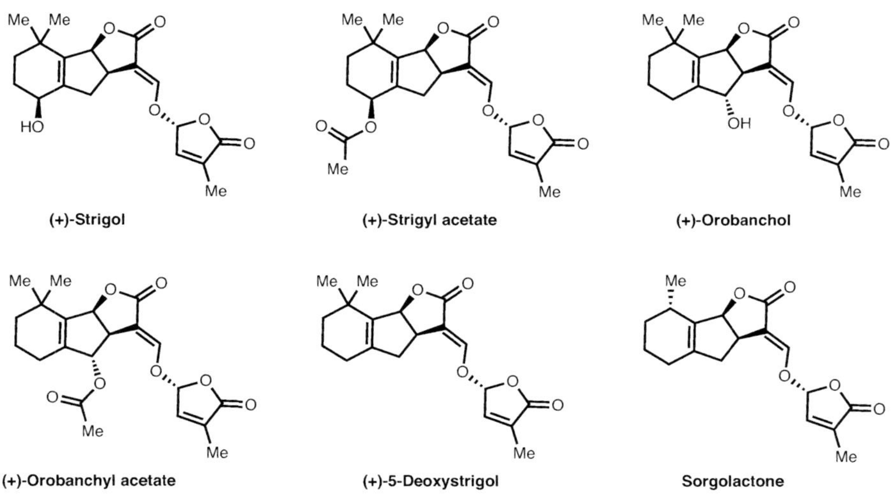

Fig. 1.2. Selected examples of canonical strigolactones.

1.3.2 Biosynthesis of canonical SLs

In the case of canonical SLs, the CYP722C enzyme subfamily is involved in the production of both strigol-type and orobanchol-type SLs (Wakabayashi *et al.*, 2020). It has been found that the VuCYP722C and SlCYP722C enzymes catalyze the conversion of carlactonoic acid (CLA) to orobanchol, presumably via the intermediate 18-OH-CLA, in cowpea (*Vigna unguiculata*) and tomato (*Solanum lycopersicum*), respectively (Wakabayashi *et al.*, 2019). Furthermore, tomato mutants lacking the non-functional SlCYP722C gene did not produce orobanchol or solanacol, a metabolite believed to be derived from orobanchol, further supporting the role of SlCYP722C in the orobanchol-type SL biosynthetic pathway (Mashiguchi *et al.*, 2021).

1.3.3 Strigol family

The strigol type of canonical SLs are naturally occurring plant-derived molecules that play a significant role in root–soil interactions and developmental processes. Initially discovered in cotton plants, strigol-type SLs are reported to induce germination in parasitic plants such as *Striga* and also to promote symbiotic interactions with arbuscular mycorrhizal fungi (AMF). Derived from β-carotene, these molecules possess a three-ringed structural skeleton and are remarkable for their ecological significance (Yoneyama and Brewer, 2021).

Canonical SLs have a distinct tricyclic structure based on A-, B-, and C-rings joined to a D-ring butenolide with an enol-ether link. The unique feature is the β-orientation of the C-ring (Flematti *et al.*, 2016). Strigol, which is the classic representative of the group, generally experiences chemical transformations such as the appending of hydroxyl or acetyl groups at the A- and B-rings and thus exhibits pronounced structural diversity. These alterations are very important in dictating the biological activity and host plant interaction specificity. The D-ring, which is structurally conserved, is essential in interacting with the receptors and facilitating downstream signaling processes (Wakabayashi *et al.*, 2021a, b).

The key members of this group are strigol, 5-deoxystrigol, and sorgomol, each having different roles to play in ecological interactions. Strigol, a typical constituent of cotton plants, is a strong germination stimulant for seeds of *Striga*. In sorghum, 5-deoxystrigol is an important intermediate in its biosynthetic pathway. Sorgomol, however, is known to stimulate symbiotic associations with mycorrhizal fungi and modulate interactions with parasitic weeds. These molecules have variations in the chemical modifications of their AB-ring structures, which have a major influence on their biological activity (Wakabayashi *et al.*, 2025).

Strigol-type SLs are derived from the precursor β-carotene via a biosynthetic process that includes CLA as a central intermediate. The conversion is catalyzed by enzymes such as D27, CCD7, and CCD8. The cytochrome P450 enzyme CYP722C is pivotal in converting CLA to 5-deoxystrigol, a central compound in the biosynthesis of strigol-type SLs in crops such as sorghum and cotton. Current research has found that there is a particular strigol-producing enzyme in cotton, and this explains the terminal processes of SL biosynthesis and its role in generating strigol. This biochemical process is crucial for rhizosphere ecological interactions, such as inducing *Striga* seed germination and symbiosis with AMF. SLs also function as plant hormones that determine growth habits and shoot branching. Variations of these biosynthetic pathways in plant species provide useful information to enhance crop attributes by using personalized biotechnological approaches (Flematti *et al.*, 2016; Wakabayashi *et al.*, 2025).

SLs of the strigol type are widely distributed in many plant taxa, especially in monocot plants such as sorghum and dicot plants such as cotton. Sorghum is reported to exude large amounts of 5-deoxystrigol and sorgomol from its roots, which greatly enhances its susceptibility to parasitic weeds such as *Striga*. Cotton, in contrast, releases mostly strigol. The diversity of SL compounds released via root exudates is essential in determining host–parasitic plant interactions. Knowledge regarding the distribution is important in the formulation of crop management strategies. For instance, it could assist in overcoming infestations by *Striga* by adjusting the pattern of secretion of SLs in sorghum, as well as improving positive associations with mycorrhizal fungi, which could result in increased nutrient uptake (Naseer *et al.*, 2024).

1.3.4 Orobanchol family

The orobanchol class is a subcategory of traditional SLs, which have an α-directed C-ring and a 2′R-configured D-ring. These plant-derived compounds play critical roles as chemical messengers in the soil ecosystem and are important for plant growth and development. First isolated from red clover in 1998, orobanchol has come to be identified as a master SL, most notably for its capacity to induce germination in parasitic plants such as *Orobanche minor*. As an apocarotenoid, orobanchol is also a plant hormone, affecting shoot branching and promoting symbiotic interactions with AMF. What distinguishes this family from strigol-type SLs is its characteristic three-dimensional molecular structure (Wakabayashi *et al.*, 2022b).

Orobanchol-type SLs have a typical three-ring ABC core structure that is linked to a butenolide D-ring through an enol-ether bond. The α-orientation of the C-ring is a characteristic feature of these compounds, distinguishing them from strigol-type SLs with a β-orientation. The D-ring maintains a uniform 2′R stereochemistry, which is critical for their biological activity. These SLs exhibit structural variation due to chemical modifications on A- and B-rings that involve hydroxylation, acetylation, and epoxidation. Alterations of these kinds, exemplified by compounds such as orobanchyl acetate and solanacol, shape how these compounds act on target receptors and affect their functions during plant development as well as communication ecologically (Yoneyama, 2020).

Other important representatives of the orobanchol group of SLs are orobanchol, orobanchyl acetate, solanacol, and fabacyl acetate. Orobanchol, present generally in legumes such as red clover, is a vital germination-inducing stimulant for parasitic *Orobanche* species. Solanacol, obtained from tomato plants, is characterized by its aromatic A-ring skeleton. Orobanchyl acetate and fabacyl acetate, which are found in cowpea and pea, respectively, have acetylated AB-ring systems that make them more efficient signaling compounds. Their structural characteristics have an impact on their effectiveness in inducing parasitic seed germination and symbiotic association with AMF (Yoneyama, 2020). Biosynthesis of orobanchol-type SLs is initiated from the carotenoid β-carotene through intermediates CL and CLA, a process catalyzed by D27, CCD7, CCD8, and CYP711A enzymes family members. In the case of plants such as tomato and cowpea, direct conversion of CLA to orobanchol is catalyzed by the enzyme CYP722C. This process is directed by a particular dirigent-like protein, stereoselective BC-ring-forming factor (SRF), which directs the stereospecific formation of an α-configured C-ring. Additionally, in tomato, the enzyme CYP712G1 is responsible for downstream orobanchol modifications resulting in the synthesis of solanacol. Stereochemical accuracy in this biosynthetic process is crucial to the biological role of orobanchol-type SLs in plant developmental and ecological signaling processes (Homma *et al.*, 2024).

1.3.5 Sorgolactone and related analogs

The SL sorgolactone is a carotenoid-derived phytohormone with a tricyclic ABC-ring system and a butenolide D-ring, characterized by a hydroxy group at C-9 of 5-deoxystrigol. Its biosynthesis starts with all-*trans*-β-carotene, processed by enzymes such as D27, CCD7, and CCD8 to form CL, which is then converted to 5-deoxystrigol and further hydroxylated by SbCYP728B35, which is a cytochrome P450 enzyme, to produce sorgomol in sorghum (*Sorghum bicolor*). The structural variability occurs due to the variations in A- and B-rings, whereas the C- and D-rings are conserved for their bioactivity. This pathway is extremely conserved among plants, with species-specific enzymes adding to structural variations (Wakabayashi *et al.*, 2021b).

Sorgolactone is a phytohormone that controls plant growth, shoot branching inhibition, and root architecture modification, as well as an AMF symbiotic rhizosphere signal. It induces parasitic weed *Striga* germination by taking advantage of SLs as host cues. Its mechanism of action includes binding to the D14 receptor and activating downstream signaling through MAX2, causing degradation of SMXL proteins in order to modulate gene expression (Shimels *et al.*, 2025).

The construction of sorgolactone and analogs frequently involves stereoselective approaches to mimic the intricate ABCD-ring system, with radical cyclization being a principal method for building the tricyclic core from precursors such as citronellate derivatives. Analogs such as GR24 and strigolactam derivatives reduce the complexity by maintaining the bioactive D-ring but altering the ABC-rings to improve stability and receptor binding. Structural modifications concentrate on changing the enol-ether bridge and D-ring substituents in order to enhance hydrolytic stability and bioactivity for parasitic weed control. The synthetic work attempts to develop scalable, inexpensive mimics for use in agriculture, such as suicidal germination (Dell'Oste *et al.*, 2021).

1.3.6 Non-canonical SLs

Non-canonical SLs are a unique class of secondary metabolites in plants synthesized from carotenoid precursors. In contrast to canonical SLs, they lack all the usual A-, B-, or C-rings that constitute the typical ABCD-ring structure. However, they do maintain the enol-ether–D-ring, which is crucial for their biological roles. Examples of these compounds include CL, CLA, and methyl carlactonoate, all of which are found in both roots and shoots (Fig. 1.3). In plant species such as maize, sunflower, and black oat, these non-canonical SLs are major contributors to the stimulation of seed germination through substances released by the roots. They are major regulators of parasitic plant seed germination in the rhizosphere, and function as signaling molecules, stimulating symbiotic relationships with AMF. They are also increasingly being recognized due to their involvement in internal plant signaling, especially regulation of shoot branching (Yoneyama *et al.*, 2018a).

Avenaol and heliolactone, discovered in black oat (*Avena strigosa*) and sunflower (*Helianthus annuus*), respectively, are newly recognized compounds that promote seed germination and are classified as non-canonical SLs (Xie, 2016). In the shoots of many flowering plants, the enzyme CYP722A catalyzes the conversion of CLA into 16-hydroxy-CLA (16-OH-CLA), a shoot-specific non-canonical SL (Zhou *et al.*, 2025). CL, a key intermediate in the SL biosynthetic process, is considered the most structurally basic member of this group (Xie, 2016).

1.3.7 Diversified SL-like compounds

Zealactones, which have been found in *Zea mays*, are an example of the wide structural diversity among SL-related compounds. Unlike the traditional (di)methylcyclohexene ring present in typical SLs, zealactones contain a 4,4-dimet hyltetrahydrofuran-2-one group as their A-ring. This unique molecule organization indicates the participation of novel biosynthetic pathways and implies that zealactones might possess novel roles in modifying plant physiology as well as interaction with neighboring organisms (Charnikhova *et al.*, 2017).

In grapevine rootstocks, limited nitrogen availability stimulates the release of a non-canonical SL known as vitislactone. This specific

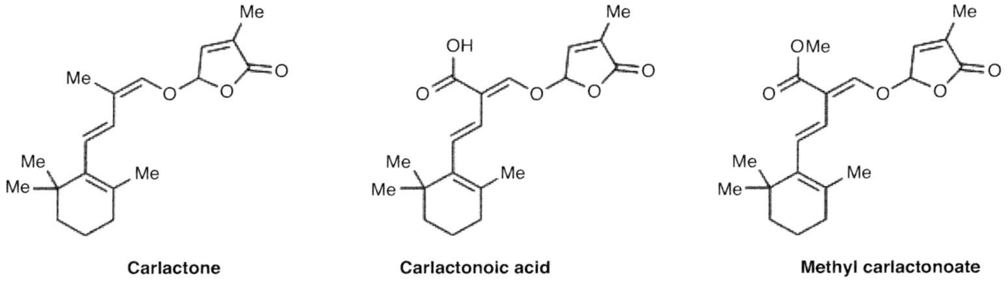

Fig. 1.3. Selected examples of non-canonical strigolactones.

compound, uncovered using LC-MS/MS and NMR techniques, appears to be unique to grapevines, as it has not been reported in other plant species. Its synthesis is linked to stronger symbiotic interactions with mycorrhizal fungi and a noticeable decrease in shoot branching, suggesting its involvement in plant responses to nutrient limitation and in shaping growth patterns (Lailheugue *et al.*, 2023).

Enzymes specific to certain species are responsible for the synthesis of various SL-like compounds. One of these enzymes, the cytochrome P450 CYP722A, is responsible for converting CLA to 16-OH-CLA, a non-canonical SL that is most prevalent in plant shoots. This molecule is important for shoot branching regulation and is a major breakthrough in the evolution of SL biosynthesis, which underlines the sophistication and diversity of these signaling compounds (Zhou *et al.*, 2025).

1.3.8 Functional and structural variations

Non-canonical SLs attain their functional diversity due to divergent biosynthetic pathways characterized by distinct biosynthetic mechanisms following the conversion of CL to an essential precursor. In contrast to the canonical SLs possessing a conserved ABCD-ring system, the non-canonical SLs follow enzymatic post-translational modifications such as MAX1 homolog-catalyzed oxidation and yield multiple metabolites with varied biological functions. Metabolites such as heliolactone and zealactone are secreted into the rhizosphere, where they help regulate symbiotic and parasitic interactions, emphasizing their ecological importance. Their simpler chemical structures often enhance their stability in the acidic environment of the rhizosphere (Yoneyama *et al.*, 2018b).

Structural differences in non-canonical SLs have a notable impact on how they interact with receptors such as DWARF14 and KAI2. These receptors identify the conserved D-ring but have different affinities in binding, enabling non-canonical SLs to subtly control traits such as shoot branching in certain plant species. For instance, in *Arabidopsis*, 16-OH-CLA controls shoot branching, highlighting its role as an endogenous hormone. This structural difference reflects the evolutionary diversification of SLs across plant lineages (Bürger and Chory, 2020).

1.3.9 Synthetic SL analogs

Synthetic analogs of SLs have proven to be useful tools in controlling parasitic weeds by inducing seed germination in the absence of a host, which eventually results in the destruction of the weeds. Such analogs, engineered to mimic the biological activity of natural SLs, possess less complex structures that are easy to synthesize without losing their efficacy. Compounds such as GR24 and strigolactams have been identified as highly active in stimulating the germination of parasitic weeds such as *Orobanche cumana* and have desirable properties such as solubility in water. They hold promise in sustainable agriculture because they reduce parasitic weed damage to crops (Lumbroso *et al.*, 2016).

Apart from their use in agriculture, synthetic SL analogs exhibit potential anticancer activities through the inhibition of cancer cell proliferation and induction of apoptosis. Studies involving the hepatocellular carcinoma cell line HepG2 have shown that the compounds TIT3 and TIT7 inhibit cell viability in a dose-response fashion without affecting normal cells. These compounds interfere with microtubule dynamics, inhibit cell migration, and trigger programmed cell death, and hence are possible lead compounds for the construction of new chemotherapeutic drugs (Hasan *et al.*, 2018).

1.3.10 Development of artificial SLs

Artificial SLs are becoming widely known as valuable players in medicine and agriculture because of their ability to control plant growth and influence specific cellular processes. Analogs such as GR24 replicate the bioactivity of natural SLs, particularly inducing suicidal germination in parasitic weeds such as *Striga*. Enhancements in the chemical structure of such compounds have made them more selective and efficient, permitting uses such as inducing hyphal branching in AMF and controlling shoot branching in plants. Such breakthroughs illustrate the potential of synthetic SLs to provide sustainable solutions to weed control and crop development (Zwanenburg and Blanco-Ania, 2018).

New SL analogs have been synthesized with innovative synthesis methods to yield products of enhanced stability and biological activity. Studies of their structure–activity relationships have found that essential structural elements such as the D-ring and enol-ether bond contribute significantly to their biological activity. Recent research has shown that such novel SL mimics can regulate photosynthesis and induce biomass development in *Chlorella sorokiniana*, enhancing the stress tolerance of the plant. These tailor-made analogs exemplify the versatility of SLs for fulfilling specific physiological needs, setting the stage for specific agricultural applications and potential therapeutic uses (Popa *et al.*, 2023).

Artificial SLs are of critical importance in medicine outside agriculture, specifically as autophagy inhibitors. Advances over the past few years have shown the potential to target colorectal cancer cells by inhibiting autophagosome and lysosome fusion, creating significant cytotoxicity and tumor inhibition in animal models. These findings point to additional therapeutic possibilities for SLs in manipulating cellular functions, paving the way for future cancer therapies. The versatile functions of synthetic SLs underscore the need for interdisciplinary research to realize their maximum therapeutic potential (Yang *et al.*, 2022).

1.3.11 Applications in agricultural and biological research

Synthetic analogs of SLs have become important tools in agricultural studies, especially for controlling parasitic weeds such as *Orobanche* and *Striga*, which significantly reduce crop productivity. These analogs promote the premature germination of weed seeds, thereby depleting the seed reservoir while leaving the host plants unaffected. Research has shown that compounds such as methyl phenlactonoates are highly effective in stimulating germination, providing a more straightforward and cost-efficient option compared with natural SLs. Their use contributes to sustainable weed management by lessening the dependence on conventional herbicides (Kountche *et al.*, 2018).

Synthetic analogs of SLs, such as GR24, have been instrumental in elucidating our knowledge of plant development in biological studies. They affect important physiological characteristics, such as root architecture, shoot branching, and environmental stress responses in plants. By imitating the behavior of natural SLs, they have enabled scientists to explore hormonal signaling mechanisms more effectively. In addition, their application has been particularly valuable in studying interactions between plants and AMF, which support improved nutrient acquisition in agricultural systems. Investigations using *Arabidopsis* have been especially informative, uncovering the relationship between chemical structure and biological activity. This knowledge is now guiding the creation of customized synthetic derivatives with targeted biological functions (Kang *et al.*, 2023).

In addition to their established role in agriculture, synthetic SL analogs are gaining attention in the medical field for their potential anticancer effects. Agents TIT3 and TIT7, for example, have been found to specifically inhibit the proliferation of hepatocellular carcinoma cells through interference with microtubule function and induction of programmed cell death without exerting much effect on normal cells. Such selective action underscores their potential as new chemotherapeutic agents and implies a more universal significance of SL research that goes beyond the biology of plants (Hasan *et al.*, 2019).

1.4 Divergent Biosynthetic Pathways of Strigolactones

D27, CCD7, and CCD8 cleave β-carotene enzymatically, initiating the biosynthesis of SLs and producing CL as a central intermediate. The cytochrome P450 enzyme MAX1 catalyzes the transformation of CL into CLA, a crucial biosynthetic derivative, as shown by Hasan *et al.* (2019), thereby causing the pathway to branch. Distinct SL structures, including canonical types with a tricyclic ABC scaffold and non-canonical forms lacking this configuration, are subsequently synthesized through modifications carried out by enzymes that vary across plant species (Al-Babili and Bouwmeester, 2015). These structural variations support physiological and ecological adaptations in SL signaling (Yoda *et al.*, 2021).

The synthesis of canonical SLs, such as 5-deoxystrigol, is facilitated in sorghum by sulfotransferases and cytochrome P450 711A (CYP711A) enzymes. In contrast, a 2-oxoglutarate-dependent dioxygenase generates the non-canonical compound lotuslactone in *Lotus japonicus*. The production of zealactone in maize is mediated by a parallel biosynthetic route involving 3-hydroxy-methylcarlactonoic acid, where *Zea mays* cytochrome P450 706C37 (ZmCYP706C37) functions as the key catalyst (Li *et al.*, 2024). These secondary routes of SL biosynthesis signify the evolutionary flexibility that is intrinsic in SL biosynthetic pathways in various plant taxa (Mori *et al.*, 2020). Regulation of plant development and environmental sensitivity are facilitated by the molecular diversity of SLs, which is critical for their specialized biological activities (Li *et al.*, 2024).

Computational modeling has revealed that SL biosynthesis is not inherently feedback-controlled, allowing plants to dynamically adjust SL levels in response to environmental conditions such as nutrient status. This flexibility is expressed in a variety of biosynthetic pathways that produce SL variants specialized for activities such as inhibiting axillary bud growth or mediating rhizosphere interactions (Bhoi *et al.*, 2021). Increased understanding of these biosynthetic pathways is opening up biotechnological strategies for the enhancement of stress tolerance and rational optimization of plant architecture in crop species (Lucido *et al.*, 2022). The diversified structural types of SLs indicate their functional promiscuity under both ecological fluctuation and developmental cues (Bhoi *et al.*, 2021).

1.4.1 General overview of plant hormone biosynthesis

Plant hormone production takes place through complex biochemical pathways that produce compounds involved in signaling the regulation of physiological development, adaptation, and growth reactions. Highly regulated enzymatic reactions produce the major hormones such as auxins, gibberellins, cytokinins, ethylene, and abscisic acid (ABA) through specific metabolic pathways (Wang *et al.*, 2020). Particular biochemical precursors such as amino acids and isoprenoid derivatives are utilized in these pathways, which are regulated by environmental factors including temperature fluctuations, light intensity, and various environmental stressors (Raza *et al.*, 2019). Understanding and elucidating these biosynthetic processes is essential for advancing agricultural strategies aimed at enhancing crop performance and improving tolerance to environmental stressors (Waadt *et al.*, 2022).

In plants, the synthesis of auxins, especially indole-3-acetic acid (IAA), is conducted mainly through a tryptophan-dependent mechanism, where critical enzymes such as YUCCA function as primary catalysts (Casanova-Sáez *et al.*, 2021). Geranylgeranyl diphosphate serves as the precursor for gibberellins, which are biosynthesized through enzymatic reactions involving ent-kaurene oxidase, thereby influencing processes such as stem elongation and seed sprouting (Binenbaum *et al.*, 2018). The production of cytokinins from adenine-based compounds is regulated by enzymes including isopentenyltransferase, which modulate cellular division (Kieber and Schaller, 2018). The integration of these hormonal pathways enables plants to adjust their growth and development in response to both intrinsic signals and environmental stimuli (Casanova-Sáez *et al.*, 2021).

The biosynthesis of ABA and ethylene plays essential roles in mediating plant responses to environmental stress and in regulating fruit maturation. Carotenoid-derived ABA is synthesized through the catalytic activity of 9-*cis*-epoxycarotenoid dioxygenase, and its concentration rises significantly during conditions such as drought and high salinity (Chen *et al.*, 2020). Ethylene biosynthesis from methionine is facilitated by the enzyme 1-aminocyclopropane-1-carboxylic acid synthase, involved in leaf senescence and stress adaptation processes (Pattyn *et al.*, 2021). The regulatory mechanisms controlling these biosynthetic processes have been uncovered by new genomic and proteomic technologies, which are the foundations for efforts directed toward engineering plant species with enhanced stress tolerance (Ku *et al.*, 2018).

1.5 Importance of Strigolactones in Plants

1.5.1 Role of SLs in parasitic weed germination

SLs are organic plant molecules produced from carotenoid precursors (Zorrilla *et al.*, 2022). They are significant in plant communication, for example in interactions with parasitic weeds such as *Striga* and *Orobanche*, which are significant agricultural pests. *Striga* particularly inflicts serious damage to key food crops such as maize, pearl millet, and sorghum in sub-Saharan Africa, frequently causing heavy yield losses and compromising food security in the region (Ito, 2023). Most parasitic species in the Orobanchaceae family have seeds that are able to survive in soil for a long time, until they sense specific chemical signals released by host plant roots around them. This extremely specialized adaptation is crucial for the survival of obligate parasites such as witchweeds (*Striga* and *Alectra*) and broomrapes (*Orobanche* and *Phelipanche*), which need to attach themselves to a host plant as soon as possible after germination in order to support their growth and development (Nelson, 2021).

Seeds of parasitic plants need a special chemical signal or germination stimulant to germinate, suggesting that there is a nearby suitable host. Upon sensing this signal, the radicle will grow within 1–2 days, usually attaining a few millimeters to 1 cm in length. Once germination occurs, the parasite then identifies other host-derived signals called haustorium-inducing factors. These substances prompt the formation of the pre-haustorium, a specialized invasive structure characterized by a swollen radicle tip and the emergence of haustorial hairs on its surface (Goyet *et al.*, 2019; Wada *et al.*, 2019). As soon as the pre-haustorium makes contact with the root of the host plant, it gives rise to specialized intrusive cells that penetrate the host's root tissues. This process results in development of the haustorium, which is an essential structure linking the vascular systems of both the parasite and its host. By this association, the parasite gains access to essential resources such as water and nutrients from the host plant directly (Brun *et al.*, 2021).

A potential method of controlling parasitic weeds is to treat the soil with a germination stimulant prior to planting the host crop. This practice causes dormant parasitic seeds to begin germination even in the absence of a host plant, preventing them from attaching and leading to their eventual death. This approach is known as the "suicidal germination strategy." To manage *Striga hermonthica* in sorghum, synthetic SL analogs are used to trigger premature seed germination (Jamil *et al.*, 2022). Unlike natural SLs, these mimics lack the enol-ether group but feature an aryloxy or aroyloxy group at the C-5 position of the D-ring, allowing them to initiate germination without supporting the parasite's growth (Zwanenburg *et al.*, 2016).

1.5.2 Regulation of plant architecture

The most well-documented function of SLs is the regulation of shoot branching. SLs work antagonistically with cytokinins by acting as inhibitors of axillary bud growth to maintain optimal branching patterns (Ferguson and Beveridge, 2009). This function was initially revealed through studies of high-branching mutants in *Arabidopsis* (*max* mutants), pea (*rms* mutants), and rice (*d* mutants), which were later found to be defective in biosynthesis or signaling of SLs (Waters *et al.*, 2017). Subsequent research has explained that SLs modulate shoot branching by having an impact on auxin transport and transcription factor activity, particularly the *BRANCHED1* (*BRC1*) gene, which regulates bud dormancy (Brewer *et al.*, 2016).

Beyond shoot architecture, SLs also influence root development by enhancing plant root growth, inhibiting lateral root formation, and stimulating root hair elongation (Kapulnik and Koltai, 2014). They also regulate stem thickness, leaf shape, and senescence, hence contributing to overall plant morphology and development (Yamada *et al.*, 2014). Recent studies have shown that SLs interact with other plant hormones such as auxin, cytokinins, ABA, and ethylene in complex and regulatory networks that influence plant growth under varying environment conditions (Waters *et al.*, 2017).

1.5.3 Roles in plant–microbe interactions

SLs play crucial roles in establishing symbiotic associations with AMF. Signaling molecules are exuded into the rhizosphere, where they attract AMF and stimulate their hyphal branching and metabolic activity. This facilitates the establishment of mutualistic relationships that enhance plant nutrient acquisition, especially phosphorus and nitrogen, as explained by Lanfranco *et al.* (2018). This function is evolutionarily conserved across land plants and is considered to be one of the ancestral roles of SLs (Waters *et al.*, 2017).

SLs not only promote symbiotic association with AMF but also trigger the germination of parasitic plants of the Orobanchaceae family such as *Striga*, *Orobanche*, and *Phelipanche* species, which are responsible for substantial crop losses worldwide (Jamil *et al.*, 2018). This dual role highlights the complex evolutionary history of SLs and presents challenges for agriculture management. Recent studies have demonstrated that developing synthetic SL analogs can actually stimulate AMF colonization without promoting parasitic plant germination, which could be a potential solution for sustainable agriculture (Yoneyama *et al.*, 2013).

1.5.4 Adaptation to environmental stress

Emerging evidence suggests that SLs are integral to plant responses to various environmental stress stimuli. Plants increase the biosynthesis of SLs under phosphate and nitrogen deficiency, which not only enhances AMF symbiosis for improved nutrient acquisition but also enhances root and shoot architecture to optimize the resource allocation (Marzec, 2016). During drought stress, SLs promote stomatal closure to reduce water loss and enhance drought tolerance (Ha *et al.*, 2014). Furthermore, SLs influence plant responses to salt stress, oxidative stress, and light conditions demonstrating their versatility as environmental response mediators (Bu *et al.*, 2014).

Recent studies have shown that genetic and biochemical connections between SL signaling and stress-responsive pathways, including ABA signaling and reactive oxygen species homeostasis, provide mechanistic insights into how these hormones enhance plant resilience (Mao *et al.*, 2020).

The role of SLs in plant development and environmental adaptation underscores their potential as targets for agriculture improvement. Manipulating the biosynthesis or signaling of SLs could lead to crops with optimized architecture, enhanced nutrient efficiency, improved stress tolerance, and resistance to parasitic weeds, hence contributing to sustainable food production in challenging environments (Wang *et al.*, 2019).

1.5.5 Evolutionary significance of SLs

The evolutionary history of SLs provides fascinating insights into plant adaptation and diversification across geological time scales. Comparative physiological and genomic studies have revealed that SLs emerged early in plant evolution, with their core biosynthetic pathway and signaling components present in most land plant lineages such as bryophytes, lycophytes, ferns, gymnosperms, and angiosperms, as discussed by Waters *et al.* (2017).

1.6 Origin and Conservation of Strigolactone Pathways

Molecular evidence suggests that the primordial function of SLs was likely related to the establishment of symbiotic relations with AMF, which facilitated the colonization of terrestrial environments by early land plants approximately 450 million years ago (Delaux *et al.*, 2012). This hypothesis is supported by the presence of genes responsible for the biosynthesis of SLs in bryophytes such as *Physcomitrella patens*, which produce and respond to these compounds despite lacking the canonical shoot branching architecture seen in higher plants (Proust *et al.*, 2011).

The core components of SL biosynthesis, particularly the carotenoid cleavage dioxygenases (CCDs) and cytochrome P450 enzymes, show considerable conservation in diverse plant lineages, validating the strong evolutionary pressure to maintain this pathway (Ruyter-Spira *et al.*, 2013). Similarly, the D14 receptor and MAX3/D3 F-box protein-like signaling components are widely conserved, although they exhibit lineage-specific adaptations that reflect the expanding functional repertoire of SLs throughout plant evolution (Nelson *et al.*, 2011).

Different plants produce distinct SL profiles, which may be adapted to specific ecological niches, developmental requirements, and symbiotic partners showing the structural diversity of SLs across plant species as an evolutionary specialization (Yoneyama *et al.*, 2013). The parasitic plants have evolved the ability to detect SLs derived from the host as germination cues, exploiting the conserved role of these compounds in establishing mycorrhizal symbiosis (Conn *et al.*, 2015).

1.6.1 Convergent evolution in SL signaling

Recent research has demonstrated the fascinating examples of convergent evolution in SL perception mechanism. While most plants utilize the α/β-hydrolase D14 as the primary SL receptor, parasitic plants of the Orobanchaceae family have evolved a divergent receptor called HTL/KAI2, which recognizes SLs derived from the host with high sensitivity (Conn *et al.*, 2015). Similarly, convergent evolution has been observed in the biosynthetic pathways of SLs, with some plant lineages developing alternative routes to produce these compounds (Walker *et al.*, 2019).

These evolutionary patterns underscore the ecological importance of SLs and demonstrate that they have been subject to diverse selective pressures throughout plant evolution. Understanding the evolutionary history of SLs provides insight into plant adaptation and diversification, as well as informing strategies for engineering these pathways for agricultural improvement (Marzec and Brewer, 2019).

1.6.2 Expansion of SL research beyond parasitic plants

SLs are a distinct set of plant hormones that affect multiple dimensions of plant growth, their earliest known function being the inhibition of shoot branching (Waters *et al.*, 2017).

SLs play important regulatory roles in enabling plants to adapt to abiotic stress, primarily because they are involved in the modulation of plant growth and development. In moss, SLs aid in the determination of growth patterns and regulation of interactions between neighboring colonies (Proust *et al.*, 2011). These compounds also participate in metabolic processes that boost plant defense against both biotic and abiotic stress agents. SL synthesis is extremely sensitive to the nature and level of environmental stress experienced at various growth stages of the plant. New evidence points to the interaction between SLs and other hormones (e.g. ABA), and proposes that SLs are an intrinsic part of hormone-controlled stress adaptation processes (Saeed *et al.*, 2017). They play a role in plant tolerance to extreme conditions such as drought, heat, and nutrient deficit, specifically in countering the impacts of phosphorus and nitrogen deficiency (Soliman *et al.*, 2022).

1.7 Conclusion

SLs are active plant hormones with important roles in regulating plant growth and assisting in interactions with the environment. Originally discovered to cause germination of parasitic plants, their functions are now known to include regulation of shoot growth, alteration of root architecture, and elicitation of association with symbiotic fungi. Their varied chemical structures and complex biosynthesis pathways are a manifestation of evolutionary adaptation to diverse ecological functions. Continuous studies endure to unveil their potential in promoting plant tolerance against environmental stress and their potential uses in agriculture, particularly in boosting crop performance and offering natural mechanisms to control parasitic weeds in challenging environments.

References

Al-Babili, S. (2024) A dirigent of the ring for strigolactone stereochemistry. *Proceedings of the National Academy of Sciences USA* 121(34), e2410953121. DOI: 10.1073/pnas.2410953121.

Al-Babili, S. and Bouwmeester, H.J. (2015) Strigolactones, a novel carotenoid-derived plant hormone. *Annual Review of Plant Biology* 66, 161–186. DOI: 10.1146/annurev-arplant-043014-114759.

Bhoi, A., Yadu, B., Chandra, J. and Keshavkant, S. (2021) Contribution of strigolactone in plant physiology, hormonal interaction and abiotic stresses. *Planta* 254(2), 28. DOI: 10.1007/s00425-021-03678-1.

Binenbaum, J., Weinstain, R. and Shani, E. (2018) Gibberellin localization and transport in plants. *Trends in Plant Science* 23(5), 410–421. DOI: 10.1016/j.tplants.2018.02.005.

Brewer, P.B., Koltai, H. and Beveridge, C.A. (2013) Diverse roles of strigolactones in plant development. *Molecular Plant* 6(1), 18–28. DOI: 10.1093/mp/sss130.

Brewer, P.B., Yoneyama, K., Filardo, F., Meyers, E., Scaffidi, A. *et al.* (2016) LATERAL BRANCHING OXIDOREDUCTASE acts in the final stages of strigolactone biosynthesis in *Arabidopsis*. *Proceedings of the National Academy of Sciences USA* 113(22), 6301–6306. DOI: 10.1073/pnas.1601729113.

Brun, G., Spallek, T., Simier, P. and Delavault, P. (2021) Molecular actors of seed germination and haustoriogenesis in parasitic weeds. *Plant Physiology* 185(4), 1270–1281. DOI: 10.1093/plphys/kiaa041.

Brun, G., Braem, L., Thoiron, S., Gevaert, K., Goormachtig, S. *et al.* (2018) Seed germination in parasitic plants: What insights can we expect from strigolactone research? *Journal of Experimental Botany* 69(9), 2265–2280. DOI: 10.1093/jxb/erx472.

Bu, Q., Lv, T., Shen, H., Luong, P., Wang, J. *et al.* (2014) Regulation of drought tolerance by the F-box protein MAX2 in *Arabidopsis*. *Plant Physiology* 164(1), 424–439. DOI: 10.1104/pp.113.226837.

Bürger, M. and Chory, J. (2020) The many models of strigolactone signaling. *Trends in Plant Science* 25(4), 395–405. DOI: 10.1016/j.tplants.2019.12.009.

Casanova-Sáez, R., Mateo-Bonmatí, E. and Ljung, K. (2021) Auxin metabolism in plants. *Cold Spring Harbor Perspectives in Biology* 13(3), a039867. DOI: 10.1101/cshperspect.a039867.

Charnikhova, T.V., Gaus, K., Lumbroso, A., Sanders, M., Vincken, J.-P. *et al.* (2017) Zealactones. Novel natural strigolactones from maize. *Phytochemistry* 137, 123–131. DOI: 10.1016/j.phytochem.2017.02.010.

Chen, K., Li, G.J., Bressan, R.A., Song, C.P., Zhu, J.K. *et al.* (2020) Abscisic acid dynamics, signaling, and functions in plants. *Journal of Integrative Plant Biology* 62(1), 25–54. DOI: 10.1111/jipb.12899.

Conn, C.E., Bythell-Douglas, R., Neumann, D., Yoshida, S., Whittington, B. *et al.* (2015) Convergent evolution of strigolactone perception enabled host detection in parasitic plants. *Science* 349(6247), 540–543. DOI: 10.1126/science.aab1140.

Cook, C.E., Whichard, L.P., Turner, B., Wall, M.E. and Egley, G.H. (1966) Germination of witchweed (*Striga lutea* Lour.): Isolation and properties of a potent stimulant. *Science* 154(3753), 1189–1190. DOI: 10.1126/science.154.3753.1189.

Cook, C.E., Whichard, L.P., Wall, M.E., Egley, G.H., Coggon, P. *et al.* (1972) Germination stimulants. II. Structure of strigol, a potent seed germination stimulant for witchweed (*Striga lutea*). *Journal of the American Chemical Society* 94(17), 6198–6199. DOI: 10.1021/ja00772a048.

Delaux, P.M., Xie, X., Timme, R.E., Puech-Pages, V., Dunand, C. *et al.* (2012) Origin of strigolactones in the green lineage. *New Phytologist* 195(4), 857–871. DOI: 10.1111/j.1469-8137.2012.04209.x.

Dell'Oste, V., Spyrakis, F. and Prandi, C. (2021) Strigolactones, from plants to human health: Achievements and challenges. *Molecules* 26(15), 4579. DOI: 10.3390/molecules26154579.

el Zahar Haichar, F., Santaella, C., Heulin, T. and Achouak, W. (2014) Root exudates mediated interactions belowground. *Soil Biology and Biochemistry* 77, 69–80. DOI: 10.1016/j.soilbio.2014.06.017.

Ferguson, B.J. and Beveridge, C.A. (2009) Roles for auxin, cytokinin, and strigolactone in regulating shoot branching. *Plant Physiology* 149(4), 1929–1944. DOI: 10.1104/pp.109.135475.

Flematti, G.R., Scaffidi, A., Waters, M.T. and Smith, S.M. (2016) Stereospecificity in strigolactone biosynthesis and perception. *Planta* 243(6), 1361–1373. DOI: 10.1007/s00425-016-2523-5.

Gomez-Roldan, V., Fermas, S., Brewer, P.B., Puech-Pagès, V., Dun, E.A. *et al.* (2008) Strigolactone inhibition of shoot branching. *Nature* 455(7210), 189–194. DOI: 10.1038/nature07271.

Goyet, V., Wada, S., Cui, S., Wakatake, T., Shirasu, K. *et al.* (2019) Haustorium inducing factors for parasitic orobanchaceae. *Frontiers in Plant Science* 10, 1056. DOI: 10.3389/fpls.2019.01056.

Guercio, A.M., Palayam, M. and Shabek, N. (2023) Strigolactones: Diversity, perception, and hydrolysis. *Phytochemistry Reviews* 22(2), 339–360. DOI: 10.1007/s11101-023-09853-4.

Ha, C.V., Leyva-González, M.A., Osakabe, Y., Tran, U.T., Nishiyama, R. *et al.* (2014) Positive regulatory role of strigolactone in plant responses to drought and salt stress. *Proceedings of the National Academy of Sciences USA* 111(2), 851–856. DOI: 10.1073/pnas.1322135111.

Hasan, M.N., Choudhry, H., Razvi, S.S., Moselhy, S.S., Kumosani, T.A. *et al.* (2018) Synthetic strigolactone analogues reveal anti-cancer activities on hepatocellular carcinoma cells. *Bioorganic & Medicinal Chemistry Letters* 28(6), 1077–1083. DOI: 10.1016/j.bmcl.2018.02.016.

Hasan, M.N., Razvi, S.S., Choudhry, H., Hassan, M.A., Moselhy, S.S. *et al.* (2019) Gene ontology and expression studies of strigolactone analogues on a hepatocellular carcinoma cell line. *Analytical Cellular Pathology* 2019, 1598182. DOI: 10.1155/2019/1598182.

Hauck, C., Müller, S. and Schildknecht, H. (1992) A germination stimulant for parasitic flowering plants from *Sorghum bicolor*, a genuine host plant. *Journal of Plant Physiology* 139(4), 474–478. DOI: 10.1016/S0176-1617(11)80497-9.

Homma, M., Wakabayashi, T., Moriwaki, Y., Shiotani, N., Shigeta, T. *et al.* (2024) Insights into stereoselective ring formation in canonical strigolactone: Identification of a dirigent domain-containing enzyme catalyzing orobanchol synthesis. *Proceedings of the National Academy of Sciences USA* 121(26), e2313683121. DOI: 10.1073/pnas.2313683121.

Ito, S. (2023) Recent advances in the regulation of root parasitic weed damage by strigolactone-related chemicals. *Bioscience, Biotechnology, and Biochemistry* 87(3), 247–255. DOI: 10.1093/bbb/zbac208.

Jamil, M., Kountche, B.A., Haider, I., Guo, X., Ntui, V.O. *et al.* (2018) Methyl phenlactonoates are efficient strigolactone analogs with simple structure. *Journal of Experimental Botany* 69(9), 2319–2331. DOI: 10.1093/jxb/erx438.

Jamil, M., Wang, J.Y., Yonli, D., Ota, T., Berqdar, L. *et al.* (2022) *Striga hermonthica* suicidal germination activity of potent strigolactone analogs: Evaluation from laboratory bioassays to field trials. *Plants* 11(8), 1045. DOI: 10.3390/plants11081045.

Jia, K.P., Baz, L. and Al-Babili, S. (2018) From carotenoids to strigolactones. *Journal of Experimental Botany* 69(9), 2189–2204. DOI: 10.1093/jxb/erx476.

Kang, Z., Yan, Y., Lu, R., Dong, X., Xu, J. *et al.* (2023) Synthesis and biological profiling of novel strigolactone derivatives for *Arabidopsis* growth and development. *Journal of Agricultural and Food Chemistry* 71(34), 12859–12874. DOI: 10.1021/acs.jafc.3c02135.

Kapulnik, Y. and Koltai, H. (2014) Strigolactone involvement in root development, response to abiotic stress, and interactions with the biotic soil environment. *Plant Physiology* 166(2), 560–569. DOI: 10.1104/pp.114.244939.

Kieber, J.J. and Schaller, G.E. (2018) Cytokinin signaling in plant development. *Development* 145(4), dev149344. DOI: 10.1242/dev.149344.

Kountche, B.A., Novero, M., Jamil, M., Asami, T., Bonfante, P. *et al.* (2018) Effect of the strigolactone analogs methyl phenlactonoates on spore germination and root colonization of arbuscular mycorrhizal fungi. *Heliyon* 4(11), e00936. DOI: 10.1016/j.heliyon.2018.e00936.

Ku, Y.S., Sintaha, M., Cheung, M.Y. and Lam, H.M. (2018) Plant hormone signaling crosstalks between biotic and abiotic stress responses. *International Journal of Molecular Sciences* 19(10), 3206. DOI: 10.3390/ijms19103206.

Lailheugue, V., Merlin, I., Boutet, S., Perreau, F., Pouvreau, J.-B. *et al.* (2023) Vitislactone, a non-canonical strigolactone exudated by grapevine rootstocks in response to nitrogen starvation. *Phytochemistry* 215, 113837. DOI: 10.1016/j.phytochem.2023.113837.

Lanfranco, L., Fiorilli, V., Venice, F. and Bonfante, P. (2018) Strigolactones cross the kingdoms: Plants, fungi, and bacteria in the arbuscular mycorrhizal symbiosis. *Journal of Experimental Botany* 69(9), 2175–2188. DOI: 10.1093/jxb/erx432.

Li, C., Haider, I., Wang, J.Y. and Quinodoz, P. (2024) OsCYP706C2 diverts rice strigolactone biosynthesis to a noncanonical pathway branch. *Science Advances* 10(35), eadq3942. DOI: 10.1126/sciadv.adq3942.

Lucido, A., Basallo, O., Sorribas, A., Marin-Sanguino, A., Vilaprinyo, E. *et al.* (2022) A mathematical model for strigolactone biosynthesis in plants. *Frontiers in Plant Science* 13, 979162. DOI: 10.3389/fpls.2022.979162.

Lumbroso, A., Villedieu-Percheron, E., Zurwerra, D., Screpanti, C., Lachia, M. *et al.* (2016) Simplified strigolactams as potent analogues of strigolactones for the seed germination induction of *Orobanche cumana* Wallr. *Pest Management Science* 72(11), 2054–2068. DOI: 10.1002/ps.4268.

MacAlpine, G.A., Raphael, R.A., Shaw, A., Taylor, A.W. and Wild, H.J. (1976) Synthesis of the germination stimulant (±)-strigol. *Journal of the Chemical Society, Perkin Transactions 1* 1(4), 410–416. DOI: 10.1039/P19760000410.

Mao, C., He, J., Liu, L., Deng, Q., Yao, X. *et al.* (2020) OsNAC2 integrates auxin and cytokinin pathways to modulate rice root development. *Plant Biotechnology Journal* 18(2), 429–442. DOI: 10.1111/pbi.13209.

Marzec, M. (2016) Perception and signaling of strigolactones. *Frontiers in Plant Science* 7, 1260. DOI: 10.3389/fpls.2016.01260.

Marzec, M. and Brewer, P. (2019) Binding or hydrolysis? How does the strigolactone receptor work? *Trends in Plant Science* 24(7), 571–574. DOI: 10.1016/j.tplants.2019.05.001.

Mashiguchi, K., Seto, Y. and Yamaguchi, S. (2021) Strigolactone biosynthesis, transport and perception. *Plant Journal* 105(2), 335–350. DOI: 10.1111/tpj.15059.

Mori, N., Nomura, T. and Akiyama, K. (2020) Identification of two oxygenase genes involved in the respective biosynthetic pathways of canonical and non-canonical strigolactones in *Lotus japonicus*. *Planta* 251(2), 40. DOI: 10.1007/s00425-019-03332-x.

Musselman, L.J. (1980) The biology of *Striga*, *Orobanche*, and other root-parasitic weeds. *Annual Review of Phytopathology* 18(1), 463–489. DOI: 10.1146/annurev.py.18.090180.002335.

Naseer, M.A., Zhang, Z.Q., Mukhtar, A., Asad, M.S., Wu, H.Y. *et al.* (2024) Strigolactones: A promising tool for nutrient acquisition through arbuscular mycorrhizal fungi symbiosis and abiotic stress tolerance. *Plant Physiology and Biochemistry* 215, 109057. DOI: 10.1016/j.plaphy.2024.109057.

Nelson, D.C. (2021) The mechanism of host-induced germination in root parasitic plants. *Plant Physiology* 185(4), 1353–1373. DOI: 10.1093/plphys/kiab043.

Nelson, D.C., Scaffidi, A., Dun, E.A., Waters, M.T., Flematti, G.R. *et al.* (2011) F-box protein MAX2 has dual roles in karrikin and strigolactone signaling in *Arabidopsis thaliana*. *Proceedings of the National Academy of Sciences USA* 108(21), 8897–8902. DOI: 10.1073/pnas.1100987108.

Nomura, T., Seto, Y. and Kyozuka, J. (2024) Unveiling the complexity of strigolactones: Exploring structural diversity, biosynthesis pathways, and signaling mechanisms. *Journal of Experimental Botany* 75(4), 1134–1147. DOI: 10.1093/jxb/erad412.

Pattyn, J., Vaughan-Hirsch, J. and van de Poel, B. (2021) The regulation of ethylene biosynthesis: A complex multilevel control circuitry. *New Phytologist* 229(2), 770–782. DOI: 10.1111/nph.16873.

Popa, D.G., Georgescu, F., Dumitrascu, F., Shova, S., Constantinescu-Aruxandei, D. *et al.* (2023) Novel strigolactone mimics that modulate photosynthesis and biomass accumulation in *Chlorella sorokiniana*. *Molecules* 28(20), 7059. DOI: 10.3390/molecules28207059.

Proust, H., Hoffmann, B., Xie, X., Yoneyama, K., Schaefer, D.G. *et al.* (2011) Strigolactones regulate protonema branching and act as a quorum sensing-like signal in the moss *Physcomitrella patens*. *Development* 138(8), 1531–1539. DOI: 10.1242/dev.058495.

Raza, A., Razzaq, A., Mehmood, S.S., Zou, X., Zhang, X. *et al.* (2019) Impact of climate change on crops adaptation and strategies to tackle its outcome: A review. *Plants* 8(2), 34. DOI: 10.3390/plants8020034.

Ruyter-Spira, C., Al-Babili, S., van der Krol, S. and Bouwmeester, H. (2013) The biology of strigolactones. *Trends in Plant Science* 18(2), 72–83. DOI: 10.1016/j.tplants.2012.10.003.

Saeed, W., Naseem, S. and Ali, Z. (2017) Strigolactones biosynthesis and their role in abiotic stress resilience in plants: A critical review. *Frontiers in Plant Science* 8, 1487. DOI: 10.3389/fpls.2017.01487.

Shimels, M.Z., Rendine, S., Ruyter-Spira, C., Rich, P.J., Ejeta, G. *et al.* (2025) The role of strigolactone structural diversity in the host specificity and control of *Striga*, a major constraint to sub-Saharan agriculture. *Plants, People, Planet* 7(2), 318–330. DOI: 10.1002/ppp3.10549.

Soliman, S., Wang, Y., Han, Z., Pervaiz, T. and El-Kereamy, A. (2022) Strigolactones in plants and their interaction with the ecological microbiome in response to abiotic stress. *Plants* 11(24), 3499. DOI: 10.3390/plants11243499.

Ueno, K., Sugimoto, Y. and Zwanenburg, B. (2015) The genuine structure of alectrol: End of a long controversy. *Phytochemistry Reviews* 14(5), 835–847. DOI: 10.1007/s11101-014-9380-2.

Umehara, M., Hanada, A., Yoshida, S., Akiyama, K., Arite, T. *et al.* (2008) Inhibition of shoot branching by new terpenoid plant hormones. *Nature* 455(7210), 195–200. DOI: 10.1038/nature07272.

Waadt, R., Seller, C.A., Hsu, P.K., Takahashi, Y., Munemasa, S. *et al.* (2022) Plant hormone regulation of abiotic stress responses. *Nature Reviews. Molecular Cell Biology* 23(10), 680–694. DOI: 10.1038/s41580-022-00479-6.

Wada, S., Cui, S. and Yoshida, S. (2019) Reactive oxygen species (ROS) generation is indispensable for haustorium formation of the root parasitic plant *Striga hermonthica*. *Frontiers in Plant Science* 10, 328. DOI: 10.3389/fpls.2019.00328.

Wakabayashi, T., Shida, K., Kitano, Y., Takikawa, H., Mizutani, M. *et al*. (2020) CYP722C from *Gossypium arboreum* catalyzes the conversion of carlactonoic acid to 5-deoxystrigol. *Planta* 251(5), 97. DOI: 10.1007/s00425-020-03390-6.

Wakabayashi, T., Hamana, M., Mori, A., Akiyama, R., Ueno, K. *et al*. (2019) Direct conversion of carlactonoic acid to orobanchol by cytochrome P450 CYP722C in strigolactone biosynthesis. *Science Advances* 5(12), eaax9067. DOI: 10.1126/sciadv.aax9067.

Wakabayashi, T., Yasuhara, R., Miura, K., Takikawa, H., Mizutani, M. *et al*. (2021a) Specific methylation of (11*R*)-carlactonoic acid by an arabidopsis SABATH methyltransferase. *Planta* 254(5), 88. DOI: 10.1007/s00425-021-03738-6.

Wakabayashi, T., Ishiwa, S., Shida, K., Motonami, N., Suzuki, H. *et al*. (2021b) Identification and characterization of sorgomol synthase in sorghum strigolactone biosynthesis. *Plant Physiology* 185(3), 902–913. DOI: 10.1093/plphys/kiaa113.

Wakabayashi, T., Moriyama, D., Miyamoto, A., Okamura, H., Shiotani, N. *et al*. (2022a) Identification of novel canonical strigolactones produced by tomato. *Frontiers in Plant Science* 13, 1064378. DOI: 10.3389/fpls.2022.1064378.

Wakabayashi, T., Ueno, K. and Sugimoto, Y. (2022b) Structure elucidation and biosynthesis of orobanchol. *Frontiers in Plant Science* 13, 835160. DOI: 10.3389/fpls.2022.835160.

Wakabayashi, T., Nakayama, M., Kitano, Y., Homma, M., Miura, K. *et al*. (2025) Discovery of strigol synthase from cotton (*Gossypium hirsutum*): The enzyme behind the first identified germination stimulant for *Striga*. *Plants People Planet* 7(2), 354–359. DOI: 10.1002/ppp3.10420.

Walker, C.H., Siu-Ting, K., Taylor, A., O'Connell, M.J. and Bennett, T. (2019) Strigolactone synthesis is ancestral in land plants, but canonical strigolactone signalling is a flowering plant innovation. *BMC Biology* 17(1), 70. DOI: 10.1186/s12915-019-0689-6.

Wang, J., Song, L., Gong, X., Xu, J. and Li, M. (2020) Functions of jasmonic acid in plant regulation and response to abiotic stress. *International Journal of Molecular Sciences* 21(4), 1446. DOI: 10.3390/ijms21041446.

Wang, J.Y., Haider, I., Jamil, M., Fiorilli, V., Saito, Y. *et al*. (2019) The apocarotenoid metabolite zaxinone regulates growth and strigolactone biosynthesis in rice. *Nature Communications* 10(1), 810. DOI: 10.1038/s41467-019-08461-1.

Waters, M.T., Gutjahr, C., Bennett, T. and Nelson, D.C. (2017) Strigolactone signaling and evolution. *Annual Review of Plant Biology* 68(1), 291–322. DOI: 10.1146/annurev-arplant-042916-040925.

Xie, X. (2016) Structural diversity of strigolactones and their distribution in the plant kingdom. *Journal of Pesticide Science* 41(4), 175–180. DOI: 10.1584/jpestics.J16-02.

Xie, X., Yoneyama, K. and Yoneyama, K. (2010) The strigolactone story. *Annual Review of Phytopathology* 48(1), 93–117. DOI: 10.1146/annurev-phyto-073009-114453.

Yamada, Y., Furusawa, S., Nagasaka, S., Shimomura, K., Yamaguchi, S. *et al*. (2014) Strigolactone signaling regulates rice leaf senescence in response to a phosphate deficiency. *Planta* 240(2), 399–408. DOI: 10.1007/s00425-014-2096-0.

Yang, S.T., Fan, J.B., Liu, T.T., Ning, S., Xu, J.H. *et al*. (2022) Development of strigolactones as novel autophagy/mitophagy inhibitors against colorectal cancer cells by blocking the autophagosome-lysosome fusion. *Journal of Medicinal Chemistry* 65(14), 9706–9717. DOI: 10.1021/acs.jmedchem.2c00275.

Yoda, A., Mori, N., Akiyama, K., Kikuchi, M., Xie, X. *et al*. (2021) Strigolactone biosynthesis catalyzed by cytochrome P450 and sulfotransferase in sorghum. *New Phytologist* 232(5), 1999–2010. DOI: 10.1111/nph.17737.

Yokota, T., Sakai, H., Okuno, K., Yoneyama, K. and Takeuchi, Y. (1998) Alectrol and orobanchol, germination stimulants for *Orobanche minor*, from its host red clover. *Phytochemistry* 49(7), 1967–1973. DOI: 10.1016/S0031-9422(98)00419-1.

Yoneyama, K. (2020) Recent progress in the chemistry and biochemistry of strigolactones. *Journal of Pesticide Science* 45(2), 45–53. DOI: 10.1584/jpestics.D19-084.

Yoneyama, K. and Brewer, P.B. (2021) Strigolactones, how are they synthesized to regulate plant growth and development? *Current Opinion in Plant Biology* 63, 102072. DOI: 10.1016/j.pbi.2021.102072.

Yoneyama, K., Awad, A.A., Xie, X., Yoneyama, K. and Takeuchi, Y. (2010) Strigolactones as germination stimulants for root parasitic plants. *Plant and Cell Physiology* 51(7), 1095–1103. DOI: 10.1093/pcp/pcq055.

Yoneyama, K., Xie, X., Kisugi, T., Nomura, T. and Yoneyama, K. (2013) Nitrogen and phosphorus fertilization negatively affects strigolactone production and exudation in sorghum. *Planta* 238(5), 885–894. DOI: 10.1007/s00425-013-1943-8.

Yoneyama, K., Xie, X., Yoneyama, K., Kisugi, T., Nomura, T. *et al.* (2018a) Which are the major players, canonical or non-canonical strigolactones? *Journal of Experimental Botany* 69(9), 2231–2239. DOI: 10.1093/jxb/ery090.

Yoneyama, K., Mori, N., Sato, T., Yoda, A., Xie, X. *et al.* (2018b) Conversion of carlactone to carlactonoic acid is a conserved function of MAX1 homologs in strigolactone biosynthesis. *New Phytologist* 218(4), 1522–1533. DOI: 10.1111/nph.15055.

Zhou, A., Kane, A., Wu, S., Wang, K., Santiago, M. *et al.* (2025) Evolution of interorganismal strigolactone biosynthesis in seed plants. *Science* 387(6731), eadp0779. DOI: 10.1126/science.adp0779.

Zorrilla, J.G., Rial, C., Varela, R.M., Molinillo, J.M.G. and Macías, F.A. (2022) Strategies for the synthesis of canonical, non-canonical and analogues of strigolactones, and evaluation of their parasitic weed germination activity. *Phytochemistry Reviews* 21(5), 1627–1659. DOI: 10.1007/s11101-022-09801-8.

Zwanenburg, B. and Blanco-Ania, D. (2018) Strigolactones: New plant hormones in the spotlight. *Journal of Experimental Botany* 69(9), 2205–2218. DOI: 10.1093/jxb/erx487.

Zwanenburg, B., Mwakaboko, A.S. and Kannan, C. (2016) Suicidal germination for parasitic weed control. *Pest Management Science* 72(11), 2016–2025. DOI: 10.1002/ps.4222.

2 Phytohormonal Crosstalk with Strigolactones in Plants

Muhammad Ilyas, Zunaira Malik, Muhammad Munawar Arshad, Ayesha, Naila Ali* and Zaib-un-Nisa

Institute of Molecular Biology and Biotechnology, University of Lahore, Lahore

Abstract

Strigolactones (SLs) are a class of carotenoid-derived phytohormones recognized for their multi-faceted roles in regulating plant growth, development, and environmental responses. Initially identified for their function in promoting symbiosis with arbuscular mycorrhizal fungi and inhibiting parasitic seed germination, SLs have since been established as key modulators of plant architecture and adaptive responses. A growing body of research reveals that SLs do not function in isolation; rather, they participate in extensive crosstalk with other phytohormones to orchestrate complex physiological and developmental processes. SL interactions with auxins are central to apical dominance and root system architecture, where SLs modulate polar auxin transport by affecting PIN protein localization. SL–cytokinin crosstalk influences axillary bud outgrowth and meristem activity, often functioning antagonistically. In abscisic acid signaling, SLs act synergistically to regulate stomatal closure, improve drought tolerance, and mediate reactive oxygen species detoxification. SLs also enhance ethylene signaling during root-hair elongation and contribute to stress adaptation under nutrient deficiencies. Additionally, interactions with gibberellins, jasmonic acid, and salicylic acid reflect the broader integration of SLs into biotic and abiotic stress signaling networks. These crosstalk events are mediated through shared signaling components such as transcription factors (e.g. BES1, BRC1), F-box proteins (e.g. MAX2), and downstream repressors (e.g. SMXLs). SLs influence gene expression both directly and through modulation of other hormone pathways, thereby coordinating growth and defense responses in a context-dependent manner. Recent advances in transcriptomics, biosensors, and mutant analyses have shed light on the spatial–temporal regulation of SL signaling and its interaction with other hormonal pathways. Understanding these mechanisms is vital for decoding how plants maintain developmental plasticity in dynamic environments. This analysis summarizes current knowledge on SL-mediated phytohormonal crosstalk, emphasizing its roles in shoot branching, root architecture, stomatal regulation, stress responses, and nutrient signaling. A detailed exploration of SL interaction networks offers valuable insights for developing novel strategies in crop improvement, particularly under conditions of abiotic stress and limited resource availability. Unraveling these integrative hormonal pathways holds promise for the design of high-yield, stress-resilient cultivars suited to the challenges of global climate change and sustainable agriculture.

Keywords: antioxidant defense system, climate change resilience, CRISPR-Cas9, downstream repressors (SMXLs), hormonal crosstalk, oxidative stress, PIN protein localization, reactive oxygen species, strigolactones, sustainable agriculture

*Corresponding author: naila.ali@imbb.uol.edu.pk

DOI: 10.1079/9781836990598.0002

2.1 Introduction

Strigolactones (SLs) are carotenoid-derived hormones acting as key integrators in plant development and stress resilience, engaging in dynamic crosstalk with abscisic acid (ABA), jasmonic acid (JA), auxin, cytokinins (CKs), ethylene, brassinosteroids (BRs), gibberellins (GAs), and salicylic acid (SA) (Korek and Marzec, 2023). SL–ABA crosstalk is extensively documented: ABA upregulates SL biosynthetic genes (*DWARF27* (*D27*), *carotenoid cleavage dioxygenase 7* (*CCD7*), and *carotenoid cleavage dioxygenase 8* (*CCD8*)) in rice, *Arabidopsis*, tomato, maize, and grapevine, enhancing SL levels under stress (Ding *et al.*, 2025). Conversely, SLs (via GR24) induce NCED ABA-biosynthesis genes (*OsNCED1/2/3*) in rice shoots and roots, elevating ABA and promoting stomatal closure and drought tolerance (Parwez *et al.*, 2022). However, antagonism also occurs: high ABA in barley and rice suppresses SL-biosynthesis transcripts, while SLs in grape berries accelerate ABA degradation (Gao *et al.*, 2022). SLs positively modulate JA and ABA synergy during cotton's defense against *Verticillium* wilt: ABA enhances SL biosynthesis, and JA (via MYC2) negatively feeds back, forming a homeostatic loop controlling SL content (Yi *et al.*, 2023). In tomato under cold conditions, ABA and JA act synergistically: ABA induces the JA-biosynthesis gene *SlLOXD/AOS/AOC/OPR3*, and JA (through MYC2 targeting *NCED2*) boosts ABA, together enhancing cold tolerance (Ding *et al.*, 2022). SLs also crosstalk with auxin, typically antagonizing auxin-mediated root branching and adventitious root formation: in melon, GR24 + IAA shifts expression of SL genes (*crtB/P/Q*, *D14*, *SMAX*) and signaling genes (*PYR/PYL*, *PP2C*, *SnRK2*, *EIN3*, *BAK1*, *BRI1*, *JAZ*, *TGA/PR-1*) to modulate architecture (Li *et al.*, 2023). SLs interact synergistically with CKs and ethylene: GR24 inhibits primary root elongation in *Arabidopsis* (CK synergy), and SL-insensitive tiller/dwarf rice mutants have elevated auxin levels (Voko *et al.*, 2024). SLs also influence SA and reactive oxygen species (ROS)-related defenses: exogenous SL activates auxin-responsive, ROS- and SA-inducible genes; SL-biosynthesis mutants show compromised tomato resistance to root-knot nematode (Xu *et al.*, 2019). Through the F-box MAX2, SL perception intersects BR signaling by targeting BZR1/BES1 for degradation, and acts in parallel to ABA in guard-cell regulation (Huang *et al.*, 2025a). JA–ABA crosstalk is mediated via JAZ and ABI3/ABI5 interactions; KEG ubiquitin ligase stabilizes JAZ12 and degrades ABI5, fine-tuning ABA/JA responses (Mei *et al.*, 2023). MYC2 integrates signals from SL, JA, and ABA, repressing SL genes (CCD7/8) in cotton and activating NCED2 in tomato (Zeng *et al.*, 2025). SL–auxin antagonism also adjusts root structural design under phosphate/nitrogen deficiency via modulation of auxin transport (Mehmood *et al.*, 2025a). SL–ABA synergy promotes drought adaptation by increasing ABA sensitivity, stomatal control, and antioxidant enzyme activity in pepper, grapevine, apple, cucumber, tomato, lettuce, and *Pennisetum* (Huang *et al.*, 2025b). SLs enhance cell elongation, chlorophyll synthesis, fruit ripening, vascular cambium activity, redox balance, and secondary growth in horticultural plants, often mediated by ABA (Kapoor *et al.*, 2022). SL–JA–SA interplay underpins systemic defense signaling against pathogens and herbivores (Ku *et al.*, 2024). In sum, SLs are central regulators orchestrating plant development (shoot/root architecture, branching, secondary growth) and stress adaptation (drought, cold, pathogens) via nuanced synergistic and antagonistic interactions with other phytohormones, with feedbacks involving biosynthesis genes (D27, CCD7/8, NCEDs), receptors/hubs (MAX2, D14, MYC2, JAZ, ABI5), and environmental triggers (Liu *et al.*, 2023).

2.2 Importance of Strigolactones as Central Signaling Molecules

SLs are increasingly recognized as central signaling molecules that integrate plant development and environmental responses. These carotenoid-derived phytohormones regulate diverse developmental processes such as shoot branching, root architecture, and leaf senescence, while also mediating responses to abiotic and biotic stresses (Sharma *et al.*, 2024a, b) . In shoot architecture, SLs suppress axillary bud outgrowth by modulating the expression of BRANCHED1 (BRC1) and auxin transport via PIN proteins (Yan *et al.*,

2020). For instance, GR24, a synthetic SL analog, inhibits lateral bud development in *Arabidopsis* and rice, even under low auxin levels (Aliche *et al.*, 2020). In roots, SLs regulate primary root elongation and lateral root density, particularly under nutrient-deficient conditions such as phosphate or nitrogen scarcity. Under drought, SLs upregulate ABA biosynthesis genes (e.g. NCEDs) and enhance ABA sensitivity in guard cells, leading to stomatal closure and improved water retention (Wei *et al.*, 2025). They also stimulate antioxidant enzyme activity, reducing ROS damage during abiotic stress (Sharma *et al.*, 2024a, b). SLs mediate beneficial rhizosphere interactions by promoting arbuscular mycorrhizal fungi (AMF) colonization, crucial for phosphorus uptake (Liu *et al.*, 2024a). In cotton, a synergistic interaction between SL, ABA, and JA enhances resistance against *Verticillium dahliae* through MYC2-mediated signaling feedback loops (Yi *et al.*, 2023). Moreover, SLs crosstalk with other hormones such as CKs and GAs to fine-tune tillering and senescence (Sharma *et al.*, 2024a, b). They also play a role in horticultural crops such as tomato and cucumber, with SLs modulating fruit ripening, vascular development, and chlorophyll biosynthesis (Wang *et al.*, 2025b). Through their receptor D14 and the F-box protein MAX2, SLs regulate downstream repressors such as D53/SMXLs, integrating developmental and stress cues (Eckardt *et al.*, 2024).

2.3 Strigolactone Signaling Pathway Recap

2.3.1 Biosynthesis and perception (D14, MAX2, and SMXL proteins)

SLs are phytohormones that regulate plant development and environmental adaptation through a well-defined biosynthesis and signaling cascade. The SL-biosynthesis pathway starts in plastids, where the isomerase DWARF27 (D27) converts all-*trans*-β-carotene to 9-*cis*-β-carotene. This is followed by the sequential action of the carotenoid cleavage dioxygenases CCD7 and CCD8, which produce carlactone, a pivotal SL precursor (Sharma *et al.*, 2024a, b). Carlactone is then exported to the cytosol and oxidized by cytochrome P450 enzymes of the MORE AXILLARY GROWTH 1 (MAX1) family into canonical SLs, including 5-deoxystrigol, orobanchol, and strigol, depending on species-specific enzymatic modifications (Taulera, 2022). Perception of SLs occurs through the receptor DWARF14 (D14), an α/β-hydrolase that binds and hydrolyzes SL molecules, causing a structural rearrangement that allows interaction with the F-box protein MAX2 (Zaidi *et al.*, 2024). This D14–MAX2 complex facilitates the ubiquitination and proteasomal degradation of SMXL (SUPPRESSOR OF MAX2 1-LIKE) proteins, notably SMXL6, SMXL7, and SMXL8 in *Arabidopsis*, which are negative regulators of SL signaling (Mukherjee *et al.*, 2023). The degradation of SMXL repressors lifts the inhibition on downstream transcriptional networks that control axillary bud outgrowth, root system architecture, and leaf senescence (Yuan *et al.*, 2023).

Mutants defective in D14 or MAX2 exhibit classic SL-deficient phenotypes, including excessive shoot branching, dwarfism, reduced apical dominance, and diminished response to phosphate starvation. For instance, *Arabidopsis d14* and *max2* mutants form additional branches due to failure to repress bud outgrowth, while in rice, *Osd14* and *Osmax2* mutants similarly show bushy phenotypes and reduced panicle length (Su *et al.*, 2024b). SMXL proteins have also been shown to mediate hormonal crosstalk: SMXL6 interacts with DELLA proteins to link SL and GA signaling, while SMXL7 modulates auxin transport genes, influencing root development (Sun *et al.*, 2024). Furthermore, in pea and tomato, SL perception regulates vascular differentiation and cambial activity, linking the D14–MAX2–SMXL module to secondary growth and mechanical support (Wang *et al.*, 2024).

Recent studies indicate that MAX2 is not only central to SL signaling but also participates in signaling pathways of other hormones, such as karrikins and ABA, suggesting convergence at the level of SCF^{MAX2}-mediated proteolysis (Temmerman *et al.*, 2022). In addition, fine-tuning of SMXL gene expression is crucial for balancing growth and stress responses. For example, under drought, MAX2-dependent SL signaling promotes stomatal closure and antioxidant responses through SMXL degradation (Kamran *et al.*, 2024). Thus, the SL pathway—comprising the biosynthetic enzymes (D27,

CCD7/8, MAX1), perception module (D14, MAX2), and regulatory repressors (SMXLs)—acts as a central role in integrating developmental and environmental cues across diverse plant species (Yuan *et al.*, 2023).

2.3.2 Transport and movement within plant tissues

SLs are transported within plant tissues to function as systemic signaling molecules, coordinating development and environmental responses. SLs move both acropetally (from roots to shoots) and laterally to target tissues to regulate shoot branching, root architecture, and symbiotic interactions (Liu *et al.*, 2024a). The transport is largely mediated by ATP-binding cassette (ABC) transporters, particularly the PLEIOTROPIC DRUG RESISTANCE 1 (PDR1) family (Khunweeraphong and Kuchler, 2021). In *Petunia hybrida*, PDR1 is localized asymmetrically in root hypodermal passage cells and xylem parenchyma, exporting SLs into the rhizosphere and facilitating shootward movement (Akhiyarova *et al.*, 2024). Similar roles of ABCG/PDR transporters have been confirmed in *Arabidopsis* and tomato, indicating evolutionary conservation of SL movement mechanisms (Mehmood *et al.*, 2025b).

Transport is also developmentally and environmentally regulated. For instance, under phosphate starvation, SL transport to the shoot increases to suppress axillary bud growth and conserve energy (Liu *et al.*, 2024a). In rice, the *OsPDR9* transporter has been implicated in controlling SL efflux from roots and long-distance movement via the xylem (Lwin *et al.*, 2025). Moreover, the polar localization of SL transporters ensures directionality: in *Petunia*, PDR1 localizes toward the stele in roots and toward axillary bud vasculature in shoots, aligning with the hormone's function in inhibiting lateral branching (Al Tahan *et al.*, 2024). In addition to transporter-mediated movement, SLs may also move via simple diffusion or through phloem loading/unloading mechanisms, although this is less well understood. Studies using synthetic SL analogs such as GR24 have shown that SLs can be translocated to aerial tissues in a MAX2-dependent manner, indicating active signaling beyond passive transport (Liu *et al.*, 2024a).

Grafting experiments in *Arabidopsis* and tomato have confirmed that SL-deficient rootstocks can restore normal shoot branching when grafted with SL-competent scions, proving the mobile nature of SLs and their physiological relevance (Lee *et al.*, 2023). SL distribution is tightly regulated to avoid mis-signaling. For example, mutants in PDR-type transporters exhibit altered shoot phenotypes, increased lateral roots, and reduced mycorrhizal colonization due to defective SL localization (Díaz *et al.*, 2025). Furthermore, environmental signals such as drought and nutrient availability modulate SL export to both the rhizosphere (to recruit symbionts) and the shoot (to adjust growth), positioning SL transport as a key mediator of adaptive plasticity (Wang *et al.*, 2025a).

2.4 Strigolactone–Auxin Crosstalk

2.4.1 Regulation of shoot branching and root development

SLs are best known for their inhibitory effect on shoot branching (tillering). In nutrient-deficient conditions (especially low phosphate), SL biosynthesis is upregulated, and the hormone is transported from roots to shoots to suppress axillary bud outgrowth. This regulation conserves energy and resources under stress (Liu *et al.*, 2024a). Perception of SLs by the D14 receptor, followed by signaling through the F-box protein MAX2, leads to degradation of SMXL6/7/8 repressors, preventing activation of branching-promoting genes such as *BRC1* (Guan *et al.*, 2025). In *Arabidopsis*, *max* mutants (e.g. *max2*) exhibit excessive branching, confirming the role of SLs in apical dominance. Similarly, *OsD14* mutants in rice produce more tillers, supporting conserved SL function across species (Wang *et al.*, 2023).

In roots, SLs function both directly and through crosstalk with auxin. SLs inhibit lateral root formation and adventitious rooting, primarily by modulating auxin transport via PIN proteins. For example, exogenous application of the SL analog GR24 reduces lateral root density by altering PIN1 localization (Sun *et al.*, 2022).

Additionally, under low nitrogen or phosphate, SLs reduce root branching to optimize foraging efficiency and help form deeper, more exploratory root systems (Lu *et al.*, 2024). SL–auxin antagonism is key here: while auxin promotes lateral root development, SLs counterbalance this by repressing auxin transport or signaling (Yu *et al.*, 2021). SLs also modulate root-hair elongation, beneficial for nutrient uptake. In tomato and rice, SL-deficient mutants show shortened root hairs, while GR24 treatment restores hair length, especially under phosphorus deficiency (Ravazzolo and Quaggiotti, 2024).

2.4.2 Transport modulation and feedback regulation

SLs exhibit a tightly regulated transport and feedback system that ensures precise hormonal signaling for plant growth and environmental adaptation. SLs are synthesized mainly in roots and translocated to shoots through the xylem, where they regulate processes such as shoot branching, root development, and stress responses. The polar and non-polar transport of SLs is mediated by ABC transporters, particularly PDR1 in petunia and tomato, which facilitates root-to-shoot SL movement (Aloni, 2021). The directional movement of SLs is influenced by nutrient status: under phosphate or nitrogen starvation, SL production and export increase, enhancing systemic signaling to suppress branching and modify root traits (Yuan *et al.*, 2023).

Feedback regulation is critical for maintaining SL homeostasis. Negative feedback loops modulate SL biosynthesis genes (e.g. *D27*, *CCD7*, *CCD8*) to avoid overaccumulation. For instance, SL accumulation represses the expression of its own biosynthetic genes via transcriptional regulators, including BRC1 in shoots and SMXL proteins in roots, which inhibit the biosynthesis machinery once developmental thresholds are met (Ranganath and Maity, 2024). Additionally, hormonal crosstalk fine-tunes this regulation: ABA and JA can upregulate SL biosynthesis under stress, while auxin and CK interactions help modulate SL transport and feedback to align root and shoot development with external cues (Ameen *et al.*, 2025). Moreover, SL perception via D14 and MAX2 not only triggers degradation of repressors (SMXL6/7/8) but also activates feedback mechanisms that reduce SL output when a sufficient response has occurred (Chang *et al.*, 2025).

2.5 Strigolactone and Abscisic Acid Interaction

2.5.1 Cooperative roles in drought and salinity tolerance

SLs and ABA are two crucial phytohormones that perform synergistic and cooperative roles in plant responses to abiotic stresses, particularly drought and salinity. Recent studies have increasingly demonstrated the crosstalk between SL and ABA pathways in regulating stress-tolerance mechanisms. Both hormones modulate root system architecture and stomatal conductance, enhancing water-use efficiency under drought stress (Jogawat *et al.*, 2021). SLs are known to promote deeper rooting and reduce shoot branching, optimizing water uptake and reducing transpiration, while ABA predominantly induces stomatal closure and activates expression of stress-responsive genes (Margay *et al.*, 2024). Under salinity conditions, SLs enhance salt tolerance by regulating ion homeostasis and antioxidant defense systems, mechanisms often co-regulated with ABA signaling (Khan *et al.*, 2023a). The SL–ABA interaction has been shown to modulate ROS-scavenging enzymes, thereby improving cellular redox balance during stress (Hashem and Khalil, 2023).

Genetic studies have revealed that mutants deficient in SL biosynthesis or signaling (e.g. *max* or *d14*) show hypersensitivity to drought and salinity, similar to ABA-insensitive mutants, suggesting overlapping downstream pathways (Liu *et al.*, 2023). Additionally, exogenous application of SL analogs such as GR24 improves ABA sensitivity, leading to enhanced drought tolerance in crops such as rice and tomato (Ghosh and Roychoudhury, 2024). Transcriptomic analyses have identified common targets of SL and ABA pathways, including genes encoding LEA proteins, dehydrins, and transcription factors (TFs) such as DREB and NAC (Kaya, 2025). These findings support the model where SLs potentiate ABA action rather than act independently. Moreover, SL-signaling components such as D14 and MAX2 interact with ABA receptors (PYR/PYL), facilitating coordinated responses (Ding *et al.*, 2025). The cooperative effect is also evident in root-to-shoot signaling, where ABA-induced expression of SL biosynthetic genes increases under osmotic stress (Kumar *et al.*, 2023). As agriculture faces escalating drought and salinity threats due to climate change, understanding SL–ABA crosstalk is critical for engineering resilient crops. Advances in clustered regularly interspaced short palindromic repeats (CRISPR)/Cas9-based gene editing are being explored to modulate SL and ABA signaling for enhanced stress adaptation (Shelake *et al.*, 2022).

2.5.2 ROS signaling and stomatal regulation

ROS signaling is significant for plant stress perception and response, particularly in the regulation of stomatal movement during abiotic stresses such as drought and salinity. Under stress conditions, plants rapidly accumulate ROS, including hydrogen peroxide and superoxide (O_2^-), which serve as secondary messengers in various signaling cascades (Khan *et al.*, 2025). ROS are generated in guard cells primarily by NADPH oxidases, also known as respiratory burst oxidase homologs (RBOHs), especially RBOHD and RBOHF (Yu *et al.*, 2023). Upon abiotic stress or ABA stimulation, RBOH-dependent ROS production increases, promoting stomatal closure to minimize water loss (Devireddy *et al.*, 2021). This process is tightly regulated through calcium (Ca^{2+}) influx, protein kinases such as OST1, and anion channels such as SLAC1, all of which are modulated by ROS. ROS signaling thus acts as a key integrator of hormonal and environmental cues in guard cells (Wang *et al.*, 2025).

Additionally, ROS interact with nitric oxide (NO) and ABA to modulate downstream components of stomatal regulation. For instance, ROS and NO act synergistically to activate ion fluxes across the plasma membrane of guard cells (Ahmad *et al.*, 2024). Antioxidant enzymes, such as superoxide dismutase (SOD), catalase (CAT), and ascorbate peroxidase (APX), also contribute to fine-tuning ROS levels, preventing oxidative damage while preserving signaling function (Rahman *et al.*, 2024). Current transcriptomic studies highlight that ROS-responsive TFs, such as ZAT12, WRKY, and NAC families, regulate genes involved in stomatal dynamics and stress acclimation (Su *et al.*, 2024a). ROS-mediated stomatal closure is a rapid and reversible process, critical for balancing photosynthetic efficiency with stress protection. Furthermore, environmental signals such as light and CO_2 also influence ROS production in guard cells, indicating that ROS signaling integrates multiple inputs (Ma and Bai, 2021). Mutants deficient in ROS production or scavenging often exhibit impaired stomatal function and increased susceptibility to stress (Yadav *et al.*, 2023). Advanced imaging and genetic tools have confirmed ROS's spatial–temporal patterns in guard cells under fluctuating environments. As the impact of climate change intensifies, leveraging ROS signaling pathways offers a promising route to improve crop water-use efficiency and stress tolerance through genetic and chemical interventions (Bao *et al.*, 2024).

2.6 Strigolactone and Cytokinin Interaction

2.6.1 Antagonistic control of meristem activity and shoot architecture

SLs and CKs are key phytohormones that act antagonistically to regulate shoot architecture and meristem activity, particularly in the control of shoot branching and apical dominance. While CKs promote cell division and the outgrowth of axillary buds by stimulating meristematic activity, SLs suppress bud outgrowth, thereby maintaining apical dominance and restricting excessive branching (Yuan *et al.*, 2023). This antagonistic interaction is fundamental to optimizing plant architecture for environmental adaptation and resource efficiency. SLs inhibit CK biosynthesis by downregulating *ISOPENTENYL TRANSFERASE* (*IPT*) genes and promote CK degradation by upregulating *CYTOKININ OXIDASE/DEHYDROGENASE* (*CKX*) genes, thereby limiting CK availability in axillary buds (Chakraborty *et al.*, 2025). Conversely, CKs can counteract SL signaling by reducing the expression of SL perception components such as *D14* and *MAX2*, thus promoting branching (Yan *et al.*, 2020).

Transcriptomic and mutant analyses have provided further evidence for this hormonal crosstalk. In *Arabidopsis*, *max* mutants, which are deficient in SL signaling, exhibit excessive branching due to elevated CK levels in axillary meristems (Claassens, 2024). In contrast, CK-deficient mutants show suppressed branching, which can partially be restored by SL inhibition. These opposing effects are mediated through shared TFs, including BRC1, which is upregulated by SLs and downregulated by CKs (Guan *et al.*, 2025). BRC1 acts as a central node that integrates SL and CK signals to regulate bud fate. Additionally, studies in crops such as rice and tomato show that exogenous SL application reduces CK-responsive gene expression in axillary tissues, leading to decreased tillering or shoot branching (Liu *et al.*, 2024b). In maize, SLs restrict inflorescence branching by suppressing CK accumulation in meristematic zones, highlighting the conserved nature of this antagonism across species (Lv *et al.*, 2023). Hormonal balance between SLs and CKs also regulates the shoot apical meristem activity. CKs maintain meristematic cell proliferation, whereas SLs act to repress excessive growth, promoting a transition from indeterminate to determinate development (Chaudhry *et al.*, 2024). This interaction is crucial under stress or nutrient-deficient conditions, where SLs promote energy-saving architectural traits, while CKs maintain growth potential. Overall, fine-tuning the SL–CK balance has significant implications for crop architecture optimization, particularly in breeding programs targeting high-density planting or limited-resource environments (Guan *et al.*, 2025).

2.6.2 Modulation of root-to-shoot signaling

Modulation of root-to-shoot signaling is critical for coordinating plant responses to environmental stimuli such as drought, nutrient deficiency, and salinity. This systemic communication ensures that information sensed in roots is relayed to the shoot to initiate appropriate physiological or developmental responses. A key mechanism of root-to-shoot signaling involves the transport of phytohormones such as ABA, SLs, CKs, and peptides through the xylem (Xia *et al.*, 2025). For example, during drought stress, roots synthesize ABA, which is rapidly transported to leaves where it induces stomatal closure to conserve water (Mukherjee *et al.*, 2023). Similarly, nitrogen deficiency triggers increased root production of SLs, which travel to the shoot to inhibit lateral bud outgrowth, thereby reallocating resources to root growth (Naseer *et al.*, 2024).

Modern studies have identified small peptides such as CLAVATA3/EMBRYO SURROUNDING REGION-related (CLE) and C-TERMINALLY ENCODED PEPTIDEs (CEPs) as crucial mediators of root-derived signals. Under nitrogen limitation, CEPs produced in roots move to shoots, where they bind to receptors that promote systemic responses, including enhanced nitrate uptake (Xiao *et al.*, 2025). In phosphate-starved conditions, root-derived SLs are also upregulated and modulate shoot architecture by limiting tillering and enhancing root elongation (Mishra, 2021). Cytokinins synthesized in roots act as positive root-to-shoot signals under nutrient-rich conditions, stimulating shoot growth and leaf expansion (Vysotskaya *et al.*, 2021). The balance of CKs and SLs forms a dynamic feedback system that integrates environmental cues with plant architecture and development (Pierik *et al.*, 2021).

Mobile RNAs and microRNAs (miRNAs) have also emerged as important modulators of root-to-shoot signaling. For instance, miR399 is produced in phosphate-deficient shoots and moves to roots to suppress PHO2, a repressor of phosphate uptake (Fahad *et al.*, 2025). Aquaporins and hydraulic signals further modulate water potential gradients between roots and shoots, impacting transpiration and growth. Root-sensed salinity stress can alter xylem sap composition, affecting shoot ion homeostasis and growth patterns (Zhou *et al.*, 2024). Moreover, current transcriptomic data reveal that systemic signals modulate gene expression in distant tissues, ensuring an integrated whole-plant response (Xia *et al.*, 2025). In crops such as rice and maize, manipulating root-to-shoot signaling pathways through genetic engineering or hormonal treatments has shown promise in improving stress resilience and nutrient-use efficiency (Khan *et al.*, 2023b). Understanding the crosstalk among hormonal, peptide, and hydraulic signals offers novel opportunities to enhance crop productivity in fluctuating environments (Aloo *et al.*, 2023).

2.7 Strigolactone and Ethylene Interplay

2.7.1 Joint regulation in root-hair elongation and stress adaptation

SLs and ethylene exhibit a synergistic interplay in regulating root-hair elongation and plant adaptation to environmental stresses, particularly in non-grass species such as *Arabidopsis thaliana*, *Medicago truncatula*, and *Brassica napus* (Raza *et al.*, 2021). Root hairs are essential for water and nutrient uptake, and their elongation is a critical adaptive response to nutrient deficiencies and abiotic stresses. SLs have been shown to enhance ethylene biosynthesis by upregulating key biosynthetic genes such as 1-aminocyclopropane-1-carboxylate synthase (*ACS*), leading to increased ethylene production and subsequent stimulation of root-hair growth (Jia *et al.*, 2022). In *Arabidopsis*, mutants deficient in SL signaling (e.g. *max2*) exhibit reduced root-hair length, which can partially be restored by exogenous application of ethylene precursors such as 1-aminocyclopropane-1-carboxylic acid (ACC), indicating that SLs function upstream of ethylene in this pathway (Aiba and Umehara, 2024).

Conversely, ethylene also enhances the sensitivity of cells to SLs by modulating SL receptor gene expression (*D14*) and signaling components such as *MAX2*, establishing a positive feedback loop that intensifies root-hair development under stress (Sharma *et al.*, 2024a, b). In *Medicago truncatula*, SL–ethylene crosstalk regulates root-hair elongation under phosphate starvation, enhancing root surface area to improve phosphorus acquisition (Hu *et al.*, 2025). Under drought stress, this interaction also supports the formation of deeper root systems with increased hair density to improve water absorption. Ethylene stabilizes the TFs EIN3/EIL1, which interact with SL-regulated transcription networks such as BZR1 and SMXLs to drive root-hair differentiation (Kang *et al.*, 2022).

In *B. napus*, combined treatment with SL and ethylene enhances tolerance to salinity by promoting the elongation of root hairs and increasing antioxidant enzyme activities (He *et al.*, 2022). This response is associated with the upregulation of stress-responsive genes such as *RD29A*, *P5CS*, and ROS-scavenging enzymes. Moreover, transcriptomic profiling in *Arabidopsis* shows that both hormones co-regulate a subset of genes involved in cell-wall loosening and ROS signaling, which are critical for root-hair expansion (Zhao *et al.*, 2020). The dual-hormone regulation ensures plasticity in root development, allowing plants to respond dynamically to fluctuating environments (Jia *et al.*, 2022).

2.7.2 Ethylene biosynthesis affected by SL levels

SLs show a pivotal performance in modulating ethylene biosynthesis, influencing plant development and stress adaptation across a variety of non-model species such as *Medicago truncatula*, *Brassica napus*, *Solanum melongena* (eggplant), and *Populus trichocarpa* (poplar) (Dong *et al.*, 2024). It can upregulate ethylene production by enhancing the expression of key biosynthetic genes, particularly *ACS* and ACC oxidase (*ACO*), which catalyze the rate-limiting steps in ethylene synthesis (Khan *et al.*, 2024a). In *Medicago truncatula*, application of GR24, a synthetic SL analog, significantly increased ethylene levels under phosphate starvation, correlating with enhanced root-hair development and improved phosphorus uptake (Kapoor *et al.*, 2024).

In *Brassica napus*, SL-induced ethylene biosynthesis supports root system remodeling during salt stress, enhancing lateral root density and improving Na^+ exclusion from shoots (Chen *et al.*, 2025). Exogenous SL treatment also elevates *BnACS6* and *BnACO2* expression in roots, suggesting that SLs modulate ethylene synthesis both transcriptionally and post-transcriptionally

(Wang *et al.*, 2021). In *Solanum melongena*, SLs promote ethylene biosynthesis during drought stress, where increased ethylene contributes to stomatal regulation and ROS detoxification, ultimately improving water-use efficiency (Luqman *et al.*, 2023). *Populus trichocarpa*, a woody perennial species, shows a strong correlation between SL accumulation and ethylene biosynthesis under hypoxic conditions. Transcriptome profiling reveals that SL treatments upregulate *PtACS1* and *PtACO4* and downregulate ethylene repressors, enhancing the tree's tolerance to flooding (Dai *et al.*, 2024). Interestingly, SLs not only affect gene expression but also modulate ethylene precursor (ACC) transport across tissues, amplifying the ethylene signal systemically (Khan *et al.*, 2024b).

The synergistic action between SLs and ethylene biosynthesis also extends to developmental processes. In *Chrysanthemum morifolium*, SL-mediated enhancement of ethylene levels accelerates flower senescence, which is vital for synchronizing bloom cycles in horticulture (Alvi *et al.*, 2022). In legumes, such as *Pisum sativum*, SL-deficient mutants show decreased ethylene production, impairing nodulation and root-hair formation—processes crucial for symbiotic nitrogen fixation (Kaniganti *et al.*, 2022). This regulatory axis allows plants to flexibly adapt to diverse environmental stresses. By fine-tuning ethylene biosynthesis, SLs ensure appropriate physiological responses without triggering stress-induced growth inhibition (Nazir *et al.*, 2024).

2.8 Interaction with Gibberellins, Brassinosteroids, and Jasmonates

2.8.1 Integration of SLs with GAs in stem elongation and dormancy

SLs integrate with GAs, BRs, and JAs to orchestrate stem elongation and dormancy regulation, especially in plant species beyond typical model crops, such as *Populus trichocarpa*, *Medicago truncatula*, *Chrysanthemum morifolium*, *Petunia hybrida*, and *Camellia sinensis* (Selwal *et al.*, 2023). In *Populus*, SLs act as key modulators of stem elongation by interacting antagonistically with GAs. SL-deficient or SL-insensitive mutants exhibit enhanced internode elongation due to upregulated GA biosynthesis genes such as *GA20ox* and *GA3ox* (Das *et al.*, 2025). Conversely, SL application represses GA signaling by promoting the accumulation of DELLA proteins, key repressors of GA-induced growth, thereby reducing stem elongation and supporting bud dormancy (Yang *et al.*, 2023a).

In *Medicago truncatula*, a similar antagonistic interaction is observed, where GR24 treatment reduces GA activity and suppresses elongation of primary stems under phosphate-limited conditions (Jhu *et al.*, 2025). Interestingly, SL–GA crosstalk is highly context-dependent: under low-nutrient or high-density planting conditions, SLs dominate to restrict elongation, whereas in nutrient-rich environments, GAs promote growth more strongly (Duc *et al.*, 2025). BRs, known for promoting cell elongation and vascular differentiation, also interact with SLs in a tissue-specific manner. In *Petunia hybrida*, SLs negatively regulate BR activity in shoot tissues by modulating the expression of *BRI1* and downstream TFs such as *BZR1*, resulting in compact plant architecture (Altamura *et al.*, 2023). The cross-regulation ensures controlled elongation in response to light and nutrient signals. In *Camellia sinensis* (tea plant), combined SL and BR signaling affects internode development during the spring flush. SL application suppresses BR-responsive genes, which limits internode elongation, resulting in more compact bushy plants ideal for leaf harvest (Jin *et al.*, 2022).

SLs also synergize with jasmonates, particularly under stress or dormancy-inducing conditions. In *Chrysanthemum morifolium*, cold-induced dormancy is associated with elevated SL and JA levels. SLs promote JA biosynthesis by upregulating *LOX* and *AOS* genes, which helps trigger growth arrest and the activation of stress-responsive gene networks (Rehman *et al.*, 2023). This integration is vital for maintaining dormancy under unfavorable conditions and preparing for reactivation during favorable seasons (Klupczyńska and Pawłowski, 2021).

Moreover, in *Populus*, SL–JA synergy plays a key role in suppressing premature bud break during winter. Transcriptomic analyses reveal that SLs upregulate JA-signaling components such as *JAZ* repressors and *MYC2*, contributing to the repression of cell-division genes in axillary meristems (Rehman *et al.*, 2023). This mechanism ensures energy conservation during seasonal dormancy. The integration of SLs with GAs, BRs, and JAs thus forms a complex hormonal web, allowing plants to balance growth and dormancy in response to environmental and developmental cues (Abd-Elsalam and Mohamed, 2024). Understanding these interactions in diverse species offers strategies to engineer desired plant forms such as compact ornamental varieties, stress-tolerant perennials, and seasonally regulated woody crops through targeted hormone regulation (Kisvarga *et al.*, 2023).

2.8.2 SL and JA/BR roles in stress responses and defense

SLs interact with JA and BRs to coordinate complex stress responses and defense mechanisms in plants, particularly in non-model species. SL–JA crosstalk is especially critical in mounting a defense against both abiotic and biotic stresses (Wang *et al.*, 2024). In *Medicago truncatula*, SLs enhance JA biosynthesis under drought conditions by upregulating key JA pathway genes such as *LOX3*, *AOS*, and *JAR1*, thereby inducing the expression of dehydration-responsive elements (DREBs) and enhancing proline accumulation. This response improves osmotic adjustment and root survival during water deficit (Zhang *et al.*, 2022c).

Similarly, in *Camellia sinensis*, co-application of SL and JA analogs increases resistance to tea blister blight (*Exobasidium vexans*) by inducing pathogenesis-related (PR) genes and enhancing activities of defensive enzymes such as peroxidase and β−1, 3-glucanase. JA acts downstream of SL in this process, with SL acting as a priming agent that sensitizes tissues to JA-mediated immune signaling (Zhang *et al.*, 2022b). In *Populus trichocarpa*, SL–JA interaction enhances tolerance to herbivory by increasing the production of defensive secondary metabolites such as phenolics and terpenoids (Jiang *et al.*, 2024). SLs induce JA-signaling regulators such as *JAZ* repressors and *MYC2*, which orchestrate transcriptional activation of defense-related genes upon insect attack (Li *et al.*, 2022). BRs, in contrast, show synergistic interaction with SLs in abiotic stress tolerance, especially under salinity and oxidative stress. In *Petunia hybrida*, combined SL and BR treatments elevate antioxidant enzyme levels (e.g. SOD, CAT, and APX) and reduce electrolyte leakage under salt stress, indicating enhanced membrane stability (Hasanuzzaman *et al.*, 2020). BR-responsive genes such as *BZR1* and *TCH4* are upregulated when SL signaling is active, suggesting a cooperative relationship that enhances stress adaptation. In *Brassica napus*, SL–BR interplay also increases heat-stress tolerance by modulating heat-shock protein gene expression and improving photosynthetic efficiency (Cao *et al.*, 2024).

Interestingly, the interaction of SLs with JA and BR is tightly regulated at both transcriptional and post-translational levels. In many species, SLs influence the expression of JA/BR receptor genes, enhancing hormone sensitivity under stress (Kumar *et al.*, 2025). In *Populus*, protein–protein interaction assays confirm that SL-signaling components such as MAX2 interact with JA co-repressors (JAZs) and BR regulators (BIN2), modulating downstream stress responses (Chang *et al.*, 2025). These findings illustrate that SLs serve as hormonal hubs, modulating JA and BR signaling networks to fine-tune defense responses depending on the nature and severity of environmental stimuli (Sharma *et al.*, 2024a, b). The integration of these hormone pathways allows plants to prioritize growth or defense as needed, enhancing survival under dual stress conditions such as pathogen attack during drought. This dynamic regulation is crucial for optimizing resource allocation in perennial and economically important species (Samanta and Roychoudhury, 2023).

2.9 Network and Systems Biology Approaches to Hormonal Crosstalk

Network and systems biology approaches have revolutionized our understanding of hormonal crosstalk, especially the integration of SL signaling with other phytohormones in non-model plant species. By employing omics-based methods of transcriptomics, proteomics, metabolomics, and phosphoproteomics, researchers are now able to map complex regulatory networks and identify signaling hubs that mediate plant development and stress responses in crops (Magdy *et al.*, 2023). These approaches enable a systems-level view of how SL interacts with GAs, BRs, JAs, ABA, and CKs (Chen *et al.*, 2023).

In *Populus trichocarpa*, integration of SL and ABA signaling under drought stress was revealed through transcriptomic and phosphoproteomic profiling, where SL-responsive TFs (e.g. DREB2, NAC72) also regulate ABA biosynthesis and perception genes such as *NCED3* and *PYL8* (Yang *et al.*, 2023b). Co-expression network analysis pinpointed MAX2 and SnRK2.6 as central nodes, suggesting shared signaling hubs between SL and ABA pathways (Chang *et al.*, 2020). Similarly, in *Medicago truncatula*, multi-omics profiling under phosphate starvation identified SL and JA as core regulators of root system architecture, with key signaling crosstalk occurring at the level of transcriptional repressors such as JAZs and SMXLs (Su *et al.*, 2024b).

Proteomic studies in *Camellia sinensis* under heat-stress conditions revealed that SL enhances JA- and BR-responsive protein networks involved in ROS scavenging, including peroxidases, glutathione *S*-transferases, and heat-shock proteins (do Carmo Santos *et al.*, 2025). Integration of metabolomic data showed SL-driven increases in flavonoids and terpenoids, important for both defense and signaling feedback. In *Brassica napus*, systems biology tools such as weighted gene co-expression network analysis helped identify SL–GA–CK crosstalk hubs that control stem elongation and meristem identity, especially under variable nitrogen conditions (Xu *et al.*, 2023). A key finding was the convergence of multiple hormonal signals on the regulation of TCP and SPL TF families (Zhou *et al.*, 2022).

High-throughput phosphoproteomics in *Chrysanthemum morifolium* revealed that SL-signaling components such as D14 and MAX2 undergo phosphorylation in response to BR and JA treatments, affecting downstream hormone response factors and transcriptional networks controlling flower senescence and stress tolerance (Arefian *et al.*, 2021). These findings illustrate post-translational regulation as a vital mechanism in hormone crosstalk. Integrative multi-omics in these species has enabled the identification of conserved signaling hubs such as MAX2, DELLA proteins, BZR1, and MYC2 that operate as key nodes connecting diverse hormonal pathways (Roychowdhury *et al.*, 2023). These nodes act as molecular switches, balancing trade-offs between growth and stress responses. For example, in *Populus*, MAX2 not only regulates SL signaling but also physically interacts with BR and JA regulators to coordinate adaptive growth under high-density planting (Samanta and Roychoudhury, 2025).

Network modeling and dynamic simulation platforms have further allowed prediction of system behavior under hormonal perturbations. Boolean network models in *Medicago* suggest that low phosphate triggers a SL–JA–ABA regulatory module to prioritize root growth and drought avoidance (Karanam *et al.*, 2021). These models are increasingly being validated through CRISPR-Cas9 gene editing and hormone biosensor imaging. Overall, systems biology enables us to move beyond single-hormone analysis to understand the interdependencies and dynamic feedback mechanisms that govern plant adaptability (Chen *et al.*, 2024). This holistic view opens up new breeding targets for climate resilience, particularly in long-lived or understudied crops (Campbell *et al.*, 2025).

2.10 Shared Transcription Factors and Regulatory Nodes

Shared TFs and regulatory nodes are pivotal components that facilitate hormonal crosstalk in plants, especially involving SLs and other key phytohormones such as JAs, ABA, BRs, CKs, and GAs (Parwez *et al.*, 2022). These TFs function as integration hubs, translating multiple hormone signals into synchronized gene expression patterns, thereby enabling plants to finely balance growth, development, and stress responses (Nishanth, 2024). With the advancement of multi-omics and systems biology platforms, several TFs and signaling components have been uncovered as shared regulatory hubs across diverse plant lineages, extending beyond classical model organisms (Abdullah-Zawawi *et al.*, 2022).

One of the most prominent shared regulators is the TF MYC2, a basic helix–loop–helix protein central to JA signaling. In *Medicago truncatula*, MYC2 integrates SL and JA signaling under herbivory stress by regulating flavonoid and terpenoid biosynthesis genes, contributing to defense responses (Ribeiro *et al.*, 2022). MYC2 activity is influenced by the SL pathway through the SCF^{MAX2} complex, facilitating crosstalk between JA and BR signaling. Likewise, WRKY family TFs such as WRKY40 and WRKY72 have emerged as key players in *Populus trichocarpa*, mediating the SL–ABA–JA regulatory axis (Wani *et al.*, 2021). These TFs control expression of stress-responsive genes such as *PtRD22* and *PtPR5* under drought conditions, often in cooperation with NAC and AP2/ERF factors (Ma *et al.*, 2024).

The TF BZR1, a central regulator in BR signaling, also acts as a convergence point for SL–BR signaling. In *Brachypodium distachyon*, BZR1 interacts with SL receptor homologs such as BdD14 and co-repressors BdSMXLs, regulating genes related to cell-wall biosynthesis and stem elongation, which affects plant architecture under low-nutrient conditions (Cao *et al.*, 2024). Additionally, TCP TFs such as TCP3 and TCP17 play a significant role in SL–CK crosstalk. In *Lotus japonicus*, TCP17 regulates axillary bud development in concert with the SL-repressing BRC1 and CK-regulated genes such as *LjLOG4* and *LjARR5*, showing SL–CK integration in shoot branching control (Zhang *et al.*, 2022a).

Another critical hub, DELLA proteins, known for repressing GA signaling, are also involved in SL crosstalk by modulating the transcription and stability of SMXL repressors. In *Glycine max* (soybean), the DELLA protein GmRGL3 physically interacts with GmSMXL8, repressing GA-promoted elongation and enhancing SL-mediated stress tolerance through regulation of antioxidant genes such as *GmCAT1* and *GmAPX2* (Sarwar *et al.*, 2023). Similarly, AP2/ERF family TFs, which mediate ethylene and ABA responses, have been shown to integrate SL signals under abiotic stress (Feng *et al.*, 2020). In *Populus euphratica*, ERF95 and ERF109 regulate genes involved in ROS detoxification and senescence such as *PeGSTU17* and *PeORE1*, integrating ABA, ethylene, and SL signals during salinity stress (Li *et al.*, 2024).

Co-expression network analysis and phosphoproteomic profiling in *Brachypodium* and *Medicago* have further revealed that these shared TFs are part of dynamic feedback loops, modulating hormone biosynthesis and signal transduction (Waititu *et al.*, 2021). For instance, post-translational modifications, such as phosphorylation of MYC2 by MAPKs and ubiquitination of SMXLs by MAX2, have been validated in *Medicago* mutants, showing how MAX2 affects not only SL targets but also JA and ethylene pathways via shared protein substrates (Luo *et al.*, 2023). Collectively, these shared TFs and nodes form tightly interconnected regulatory hubs, demonstrating remarkable evolutionary conservation and adaptability across species (Roy *et al.*, 2022). Their plasticity and cross-hormonal control make them promising candidates for targeted manipulation in underutilized crops to improve resilience, biomass, and reproductive efficiency under environmental stresses (Rosero *et al.*, 2020).

2.11 Agronomic Implications and Manipulation

2.11.1 Hormonal balance optimization in crop improvement

Hormonal balance optimization through manipulation of SLs and other hormone pathways presents a promising strategy for improving agronomic traits such as yield, stress tolerance, nutrient-use efficiency, and plant architecture (Manzoor *et al.*, 2023). Modern advancement shows that targeted modulation of hormonal crosstalk, particularly involving SLs, CKs, ABA, GAs, and auxins, can improve crop productivity under suboptimal conditions (Kaniganti *et al.*, 2022). SLs, originally characterized for their role in suppressing shoot branching, are now understood to function as systemic regulators of plant development and environmental response. Fine-tuning their levels or signaling components such as D14, MAX2, and SMXLs has shown promise in enhancing shoot–root balance and optimizing resource allocation (Mashiguchi *et al.*, 2021).

In crops such as *Brassica napus*, overexpression of SL-biosynthesis genes such as *CCD7* and *CCD8* leads to reduced branching and increased main stem biomass, contributing to greater harvest index under limited nitrogen supply (Mukherjee *et al.*, 2023). Conversely, CRISPR/Cas9-mediated disruption of SL receptors has improved biomass allocation toward reproductive tissues in *Glycine max*, without compromising overall yield (Kumawat *et al.*, 2025). Similarly, modulation of SL signaling in combination with CKs has been exploited to create compact yet productive canopies in legumes and oilseeds, enhancing light penetration and photosynthetic efficiency (van Niekerk, 2020).

Integration of SL–ABA signaling is particularly beneficial under drought conditions. By enhancing SL biosynthesis or sensitivity, researchers have induced more efficient stomatal closure and deeper root growth, leading to improved drought resilience in various crops (Wei *et al.*, 2025). Foliar application of SL analogs such as GR24 has also been shown to reduce ABA thresholds for stomatal closure, conserving water without compromising CO_2 uptake. Meanwhile, dual manipulation of SL and GA pathways has been used to balance stem elongation and lodging resistance, especially in semi-dwarf cultivars of barley and sorghum (van Weringh, 2023). Furthermore, multi-hormonal manipulation has been incorporated into breeding programs through marker-assisted selection and genomic selection approaches. Expression quantitative trait locus mapping has identified key loci that regulate SL-responsive genes involved in yield and root development, facilitating selection of elite genotypes with optimized hormonal profiles (Sheoran *et al.*, 2022). Synthetic biology platforms have also been used to engineer inducible hormone circuits that regulate SL biosynthesis in response to environmental signals, offering precision agriculture solutions for variable field conditions (Singla, 2020).

Biostimulants and hormone mimics are also gaining ground. Application of synthetic SLs in combination with beneficial microbes such as AMF has shown synergistic effects on phosphorus uptake and biomass accumulation in nutrient-poor soils (Bisht *et al.*, 2025). Additionally, co-application of SL analogs with BRs or auxins has been shown to accelerate regeneration in tissue culture and promote uniform growth during vegetative propagation. In supposition, manipulation of hormonal balance involving SLs is becoming a cornerstone in next-generation crop improvement strategies (Pasternak and Steinmacher, 2024). By leveraging gene editing, chemical treatments, and systems biology insights, plant breeders and agronomists are unlocking new opportunities to enhance crop resilience, productivity, and sustainability under increasingly variable climates and marginal soils (Ge and Wang, 2025).

2.11.2 Hormone interaction manipulation for targeted traits

Hormone interaction manipulation has emerged as a powerful tool for improving targeted agronomic traits in crops, including yield components, stress resilience, nutrient uptake, and architecture optimization (Li *et al.*, 2025). Central to this approach is the strategic alteration of crosstalk between hormones such as SLs, CKs, ABA, GAs, JAs, and auxins (Bhatt *et al.*, 2020). Current investigations indicate that manipulating these interactions, rather than modifying individual hormone pathways in isolation, leads to more stable and context-specific outcomes across diverse environments (Bhatt *et al.*, 2020). For example, in *Brassica napus*, overexpression of the SL biosynthetic gene *CCD7* coupled with downregulation of CK biosynthesis genes (*IPT3*) resulted in reduced branching and enhanced apical dominance, leading to more efficient light use and yield improvement (Jardim-Messeder *et al.*, 2025). In *Solanum tuberosum*, external application of SL analog GR24 in combination with GA inhibitors helped improve tuberization under short-day stress by modulating auxin–GA dynamics (Kondhare *et al.*, 2021).

In *Medicago truncatula*, dual manipulation of SL and ABA pathways improved drought resilience by enhancing root growth and promoting stomatal closure, without a negative impact on biomass (Kou *et al.*, 2024). Similarly, in *Glycine max*, CRISPR-based disruption of *SMXL* repressors altered SL–JA interaction to enhance insect resistance while maintaining pod set and seed size (Yu *et al.*, 2023). Targeted hormonal manipulation has also improved nutrient efficiency. In *Vigna unguiculata*, SL–CK interaction modulation improved phosphorus use efficiency by favoring root-hair development and mycorrhizal colonization (Sharma *et al.*, 2024a, b). In *Capsicum annuum*, fine-tuning the ABA–SL axis improved salinity tolerance by stabilizing ion homeostasis and enhancing antioxidant enzyme activity (Corpas *et al.*, 2025).

Furthermore, synthetic hormone combinations are being applied in precision agriculture. In *Hordeum vulgare*, exogenous SLs applied with BR analogs improved stem robustness and reduced lodging while supporting high grain yield (Ahammed *et al.*, 2021). These approaches are now supported by omics-guided selection, enabling breeders to map hormonal responsiveness to traits at the transcriptomic and metabolomic levels, as demonstrated in *Phaseolus vulgaris* (Cortés *et al.*, 2023). Ultimately, manipulating hormone interactions allows trait stacking such as combining drought tolerance with yield maintenance, or pest resistance with growth vigor, by tuning complex regulatory networks rather than single traits (Singh *et al.*, 2024).

2.12 Conclusion and Research Outlook

Hormonal crosstalk dynamics in plants involve complex, context-dependent interactions among various phytohormones, including SLs, auxins, CKs, ABA, GAs, BRs, JAs, and ethylene (Harsha *et al.*, 2023). These interactions enable plants to balance growth, development, and stress responses through shared signaling components, transcriptional regulators, and feedback mechanisms (Aerts *et al.*, 2021). Current breakthroughs emphasize that systems biology and omics technologies have revealed how these hormones operate not in isolation but through dynamic regulatory networks (Kaya, 2024).

SLs serve as central integrators, interacting antagonistically with CKs to control shoot branching, while working cooperatively with auxins in root development and gravitropism (Sharma *et al.*, 2021). In drought conditions, SLs enhance ABA sensitivity, contributing to improved stomatal regulation and root system plasticity (Khalid *et al.*, 2024). Similarly, SL–GA interplay regulates shoot elongation and bud dormancy, especially under nutrient-deficient conditions where energy conservation is crucial (Liu *et al.*, 2024b). SL–JA and SL–BR crosstalk are increasingly recognized in stress defense, including pathogen resistance and oxidative stress mitigation. The TFs MYC2 and BZR1, for example, are shared regulators modulated by SL signals to fine-tune downstream genes (Gilroy and Breen, 2022). Ethylene and SL interactions control root-hair elongation and nutrient uptake, demonstrating synergistic roles under low-phosphorus or salinity stress (Wang *et al.*, 2025). At the molecular level, shared signaling components such as the SCF^{MAX2} complex and DELLA proteins act as central hubs. These regulators mediate degradation of repressors (e.g. SMXLs) or alter transcriptional responses in response to multi-hormonal cues (Huang *et al.*, 2025b). Hormone crosstalk is also modulated spatially and temporally, with organ-specific responses allowing localized adaptation, such as SL–CK balancing in axillary buds versus SL–ABA coordination in roots (Dhal *et al.*, 2023).

Contemporary omics analyses such as transcriptomics, proteomics, and metabolomics have mapped global hormone-responsive gene networks and identified shared TFs such as WRKYs, NACs, and TCPs that facilitate crosstalk (Berry, 2022). These nodes enable flexible and redundant regulation, enhancing plant adaptability to fluctuating environments. Ultimately, hormonal crosstalk functions as a dynamic, multi-layered regulatory system (Chen *et al.*, 2025). Understanding these interactions enables targeted manipulation of signaling pathways for crop improvement, integrating environmental responsiveness with developmental optimization (Thilakarathne *et al.*, 2025).

2.12.1 Emerging areas of SL interaction research

State-of-the-art advancements in SL research have uncovered a wide spectrum of emerging functions that transcend traditional roles in plant development and parasitic weed control (Srivastava *et al.*, 2022). A particularly exciting area is the regulation of rhizosphere microbiota by SLs. While historically associated with stimulating AMF, SLs have now been shown to selectively influence bacterial populations in the root zone (Boyno *et al.*, 2023). For instance, in *Malus domestica*, SL application modified microbial composition by increasing beneficial taxa such as *Burkholderia* and *Pseudomonas*, which can enhance nutrient uptake and disease resistance (Bokszczanin *et al*., 2023). This functional extension of SLs suggests they may serve as rhizosphere engineering signals, helping plants adapt to stress through indirect microbial mechanisms (Priya *et al*., 2024).

In addition, genome editing using CRISPR/Cas9 is revolutionizing the potential of SL pathway manipulation. Researchers have targeted SL biosynthesis (e.g. *D27*, *MAX1*) and transporter (e.g. *PDR1a/b* in sorghum) genes to fine-tune SL exudation (Barman *et al*., 2020). Liu *et al*. (2024b) demonstrated that editing these genes led to a dramatic reduction in *Striga hermonthica* germination cues, effectively lowering field infestation rates without negatively affecting plant biomass or yield. These precise edits allow dual optimization of crop defense and symbiosis (Kimutai *et al*., 2024). Another frontier lies in synthetic biology and microbial platforms. Scientists have reconstructed SL biosynthetic pathways in *Escherichia coli* and yeast, allowing scalable production of key SL intermediates such as carlactone and methyl carlactonoate (Feng *et al*., 2022). This approach facilitates the exploration of new SL analogs that may possess enhanced bioactivity or specificity for certain signaling pathways, including SL–ABA or SL–JA crosstalk (Figueroa-Macías *et al*., 2021).

Furthermore, integrated omics approaches combining transcriptomics, metabolomics, and proteomics are providing a systems-level view of SL functions. In *Populus trichocarpa*, SL-regulated genes were found to include those associated with secondary metabolism, ROS scavenging, and stress hormone synthesis (Kangi *et al*., 2024). Similarly, in *Glycine max*, SLs were shown to upregulate nodulation-related genes while enhancing drought resistance through ABA synergy. These findings illustrate how SLs orchestrate complex hormonal and environmental networks beyond their primary signaling pathways (Ahmed *et al*., 2025).

The development of biosensors has enabled real-time tracking of SL levels in living tissues. Fluorescence-based SL sensors respond dynamically to nutrient deficiencies such as phosphate starvation, revealing localized SL accumulation in root tips and axillary buds (Huang and Jin, 2024). This *in vivo* approach has immense value in studying temporal dynamics of SL signaling and guiding precision agriculture (Aarif *et al*., 2025). Finally, new roles for SLs in abiotic stress and development are emerging, including their involvement in thermomorphogenesis, photoperiod sensing, and heavy-metal detoxification (Hossain *et al*., 2022). Although still nascent, research indicates that SLs may modulate key TFs involved in redox balance and hormonal crosstalk under metal stress and high temperatures. These multi-faceted functions underscore the potential of SL-targeted interventions in crop systems facing climatic challenges (Waidi *et al*., 2024).

References

Aarif, K.O.M., Alam, A. and Hotak, Y. (2025) Smart sensor technologies shaping the future of precision agriculture: Recent advances and future outlooks. *Journal of Sensors* 2025(1), 2460098. DOI: 10.1155/js/2460098.

Abd-Elsalam, K.A. and Mohamed, H.I. (eds) (2024) *Plant Growth Regulators to Manage Biotic and Abiotic Stress in Agroecosystems*. CRC Press. DOI: 10.1201/9781003389507.

Abdullah-Zawawi, M.R., Govender, N., Harun, S., Muhammad, N.A.N., Zainal, Z. *et al*. (2022) Multi-omics approaches and resources for systems-level gene function prediction in the plant kingdom. *Plants* 11(19), 2614. DOI: 10.3390/plants11192614.

Aerts, N., Pereira Mendes, M. and van Wees, S.C.M. (2021) Multiple levels of crosstalk in hormone networks regulating plant defense. *Plant Journal* 105(2), 489–504. DOI: 10.1111/tpj.15124.

Ahammed, G.J., Sharma, A. and Yu, J. (eds) (2021) *Brassinosteroids in Plant Developmental Biology and Stress Tolerance*. Academic Press, UK.

Ahmad, B., Mukarram, M., Choudhary, S., Petrík, P., Dar, T.A. *et al.* (2024) Adaptive responses of nitric oxide (NO) and its intricate dialogue with phytohormones during salinity stress. *Plant Physiology and Biochemistry* 208, 108504. DOI: 10.1016/j.plaphy.2024.108504.

Ahmed, N., Li, J., Li, Y., Deng, L., Deng, L. *et al.* (2025) Symbiotic synergy: How arbuscular mycorrhizal fungi enhance nutrient uptake, stress tolerance, and soil health through molecular mechanisms and hormonal regulation. *IMA Fungus* 16, e144989. DOI: 10.3897/imafungus.16.144989.

Aiba, H. and Umehara, M. (2024) Strigolactones interplay with signaling molecules of plant. In: Bashri, G., Hayat, S. and Bajguz, A. (eds) *Strigolactones: Synthesis, Application and Role in Plants*. Academic Press, UK, pp. 127–145.

Akhiyarova, G., Finkina, E.I., Zhang, K., Veselov, D., Vafina, G. *et al.* (2024) The long-distance transport of some plant hormones and possible involvement of lipid-binding and transfer proteins in hormonal transport. *Cells* 13(5), 364. DOI: 10.3390/cells13050364.

Al Tahan, M.A., Michaelides, K., Somasekharan Nair, S., AlShatti, S., Russell, C. *et al.* (2024) Mesoporous silica microparticle-protein complexes: Effects of protein size and solvent properties on diffusion and loading efficiency. *British Journal of Biomedical Science* 81, 13595. DOI: 10.3389/bjbs.2024.13595.

Aliche, E.B., Screpanti, C., de Mesmaeker, A., Munnik, T. and Bouwmeester, H.J. (2020) Science and application of strigolactones. *New Phytologist* 227(4), 1001–1011. DOI: 10.1111/nph.16489.

Aloni, R. (2021) The hormonal signals that regulate plant vascular differentiation. In: *Vascular Differentiation and Plant Hormones*. Springer, Cham, Switzerland, pp. 55–96. DOI: 10.1007/978-3-030-53202-4_3.

Aloo, B.N., Dessureault-Rompré, J., Tripathi, V., Nyongesa, B.O. and Were, B.A. (2023) Signaling and crosstalk of rhizobacterial and plant hormones that mediate abiotic stress tolerance in plants. *Frontiers in Microbiology* 14, 1171104. DOI: 10.3389/fmicb.2023.1171104.

Altamura, M.M., Piacentini, D., Della Rovere, F., Fattorini, L., Falasca, G. *et al.* (2023) New paradigms in brassinosteroids, strigolactones, sphingolipids, and nitric oxide interaction in the control of lateral and adventitious root formation. *Plants* 12(2), 413. DOI: 10.3390/plants12020413.

Alvi, A.F., Sehar, Z., Fatma, M., Masood, A. and Khan, N.A. (2022) Strigolactone: An emerging growth regulator for developing resilience in plants. *Plants* 11(19), 2604. DOI: 10.3390/plants11192604.

Ameen, F., Hussain, I., Afzal, S., Rasheed, R., Ashraf, M.A. *et al.* (2025) Crosstalk between strigolactones and major hormones in plants under abiotic stresses. *South African Journal of Botany* 177, 187–200. DOI: 10.1016/j.sajb.2024.12.004.

Arefian, M., Bhagya, N. and Prasad, T.S.K. (2021) Phosphorylation-mediated signalling in flowering: Prospects and retrospects of phosphoproteomics in crops. *Biological Reviews of the Cambridge Philosophical Society* 96(5), 2164–2191. DOI: 10.1111/brv.12748.

Bao, L., Liu, J., Mao, T., Zhao, L., Wang, D. *et al.* (2024) Nanobiotechnology-mediated regulation of reactive oxygen species homeostasis under heat and drought stress in plants. *Frontiers in Plant Science* 15, 1418515. DOI: 10.3389/fpls.2024.1418515.

Barman, A., Deb, B. and Chakraborty, S. (2020) A glance at genome editing with CRISPR-Cas9 technology. *Current Genetics* 66(3), 447–462. DOI: 10.1007/s00294-019-01040-3.

Berry, H.M. (2022) Characterizing the role of plant hormones during plant development and plant immunity in *Arabidopsis thaliana* and *Solanum lycopersicum* cv. Micro-Tom. PhD thesis, Colorado State University, Fort Collins, Colorado.

Bhatt, D., Nath, M., Sharma, M., Bhatt, M.D., Bisht, D.S. *et al.* (2020) Role of growth regulators and phytohormones in overcoming environmental stress. In: Roychoudhury, A. and Tripathi, D.K. (eds) *Protective Chemical Agents in the Amelioration of Plant Abiotic Stress: Biochemical and Molecular Perspectives*. Wiley, Hoboken, New Jersey, pp. 254–279.

Bisht, N., Singh, T., Ansari, M.M. and Chauhan, P.S. (2025) The hidden language of plant-beneficial microbes: Chemo-signaling dynamics in plant microenvironments. *World Journal of Microbiology and Biotechnology* 41(2), 35. DOI: 10.1007/s11274-025-04253-6.

Bokszczanin, K.Ł., Przybyłko, S., Molska-Kawulok, K. and Wrona, D. (2023) Tree root-associated microbial communities depend on various floor management systems in an intensive apple (*Malus*× *domestica* Borkh.) orchard. *International Journal of Molecular Sciences* 24(12), 9898. DOI: 10.3390/ijms24129898.

Boyno, G., Rezaee Danesh, Y., Demir, S., Teniz, N., Mulet, J.M. *et al.* (2023) The complex interplay between arbuscular mycorrhizal fungi and strigolactone: Mechanisms, sinergies, applications and future directions. *International Journal of Molecular Sciences* 24(23), 16774. DOI: 10.3390/ijms242316774.

Campbell, Q., Bedford, J.A., Yu, Y., Halpin-McCormick, A., Castaneda-Alvarez, N. *et al.* (2025) Agricultural landscape genomics to increase crop resilience. *Plant Communications* 6(2), 101260. DOI: 10.1016/j.xplc.2025.101260.

Cao, X., Wei, Y., Shen, B., Liu, L. and Mao, J. (2024) Interaction of the transcription factors BES1/BZR1 in plant growth and stress response. *International Journal of Molecular Sciences* 25(13), 6836. DOI: 10.3390/ijms25136836.

Chakraborty, R., Rehman, R.U., Siddiqui, M.W., Liu, H. and Seth, C.S. (2025) Phytohormones: Heart of plants' signaling network under biotic, abiotic, and climate change stresses. *Plant Physiology and Biochemistry* 223, 109839. DOI: 10.1016/j.plaphy.2025.109839.

Chang, G.R., Li, R.J., Zhang, Q., Zhang, Y.M., Han, Y.H. *et al.* (2020) Identification of co-expression genes related to endogenous abscisic acid in response to the stress of *Sclerospora graminicola* by WGCNA in foxtail millet. Scientia Agricultura Sinica 53(16), 3280–3293. DOI: 10.3864/j.issn.0578-1752.2020.16.007.

Chang, S.H., George, W. and Nelson, D.C. (2025) Transcriptional regulation of development by SMAX1-LIKE proteins, targets of strigolactone and karrikin/KAI2 ligand signaling. *Journal of Experimental Botany* 76(7), 1888–1906. DOI: 10.1093/jxb/eraf027.

Chaudhry, A., Chen, Z. and Gallavotti, A. (2024) Hormonal influence on maize inflorescence development and reproduction. *Plant Reproduction* 37(4), 393–407. DOI: 10.1007/s00497-024-00510-0.

Chen, F., Chen, L., Yan, Z., Xu, J., Feng, L. *et al.* (2024) Recent advances of CRISPR-based genome editing for enhancing staple crops. *Frontiers in Plant Science* 15, 1478398. DOI: 10.3389/fpls.2024.1478398.

Chen, X., Chen, Z., Watts, R. and Luo, H. (2025) Non-coding RNAs in plant stress responses: Molecular insights and agricultural applications. *Plant Biotechnology Journal* 23(8), 3195–3233. DOI: 10.1111/pbi.70134.

Chen, Y., Wang, Y., Liang, X., Zhang, Y. and Fernie, A.R. (2023) Mass spectrometric exploration of phytohormone profiles and signaling networks. *Trends in Plant Science* 28(4), 399–414. DOI: 10.1016/j.tplants.2022.12.006.

Claassens, A.P. (2024) Role of strigolactones in adventitious rooting of *Arabidopsis thaliana* callus *in vitro*. PhD thesis, Stellenbosch University, Stellenbosch, South Africa.

Corpas, F.J., Taboada, J., Rivero, R.M., Reiter, R.J. and Palma, J.M. (2025) Functions of endogenously produced and exogenously applied melatonin in higher plants. *Antioxidants & Redox Signaling* 43(1–3), 151–188. DOI: 10.1089/ars.2024.0889.

Cortés, A.J., Castillejo, M.Á. and Yockteng, R. (2023) Omics' approaches for crop improvement. *Agronomy* 13(5), 1401. DOI: 10.3390/agronomy13051401.

Dai, L., Chaudhari, B., Wang, T., Li, X., Zeng, D. *et al.* (2024) A transcriptomic analysis sheds light on the molecular regulation of wood formation in *Populus trichocarpa* during drought stress. *Forests* 15(5), 873. DOI: 10.3390/f15050873.

Das, S., Shil, S., Rime, J., Alice, A.K., Yumkhaibam, T. *et al.* (2025) Phytohormonal signaling in plant resilience: Advances and strategies for enhancing abiotic stress tolerance. *Plant Growth Regulation* 105(2), 329–360. DOI: 10.1007/s10725-025-01279-6.

Devireddy, A.R., Zandalinas, S.I., Fichman, Y. and Mittler, R. (2021) Integration of reactive oxygen species and hormone signaling during abiotic stress. *Plant Journal* 105(2), 459–476. DOI: 10.1111/tpj.15010.

Díaz, V., Villalobos, M., Arriaza, K., Flores, K., Hernández-Saravia, L.P. *et al.* (2025) Decoding the dialog between plants and arbuscular mycorrhizal fungi: A molecular genetic perspective. *Genes* 16(2), 143. DOI: 10.3390/genes16020143.

Ding, F., Wang, X., Li, Z. and Wang, M. (2022) Jasmonate positively regulates cold tolerance by promoting ABA biosynthesis in tomato. *Plants* 12(1), 60. DOI: 10.3390/plants12010060.

Ding, F., Fan, X., Tian, R., Wang, M. and Sun, Z. (2025) Crosstalk of abscisic acid with other hormones and signaling molecules in tomato cold stress tolerance. *Horticulturae* 11(6), 647. DOI: 10.3390/horticulturae11060647.

do Carmo Santos, M.L., Pereira Silva Santos, A., Novais, D., Dos Santos Lopes, N., Pirovani, C.P. *et al.* (2025) The family of glutathione peroxidase proteins and their role against biotic stress in plants: A systematic review. *Frontiers in Plant Science* 16, 1425880. DOI: 10.3389/fpls.2025.1425880.

Dong, J., Ding, C., Chen, H., Fu, H., Pei, R. *et al.* (2024) Functions of exogenous strigolactone application and strigolactone biosynthesis genes GhMAX3/GhMAX4b in response to drought tolerance in cotton (*Gossypium hirsutum* L.). *BMC Plant Biology* 24(1), 1008. DOI: 10.1186/s12870-024-05726-w.

Duc, N.T., Harika, A., Raju, D., Kumar, S., Pandey, R. *et al.* (2025) Maximizing nitrogen stress tolerance through high-throughput phenotyping in rice. *Plant Stress* 15, 100764. DOI: 10.1016/j.stress.2025.100764.

Eckardt, N.A., Avin-Wittenberg, T., Bassham, D.C., Chen, P., Chen, Q. *et al.* (2024) The lowdown on breakdown: Open questions in plant proteolysis. *Plant Cell* 36(9), 2931–2975. DOI: 10.1093/plcell/koae193.

Fahad, M., Tariq, L., Li, W. and Wu, L. (2025) MicroRNA gatekeepers: Orchestrating rhizospheric dynamics. *Journal of Integrative Plant Biology* 67(3), 845–876. DOI: 10.1111/jipb.13860.

Dhal, P.B., Devi, R.S., Bhadouria, R. and Tripathi, S. (2023) Phytohormones: Role in ecophysiological responses of tropical plants to varying resource availability. In: Tripathi, S., Bhadouria, R., Srivastava, P., Singh, R. and Devi, R.S. (eds) *Ecophysiology of Tropical Plants*. CRC Press, Boca Raton, Florida, pp. 265–283.

Feng, C., Chen, J., Ye, W., Liao, K., Wang, Z. *et al.* (2022) Synthetic biology-driven microbial production of resveratrol: Advances and perspectives. *Frontiers in Bioengineering and Biotechnology* 10, 833920. DOI: 10.3389/fbioe.2022.833920.

Feng, K., Hou, X.-L., Xing, G.-M., Liu, J.-X., Duan, A.-Q. *et al.* (2020) Advances in AP2/ERF superfamily transcription factors in plant. *Critical Reviews in Biotechnology* 40(6), 750–776. DOI: 10.1080/07388551.2020.1768509.

Figueroa-Macías, J.P., García, Y.C., Núñez, M., Díaz, K., Olea, A.F. *et al.* (2021) Plant growth-defense trade-offs: Molecular processes leading to physiological changes. *International Journal of Molecular Sciences* 22(2), 693. DOI: 10.3390/ijms22020693.

Gao, Y., Wang, X., Liu, X. and Liang, Z. (2022) Hormone biosynthesis and metabolism members of 2OGD superfamily are involved in berry development and respond to MeJA and ABA treatment of *Vitis vinifera* L. *BMC Plant Biology* 22(1), 427. DOI: 10.1186/s12870-022-03810-7.

Ge, A.H. and Wang, E. (2025) Exploring the plant microbiome: A pathway to climate-smart crops. *Cell* 188(6), 1469–1485. DOI: 10.1016/j.cell.2025.01.035.

Ghosh, P. and Roychoudhury, A. (2024) Molecular basis of salicylic acid–phytohormone crosstalk in regulating stress tolerance in plants. *Brazilian Journal of Botany* 47(3), 735–750. DOI: 10.1007/s40415-024-00983-3.

Gilroy, E. and Breen, S. (2022) Interplay between phytohormone signalling pathways in plant defence – other than salicylic acid and jasmonic acid. *Essays in Biochemistry* 66(5), 657–671. DOI: 10.1042/EBC20210089.

Guan, Y., Li, L., Wang, D., Zhou, J., Qi, W. *et al.* (2025) Multifaceted functions of strigolactones in annual and perennial plants: Developmental regulation, phytohormone crosstalk and abiotic stresses. *Plant Stress* 16, 100895. DOI: 10.1016/j.stress.2025.100895.

Harsha, S.G., Sheela, H.S., Shivashankara, K.S., Laxman, R.H. and Prathibha, M.D. (2023) Molecular understanding of crosstalk between plant growth hormones and plant growth regulators under multiple and combined abiotic stress tolerance. In: Gupta, N.K., Shavrukov, Y., Singhal, R.K. and Borisjuk, N. (eds) *Multiple Abiotic Stress Tolerances in Higher Plants: Addressing the Growing Challenges*. CRC Press, Boca Raton, Florida, pp. 131–142. DOI: 10.1201/9781003300564-9.

Hasanuzzaman, M., Bhuyan, M.H.M.B., Zulfiqar, F., Raza, A., Mohsin, S.M. *et al.* (2020) Reactive oxygen species and antioxidant defense in plants under abiotic stress: Revisiting the crucial role of a universal defense regulator. *Antioxidants* 9(8), 681. DOI: 10.3390/antiox9080681.

Hashem, H.A. and Khalil, R. (2023) Insight into the interaction of strigolactones, abscisic acid, and reactive oxygen species signals. In: Faizan, M., Hayat, S. and Ahmed, S.M. (eds) *Reactive Oxygen Species: Prospects in Plant Metabolism*. Springer Nature Singapore, Singapore, pp. 179–211. DOI: 10.1007/978-981-19-9794-5_11.

He, X., Wan, Z., Jin, N., Jin, L., Zhang, G. *et al.* (2022) Enhancement of cucumber resistance under salt stress by 2, 4-epibrassinolide lactones. *Frontiers in Plant Science* 13, 1023178. DOI: 10.3389/fpls.2022.1023178.

Hossain, A., Pamanick, B., Venugopalan, V.K., Ibrahimova, U., Rahman, M.A. *et al.* (2022) Emerging roles of plant growth regulators for plants adaptation to abiotic stress-induced oxidative stress. In: Aftab, T. and Naeem, M. (eds) *Emerging Plant Growth Regulators in Agriculture*. Academic Press, UK, pp. 1–72. DOI: 10.1016/B978-0-323-91005-7.00010-2.

Huang, S. and Jin, S. (2024) Enhancing drought tolerance in horticultural plants through plant hormones: A strategic coping mechanism. *Frontiers in Plant Science* 15, 1502438. DOI: 10.3389/fpls.2024.1502438.

Huang, M., Liu, W., Su, D., Abid, G., Huang, B. *et al.* (2025a) The BES1/BZR1 transcriptional factor SlBES2 regulates photosynthetic apparatus in tomato fruit. *BMC Plant Biology* 25(1), 70. DOI: 10.1186/s12870-025-06085-w.

Huang, Q., Lin, X., Nie, D., Wang, J., Fan, K. *et al.* (2025b) Enhanced electrochemical and photochemical sensors for *in vivo* biological monitoring based on carbon nanomaterials. *Advanced Functional Materials* 35(40), 2503027. DOI: 10.1002/adfm.202503027.

Hu, D., Zhang, J., Yang, Y., Yu, D., Zhang, H. *et al.* (2025) Molecular mechanisms underlying plant responses to low phosphate stress and potential applications in crop improvement. *New Crops* 2, 100064. DOI: 10.1016/j.ncrops.2024.100064.

Jardim-Messeder, D., de Souza-Vieira, Y. and Sachetto-Martins, G. (2025) Dressed up to the nines: The interplay of phytohormones signaling and redox metabolism during plant response to drought. *Plants* 14(2), 208. DOI: 10.3390/plants14020208.

Jhu, M.Y., Souza, V.H.M. and Schiessl, K. (2025) From hosts to parasites: Hormones driving symbiosis-induced *de novo* organogenesis. In: *Trends in Plant Science*, Vol. 30 (12). pp. 1372–1391. DOI: 10.1016/j.tplants.2025.05.015.

Jiang, J., Fan, G., Wang, R., Yao, W., Zhou, B. *et al.* (2024) Multi-omics analysis of *Populus simonii* × *P. nigra* leaves under *Hyphantria cunea* stress. *Frontiers in Plant Science* 15, 1392433. DOI: 10.3389/fpls.2024.1392433.

Jia, Z., Giehl, R.F.H. and von Wirén, N. (2022) Nutrient–hormone relations: Driving root plasticity in plants. *Molecular Plant* 15(1), 86–103. DOI: 10.1016/j.molp.2021.12.004.

Jin, Q., Wang, Z., Chen, Y., Luo, Y., Tian, N. *et al.* (2022) Transcriptomics analysis reveals the signal transduction mechanism of brassinolides in tea leaves and its regulation on the growth and development of *Camellia sinensis*. *BMC Genomics* 23(1), 29. DOI: 10.1186/s12864-021-08179-9.

Jogawat, A., Yadav, B., Chhayya, 0.0, lakra, N., Singh, A.K. *et al.* (2021) Crosstalk between phytohormones and secondary metabolites in the drought stress tolerance of crop plants: A review. *Physiologia Plantarum* 172(2), 1106–1132. DOI: 10.1111/ppl.13328.

Kamran, M., Melville, K.T. and Waters, M.T. (2024) Karrikin signalling: Impacts on plant development and abiotic stress tolerance. *Journal of Experimental Botany* 75(4), 1174–1186. DOI: 10.1093/jxb/erad476.

Kangi, E., Brzostek, E.R., Bills, R.J., Callister, S.J., Zink, E.M. *et al.* (2024) A multi-omic survey of black cottonwood tissues highlights coordinated transcriptomic and metabolomic mechanisms for plant adaptation to phosphorus deficiency. *Frontiers in Plant Science* 15, 1324608. DOI: 10.3389/fpls.2024.1324608.

Kang, J., Peng, Y. and Xu, W. (2022) Crop root responses to drought stress: Molecular mechanisms, nutrient regulations, and interactions with microorganisms in the rhizosphere. *International Journal of Molecular Sciences* 23(16), 9310. DOI: 10.3390/ijms23169310.

Kaniganti, S., Bhattacharya, J., Petla, B.P. and Reddy, P.S. (2022) Strigolactone, a neglected plant hormone, with a great potential for crop improvement: Crosstalk with other plant hormones. *Environmental and Experimental Botany* 204, 105072. DOI: 10.1016/j.envexpbot.2022.105072.

Kapoor, L., Simkin, A.J., Doss, C.G.P. and Siva, R. (2022) Fruit ripening: Dynamics and integrated analysis of carotenoids and anthocyanins. *BMC Plant Biology* 22(1), 27. DOI: 10.1186/s12870-021-03411-w.

Kapoor, R.T., Alam, P., Chen, Y. and Ahmad, P. (2024) Strigolactones in plants: From development to abiotic stress management. *Journal of Plant Growth Regulation* 43(3), 903–919. DOI: 10.1007/s00344-023-11148-z.

Karanam, A., He, D., Hsu, P.-K., Schulze, S., Dubeaux, G. *et al.* (2021) Boolink: A graphical interface for open access Boolean network simulations and use in guard cell CO_2 signaling. *Plant Physiology* 187(4), 2311–2322. DOI: 10.1093/plphys/kiab344.

Kaya, C. (2024) Microbial modulation of hormone signaling, proteomic dynamics, and metabolomics in plant drought adaptation. *Food and Energy Security* 13(1), e513. DOI: 10.1002/fes3.513.

Kaya, C. (2025) Molecular crosstalk of transcription factors and hormones in drought tolerance genes in plants. *Turkish Journal of Botany* 49(3), 145–164. DOI: 10.55730/1300-008X.2852.

Khalid, M.F., Shafqat, W., Khan, R.I., Jawaid, M.Z., Hussain, S. *et al.* (2024) Unveiling the resilience mechanism: Strigolactones as master regulators of plant responses to abiotic stresses. *Plant Stress* 12, 100490. DOI: 10.1016/j.stress.2024.100490.

Khan, M.I.R., Singh, A. and Poor, P. (eds) (2023b) *Plant Hormones in Crop Improvement*. Academic Press, Cambridge, Massachusetts.

Khan, M.T.A., Saeed, T., Ahmad, A., Fariduddin, Q. and Yusuf, M. (eds) (2025) *Hydrogen Peroxide: Signaling Mechanisms and Crosstalk in Plant Development and Stress Responses*. CRC Press, Boca Raton, Florida.

Khan, S., Alvi, A.F. and Khan, N.A. (2024a) Role of ethylene in the regulation of plant developmental processes. *Stresses* 4(1), 28–53. DOI: 10.3390/stresses4010003.

Khan, S., Alvi, A.F., Saify, S., Iqbal, N. and Khan, N.A. (2024b) The ethylene biosynthetic enzymes, 1-am inocyclopropane-1-carboxylate (ACC) synthase (ACS) and ACC oxidase (ACO): The less explored players in abiotic stress tolerance. *Biomolecules* 14(1), 90. DOI: 10.3390/biom14010090.

Khan, S., Sehar, Z., Albaqami, M. and Khan, N.A. (2023a) Ethylene crosstalk with isoprenoid-derived signaling molecules in the context of salinity tolerance. *Environmental and Experimental Botany* 212, 105379. DOI: 10.1016/j.envexpbot.2023.105379.

Khunweeraphong, N. and Kuchler, K. (2021) Multidrug resistance in mammals and fungi – from MDR to PDR: A rocky road from atomic structures to transport mechanisms. *International Journal of Molecular Sciences* 22(9), 4806. DOI: 10.3390/ijms22094806.

Kimutai, J.J.C., Makumbi, D., Burgueño, J., Pérez-Rodríguez, P., Crossa, J. *et al.* (2024) Genomic prediction of the performance of tropical doubled haploid maize lines under artificial *Striga hermonthica* (Del.) Benth. *Infestation* G3 14(10), jkae186. DOI: 10.1093/g3journal/jkae186.

Kisvarga, S., Horotán, K., Wani, M.A. and Orlóci, L. (2023) Plant responses to global climate change and urbanization: Implications for sustainable urban landscapes. *Horticulturae* 9(9), 1051. DOI: 10.3390/horticulturae9091051.

Klupczyńska, E.A. and Pawłowski, T.A. (2021) Regulation of seed dormancy and germination mechanisms in a changing environment. *International Journal of Molecular Sciences* 22(3), 1357. DOI: 10.3390/ijms22031357.

Kondhare, K.R., Patil, A.B. and Giri, A.P. (2021) Auxin: An emerging regulator of tuber and storage root development. *Plant Science* 306, 110854. DOI: 10.1016/j.plantsci.2021.110854.

Korek, M. and Marzec, M. (2023) Strigolactones and abscisic acid interactions affect plant development and response to abiotic stresses. *BMC Plant Biology* 23(1), 314. DOI: 10.1186/s12870-023-04332-6.

Kou, C., Song, F., Li, D., Xu, H., Zhang, S. *et al.* (2024) A necessary considering factor for crop resistance: Precise regulation and effective utilization of beneficial microorganisms. *New Crops* 1, 100023. DOI: 10.1016/j.ncrops.2024.100023.

Kumawat, S., Pragati, K., Garkoti, P., Singh, A. and Senthilkumar, V. (2025) Next-gen seeds: Harnessing genetic engineering for improving seed quality traits. In: Roy, B., Shukla, G., Dunna, V., Sharma, P. and Shukla, P.S. (eds) *Advances in Seed Quality Evaluation and Improvement*. Springer, Singapore, pp. 273–307. DOI: 10.1007/978-981-96-5178-8_11.

Kumar, P., Kumar, P., Suniti, K., Avni, U. and Mann, A. (2023) Transcriptional regulatory network involved in drought and salt stress response in rice. In: Kumar, A., Dhansu, P. and Mann, A. (eds) *Salinity and Drought Tolerance in Plants: Physiological Perspectives*. Springer Nature Singapore, Singapore, pp. 237–274. DOI: 10.1007/978-981-99-4669-3_13.

Kumar, A., Partap, M. and Warghat, A.R. (2025) Jasmonic acid: A versatile phytohormone regulating growth, physiology, and biochemical responses. *Journal of Plant Growth Regulation* 44(1), 131–154. DOI: 10.1007/s00344-024-11376-x.

Ku, Y.S., Liao, Y.J., Chiou, S.P., Lam, H.M. and Chan, C. (2024) From trade-off to synergy: Microbial insights into enhancing plant growth and immunity. *Plant Biotechnology Journal* 22(9), 2461–2471. DOI: 10.1111/pbi.14360.

Lee, C., Harvey, J.T., Nagila, A., Qin, K. and Leskovar, D.I. (2023) Thermotolerance of tomato plants grafted onto wild relative rootstocks. *Frontiers in Plant Science* 14, 1252456. DOI: 10.3389/fpls.2023.1252456.

Li, C., Xu, M., Cai, X., Han, Z., Si, J. *et al.* (2022) Jasmonate signaling pathway modulates plant defense, growth, and their trade-offs. *International Journal of Molecular Sciences* 23(7), 3945. DOI: 10.3390/ijms23073945.

Li, J., Fan, M., Zhang, Q., Lü, G., Wu, X. *et al.* (2023) Transcriptome analysis reveals that auxin promotes strigolactone-induced adventitious root growth in the hypocotyl of melon seedlings. *Frontiers in Plant Science* 14, 1192340. DOI: 10.3389/fpls.2023.1192340.

Li, D., Si, J. and Ren, X. (2024) Coordination and adaptation of water processes in *Populus euphratica* in response to salinity. *Frontiers in Plant Science* 15, 1443444. DOI: 10.3389/fpls.2024.1443444.

Li, J., Forghieri, G., Geelen, D., Jardin, P.D. and Brown, P.H. (2025) The optimization of crop response to climatic stress through modulation of plant stress response mechanisms. Opportunities for biostimulants and plant hormones to meet climate challenges. *New Phytologist* 249(1), 130–151. DOI: 10.1111/nph.70701.

Liu, M., Shan, Q., Ding, E., Gu, T. and Gong, B. (2023) Karrikin increases tomato cold tolerance via strigolactone and the abscisic acid signaling network. *Plant Science* 332, 111720. DOI: 10.1016/j.plantsci.2023.111720.

Liu, Z., Ruonala, R. and Helariutta, Y. (2024a) Control of phloem unloading and root development. *Journal of Plant Physiology* 295, 154203. DOI: 10.1016/j.jplph.2024.154203.

Liu, Y., Chen, S., Pal, S., Yu, J., Zhou, Y. *et al.* (2024b) The hormonal, metabolic, and environmental regulation of plant shoot branching. *New Crops* 1, 100028. DOI: 10.1016/j.ncrops.2024.100028.

Lu, H., Ren, M., Lin, R., Jin, K. and Mao, C. (2024) Developmental responses of roots to limited phosphate availability: Research progress and application in cereals. *Plant Physiology* 196(4), 2162–2174. DOI: 10.1093/plphys/kiae495.

Luo, L., Wang, Y., Qiu, L., Han, X., Zhu, Y. *et al.* (2023) MYC2: A master switch for plant physiological processes and specialized metabolite synthesis. *International Journal of Molecular Sciences* 24(4), 3511. DOI: 10.3390/ijms24043511.

Luqman, M., Shahbaz, M., Maqsood, M.F., Farhat, F., Zulfiqar, U. *et al.* (2023) Effect of strigolactone on growth, photosynthetic efficiency, antioxidant activity, and osmolytes accumulation in different maize (*Zea mays* L.) hybrids grown under drought stress. *Plant Signaling & Behavior* 18(1), 2262795. DOI: 10.1080/15592324.2023.2262795.

Lv, Z., Zhao, W., Kong, S., Li, L. and Lin, S. (2023) Overview of molecular mechanisms of plant leaf development: A systematic review. *Frontiers in Plant Science* 14, 1293424. DOI: 10.3389/fpls.2023.1293424.

Lwin, C.S., Jung, H., Kim, M.-S., Lee, E.-J. and Lee, T.-G. (2025) Geochemical speciation, uptake, and transportation mechanisms of arsenic, cadmium, and lead in soil–rice systems: Additional aspects and challenges. *Antioxidants* 14(5), 607. DOI: 10.3390/antiox14050607.

Magdy, M., Mostofa, M.G. and Rahimi, M. (2023) Abiotic stress alleviation in plants: Morpho-physiological and molecular aspects. *Frontiers in Plant Science* 14, 1295638. DOI: 10.3389/fpls.2023.1295638.

Ma, X. and Bai, L. (2021) Elevated CO_2 and reactive oxygen species in stomatal closure. *Plants* 10(2), 410. DOI: 10.3390/plants10020410.

Manzoor, M.A., Xu, Y., lv, Z., Xu, J., Wang, Y. *et al.* (2023) Fruit crop abiotic stress management: A comprehensive review of plant hormones mediated responses. *Fruit Research* 3, 30. DOI: 10.48130/frures-2023-0030.

Margay, A.R., Mehmood, A. and Bashir, L. (2024) Review on hormonal regulation of drought stress response in plants. *International Journal of Plant & Soil Science* 36(8), 902–916. DOI: 10.9734/ijpss/2024/v36i84921.

Mashiguchi, K., Seto, Y. and Yamaguchi, S. (2021) Strigolactone biosynthesis, transport and perception. *Plant Journal* 105(2), 335–350. DOI: 10.1111/tpj.15059.

Ma, Z., Hu, L. and Jiang, W. (2024) Understanding AP2/ERF transcription factor responses and tolerance to various abiotic stresses in plants: A comprehensive review. *International Journal of Molecular Sciences* 25(2), 893. DOI: 10.3390/ijms25020893.

Mehmood, I., Wani, K.I. and Aftab, T. (2025a) Nature, structural diversity, biosynthetic pathway, and strigolactone transport in plants. In: Aftab, T. and Wani, K.I. (eds) *Strigolactones: Emerging Plant Hormones*. Scrivener Publishing, Beverly, Massachusetts, pp. 13–34. DOI: 10.1002/9781394302826.ch2.

Mehmood, I., Wani, K.I. and Aftab, T. (2025b) Strigolactone interplay with other phytohormones under stressed and normal conditions. In: Aftab, T. and Wani, K.I. (eds) *Strigolactones: Emerging Plant Hormones*. Scrivener Publishing, Beverly, Massachusetts, pp. 161–191.

Mei, S., Zhang, M., Ye, J., Du, J., Jiang, Y. *et al.* (2023) Auxin contributes to jasmonate-mediated regulation of abscisic acid signaling during seed germination in *Arabidopsis*. *Plant Cell* 35(3), 1110–1133. DOI: 10.1093/plcell/koac362.

Mishra, S. (2021) Physiological and molecular analysis of vascular tissues in *Plantago major* in response to sole or combined deficiencies to nitrogen and phosphorus. PhD thesis, Purdue University Graduate School, West Lafayette, Indiana.

Mukherjee, A., Dwivedi, S., Bhagavatula, L. and Datta, S. (2023) Integration of light and ABA signaling pathways to combat drought stress in plants. *Plant Cell Reports* 42(5), 829–841. DOI: 10.1007/s00299-023-02999-7.

Naseer, M.A., Zhang, Z.Q., Mukhtar, A., Asad, M.S., Wu, H.Y. *et al.* (2024) Strigolactones: A promising tool for nutrient acquisition through arbuscular mycorrhizal fungi symbiosis and abiotic stress tolerance. *Plant Physiology and Biochemistry* 215, 109057. DOI: 10.1016/j.plaphy.2024.109057.

Nazir, F., Peter, P., Gupta, R., Kumari, S., Nawaz, K. *et al.* (2024) Plant hormone ethylene: A leading edge in conferring drought stress tolerance. *Physiologia Plantarum* 176(1), e14151. DOI: 10.1111/ppl.14151.

Nishanth, M.J. (2024) Transcriptome meta-analysis-based identification of hub transcription factors and RNA-binding proteins potentially orchestrating gene regulatory cascades and crosstalk in response to abiotic stresses in *Arabidopsis thaliana*. *Journal of Applied Genetics* 65(2), 255–269. DOI: 10.1007/s13353-024-00837-4.

Parwez, R., Aftab, T., Gill, S.S. and Naeem, M. (2022) Abscisic acid signaling and crosstalk with phytohormones in regulation of environmental stress responses. *Environmental and Experimental Botany* 199, 104885. DOI: 10.1016/j.envexpbot.2022.104885.

Pasternak, T.P. and Steinmacher, D. (2024) Plant growth regulation in cell and tissue culture *in vitro*. *Plants* 13(2), 327. DOI: 10.3390/plants13020327.

Pierik, R., Fankhauser, C., Strader, L.C. and Sinha, N. (2021) Architecture and plasticity: Optimizing plant performance in dynamic environments. *Plant Physiology* 187(3), 1029–1032. DOI: 10.1093/plphys/kiab402.

Priya, S., Somenahally, A.C., Obayomi, O., Gentry, T.J., Sarker, T.C. *et al.* (2024) Exploring the application of signaling compounds and soil amendments to modulate plant–microbe interactions for improved plant salinity tolerance. *Plant and Soil* 501(1–2), 267–288. DOI: 10.1007/s11104-024-06512-1.

Rahman, M.M., Ghosh, P.K., Akter, M., Al Noor, M.M., Rahman, M.A. *et al.* (2024) Green vanguards: Harnessing the power of plant antioxidants, signal catalysts, and genetic engineering to combat reactive oxygen species under multiple abiotic stresses. *Plant Stress* 13, 100547. DOI: 10.1016/j.stress.2024.100547.

Ranganath, V.A. and Maity, I. (2024) Artificial homeostasis systems based on feedback reaction networks: Design principles and future promises. *Angewandte Chemie International Edition* 63(14), e202318134. DOI: 10.1002/anie.202318134.

Ravazzolo, L. and Quaggiotti, S. (2024) Strigolactones in root growth and development. In: Beeckman, T. and Eshel, A. (eds) *Plant Roots: The Hidden Half*, 5th edn. CRC Press, Boca Raton, Florida, pp. 261–274. DOI: 10.1201/b23126-20.

Raza, A., Javed, R., Zahid, Z., Sharif, R., Hafeez, M.B. *et al.* (2021) Strigolactones for sustainable plant growth and production under adverse environmental conditions. In: Husen, A. (ed.) *Plant Performance Under Environmental Stress: Hormones, Biostimulants and Sustainable Plant Growth Management*. Springer Nature, Cham, Switzerland, pp. 129–166. DOI: 10.1007/978-3-030-78521-5.

Rehman, M., Saeed, M.S., Fan, X., Salam, A., Munir, R. *et al.* (2023) The multifaceted role of jasmonic acid in plant stress mitigation: An overview. *Plants* 12(23), 3982. DOI: 10.3390/plants12233982.

Ribeiro, B., Erffelinck, M.-L., Lacchini, E., Ceulemans, E., Colinas, M. *et al.* (2022) Interference between ER stress-related bZIP-type and jasmonate-inducible bHLH-type transcription factors in the regulation of triterpene saponin biosynthesis in *Medicago truncatula*. *Frontiers in Plant Science* 13, 1103943. DOI: 10.3389/fpls.2022.1103943.

Rosero, A., Granda, L., Berdugo-Cely, J.A., Šamajová, O., Šamaj, J. *et al.* (2020) A dual strategy of breeding for drought tolerance and introducing drought-tolerant, underutilized crops into production systems to enhance their resilience to water deficiency. *Plants* 9(10), 1263. DOI: 10.3390/plants9101263.

Roychowdhury, R., Das, S.P., Gupta, A., Parihar, P., Chandrasekhar, K. *et al.* (2023) Multi-omics pipeline and omics-integration approach to decipher plant's abiotic stress tolerance responses. *Genes* 14(6), 1281. DOI: 10.3390/genes14061281.

Roy, S., Ghosh, P., Ghosh, N. and Das, S.K. (2022) Transcriptional regulatory network topology with applications to bio-inspired networking: A survey. *ACM Computing Surveys* 54(8), 166. DOI: 10.1145/3468266.

Samanta, S. and Roychoudhury, A. (2023) Crosstalk of melatonin with major phytohormones and growth regulators in mediating abiotic stress tolerance in plants. *South African Journal of Botany* 163, 201–216. DOI: 10.1016/j.sajb.2023.10.040.

Samanta, S. and Roychoudhury, A. (2025) Molecular crosstalk of jasmonate with major phytohormones and plant growth regulators during diverse stress responses. *Journal of Plant Growth Regulation* 44(1), 62–88. DOI: 10.1007/s00344-024-11412-w.

Sarwar, R., Zhu, K.M., Jiang, T., Ding, P., Gao, Y. *et al.* (2023) DELLAs directed gibberellins responses orchestrate crop development: A brief review. *Crop Science* 63(1), 1–28. DOI: 10.1002/csc2.20825.

Selwal, N., Wani, A.K., Akhtar, N., Kaur, M. and Jassal, P.S. (2023) Molecular insights of strigolactone biosynthesis, signalling pathways, regulatory roles, and hormonal crosstalks in plant systems. *South African Journal of Botany* 160, 9–22. DOI: 10.1016/j.sajb.2023.06.046.

Sharma, M., Singh, D., Saksena, H.B., Sharma, M., Tiwari, A. *et al.* (2021) Understanding the intricate web of phytohormone signalling in modulating root system architecture. *International Journal of Molecular Sciences* 22(11), 5508. DOI: 10.3390/ijms22115508.

Sharma, P., Jha, A.B. and Dubey, R.S. (2024a) Strigolactones: Coordination with other phytohormones and enhancement of abiotic stress responses. *Environmental and Experimental Botany* 223, 105782. DOI: 10.1016/j.envexpbot.2024.105782.

Sharma, L., Roy, S., Satya, P., Alam, N.M., Goswami, T. *et al.* (2024b) Exogenous ascorbic acid application ameliorates drought stress through improvement in morpho-physiology, nutrient dynamics, stress metabolite production and antioxidant activities recovering cellulosic fibre production in jute (*Corchorus olitorius* L.). *Industrial Crops and Products* 217, 118808. DOI: 10.1016/j.indcrop.2024.118808.

Shelake, R.M., Kadam, U.S., Kumar, R., Pramanik, D., Singh, A.K. *et al.* (2022) Engineering drought and salinity tolerance traits in crops through CRISPR-mediated genome editing: Targets, tools, challenges, and perspectives. *Plant Communications* 3(6), 100417. DOI: 10.1016/j.xplc.2022.100417.

Sheoran, S., Gupta, M., Kumari, S., Kumar, S. and Rakshit, S. (2022) Meta-QTL analysis and candidate genes identification for various abiotic stresses in maize (*Zea mays* L.) and their implications in breeding programs. *Molecular Breeding* 42(5), 26. DOI: 10.1007/s11032-022-01294-9.

Singh, M., Kumar, N. and Singh, H. (2024) Plant functional traits assisted crop adaptation to abiotic and biotic stress. In: Kumar, N. and Singh, H. (eds) *Plant Functional Traits for Improving Productivity*. Springer Nature Singapore, Singapore, pp. 239–255.

Singla, P. (2020) Plant synthetic biology: A paradigm shift targeting stress mitigation, reduction of ecological footprints and sustainable transformation in agriculture. In: Giri, B. and Sharma, M.P. (eds) *Plant Stress Biology*. Springer, Singapore.

Srivastava, R., Bajpai, R., Khan, Z., Singh, S.P., Mehrotra, R. *et al.* (2022) Insight into strigolactone hormone functions in plant parasitic weeds: A regulatory perspective. *Indian Journal of Experimental Biology* 60(9), 659–666. DOI: 10.56042/ijeb.v60i09.65154.

Su, J., Liu, Y., Han, F., Gao, F., Gan, F. *et al.* (2024a) ROS, an important plant growth regulator in root growth and development: Functional genes and mechanism. *Biology* 13(12), 1033. DOI: 10.3390/biology13121033.

Su, X., Zheng, J., Diao, X., Yang, Z., Yu, D. *et al.* (2024b) MtTCP18 regulates plant structure in *Medicago truncatula*. *Plants* 13(7), 1012. DOI: 10.3390/plants13071012.

Sun, H., Li, W., Burritt, D.J., Tian, H., Zhang, H. *et al.* (2022) Strigolactones interact with other phytohormones to modulate plant root growth and development. *Crop Journal* 10(6), 1517–1527. DOI: 10.1016/j.cj.2022.07.014.

Sun, Y., Tian, Z., Zuo, D., Cheng, H., Wang, Q. *et al.* (2024) Strigolactone-induced degradation of SMXL7 and SMXL8 contributes to gibberellin- and auxin-mediated fiber cell elongation in cotton. *Plant Cell* 36(9), 3875–3893. DOI: 10.1093/plcell/koae212.

Taulera, Q. (2022) Perception and responses to strigolactones in arbuscular mycorrhizal fungi. PhD thesis, Université Paul Sabatier-Toulouse III, Toulouse, France.

Temmerman, A., Guillory, A., Bonhomme, S., Goormachtig, S. and Struk, S. (2022) Masks start to drop: Suppressor of MAX2 1-like proteins reveal their many faces. *Frontiers in Plant Science* 13, 887232. DOI: 10.3389/fpls.2022.887232.

Thilakarathne, A.S., Liu, F. and Zou, Z. (2025) Plant signaling hormones and transcription factors: Key regulators of plant responses to growth, development, and stress. *Plants* 14(7), 1070. DOI: 10.3390/plants14071070.

van Niekerk, G.L. (2020) Evaluation of the effect of selected plant growth regulators on soybean yield parameters in South Africa. Doctoral dissertation, North-West University (South Africa).

van Weringh, A. (2023) Time course transcriptome analysis of guard cells during a progressive drought in *Arabidopsis thaliana* using an optimized INTACT protocol for leaf tissue. PhD thesis, University of Toronto, Toronto, Canada.

Voko, M.P., Aremu, A.O., Makunga, N.P., Nisler, J., Doležal, K. *et al.* (2024) The potential applications of cytokinins and cytokinin oxidase/dehydrogenase inhibitors for mitigating abiotic stresses in model and non-model plant species. *Current Plant Biology* 40, 100398. DOI: 10.1016/j.cpb.2024.100398.

Vysotskaya, L., Timergalina, L., Akhiyarova, G., Korobova, A., Fedyaev, V. *et al.* (2021) Association of barley root elongation with ABA-dependent transport of cytokinins from roots and shoots under supra-optimal concentrations of nitrates and phosphates. *Cells* 10(11), 3110. DOI: 10.3390/cells10113110.

Waidi, Y.O., Debnath, S., Datta, S. and Chatterjee, K. (2024) 3D-printed silk proteins for bone tissue regeneration and associated immunomodulation. *Biomacromolecules* 25(9), 5512–5540. DOI: 10.1021/acs.biomac.4c00540.

Waititu, J.K., Cai, Q., Sun, Y., Sun, Y., Li, C. *et al.* (2021) Transcriptome profiling of maize (*Zea mays* L.) leaves reveals key cold-responsive genes, transcription factors, and metabolic pathways regulating cold stress tolerance at the seedling stage. *Genes* 12(10), 1638. DOI: 10.3390/genes12101638.

Wang, J., Wang, D., Zhu, M. and Li, F. (2021) Exogenous 6-benzyladenine improves waterlogging tolerance in maize seedlings by mitigating oxidative stress and upregulating the ascorbate-glutathione cycle. *Frontiers in Plant Science* 12, 680376. DOI: 10.3389/fpls.2021.680376.

Wang, X., Wang, Q., Yan, L., Hao, Y., Lian, X. *et al.* (2023) *PpTCP18* is upregulated by lncRNA5 and controls branch number in peach (*Prunus persica*) through positive feedback regulation of strigolactone biosynthesis. *Horticulture Research* 10(1), uhac224. DOI: 10.1093/hr/uhac224.

Wang, K., Guo, H. and Yin, Y. (2024) AP2/ERF transcription factors and their functions in *Arabidopsis* responses to abiotic stresses. *Environmental and Experimental Botany* 222, 105763. DOI: 10.1016/j.envexpbot.2024.105763.

Wang, K., Yu, J. and Zhou, J. (2025) Hormonal regulation of stress tolerance in vegetable crops. In: Ahammed, G.J. and Zhou, J. (eds) *Growth Regulation and Quality Improvement of Vegetable Crops: Physiological and Molecular Features*. Springer Nature Singapore, Singapore, pp. 349–388. DOI: 10.1007/978-981-96-0169-1_13.

Wang, X., Liu, X., Su, Y. and Shen, H. (2025a) Rice responses to abiotic stress: Key proteins and molecular mechanisms. *International Journal of Molecular Sciences* 26(3), 896. DOI: 10.3390/ijms26030896.

Wang, X., Wen, H., Suprun, A. and Zhu, H. (2025b) Ethylene signaling in regulating plant growth, development, and stress responses. *Plants* 14(3), 309. DOI: 10.3390/plants14030309.

Wani, S.H., Anand, S., Singh, B., Bohra, A. and Joshi, R. (2021) WRKY transcription factors and plant defense responses: Latest discoveries and future prospects. *Plant Cell Reports* 40(7), 1071–1085. DOI: 10.1007/s00299-021-02691-8.

Wei, Y.S., Javed, T., Liu, T.T., Ali, A. and Gao, S.J. (2025) Mechanisms of abscisic acid (ABA)-mediated plant defense responses: An updated review. *Plant Stress* 15, 100724. DOI: 10.1016/j.stress.2024.100724.

Xia, C., Miranda, J., Mendoza-Cozatl, D., Ham, B.K., Ma, J. *et al.* (2025) Decoding long-distance communication under mineral stress: Advances in vascular signalling and molecular tools for plant resilience. *Plant, Cell & Environment* 48(7), 4781–4802. DOI: 10.1111/pce.15475.

Xiao, F., Zhou, H. and Lin, H. (2025) Decoding small peptides: Regulators of plant growth and stress resilience. *Journal of Integrative Plant Biology* 67(3), 596–631. DOI: 10.1111/jipb.13873.

Xu, G., Li, L., Zhou, J., He, M., Lyu, D. *et al.* (2023) Integrated transcriptomics and metabolomics analyses reveal key genes and essential metabolic pathways for the acquisition of cold tolerance during dormancy in apple. *Environmental and Experimental Botany* 213, 105413. DOI: 10.1016/j.envexpbot.2023.105413.

Xu, X., Fang, P., Zhang, H., Chi, C., Song, L. *et al.* (2019) Strigolactones positively regulate defense against root-knot nematodes in tomato. *Journal of Experimental Botany* 70(4), 1325–1337. DOI: 10.1093/jxb/ery439.

Yadav, R.K., Tripathi, M.K., Tiwari, S., Tripathi, N., Asati, R. *et al.* (2023) Genome editing and improvement of abiotic stress tolerance in crop plants. *Life* 13(7), 1456. DOI: 10.3390/life13071456.

Yan, Y., Zhao, N., Tang, H., Gong, B. and Shi, Q. (2020) Shoot branching regulation and signaling. *Plant Growth Regulation* 92(2), 131–140. DOI: 10.1007/s10725-020-00640-1.

Yang, L., Luo, S., Jiao, J., Yan, W., Zeng, B. *et al.* (2023a) Integrated transcriptomic and metabolomic analysis reveals the mechanism of gibberellic acid regulates the growth and flavonoid synthesis in *Phellodendron chinense* schneid seedlings. *International Journal of Molecular Sciences* 24(22), 16045. DOI: 10.3390/ijms242216045.

Yang, M., Wang, L., Wang, X., Li, Y. and Huang, H. (2023b) Transcriptomic response to drought stress in *Populus davidiana* dode. *Forests* 14(7), 1465. DOI: 10.3390/f14071465.

Yi, F., Song, A., Cheng, K., Liu, J., Wang, C. *et al.* (2023) Strigolactones positively regulate *Verticillium wilt* resistance in cotton via crosstalk with other hormones. *Plant Physiology* 192(2), 945–966. DOI: 10.1093/plphys/kiad053.

Yuan, Y., Khourchi, S., Li, S., Du, Y. and Delaplace, P. (2023) Unlocking the multifaceted mechanisms of bud outgrowth: Advances in understanding shoot branching. *Plants* 12(20), 3628. DOI: 10.3390/plants12203628.

Yu, H., Zhao, J., Chen, L., Wu, T., Jiang, B. *et al.* (2023) CRISPR/Cas9-mediated targeted mutagenesis of *GmEOD1* enhances seed size of soybean. *Agronomy* 13(9), 2359. DOI: 10.3390/agronomy13092359.

Yu, G.J., Sharma, A. and Ahammed., J. (eds) (2021) *Brassinosteroids in Plant Developmental Biology and Stress Tolerance*. Academic Press, UK.

Zaidi, S., Arif, Y., Imtiaz, H., Shiraz, M. and Hayat, S. (2024) Structural chemistry, biosynthesis, and signaling of multifaceted plant growth regulator: Strigolactone. *Journal of Plant Growth Regulation* 43(8), 2489–2502. DOI: 10.1007/s00344-024-11285-z.

Zeng, T., Su, H., Wang, M., He, J., Gu, L. *et al.* (2025) The role of MYC2 transcription factors in plant secondary metabolism and stress response mechanisms. *Plants* 14(8), 1255. DOI: 10.3390/plants14081255.

Zhang, L., Fang, W., Chen, F. and Song, A. (2022a) The role of transcription factors in the regulation of plant shoot branching. *Plants* 11(15), 1997. DOI: 10.3390/plants11151997.

Zhang, Q., Guo, N., Zhang, Y., Yu, Y. and Liu, S. (2022b) Genome-wide characterization and expression analysis of pathogenesis-related 1 (PR-1) gene family in tea plant (*Camellia sinensis* (L.) O. Kuntze) in response to blister-blight disease stress. *International Journal of Molecular Sciences* 23(3), 1292. DOI: 10.3390/ijms23031292.

Zhang, X., Sun, Y., Qiu, X., Lu, H., Hwang, I. *et al.* (2022c) Tolerant mechanism of model legume plant *Medicago truncatula* to drought, salt, and cold stresses. *Frontiers in Plant Science* 13, 847166. DOI: 10.3389/fpls.2022.847166.

Zhao, C., Zhang, H., Song, C., Zhu, J.K. and Shabala, S. (2020) Mechanisms of plant responses and adaptation to soil salinity. *Innovation* 1(1), 100017. DOI: 10.1016/j.xinn.2020.100017.

Zhou, H., Hwarari, D., Ma, H., Xu, H., Yang, L. *et al.* (2022) Genomic survey of TCP transcription factors in plants: Phylogenomics, evolution and their biology. *Frontiers in Genetics* 13, 1060546. DOI: 10.3389/fgene.2022.1060546.

Zhou, H., Shi, H., Yang, Y., Feng, X., Chen, X. *et al.* (2024) Insights into plant salt stress signaling and tolerance. *Journal of Genetics and Genomics* 51(1), 16–34. DOI: 10.1016/j.jgg.2023.08.007.

3 Coordination of Plant Hormone Pathways by Strigolactones

Erdogan E. Hakki[1], Muhammad Anas[1,2], Waseem Ahmed Khattak[2], Amjid Khan[3], Sameel Hassan[4], Javed Iqbal[1,5]*, Banzeer Ahsan Abbasi[6] and Razia Kalsoom[7]

[1]Department of Soil Science & Plant Nutrition, Selcuk University Campus, Konya, Turkey; [2]Department of Plant Sciences, Faculty of Biological Sciences, Quaid-I-Azam University, Islamabad, Pakistan; [3]College of Graduate Studies, University of South Africa, Pretoria, South Africa; [4]Department of Agriculture & Forest Science (DAFNE), University of Tuscia, Viterbo, Italy; [5]Department of Botany, Bacha Khan University, Charsadda, Khyber, Pakhtunkhwa, Pakistan; [6]Department of Botany, Rawalpindi Women University, Satellite Town, Rawalpindi, Pakistan; [7]Pakistan Council of Scientific and Industrial Research (PCSIR) Labs, Sector H-9, Islamabad, Pakistan

Abstract

Strigolactones (SLs) are a novel class of carotenoid-derived phytohormones that play crucial roles in regulating plant development, architecture, and stress responses. Initially identified as germination stimulants for parasitic weeds, SLs have since been recognized as key endogenous hormones that modulate shoot branching, root-system architecture, seed germination, and photomorphogenesis. This chapter comprehensively explores the multi-faceted roles of SLs and their intricate crosstalk with other phytohormones—including auxin, cytokinins, abscisic acid, ethylene, gibberellins, jasmonates, and salicylic acid—in coordinating plant growth and adaptation. The regulatory influence of SLs extends to abiotic stress tolerance, such as drought, salinity, and temperature extremes, where they interact synergistically with abscisic acid and other hormones to enhance plant resilience. SLs also mediate nutrient-deficiency adaptations by promoting root architectural changes and facilitating beneficial symbioses with arbuscular mycorrhizal fungi, thereby improving nutrient uptake efficiency. Furthermore, SLs contribute to biotic stress defense by priming immune responses and shaping the rhizosphere microbiome, although their role in parasitic weed germination presents agricultural challenges. This chapter highlights the potential of synthetic SL analogs as agrochemicals for controlling parasitic weeds, regulating plant architecture, and enhancing stress tolerance. Advances in genetic and hormonal manipulation strategies targeting SL pathways offer promising avenues for crop improvement aimed at sustainable agriculture. Despite significant progress, further research is needed to unravel the molecular mechanisms underlying SL signaling and crosstalk, optimize the use of SL analogs in field conditions, and assess ecological impacts. Overall, this study underscores the centrality of strigolactones in plant biology and their promising applications in improving crop productivity and resilience.

Keywords: abiotic stress tolerance, crop improvement, phytohormonal crosstalk, plant development, strigolactones

*Corresponding author: javed89qau@gmail.com

DOI: 10.1079/9781836990598.0003

3.1 Introduction

Strigolactones (SLs) are bioactive molecules derived from carotenoids that are a distributed array of phytohormones and secondary metabolites involved in various aspects of plant growth, development, and microbiota management in the rhizosphere (El Sabagh *et al.*, 2022; Tariq *et al.*, 2023). Subsequently discovered as potent germination stimulants for root-parasitic weeds of the genera *Striga* and *Orobanche*, and other root-parasitic weeds in the 1960s, they were later established to play major roles in regulation of plant architecture, for example by suppressing branching of the shoot (apical dominance), modulating root branching, promoting leaf senescence, modulating internode elongation, and consequently, affecting the overall form and plasticity of the plant. There is an agreed-upon biosynthetic pathway in synthesizing SLs from carotenoids, which includes four enzymes, D27, CCD7, CCD8, and MAX1, whereby β-carotene is first isomerized and then enzymatically cleaved to produce carlactone, which is subsequently converted into diverse bioactive SLs, some of which remain structurally uncharacterized in certain plant species (Chesterfield, 2020; En-Nahli *et al.*, 2023; Jain *et al.*, 2025). These are usually found at only picomolar to nanomolar concentrations in root exudates of both monocots and dicots as they have strong biological properties, and there is competition for them. In this context, SLs are used as endogenous hormones by integrating with other phytohormones such as auxin, cytokinins (CKs), abscisic acid (ABA), ethylene, gibberellins (GAs), and jasmonates (JAs) for modulating plant developmental processes and various stress factors including nutrients deficiencies and other abiotic stress factors, which may include drought, salinity, and high or low temperature stresses. For example, SLs are induced by phosphate and nitrogen deficiency, and as well as modulating shoot branching for resource conservation, they also help in promoting root growth for better nutrient searching ability of the plant. Functioning of SLs is important for the plant in the rhizosphere as they are active signaling molecules in the communication between plants and soil organisms (Pandey *et al.*, 2016; Arshad *et al.*, 2023). They act as signals for arbuscular mycorrhizal fungi (AMF), which form mutualistic associations that have improved absorption of water as well as inorganic phosphate, particularly by land plants, for 400 million years. This same signaling system is used by parasitic weeds belonging to the Orobanchaceae family because it allows them to germinate only when there is a host plant nearby such as sorghum, maize, rice, or tomatoes, especially in regions of Africa, Asia, and the Mediterranean, thereby causing damage to crops. The ability of SLs to stimulate beneficial mycorrhizal associations but at the same time fulfill the needs of parasitic nematodes shows that plants have evolved to balance biosynthesis and release of these signaling molecules under the selective pressures from both mutualistic and parasitic organisms (Boyno *et al.*, 2023). As well as plant–microbe and plant–plant interactions, SLs have also been found to play a role in the defense mechanism of the plant, with an increased susceptibility of SL biosynthesis or signaling mutants to specific bacterial pathogens, resulting in changes in stomatal conductance, reactive oxygen species tolerance, and hormonal homeostasis (Jogawat *et al.*, 2021). SLs are perceived by the D14 receptor, which, in turn, interacts with the F-box protein MAX2 activating further signaling processes leading to changes in gene expression and ultimately developmental processes. Specific to parasitic plants, it is proposed that a divergent KAI2 receptor different from that of the corresponding receptor of plants has been tailored to perceive SLs and not related molecules such as karrikins. Commonly, SLs are made up of tricyclic lactone rings (ABC-rings) linked to a butenolide (D-ring) though an enol-ether bridge, and naturally occurring SLs exhibit variation in their structural features, with a plant typically producing more than one SL with differences in their efficacy and effectiveness toward different rhizosphere partners and parasites (Hossain *et al.*, 2022). This also helps explain the sophistication of signal transduction involving SLs and the ability of these systems to evolve over time. Despite their roles in plant biology, there are also studies and research about possible biomedical uses for SLs and their mimics, mainly in cancer, inflammation, and virus infection, but these are still experimental and are not part of their natural purpose. Interest in the study of SLs remains sustained by their relevance in plant adaptation, in view of their role in parasitic weed growth, in global food security and in their applications for the innovations in agriculture, such as the use of SL homologs for the modulation of plant architecture, stress responses to abiotic stresses, and control of parasitic weeds (Mutinda *et al.*, 2025). None the less, several questions such as how biosynthesis of SLs is regulated, the specificity of their signaling pathways, and their integration with other hormones, as well as environmental factors and their functional potential in plant–microbe and plant–pathogen interactions, remain unknown, and are areas that have constant research interest in terms of convergence of physiology, ecology, and biotechnology (Schuetz, 2019).

3.2 Biosynthesis and Signaling Pathways

SLs are a group of plant hormones and soil-borne signaling agents formed from carotenes, and are involved in control of plant growth and development, mycorrhiza formation and recognition, and responses to the external environment (Janeeshma *et al.*, 2024). This process stems from isomerization of all-*trans*-β-carotene to 9-*cis*-β-carotene (CISB) through the enzyme DWARF27 (D27). This is succeeded by the action of carotenoid cleavage dioxygenase 7 (CCD7), which cuts CISB to produce 9-*cis*-β-apo-10′-carotenal; this is then degraded by carotenoid cleavage dioxygenase 8 (CCD8) into carlactone, which is an essential intermediate in SL synthesis. From this point, the pathway branches depending on the type of plant and the enzymatic activity: thus, proteins of the cytochrome P450 family, and in particular the MAX1 family, are responsible for the transformation of carlactone into other SLs, such as carlactonoic acid or orobanchol, 5-deoxystrigol and others (Harrison *et al.*, 2015; Zhang *et al.*, 2018). The exact process and the enzymes involved in the synthesis of carlactone and its relatives from SLs has not been fully elucidated, but the most recent research describes orobanchol synthase to be responsible for the biosynthesis of orobanchol from carlactonoic acid, thus relating SL biosynthesis both to beneficial related plant symbiosis and to the promotion of parasitic plant germination. It has been suggested that the biosynthesis regulation of SL is supply driven and does not depend on feedback inhibition by SLs, which has been reinforced by the conclusions from mathematical modeling and experiments; hence, changes in the flux of β-carotene into the pathway, as well as altering the activity of enzymes after branching points in the pathway, affect the yield of total and individual types of SLs (Fig. 3.1) (Al-Babili and Bouwmeester, 2015).

Unfortunately, there is a lack of information about the perception of SL molecules by the α/β-hydrolase D14, the SL receptor. When bound to SLs, D14 changes its conformation to interact with the F-box protein MORE AXILLARY GROWTH 2 (MAX2), which is part of the SCF (SKP1–CUL1–F-box) ubiquitin ligase complex (Wani *et al.*, 2021). This receptor–co-receptor complex then recruits in addition downstream repressors such as D53/SMXL proteins for ubiquitination and degradation by the 26S proteasome. Liberation of these repressors provides the proper forum for the occurrence of SL-responsive gene transcription and its physiological consequences, including suppression of shoot branching, regulation of root-system configuration, and improved AMF symbiosis (Pandey *et al.*, 2016). The SL signaling pathway is also interconnected with other hormonal and environmental signaling to enable the plants to regulate growth and developmental change according to nutrient status, stress, and biotic factors (Zhao *et al.*, 2021). Recent studies have also identified that the structural differences in SLs due to the differences in biosynthetic pathway and enzyme specificity can be an important factor for their function in endogenous signaling or in the regulation of biotic interactions in the rhizosphere. Thus, SLs are synthesized from carotenoids in a multi-step process through the activity of multiple enzymes and form a structurally diverse group of hormones. Their signaling involves a receptor-mediated system that results in protein degradation and subsequent changes in gene expression, thus leading to adaptive mechanisms of growth, development, and interaction with other organisms in response to internal and external stimuli (Antar *et al.*, 2021).

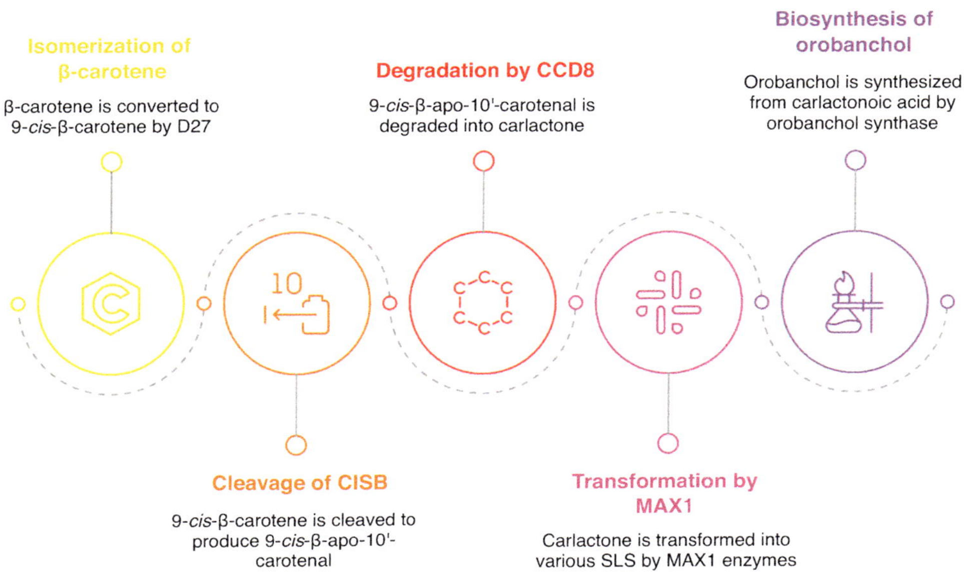

Fig. 3.1. Biosynthesis pathway of strigolactones. CCD8, carotenoid cleavage dioxygenase 8; CISB, 9-*cis*-β-carotene; SL, strigolactone.

3.3 Strigolactone Crosstalk with Other Phytohormones

3.3.1 Interaction with auxin

Both SLs and auxin exist as fundamental phytohormones that intensely control multiple developmental processes involving root architecture and shoot branching. Auxin increases the formation of adventitious roots, which results from SL exposure according to recent research on melon seedlings (Sharma *et al.*, 2021). The total output of adventitious roots improved significantly when seedlings received synthetic SL (GR24) and auxin (indole-3-acetic acid) treatments rather than just one of the hormones. The root regeneration process between auxin and SLs works in synergy due to changes in hormone signaling gene expression that enhances both compounds' existing functions essential for plant survival when transplanted or under abiotic stress (Li *et al.*, 2023). Under extreme phosphate exposure, *Medicago truncatula* displays enhanced lateral root plasticity because auxin treatment activates both SL biosynthetic and signaling markers (including MAX2 and MAX3) in the roots. The auxin–SL interaction mechanism allows plants to modify their root systems effectively when acquiring nutrients in difficult ecological conditions. Stem and root thickening receive coordinated regulation through the vascular cambium from SLs and auxin signaling, while maintaining SL signaling as a prerequisite for complete auxin-dependent activation of cambial activity between the two hormones. The transport of auxin depends on SLs through their ability to control the activity of auxin efflux carriers (PIN proteins) and block auxin distribution toward lateral buds, which strengthens apical dominance (Herold, 2009; Borghi *et al.*, 2016). The feedback mechanism regulates plant architecture by how auxin controls SL biosynthetic gene expression. The community of plant hormones functioning together displays similar behavior in plant–fungal symbiotic relationships through a relationship between reduced auxin production and diminished SL quantities, which affects both fungal and plant health outcomes. SLs and auxin interact through a complex regulatory network that enables plants to integrate environmental, physiological, and developmental signals for survival and productivity purposes (Sharma *et al.*, 2024).

3.3.2 Interaction with CKs

Shoot branching, bud outgrowth, and plant architecture are controlled by complex and often antagonistic hormonal crosstalk between SLs and CKs. In rice and other cereals, SLs are key inhibitors of axillary bud growth (tillers) acting together with CKs, which are positive regulators for activating the buds and sprouting of shoots (Lakehal and Bellini, 2019). It has been elucidated in recent research that SLs inhibit shoot branching, at least in part, by modulating CK homeostasis when they upregulate CYTOKININ OXIDASE/DEHYDROGENASE 9 (OsCKX9) expression to lower local CK levels at the shoot base and axillary buds. In rice, the loss of function in SL signaling (as in d53 mutants) results in increased levels of CK, increased tillering, and altered shoot architecture, and overexpression of OsCKX9 reduces CK levels and represses bud outgrowth (Ahmed *et al.*, 2025). Additionally it has been demonstrated by transcriptomic studies that SLs can, at the level of both CK abundance and signal transduction, suppress type-A Arabidopsis response regulator (*A-ARR*) genes, the central hub of CK signaling in buds. However, CKs can counteract SL action and promote bud growth by downregulating the components of the SL signaling pathway and potentially opposing the status of a bud as dormant or growing out. However, this antagonism is under context-dependent environmental control, as plants often use nutrients, light, and stress cues to modulate architecture adaptively to maximize resource use (Mazzoni-Putman *et al.*, 2021). In addition to shoot branching, SL–CK crosstalk affects root development, secondary growth, and the response to environmental change, emphasizing the central control exerted by this arm in plant developmental plasticity. In general, the interplay between SLs and CKs makes plant growth integrate signals of internal developmental with those external cues of the environment to coordinate regulation of plant growth and form (Ahmed *et al.*, 2025).

3.3.3 Interaction with ABA

SLs and ABA are two phytohormones that are related by multiple connections and work in several steps to control the plant response to abiotic stress, drought, and osmotic stress (Sharma *et al.*, 2024). It is noteworthy that both of these hormones are derived from carotenoid precursors and are therefore regulated coordinately during stress adaptation. Thus, under normal growth conditions, there is an optimal ratio of SL and ABA content to promote plant development; however, under situations such as water deficit or salinity stress, SL content in the roots is usually reduced, while ABA content is increased, and the plant activates a defense mechanism such as closure of stomata to minimize water loss (Visentin *et al.*, 2016). This is an indicated interaction between SL biosynthesis and ABA in plants where SL enhances the plant response to ABA, especially in guard cells, thus allowing better stomatal closure and improved drought tolerance. Some evidence from *Arabidopsis*, tomato, and *Lotus japonicus* reveals that mutants that are deficient in SL biosynthesis or signaling have lower ABA levels and exhibit a strain in the ABA-induced stomatal response under stress, thus underlining the role of SLs in regulating ABA-mediated pathways (Sharma *et al.*, 2024). Notably, SLs also regulate stomata closing through other pathways as well as ABA, confirming the layers of regulation in the SL–ABA system. Correspondingly, the expression of major SL biosynthetic genes is activated in shoots under drought-stress conditions, further supporting the fact that ABA levels are also elevated under drought-stress condition. In contrast, SL-deficient mutants are known to exhibit enhanced accumulation of ABA under normal conditions, suggesting that they are interlinked in a way that is regulated at the feedback level. Furthermore, both SLs and ABA regulate antioxidant systems and stress-responsive genes involving heat-shock and transcription factor proteins that improve the ability of plants to tolerate heat stress, drought, and salinity conditions (Hussain *et al.*, 2021). Bennett and Leyser (2014) and others have outlined how SLs can be perceived and imported into the cell to modulate other hormone pathways such as ABA in guard cells; however, the signaling pathways that occur in guard cells are not well understood. In summary, the relationship between SLs and ABA is an important hormonal network that conveys environmental information to adjust photosynthetic water-use efficiency and growth, and to enhance plant tolerance to stress in response to water deficit, making this signaling pathway a candidate for enhancing stress tolerance of crop plants through biotechnology (Fig. 3.2) (Sharma *et al.*, 2024).

3.3.4 Interaction with ethylene

SLs and ethylene are two hormones that have been reported to regulate processes such as leaf senescence and root-hair development. In *Arabidopsis*, evidence has demonstrated that both SLs and ethylene have the ability to induce leaf senescence, especially under dark-imposed conditions (Yamada and Umehara, 2015). Carbohydrate content decreases first before the synthesis of ethylene begins and is followed by that of SLs, and both hormones are pivotal for the promotion of senescence. SL-biosynthesis mutants exhibit longevity akin to plants not treated with ethylene, regardless of exposure to ethylene, but treatment with SLs alongside ethylene enhances chlorophyll breakdown and leaf yellowing (Altaf *et al.*, 2024). A similar interaction between SLs and ethylene in mediating leaf senescence has also been noted in perennial ryegrass; SLs and ethylene are involved in dark-induced alteration of chlorophyll degradation genes, and use of ethylene antagonists also negates the effects of SLs (Hu *et al.*, 2021). However, there is evidence that ethylene and SLs work cooperatively in controlling root-hair elongation in *Arabidopsis* where both hormones are needed for the growth of root hair and their signalling pathways are involved in regulating this developmental process (Kamran *et al.*, 2024). More specifically, in plant–pathogen interactions including *Pythium irregulare* and *Fusarium oxysporum*, there is evidence that although the ethylene signal transduction pathway is important for resistance against these pathogens, SL are not implicated in these disease responses or depolymerization of non-green-leaf reporters, and have no direct effects on pathogen growth and plant permissiveness in the tested systems (Foo *et al.*, 2016). Thus, there is a context dependency in the interaction of SLs with ethylene, where they play a crucial role in developmental processes such as aging and root-hair development but may not be as influential in some cases of pathogen attack (Chien and Yoon, 2024).

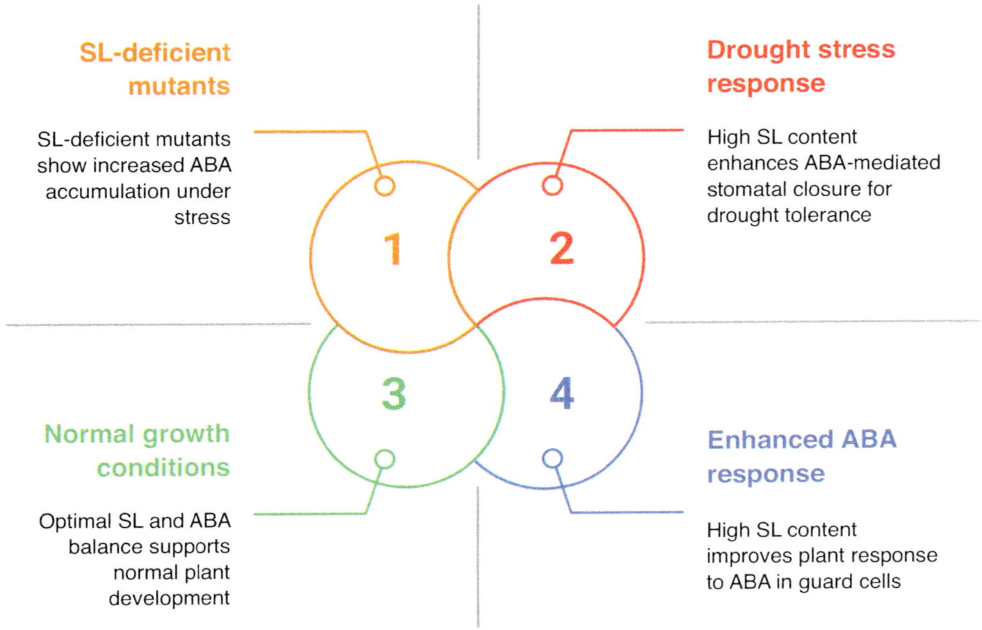

Fig. 3.2. Hormonal regulation in the plant stress response. ABA, abscisic acid; SL, strigolactone.

3.3.5 Interaction with GAs

Plants use both SLs and GAs as major messengers that control their development stages and environmental adjustments. Both groups of hormones employ α/β-hydrolase receptors, which activate proteasomal breakdown of inhibitory proteins (DELLA and D53 for GA and SL perception, respectively) through enzymatic activity to trigger gene-expression changes (Sharma *et al.*, 2024). Scientific research shows that GA signaling controls the production of SLs in plants. In particular, the study of cotton shows that GA triggers *GhGRF4* gene expression, which then activates DWARF27 (D27) to increase SL synthesis and trigger plant growth changes. Research using rice and *Arabidopsis* GA- and SL-deficient mutants shows that these hormones jointly control a group of target genes that they influence by joint regulation at the level of gene transcription, rather than by tweaking specific signaling machinery (Sun *et al.*, 2024). SLs and GAs collaborate during plant development to control growth patterns including tip branching and root growth alongside cellulose fiber elongation. GAs usually initiate SL production, which then leads to particular stage transformations. Under seed conditioning, GAs trigger *Striga hermonthica* plant seeds to produce more SL receptor molecules, which enable the seeds to detect and respond to SL signals from host plants. Although the hormones SL and GA have similar growing effects, they only occasionally depend on each other, as their signals do not directly impact DELLA protein changes. SLs and GAs influence overlapping genes in plant networks when they regulate the growth and stress tolerance response to cold temperatures (Altamura *et al.*, 2023; Yap and Tsuchiya, 2023). Research confirms that SLs and GAs regulate plant development together as a balanced system that controls specific signaling targets (Thilakarathne *et al.*, 2025).

3.3.6 Interaction with JAs

SLs along with JAs show a continuous pattern of interaction that affects multiple plant developmental aspects, as well as protection systems and stress adaptability. These two regulatory compounds operate at the core of maintaining growth and defense stability, especially when plants face stress events. The lack of SLs in rice plants has been proven to elevate JA signaling strength, which leads to greater production of defense compounds including phytoalexins and better resistance to *Pyricularia oryzae* blast fungus infection (Torres Vera, 2017). The defense enhancement requires operable JA signaling because defense priming by SL depends on JA pathways, as demonstrated by mutants that show no increase in resistance or defense metabolite accumulation when JA signaling and biosynthesis functions are impaired. This signaling recruitment depends on hormone regulation together with the controlled interactions between key components such as JAZ repressors and MYC2 transcription factors, which serve as the main elements in JA signaling pathways. SLs control component quantities and activities that lead to modulation of defense and stress adaptation gene outputs (Mazzoni-Putman *et al.*, 2021).

SLs and JAs affect developmental processes through their interactive relationship, which extends beyond defense roles. The interaction of SLs with JAs produces either cooperative or competing effects that control both root and shoot shape development and environmental signal responses such as nutrient acquisition and pathogenic conditions (Rasheed *et al.*, 2023). The hormonal signals between plants coordinate energy management to direct resources toward defense systems or toward developmental growth, based on environmental factors. The accumulation of JA and the corresponding defense response in rice plants rises after SL deficiency leads to higher soluble sugar levels because these levels enhance JA biosynthesis. Through SL–JA crosstalk, plants achieve adaptive coordination between developmental regulation and defense mechanisms by way of their complex hormonal communication networks (Samanta and Roychoudhury, 2025).

3.3.7 Interaction with salicylic acid

The close interaction between SLs and salicylic acid (SA) modulates the plant defense response, particularly systemic acquired resistance against pathogens. SLs prime *Arabidopsis thaliana* plants to exhibit increases of SA-mediated disease resistance without inducing defense genes themselves before pathogen attack (Kamle *et al.*, 2020). In contrast, treatment with the synthetic SL analog GR24 does not activate defense-related genes by itself but renders a faster and stronger induction of SA-responsive genes upon pathogenic infection, suggesting a priming effect of SLs on the SA signaling pathway. The signaling-deficient mutants *max1*, *max3*, *max4*, and especially *max2*, are defective in SL biosynthesis and/or signaling and exhibit altered basal immune function by showing higher bacterial growth and lower endogenous SA levels and SA-responsive gene expression compared with wild-type plants (Kumar *et al.*, 2025). Thus, functional SL signaling is required for basal immunity and normal SA accumulation. Furthermore, GR24 promotes disease resistance in both wild-type and SL-biosynthesis mutants, and disease resistance is lost in *max2* mutants, confirming that SL signaling via MAX2 is required for this priming. Suppression of disease resistance in wild-type plants following treatment with SL-biosynthesis inhibitors also supports a positive role of SLs in SA-mediated defense. Together, these findings indicate that SLs are not direct inducers of defense; rather, they sensitize or prime the plant immune system for faster, stronger, and ultimately more robust SA-dependent defense responses upon challenge by a pathogen (de Villiers, 2015). The integration of developmental and environmental signals via this SL–SA crosstalk serves to optimize plant immunity and establishes SLs as key mediators of hormone-mediated disease-resistance pathways. Additionally, the interaction highlights the possibility of controlling SL signaling to promote SOS development and strengthen SA-based defenses in crops that will enhance its resilience against pathogens (Roychowdhury *et al.*, 2024).

3.4 Role of Strigolactone Crosstalk in Plant Development

SLs regulate major aspects of growth and morphogenesis, as well as adaptation to the environment through complex crosstalk with other hormone pathways. Their discovery as root exudates in which parasitic weeds germinate proved to be just the tip of the iceberg in the many roles that they play in shaping shoot and root architecture, controlling seed germination and photomorphogenesis, and integrating plant responses to nutrient availability and stress (Cheng *et al.*, 2013). Far from acting in isolation, SLs play out their roles by functioning as part of complex hormonal networks, integrating themselves with auxin, CKs, ABA, ethylene and GAs to fine-tune developmental processes and allocation of resources. Crosstalk is essential for plants to balance growth and survival, especially under suboptimal conditions, and occurs both synergistically and antagonistically at molecular, cellular, and physiological levels (Parwez *et al.*, 2022).

3.4.1 Regulation of shoot branching

Axillary branching is one of the most important architectural features and productiveness characteristics of the plant shoot starting with the fates of axillary buds—buds that either remain dormant or grow out into branches. They are now known as the factors responsible for inhibition of axillary bud growth while working in conjunction with auxin but antagonistically to CKs as well as sugars (Kaniganti *et al.*, 2022). SL biosynthesis is induced under nutrient-stress conditions, especially low phosphorus or nitrogen availability, to ensure the plant conserves resources through reduced branch initiation. Molecularly, SLs regulate the key regulatory genes including *BRANCHED1* (*BRC1*) in *Arabidopsis* and its homologous genes in other plants, which are hub genes that integrate hormonal and environmental signals for bud activity (Chaudhary, 2013). SLs are translocated from roots to shoots where they repress the growth of buds through a reduction in CK levels via activation of cytokinin oxidase/dehydrogenase genes such as *OsCKI9* in rice and also through modulation of auxin transport and signaling. This antagonism is important to manage: CKs are involved into the activation of buds and SLs in CK deactivation, so only those branches that will benefit a plant are developed. In addition, SLs function with sugars and their signaling networks; for example, a high sugar concentration interferes with SL inputs and stimulates branching through CKs, indicating the homeostasis between energy status and hormonal controls (Salam *et al.*, 2021). Thus, the complex relationship between SLs, auxin and CKs is complementary, while SLs act both directly and indirectly on the structures contributing to shoot branching, plant architecture and development, improved plant form, and better adaptation to environmental conditions. Plants with reduced capacity of SL biosynthesis or signaling such as *max* mutants in *Arabidopsis* or *d* mutants in rice have an enhanced branch number and different shoot architecture, indicating the critical function of SLs in this procedure. Compatibility of SLs with other hormonal systems guarantees that branching of the shoot is not an unalterable feature of plant development but rather a reaction to a particular stimulus regulating the choice of positions for reproduction and more efficient resource management (Fig. 3.3) (Kelly *et al.*, 2023; Liu *et al.*, 2024).

3.4.2 Root development and architecture

Root-system architecture is another domain known to be closely related to SLs and their interaction with hormones, which directly concerns nutrient acquisition, stress tolerance, and mutualistic associations. SLs are involved in controlling several processes concerning root growth and development, which include primary root growth, lateral root (LR) formation, adventitious root (AR) development and root-hair density (Brewer *et al.*, 2013). Nutrient-starvation conditions, particularly lack of phosphorus and nitrogen, leads to upregulation of SL biosynthesis and consequently changes in root direction in the search for nutrients in the soil. For instance, positive regulators of SLs are involved in the regulation of primary root elongation in rice, wheat, and tomato, and also in controlling the formation of excessive ARs and regulating LRs for the purpose of maintaining a maximum balance of root proliferation and resource exploitation (Shinohara *et al.*, 2013). These processes involve the interaction between SLs and auxin: SLs impact auxin transport and its localization by regulating the genes of auxin efflux carriers (PIN proteins). In addition, SLs act synergistically with ethylene, which enhances the development of root hairs, a useful attribute as this enhances the absorptive area of roots under conditions of low nutrient supply.

Fig. 3.3. Hormonal regulation of axillary bud growth.

An interaction with CKs also occurs as both SLs and CKs are synergistic or inhibitory to each other, depending on the processes involved in root development; they form an active synergy in dictating LR development but may prohibit each other in AR formation (Kohli *et al.*, 2013). Furthermore, SLs act as root-borne molecules that regulate AMF formation on roots to improve nutrient and stress acquisition. They provide the plant with a defense mechanism to protect it from parasitic and pathogenic infections and their adverse effects on plant root architecture, microbial interactions, and nutrient changes in the soil environment. A number of mutants that affect SL biosynthesis or signaling show altered root systems, and reduced mycorrhizal association and nutrient acquisition ability, thus establishing the indispensability of SLs and their interactions with other hormones in root-system formation and soil adaptation (Bhoi *et al.*, 2021).

3.4.3 Seed germination and photomorphogenesis

Recent research shows that SLs contribute to multiple plant functions, including seed sprouting and light response, as well as their impact on shoot and root systems. Research first revealed SLs as root signals that trigger parasitic weed seeds, and then discovered their regulation of seed dormancy in normal plant life (Shu *et al.*, 2016). Through their interaction with ABA and GAs, SLs help determine seed germination readiness by enabling growth during favorable conditions. The interaction among hormones helps seeds wait for ideal conditions before starting their growth as young plants. For example, light-sensing processes interact with SLs to control the length of the hypocotyl stem and growth of cotyledons, while also adding chlorophyll to *Arabidopsis* seedlings (Ahammed *et al.*, 2020). In darkness, SLs reduce hypocotyl growth by their interaction with ethylene and auxin signaling components, which helps to switch from dark growth to light growth. Control of seedling development to help them grow optimally and use energy at the start of their life above ground. SLs help plants develop functional structures and undergo photosynthesis by linking hormone responses to external stimuli during seedling transformation. Hormone relationships with SLs in these settings depend on multiple factors influencing direct effects on gene expression and also modifying hormone production and detection pathways (Cheng *et al.*, 2013). The progeny of plants that produce modified SLs need help with initial seed growth and have worse reactions to light patterns and nutrient availability, suggesting that SLs set important rules for plant progress. Plants use internal signals plus information from their surroundings to manipulate their development while distributing resources better and building defense against stress. Understanding how SL signals work with other hormones improves plant research while creating new ways to modify crop development and stress tolerance in different environments (Saeed *et al.*, 2017).

3.5 Strigolactone and Phytohormonal Crosstalk in Stress Responses

SL phytohormones originated from carotenoids to establish themselves as essential agents that control the plant response to environmental challenges. SLs are part of advanced communication networks with other hormones that control how plants respond to various abiotic and biotic stressors. The following section analyzes the complex relationships between SLs and other phytohormones that control abiotic stress tolerance along with nutrient deficiencies and biotic stress defense mechanisms (Sharma *et al.*, 2024).

3.5.1 Abiotic stress tolerance (drought, salinity, temperature)

Drought, salinity, and extreme temperatures are major abiotic stresses that constrain plant growth and productivity. Recent reports confirm SLs as positive regulators of plant tolerance to these stresses, operating by both direct physiological effects and complex hormonal crosstalk (Bhoi *et al.*, 2021). *Arabidopsis* SL-deficient and SL signaling mutants (*max* series) are hypersensitive to drought and salt stress, indicated by lower survival rates, higher leaf water loss, higher stomatal density, and impaired ABA-induced stomatal closure compared with wild-type plants. It has been demonstrated that the SLs are crucial for stress adaptation by exogenous application of SLs rescuing the stress-sensitive phenotype of SL-deficient mutants (Sharma *et al.*, 2024). Abiotic stress resistance is enhanced by SLs as they function to integrate with the ABA signaling pathways. SLs promote ABA responsiveness under drought and salinity conditions, mainly affecting stomatal regulation: stomata are closed more efficiently and there is less water loss. SLs have physiological effects on ABA-regulated and CK metabolism-regulated genes at the transcriptomic level, thus aiding in cross-regulation at broader hormonal network levels. In addition to their role in protecting the photosynthetic machinery, SLs benefit antioxidant capacity and boost antioxidant activity, which in turn ameliorate oxidative damage under stress conditions (Ali *et al.*, 2023). Furthermore, they are able to enter into a symbiotic relationship with AMF, enhancing the plant's ability to absorb water and nutrients, further augmenting resilience of the plant. SL biosynthesis and signaling is also triggered by temperature extremes such as heat and cold stress (Boyno *et al.*, 2023). Photosynthetic efficiency and the expression of stress-responsive genes are maintained by SLs, sometimes in coordination with other hormones such as GAs and CKs. The multi-faceted role of SLs in abiotic stress tolerance makes them targets for crop improvement strategies to increase stress resilience, especially for crop production in arid and stressed climates (Sharma *et al.*, 2024).

3.5.2 Nutrient-deficiency adaptations

Plant growth worldwide faces strict limitations due to broad abiotic stress that exists in nutrient-deficient soil primarily reduced nitrogen and phosphorous. SLs increase quickly when plants face nutrient shortages and help the plant adjust its response. Researchers have shown that SLs help plants form beneficial connections with AMF (Yuan *et al.*, 2023). When plants lack phosphorus, they boost SL discharge from their roots to attract beneficial AMF in the root area. The fungi live in plant roots and help the plants obtain more phosphorus, nitrogen, and water. SLs help plants grow better roots to find nutrients more effectively. They help roots grow straight up and control side root and hair root growth, making a larger surface area for absorption of nutrients. Through interactions with different hormones and auxins, the plant system optimizes root development to access better nutrients (Mitra *et al.*, 2021). Through SLs, the plant influences microbial recruitment and specifically creates a suitable environment for nitrogen-fixing bacteria to thrive, which improves both nutrient availability and stress defense (Andreo-Jimenez *et al.*, 2015). When SL detection occurs under nutrient shortage, the plant activates its genetic networks to modify both metabolism and resource use. SLs work with other signaling pathways to help plants respond better to poor nutrient supplies without losing their ability to grow and bear crops. SL-based adaptation helps plants survive in dry soils that need rehabilitation and supports sustainable crop production techniques (Ahmed *et al.*, 2025).

3.6 Biotic Stress and Defense Mechanisms

Plants use SLs as defense compounds against both microbial pathogens and parasitic organisms. The best-known property of SLs involves stimulating parasitic weed germination, particularly of *Striga* and *Orobanche*, but researchers are now recognizing their dual role in plant protection against pathogens and in shaping microbial communities (Marro *et al.*, 2021). When plants face pathogen attack, SLs activate plant defense mechanisms through modifications of defense-related gene expression and secondary metabolite production. The priming effect occurs through communication between SL signaling pathways and two vital defense networks consisting of SA and JAs. These defense networks lead to systemic acquired resistance and induced systemic resistance responses (Kusajima *et al.*, 2022). The root microbiota's structure and functionality change due to the action of SLs, while attracting microorganisms that defend plants through suppression or defensive mechanisms. Soil microorganisms including AMF and other supportive species align with SLs, which block pathogens by competing against soil-based pathogens, thus creating defensive protection. When plants fight infections, SLs maintain hormonal equilibrium through their influence on ABA, while also regulating ethylene and CK actions to protect growth and development from damage (Zhao and Li, 2021). The exact function of SLs as protective compounds against biotic stress remains intricate due to situational variables. The germination-stimulating property of SLs toward parasitic plants creates a severe drawback, particularly in agricultural farming operations, because the system faces increased susceptibility. The combination of SL signaling control with studies of its hormone network connections could develop new protective approaches for disease control and plant defense that optimizes both health advantages and parasitic defense vulnerabilities (Tariq *et al.*, 2023).

3.7 Agricultural and Biotechnological Applications

SLs have attracted significant attention in the plant science community within the last 5 years as they have demonstrated diverse roles in organization of plant structure, plant symbiosis and pathosystems, and stress responses. Due to these special characteristics of acting *in situ* within the plant tissues and being released root exudates, they have been viewed as attractive targets for crop improvement and biotechnological applications. The subsequent sections describe the application of SL analogs and other strategies for potential future crop enhancement by hormonal regulation (Pandey *et al.*, 2016).

3.7.1 Use of SL analogs

Natural strigolactones are produced by plants in extremely small quantities; therefore, their extraction for commercial agricultural use is not feasible. However, large-scale synthesis of chemical compounds is also problematic and expensive due to their complex structure (Santana *et al.*, 2015). To address these drawbacks, researchers have synthesized several different synthetic SL analogs and mimics of SLs such as GR24, Nijmegen-1 and several methyl-phenlactonoates which possess biological activities like those of natural SLs but are easier to synthesize and more cost-effective. These analogs are intended to reproduce the principal D-ring moiety of SLs, which is vital for manifestation of their biological effects, while also introducing potential modifications that can improve their stability and selectivity, or facilitate synthetic production (Jamil *et al.*, 2022).

3.7.2 Applications in parasitic weed management

The most important agricultural use for SL analogs is the control of root-parasitic weeds, particularly *Striga* and *Orobanche* species, which pose serious yield threats to key food crops in African, Asian, and Mediterranean regions. These weeds use SLs secreted by host roots to stimulate germination of weeds. Thus, synthetic SL analogs can be applied to affected areas to germinate the parasitic seeds in the absence of a host so that they die, hence eliminating the seed bank and subsequent infestation (Kawada *et al.*, 2022). Analogs such as GR24 and the newly synthesized methyl-phenlactonoates including MP16 have been identified to exert stronger germination efficacy on *Striga* and *Phelipanche* seeds, where field tests have shown their viability for large-scale weed management. Thus, the simplicity of synthesis and high activity of some of these analogs, especially MP16, make their use appropriate in areas infested with parasitic weeds (López-Ráez *et al.*, 2009).

3.7.3 Regulation of plant architecture and yield

SLs are also considered as functional agrochemicals for controlling tillering of shoots, root development, and leaf senescence in plants. Thus, control and regulation of SL signaling could be used to maximize light interception, resource distribution, and yield regeneration in plants. For instance, using of SL analogs can control excessive tillering in cereals, hence developing stronger main stems and leading to increased grain yield. In woody crops, SL analogs may be used to increase wood biomass yield and change bark and branch characteristics to enhance crop management and harvesting (Kaniganti *et al.*, 2022).

3.7.4 Enhancement of symbiotic interactions

SLs are vital in terms of regulation of plant–AMF communication, which in turn help in nutrient acquisition, especially phosphorus, and stress tolerance. Synthetic SL analogs can enhance AMF colonization of plants, which consequently optimizes nutrient acquisition and leads to decreased number of chemical fertilizers. This application is of particular use where the rates of input for agriculture are minimum, and reducing the input and imposing techniques for maintaining ecosystems is more appropriate for agricultural purposes (Naseer *et al.*, 2024).

3.7.5 Crop establishment and stress tolerance

Some current findings show that SL analogs can be used to enhance crop germination under poor conditions such as drought, salt stress, and low nutrient availability. SL analogs can promote better stress responses of plants and stimulation of the root system, which leads to more homogenized crop stands and yield (Pandey *et al.*, 2016).

3.7.6 Challenges and future directions

Despite their encouraging uses, other important difficulties continue to block the use of SL analogs. Evaluations must take place to analyze their environmental durability and selectivity, along with cost-performance ratios and any unintended consequences. Researchers in this field continue to investigate specific crop-focused analog development and regulatory framework approval procedures. Hundreds of SL analogs are being developed, alongside better comprehension of their mode of action, creating the foundation for their widespread use in agricultural sectors.

3.8 Use of Hormones as a Mechanism to Enhance Crop Characteristics

3.8.1 Genetic manipulation of SL pathways

Scientific breakthroughs in molecular genetics and genome-editing technology have allowed scientists to precisely handle SL biosynthesis and signaling pathways in agricultural plants. Breeding efforts have developed crop varieties with optimized root and stress characteristics through manipulating expression of the genes *D27*, *CCD7*, *CCD8*, and *MAX1* for SL production, and *D14* and *MAX2* for perception. By changing the levels of SL production, cereal tillering can be controlled, while SL signaling enhancement helps crops maintain desired architectures and stress-management functions.

3.8.2 Integration with other hormonal pathways

Plant hormones coordinate plant growth and development by interacting in complex networks, involving auxin (indole-3-acetic acid), CKs, ABA, GAs, and ethylene. These interactions regulate processes such as cell division, elongation, differentiation, and responses to environmental stresses. A combination of managing SL regulatory mechanisms with key hormones creates powerful possibilities to enhance agricultural output. The combination of SL analogs and auxin inhibitors or CK modulators allows researchers to adjust shoot and root structures for precise agricultural objectives. The combination of SL manipulation with ABA signaling achieves better drought resistance, and adding cross-pathway interactions of SL with JA and SA enhances disease protection.

3.8.3 Breeding for enhanced symbiotic capacity

Current breeding initiatives choose genotypes that produce SLs at optimal levels because SLs serve as important agents in establishing beneficial symbioses between AMF and nitrogen-fixing bacteria. Production lines of crops with this enhanced capability use soil microbial communities more effectively, thus maximizing nutrient efficiency without synthetic inputs. Marker-assisted selection together with genomic prediction tools facilitate the quick development of crops that show better SL-mediated symbiotic characteristics.

3.8.4 Biotechnological approaches and synthetic biology

The scientific field of synthetic biology investigates how to develop microbial systems that produce SL derivatives and precursors suitable for agricultural applications. Researchers aim to develop transgenic crops that possess modified SL pathways for the purpose of bestowing traits such as superior yield potential together with stress resistance and parasitic weed resistance. Strategies need to undergo thorough analysis regarding biosafety concerns, alongside regulatory requirements as well as acceptability by the public.

3.8.5 Sustainable agriculture and environmental benefits

SL analogs together with hormonal manipulation methods follow sustainable agricultural practices because they eliminate dependency on herbicides along with fertilizers and irrigation systems. These technologies increase plant defensive capabilities while achieving optimal resource management and encouraging beneficial soil-based relationships to develop environmentally friendly and resilient agricultural cultivation systems. Parasitic weed-afflicted areas together with areas restricted by available resources can find potential agricultural solutions in SL-based approaches to enhance food security alongside farmer economic growth.

3.8.6 Regulatory and commercialization considerations

The market rollout of SL analogs together with SL-manipulated crops needs comprehensive regulatory systems that will confirm product security and operational effectiveness. Three essential elements determining the success of such uses are intellectual property rights together with production scalability and farmer adoption. The successful use of SL-based agricultural innovations depends on continuous field testing and active stakeholder collaboration, as well as multi-disciplinary research to solve regulatory issues and enhance benefits.

3.9 Future Perspectives and Research Directions

SLs underwent a rapid transformation from their initial classification as botanical curiosities to emerge as a vital subject in plant development research and agricultural science because they affect various aspects of plant growth and stress responses and rhizosphere dynamics. SLs function as phytohormone variants derived from carotenoids to control shoot development, root morphology, and seedling behavior, as well as fungal symbiosis and parasitic seed germination and stress response capacity. A further understanding of SLs requires solving many remaining questions and exploiting research opportunities, which will determine future paths in this scientific discipline. Research should focus primarily on creating a clearer view of SL signaling pathways, together with their dual networking effects with additional phytohormone systems. SLs interact with five main phytohormones, namely auxin, CKs, ABA, ethylene, and GAs to coordinate both development and stress responses, yet the fundamental mechanisms behind these reactions remain unclear. Advanced genetic examination tools, gene-expression analysis methodology, and protein investigation techniques are required for researchers to study the downstream signaling pathways and discover essential regulatory components while creating a dynamic hormonal system model that explains how SLs contribute to plant adaptive methods. Knowledge about plant hormone mechanisms is essential to create precise crop improvement approaches, as breeders can use targeted hormonal engineering for optimizing plant structure, resource management, and stress tolerance.

The development of synthetic SL analogs for agricultural use requires immediate investigation, together with their subsequent field-based confirmation. The compounds GR24 and Nijmegen-1 exhibit efficacy in laboratory studies and greenhouses, but their production costs and restricted application at the field scale represent major obstacles. With ongoing research, scientists aim to develop affordable SL analogs suitable for specific crops that effectively combat parasitic weeds such as *Striga* and *Orobanche* while improving mycorrhizal symbiosis and crop growth. Safe and sustainable agricultural use of synthetic SLs requires a complete risk evaluation to study their environmental spread, together with their persistence and effects on unauthorized targets and non-target soil microorganisms. The investigation of SLs through plant–microbe and plant–soil relationships represents a rising investigation direction. Plant hormones act as vital signaling molecules that convey plant microbial messages to AMF, and researchers are just starting to study their complete impact on soil microbial systems and whole ecosystem performance.

Researchers should study how soil-beneficial microorganisms respond to SL signaling to understand their role in shaping microbial communities, as well as influencing nutrient cycles and promoting soil health and plant productivity. The acquired understanding about SL signaling could lay the foundation for agricultural methods that use beneficial soil–microbial interactions to achieve sustainable crop cultivation and ecosystem sustainability. The simultaneous action of SLs as symbiotic facilitators and seed germinators for parasitic weeds demonstrates the ecological compromises that plants need to manage in their environment. Studies conducted on various plant species throughout different environments can reveal the natural-selection forces that mold SL signaling pathways for improved agricultural strategies that protect crops without losing their protective capabilities. The continuous search for plant metabolites with SL-related actions and phytohormones of new classes continues to flourish as investigators explore unrevealed regulatory aspects and practical applications. Interdisciplinary collaboration between plant researchers, chemists, agronomists, regulatory personnel, and industry organizations will be essential to transform SL knowledge into practical agricultural solutions. New vigilance programs need to build standards regulating SL-based product release while educating society about these products and their appropriate utilization. Field-trial implementation together with demonstration projects and farmer participation represent key elements to establish a connection between supporting laboratory discoveries and practical outcomes. Scientists have initiated new studies on the biomedical uses of SLs to understand their anticancer and anti-inflammatory properties, which creates opportunities for translational research between plant biology and human health. These applications are at the infancy stage but demonstrate how SLs function as a model system to study signaling systems and create new bioactive compounds.

3.10 Conclusion

SLs are important plant hormones shaping growth, development, and environmental adaptation of plants. They were first discovered to be germination stimulants of parasitic weeds but are now recognized as versatile regulators that integrate developmental cues with external signals to optimize plant architecture, resource allocation, and the stress response. It is their unique capacity to participate both as mediators of endogenous hormonal signaling and as transceivers of rhizosphere communication that makes them so central to plant biology, and so poised for development of agricultural innovations. Shoot branching represents one of the most important contributions of SLs, as they function as major inhibitors of axillary bud outgrowth. The roles of SLs in shape determination involve modulating the balance between auxin, CKs, and sugars so that the plant form is best optimized for reproductive success, or to enhance resource-use efficiency. Excessive branching lowers yield and elevates crop susceptibility to stresses, and therefore regulation of this trait is important for crop productivity. Likewise, SLs impact root-system architecture, for instance by promoting primary root elongation, regulation of LR and AR development, and increasing root-hair development. Under nutrient-deficient or stressful conditions, nutrient and water uptake can be improved, thus elucidating the adaptive significance of SL-mediated root plasticity. In addition, SLs participate in seed germination and photomorphogenesis, and intertwine with hormones such as ABA and GAs to control dormancy and seedling growth. This involvement in these early stages of plant life highlights their integrative function across the life cycle of the plant. As well as the impact on plant development, SLs function to determine plant resilience to abiotic stresses including drought, salinity, and adverse temperature extremes, in large part by interacting with ABA and other stress hormones. Moreover, they prime defense responses, providing increased biotic stress tolerance and structure-beneficial microbial interactions, but their role in germination of parasitic weeds is a complex problem for agriculture. SLs have fast developed to be used in agricultural and biotechnological applications. Synthetic SL analogs thus have potential as tools to trigger suicidal germination of parasites, regulate crop architecture for maximizing yield, and modulate symbiotic interactions needed for reducing the need for fertilizers of crops. Next-generation genetic engineering and hormonal manipulation can be used to develop crop varieties with genetically engineered SL profiles that promote improved stress tolerance and resource-use efficiency. These innovations are in line with sustainable agriculture goals by providing environmentally friendly strategies to enhance productivity and resilience. However, many questions remain unanswered. Further elucidation of the molecular details of SL signaling and their integration with other hormonal pathways are required. Manipulation of SLs has ecological implications, particularly regarding the soil microbiome and parasitic plants, that needs to be carefully assessed. Furthermore, the development of SL analogs remains limited by production at scale and cost factors, but their deployment in various agroecosystems is safe.

References

Ahammed, G.J., Gantait, S., Mitra, M., Yang, Y. and Li, X. (2020) Role of ethylene crosstalk in seed germination and early seedling development: A review. *Plant Physiology and Biochemistry* 151, 124–131. DOI: 10.1016/j.plaphy.2020.03.016.

Ahmed, N., Li, J., Li, Y., Deng, L., Deng, L. *et al.* (2025) Symbiotic synergy: How arbuscular mycorrhizal fungi enhance nutrient uptake, stress tolerance, and soil health through molecular mechanisms and hormonal regulation. *IMA Fungus* 16, e144989. DOI: 10.3897/imafungus.16.144989.

Al-Babili, S. and Bouwmeester, H.J. (2015) Strigolactones, a novel carotenoid-derived plant hormone. *Annual Review of Plant Biology* 66(1), 161–186. DOI: 10.1146/annurev-arplant-043014-114759.

Ali, A., Shah, T., Haider, G., Awan, M.I., Gohar, M. *et al.* (2023) Strigolactone-mediated oxidative stress alleviation in *Brassica rapa* through upregulating antioxidant system under water deficit conditions. *Journal of Plant Growth Regulation* 42(8), 4675–4687. DOI: 10.1007/s00344-023-10925-0.

Altaf, F., Parveen, S., Farooq, S., Lone, M.L., Haq, A.U. *et al.* (2024) Enigmas of senescence: A reappraisal on the hormonal crosstalk and the molecular mechanisms. *Theoretical and Experimental Plant Physiology* 36(1), 51–81. DOI: 10.1007/s40626-024-00308-5.

Altamura, M.M., Piacentini, D., Della Rovere, F., Fattorini, L., Falasca, G. *et al.* (2023) New paradigms in brassinosteroids, strigolactones, sphingolipids, and nitric oxide interaction in the control of lateral and adventitious root formation. *Plants* 12(2), 413. DOI: 10.3390/plants12020413.

Andreo-Jimenez, B., Ruyter-Spira, C., Bouwmeester, H.J. and Lopez-Raez, J.A. (2015) Ecological relevance of strigolactones in nutrient uptake and other abiotic stresses, and in plant–microbe interactions below-ground. *Plant and Soil* 394(1–2), 1–19. DOI: 10.1007/s11104-015-2544-z.

Antar, M., Gopal, S., Msimbira, L.A., Naamala, J., Overbeek, W. *et al.* (2021) Inter-organismal signaling in the rhizosphere. In: Gupta, V.V.S.R. and Sharma, A.K. (eds) *Rhizosphere Biology: Interactions Between Microbes and Plants*. Springer, Singapore, pp. 255–293. DOI: 10.1007/978-981-15-6125-2_13.

Arshad, A., Mushtaq, N., Sajjad, M., Ahad, A., Ilyas, M. *et al.* (2023) Role of exogenous phytohormones in mitigating stress in plants. In: Ozturk, M., Bhat, R.A., Ashraf, M., Tonelli, F.M.P., Unal, B.T. *et al.* (eds) *Phytohormones and Stress Responsive Secondary Metabolites*. Academic Press, Cambridge, Massachusetts, pp. 111–131. DOI: 10.1016/B978-0-323-91883-1.00020-6.

Bennett, T. and Leyser, O. (2014) Strigolactone signalling: Standing on the shoulders of DWARFs. *Current Opinion in Plant Biology* 22, 7–13. DOI: 10.1016/j.pbi.2014.08.001.

Bhoi, A., Yadu, B., Chandra, J. and Keshavkant, S. (2021) Contribution of strigolactone in plant physiology, hormonal interaction and abiotic stresses. *Planta* 254(2), 28. DOI: 10.1007/s00425-021-03678-1.

Borghi, L., Liu, G.-W., Emonet, A., Kretzschmar, T. and Martinoia, E. (2016) The importance of strigolactone transport regulation for symbiotic signaling and shoot branching. *Planta* 243(6), 1351–1360. DOI: 10.1007/s00425-016-2503-9.

Boyno, G., Rezaee Danesh, Y., Demir, S., Teniz, N., Mulet, J.M. *et al.* (2023) The complex interplay between arbuscular mycorrhizal fungi and strigolactone: Mechanisms, sinergies, applications and future directions. *International Journal of Molecular Sciences* 24(23), 16774. DOI: 10.3390/ijms242316774.

Brewer, P.B., Koltai, H. and Beveridge, C.A. (2013) Diverse roles of strigolactones in plant development. *Molecular Plant* 6(1), 18–28. DOI: 10.1093/mp/sss130.

Chaudhary, N. (2013) The interaction of cytokinin and strigolactone in controlling shoot branching in *Arabidopsis thaliana*. PhD thesis, Imperial College, London.

Cheng, X., Ruyter-Spira, C. and Bouwmeester, H. (2013) The interaction between strigolactones and other plant hormones in the regulation of plant development. *Frontiers in Plant Science* 4, 199. DOI: 10.3389/fpls.2013.00199.

Chesterfield, R. (2020) A synthetic biology toolbox for examining and engineering strigolactone biosynthesis. PhD thesis, Australian Institute for Bioengineering and Nanotechnology, St Lucia, Australia.

Chien, Y.C. and Yoon, G.M. (2024) Subcellular dynamics of ethylene signaling drive plant plasticity to growth and stress: Spatiotemporal control of ethylene signaling in *Arabidopsis*. *BioEssays* 46(6), e2400043. DOI: 10.1002/bies.202400043.

de Villiers, R.M. (2015) Reverse genetic analysis of gene pp1s148_40v6 in physcomitrella patens: An Atmax2 orthologue? PhD thesis, Stellenbosch University, Stellenbosch, South Africa.

El Sabagh, A., Islam, M.S., Hossain, A., Iqbal, M.A., Mubeen, M. *et al.* (2022) Phytohormones as growth regulators during abiotic stress tolerance in plants. *Frontiers in Agronomy* 4, 765068. DOI: 10.3389/fagro.2022.765068.

En-Nahli, Y., Abbes, Z., Trabelsi, I., Makaza, W., Kharrat, M. *et al.* (2023) Host–parasitic plant interaction: The key role of strigolactones. In: Mukherjee, S. and Aftab, T. (eds) *Strigolactones, Alkamides and Karrikins in Plants*. CRC Press, Boca Raton, Florida, pp. 87–99. DOI: 10.1201/9781003224525-9.

Foo, E., Blake, S.N., Fisher, B.J., Smith, J.A. and Reid, J.B. (2016) The role of strigolactones during plant interactions with the pathogenic fungus *Fusarium oxysporum*. *Planta* 243(6), 1387–1396. DOI: 10.1007/s00425-015-2449-3.

Harrison, P.J., Newgas, S.A., Descombes, F., Shepherd, S.A., Thompson, A.J. *et al.* (2015) Biochemical characterization and selective inhibition of b-carotene *cis-trans* isomerase D27 and carotenoid cleavage dioxygenase CCD8 on the strigolactone biosynthetic pathway. *FEBS Journal* 282(20), 3986–4000. DOI: 10.1111/febs.13400.

Herold, S. (2009) The role of strigolactones in the regulation of secondary growth in *Arabidopsis thaliana*. PhD thesis, Universität Wien, Vienna. DOI: 10.25365/thesis.6245.

Hossain, A., Pamanick, B., Venugopalan, V.K., Ibrahimova, U., Rahman, M.A. *et al.* (2022) Emerging roles of plant growth regulators for plants adaptation to abiotic stress-induced oxidative stress. In: Aftab, T. and Naeem, M. (eds) *Emerging Plant Growth Regulators in Agriculture*. Academic Press, UK, pp. 1–72. DOI: 10.1016/B978-0-323-91005-7.00010-2.

Hu, Q., Ding, F., Li, M., Zhang, X., Zhang, S. *et al.* (2021) Strigolactone and ethylene inhibitor suppressing dark-induced leaf senescence in perennial ryegrass involving transcriptional downregulation of chlorophyll degradation. *Journal of the American Society for Horticultural Science* 146(2), 79–86. DOI: 10.21273/JASHS04933-20.

Hussain, Q., Asim, M., Zhang, R., Khan, R., Farooq, S. *et al.* (2021) Transcription factors interact with ABA through gene expression and signaling pathways to mitigate drought and salinity stress. *Biomolecules* 11(8), 1159. DOI: 10.3390/biom11081159.

Jain, S., Kumar, R., Gunsola, D., Živković, S., Vasić, T. *et al.* (2025) Strigolactones and control of parasitic weeds. In: Aftab, T. and Wani, K.I. (eds) *Strigolactones: Emerging Plant Hormones*. Scrivener Publishing, Beverly, Massachusetts, pp. 261–279. DOI: 10.1002/9781394302826.

Jamil, M., Wang, J.Y., Yonli, D., Ota, T., Berqdar, L. *et al.* (2022) *Striga hermonthica* suicidal germination activity of potent strigolactone analogs: Evaluation from laboratory bioassays to field trials. *Plants* 11(8), 1045. DOI: 10.3390/plants11081045.

Janeeshma, E., Habeeb, H., Shackira, A.M., Sinisha, A.K., Mirshad, P.P. *et al.* (2024) Strigolactone and analogues: A new generation of plant hormones with multifactorial benefits in environmental sustainability. *Environmental and Experimental Botany* 223, 105775. DOI: 10.1016/j.envexpbot.2024.105775.
Jogawat, A., Yadav, B., Chhaya, Lakra, N., Singh, A.K. *et al.* (2021) Crosstalk between phytohormones and secondary metabolites in the drought stress tolerance of crop plants: A review. *Physiologia Plantarum* 172(2), 1106–1132. DOI: 10.1111/ppl.13328.
Kamle, M., Borah, R., Bora, H., Jaiswal, A.K., Singh, R.K. *et al.* (2020) Systemic acquired resistance (SAR) and induced systemic resistance (ISR): Role and mechanism of action against phytopathogens. In: Hesham, AL., Upadhyay, R., Sharma, G., Manoharachary, C. and Gupta, V. (eds) *Fungal Biotechnology and Bioengineering*. Springer, Cham, Switzerland, pp. 457–470. DOI: 10.1007/978-3-030-41870-0_20.
Kamran, M., Melville, K.T. and Waters, M.T. (2024) Karrikin signalling: Impacts on plant development and abiotic stress tolerance. *Journal of Experimental Botany* 75(4), 1174–1186. DOI: 10.1093/jxb/erad476.
Kaniganti, S., Bhattacharya, J., Petla, B.P. and Reddy, P.S. (2022) Strigolactone, a neglected plant hormone, with a great potential for crop improvement: Crosstalk with other plant hormones. *Environmental and Experimental Botany* 204, 105072. DOI: 10.1016/j.envexpbot.2022.105072.
Kawada, K., Koyama, T., Takahashi, I., Nakamura, H. and Asami, T. (2022) Emerging technologies for the chemical control of root parasitic weeds. *Journal of Pesticide Science* 47(3), 101–110. DOI: 10.1584/jpestics.D22-045.
Kelly, J.H., Tucker, M.R. and Brewer, P.B. (2023) The strigolactone pathway is a target for modifying crop shoot architecture and yield. *Biology* 12(1), 95. DOI: 10.3390/biology12010095.
Kohli, A., Sreenivasulu, N., Lakshmanan, P. and Kumar, P.P. (2013) The phytohormone crosstalk paradigm takes center stage in understanding how plants respond to abiotic stresses. *Plant Cell Reports* 32(7), 945–957. DOI: 10.1007/s00299-013-1461-y.
Kumar, P., Pandey, S. and Pati, P.K. (2025) Interaction between pathogenesis-related (PR) proteins and phytohormone signaling pathways in conferring disease tolerance in plants. *Physiologia Plantarum* 177(2), e70174. DOI: 10.1111/ppl.70174.
Kusajima, M., Fujita, M., Soudthedlath, K., Nakamura, H., Yoneyama, K. *et al.* (2022) Strigolactones modulate salicylic acid-mediated disease resistance in *Arabidopsis thaliana*. *International Journal of Molecular Sciences* 23(9), 5246. DOI: 10.3390/ijms23095246.
Lakehal, A. and Bellini, C. (2019) Control of adventitious root formation: Insights into synergistic and antagonistic hormonal interactions. *Physiologia Plantarum* 165(1), 90–100. DOI: 10.1111/ppl.12823.
Li, J., Fan, M., Zhang, Q., Lü, G., Wu, X. *et al.* (2023) Transcriptome analysis reveals that auxin promotes strigolactone-induced adventitious root growth in the hypocotyl of melon seedlings. *Frontiers in Plant Science* 14, 1192340. DOI: 10.3389/fpls.2023.1192340.
Liu, Y., Chen, S., Pal, S., Yu, J., Zhou, Y. *et al.* (2024) The hormonal, metabolic, and environmental regulation of plant shoot branching. *New Crops* 1, 100028. DOI: 10.1016/j.ncrops.2024.100028.
López-Ráez, J.A., Matusova, R., Cardoso, C., Jamil, M., Charnikhova, T. *et al.* (2009) Strigolactones: Ecological significance and use as a target for parasitic plant control. *Pest Management Science* 65(5), 471–477. DOI: 10.1002/ps.1692.
Marro, N., Caccia, M. and López-Ráez, J.A. (2021) Are strigolactones a key in plant–parasitic nematodes interactions? An intriguing question. *Plant and Soil* 462(1–2), 591–601. DOI: 10.1007/s11104-021-04862-8.
Mazzoni-Putman, S.M., Brumos, J., Zhao, C., Alonso, J.M. and Stepanova, A.N. (2021) Auxin interactions with other hormones in plant development. *Cold Spring Harbor Perspectives in Biology* 13(10), a039990. DOI: 10.1101/cshperspect.a039990.
Mitra, D., Sierra, B.E.G., Khoshru, B., de los Santos Villalobos, S., Belz, C. *et al.* (2021) Impacts of arbuscular mycorrhizal fungi on rice growth, development, and stress management with a particular emphasis on strigolactone effects on root development. *Communications in Soil Science and Plant Analysis* 52(14), 1591–1621. DOI: 10.1080/00103624.2021.1892728.
Mutinda, S., Jamil, M., Wang, J.Y., Berqdar, L., Ateka, E. *et al.* (2025) Strigolactone biosynthesis *lgs1* mutant alleles mined from the sorghum accession panel are a promising resource of resistance to witchweed (*Striga*) parasitism. *Plants, People, Planet* 7(2), 382–395. DOI: 10.1002/ppp3.10442.
Naseer, M.A., Zhang, Z.Q., Mukhtar, A., Asad, M.S., Wu, H.Y. *et al.* (2024) Strigolactones: A promising tool for nutrient acquisition through arbuscular mycorrhizal fungi symbiosis and abiotic stress tolerance. *Plant Physiology and Biochemistry* 215, 109057. DOI: 10.1016/j.plaphy.2024.109057.

Pandey, A., Pandey, M. and Pandey, G.K. (2016) Emerging roles of strigolactones in plant responses to stress and development. *Frontiers in Plant Science* 7, 434. DOI: 10.3389/fpls.2016.00434.

Parwez, R., Aftab, T., Gill, S.S. and Naeem, M. (2022) Abscisic acid signaling and crosstalk with phytohormones in regulation of environmental stress responses. *Environmental and Experimental Botany* 199, 104885. DOI: 10.1016/j.envexpbot.2022.104885.

Rasheed, R., Hafeez, A., Ashraf, M.A. and Hussain, I. (2023) Signaling crosstalk between gibberellins–jasmonates–strigolactones in plant defense, growth, and development. In: Husen, A. and Zhang, W. (eds) *Hormonal Cross-Talk, Plant Defense and Development: Plant Biology, Sustainability and Climate Change*. Academic Press, UK, pp. 171–194. DOI: 10.1016/B978-0-323-95375-7.00007-0.

Roychowdhury, R., Mishra, S., Anand, G., Dalal, D., Gupta, R. *et al.* (2024) Decoding the molecular mechanism underlying salicylic acid (SA)-mediated plant immunity: An integrated overview from its biosynthesis to the mode of action. *Physiologia Plantarum* 176(3), e14399. DOI: 10.1111/ppl.14399.

Saeed, W., Naseem, S. and Ali, Z. (2017) Strigolactones biosynthesis and their role in abiotic stress resilience in plants: A critical review. *Frontiers in Plant Science* 8, 1487. DOI: 10.3389/fpls.2017.01487.

Salam, B.B., Barbier, F., Danieli, R., Teper-Bamnolker, P., Ziv, C. *et al.* (2021) Sucrose promotes stem branching through cytokinin. *Plant Physiology* 185(4), 1708–1721. DOI: 10.1093/plphys/kiab003.

Samanta, S. and Roychoudhury, A. (2025) Molecular crosstalk of jasmonate with major phytohormones and plant growth regulators during diverse stress responses. *Journal of Plant Growth Regulation* 44(1), 62–88. DOI: 10.1007/s00344-024-11412-w.

Santana, A., Molinillo, J.M.G. and Macías, F.A. (2015) Trends in the synthesis and functionalization of guaianolides. *European Journal of Organic Chemistry* 2015(10), 2093–2110. DOI: 10.1002/ejoc.201403244.

Schuetz, S. (2019) Characterizing the basis of strigolactone perception by HYPOSENSITIVE TO LIGHT/KARRIKIN INSENSITIVE 2. MSc thesis, University of Toronto, Toronto, Canada.

Sharma, M., Singh, D., Saksena, H.B., Sharma, M., Tiwari, A. *et al.* (2021) Understanding the intricate web of phytohormone signalling in modulating root system architecture. *International Journal of Molecular Sciences* 22(11), 5508. DOI: 10.3390/ijms22115508.

Sharma, P., Jha, A.B. and Dubey, R.S. (2024) Strigolactones: Coordination with other phytohormones and enhancement of abiotic stress responses. *Environmental and Experimental Botany* 223, 105782. DOI: 10.1016/j.envexpbot.2024.105782.

Shinohara, N., Taylor, C. and Leyser, O. (2013) Strigolactone can promote or inhibit shoot branching by triggering rapid depletion of the auxin efflux protein PIN1 from the plasma membrane. *PLOS Biology* 11(1), e1001474. DOI: 10.1371/journal.pbio.1001474.

Shu, K., Liu, X.-D., Xie, Q. and He, Z.-H. (2016) Two faces of one seed: Hormonal regulation of dormancy and germination. *Molecular Plant* 9(1), 34–45. DOI: 10.1016/j.molp.2015.08.010.

Sun, Y., Tian, Z., Zuo, D., Cheng, H., Wang, Q. *et al.* (2024) Strigolactone-induced degradation of SMXL7 and SMXL8 contributes to gibberellin- and auxin-mediated fiber cell elongation in cotton. *Plant Cell* 36(9), 3875–3893. DOI: 10.1093/plcell/koae212.

Tariq, A., Ullah, I., Sardans, J., Graciano, C., Mussarat, S. *et al.* (2023) Strigolactones can be a potential tool to fight environmental stresses in arid lands. *Environmental Research* 229, 115966. DOI: 10.1016/j.envres.2023.115966.

Thilakarathne, A.S., Liu, F. and Zou, Z. (2025) Plant signaling hormones and transcription factors: Key regulators of plant responses to growth, development, and stress. *Plants* 14(7), 1070. DOI: 10.3390/plants14071070.

Torres Vera, R. (2017) Role of strigolactones in plant defense: Hormonal cross-talk and implication in arbuscular mycorrhizal symbiosis. PhD thesis, Universidad de Granada, Granada, Spain.

Visentin, I., Vitali, M., Ferrero, M., Zhang, Y., Ruyter-Spira, C. *et al.* (2016) Low levels of strigolactones in roots as a component of the systemic signal of drought stress in tomato. *New Phytologist* 212(4), 954–963. DOI: 10.1111/nph.14190.

Wani, K.I., Zehra, A., Choudhary, S., Naeem, M., Khan, M.M.A. *et al.* (2021) Mechanistic insights into strigolactone biosynthesis, signaling, and regulation during plant growth and development. *Journal of Plant Growth Regulation* 40(5), 1836–1852. DOI: 10.1007/s00344-020-10234-w.

Yamada, Y. and Umehara, M. (2015) Possible roles of strigolactones during leaf senescence. *Plants* 4(3), 664–677. DOI: 10.3390/plants4030664.

Yap, J.X. and Tsuchiya, Y. (2023) Gibberellins promote seed conditioning by up-regulating strigolactone receptors in the parasitic plant *Striga hermonthica*. *Plant & Cell Physiology* 64(9), 1021–1033. DOI: 10.1093/pcp/pcad056.

Yuan, K., Zhang, H., Yu, C., Luo, N., Yan, J. *et al.* (2023) Low phosphorus promotes NSP1–NSP2 heterodimerization to enhance strigolactone biosynthesis and regulate shoot and root architecture in rice. *Molecular Plant* 16(11), 1811–1831. DOI: 10.1016/j.molp.2023.09.022.

Zhang, Y., Cheng, X., Wang, Y., Díez-Simón, C., Flokova, K. *et al.* (2018) The tomato *MAX1* homolog, *SLMAX1*, is involved in the biosynthesis of tomato strigolactones from carlactone. *New Phytologist* 219(1), 297–309. DOI: 10.1111/nph.15131.

Zhao, S. and Li, Y. (2021) Current understanding of the interplays between host hormones and plant viral infections. *PLOS Pathogens* 17(2), e1009242. DOI: 10.1371/journal.ppat.1009242.

Zhao, B., Liu, Q., Wang, B. and Yuan, F. (2021) Roles of phytohormones and their signaling pathways in leaf development and stress responses. *Journal of Agricultural and Food Chemistry* 69(12), 3566–3584. DOI: 10.1021/acs.jafc.0c07908.

4 Interaction of Strigolactones with Reactive Oxygen Species in Plants

Ansa Asghar[1*], Rizwana Kanwal[1], Usman Tufail[1*], Farooq Ahmad[1], Sana Basharat[2], Mansoor Hameed[1], Muhammad Sufyan[3], Muhammad Sajid Aqeel Ahmad[1], Asima Asghar[4], Syed Mohsan Raza Shah[5] and Zahida Parveen[1]

[1]Department of Botany, University of Agriculture Faisalabad, Faisalabad, Pakistan; [2]School of Breeding and Multiplication (Sanya Institute of Breeding and Multiplication), College of Tropical Agriculture and Forestry, Hainan University, Sanya, China; [3]College of Biological Sciences, China Agricultural University, Beijing, China; [4]Department of Zoology, Lahore College for Women University, Lahore, Pakistan; [5]Department of Plant Pathology, Bahauddin Zakariya University, Multan, Pakistan

Abstract

Strigolactones (SLs), an innovative family of phytohormones generated from carotenoids, are gaining popularity because of their diverse functions in plant growth and stress-response mechanisms. Reactive oxygen species (ROS), which were previously considered to be harmful by-products, are now understood as crucial signaling molecules that help plants adapt to both biotic and abiotic stressors. The relationship between SLs and ROS is complicated, according to recent research. It has been noted that SLs control the generation of ROS and initiate antioxidant defense pathways; ROS then influence SL signaling and biosynthesis. However, through feedback processes, elevated ROS produced under stress might impact SL metabolism and signaling pathways. Redox equilibrium, stress tolerance, and plant developmental plasticity all depend on this reciprocal communication. New tactics to improve plant resistance will be developed with a better knowledge of the SL–ROS interaction.

Keywords: antioxidant defense, carlactone, detoxification of ROS, phytohormones, strigolactones

4.1 Introduction

Plants have developed highly complex signaling networks to respond to environmental conditions. Within these networks, strigolactones (SLs) and reactive oxygen species (ROS) perform central roles in plant stress physiology and expansion (Mansoor *et al.*, 2024). SLs, originally discovered as carotenoid-derived phytohormones that induce the germination of parasitic plant seeds, are now recognized for their broader role in regulating various plant development processes, including shoot branching, root system architecture, and seed germination (Nelson, 2021). SLs have also been shown to shape plant responses against nutrient deficiency and drought. Likewise, ROS are constantly generated as derivatives of essential

*Corresponding authors: ansaasghar1122@gmail.com and usman3383ar@gmail.com

DOI: 10.1079/9781836990598.0004

metabolic activities, including photosynthesis and respiration (Khorobrykh *et al.*, 2020). ROS, which damage cellular structures when accumulated at high levels, function as signaling molecules at low and controlled levels, regulating growth, development, and stress responses (Mittler *et al.*, 2022). Plants activate antioxidant defense systems such as catalase, superoxide dismutase, ascorbate peroxidase, glutathione, and ascorbate to maintain this balance (Fujita and Hasanuzzaman, 2022).

In this chapter, the molecular basis of the SL–ROS interaction is discussed: how SLs affect ROS manufacture and detoxification, the effects of ROS on SL metabolism, and the integrative role of this interaction on stress tolerance and development. Understanding this oxidative–hormonal interface is of great importance for the development of plants resistant to environmental stresses (Sze *et al.*, 2021).

4.1.1 Chemical structure of SLs

To date, numerous members have been tentatively identified; however, over 25 members have been structurally characterized, contingent upon the occurrence or non-appearance of the complete ABC-ring structure, categorized into canonical and non-canonical SL forms (Bansal *et al.*, 2022). The C-2′ configurations of obviously occurring SLs that have been identified to date are predominantly of the 2′R form. The enol-ether bridge and the D-ring components have been demonstrated to be important for their biological activities, particularly in plant systems (Dembitsky, 2023).

4.1.2 SLs: Biosynthesis, transport, and functions

Plants exhibit responses to various abiotic stimuli through anatomical, cellular, morphological, and physiological changes (Li *et al.*, 2023). Phytohormones, which control development, development, growth, and nutrient allocation, have a major impact on a plant's capacity to adapt to different surroundings (El Sabagh *et al.*, 2022). Often known as plant growth regulators, these signal molecules are produced by the plant. A unique class of phytohormones known as SLs has an ecological function in reducing a range of stressors (Bhoi *et al.*, 2021). SLs are produced from precursor carotenoids in the roots of plants (Tables 4.1 and 4.2) and are important for bud development, shoot splitting, nodulation, and

Table 4.1. The SL biosynthesis pathway. Plasmid-localized enzymes enable the sequential processes. Carlactone undergoes additional oxidation by the CYP711A family, resulting in the formation of carlactonoic acid. Certain members of these families are capable of synthesizing the canonical strigolactone orobranchol from conjugated linoleic acid.

Enzyme	Substrate → product	Organism(s)	Reference
D27	All-*trans*-β-carotene → 9-*cis*-β-carotene	*Arabidopsis*, rice	Yang *et al.* (2023b)
CCD7 (MAX3)	9-*cis*-β-Carotene → intermediate	*Arabidopsis*, pea	Liu *et al.* (2021)
CCD8 (MAX4)	Intermediate → carlactone	*Arabidopsis*, pea	Mehmood *et al.* (2025)
MAX1 (CYP711A1)	Carlactone → carlactonoic acid	*Arabidopsis*	Sigalas *et al.* (2023)
CYP711A2 (Os900)	Carlactone → 4-deoxyorobanchol	Rice	Chen *et al.* (2023)
CYP711A3 (Os1400)	4-Deoxyorobanchol → orobranchol	Rice	Chen *et al.* (2023)
CYP711A17/18	Carlactone or carlactonoic acid → diverse SLs	*Selaginella moellendorffii*, maize	Mashiguchi *et al.* (2021)

Table 4.2. Chemical type and role of compounds.

Compound	Function/role	Chemical type	Reference
All-*trans*-β-carotene	Starting carotenoid	Precursor	Sun *et al.* (2021)
Carlactone	Core biosynthetic intermediate	SL backbone	Homma *et al.* (2023)
Carlactonoic acid	Precursor to various strigolactones (SLs)	Oxidized form of carlactone	Mashiguchi *et al.* (2022)
4-Deoxyorobanchol (4DO)	Canonical SL intermediate in rice	SL	Mizutori *et al.* (2025)
Orobranchol (ORO)	Fully functional canonical SL	Final bioactive compound	Pérez-Escalante *et al.* (2022)

photomorphogenesis (Jain *et al.*, 2025). Because of their function in physiological and molecular responses to abiotic stressors, SLs have recently attracted interest. SLs contribute to stress reduction via a number of molecular mechanisms and interact with other phytohormones (Bhoi *et al.*, 2021).

Numerous processes, including germination of seeds and early seedling growth, leaf senescence, and the control of main and lateral root or root-hair extension, have been shown to be impacted by SLs in experimental investigations (Kohli *et al.*, 2022). Evidence is mounting that suggests SLs are involved in the plant's reaction to a variety of biotic and abiotic stressors alongside these developmental processes (Imaduwage and Hewadikaram, 2024). The plant's reaction to less-than-ideal environmental circumstances, such as drought, salt, temperature fluctuations, nutritional shortages, oxidative stress, and changes in light intensity and quality, has been linked to SL activity (Lee *et al.*, 2024).

4.1.3 Transport mechanisms and perception

Grafting experiments using hypocotyl grafts between SL-deficient rootstocks and wild-type shoots, as well as the reciprocal arrangement, have shown that a branching inhibitor can move from the roots to the shoots (Yan *et al.*, 2020). However, these studies also reveal that root-to-shoot transport is not essential for suppressing branching, as local synthesis within the shoot can achieve the same effect (Dun *et al.*, 2023). Even so, the movement of SLs remains an important regulatory mechanism, enabling fine-tuned adjustment of the root-to-shoot branching ratio. In conditions of phosphate deprivation, there was an observed increase in the concentration of orobanchol, aligning with the established understanding that SLs play a significant role in the suppression of branching and tillering in phosphate-limited environments (Mohapatra, 2021).

4.2 Roles in Shoot Branching, Root Architecture, Seed Germination, and Symbiosis

SLs are phytohormones generated from carotenoids that are elaborate in a variety of physiological and developmental processes (Wu *et al.*, 2022). According to Mareri *et al.* (2022), they have an impact on the development of the root and shoot, leaf senescence, and the reaction to environmental stressors. They encourage the height of the plant, root-hair elongation, additional growth, and the onset of senescence, but they prevent the production of lateral roots (Li *et al.*, 2018).

4.2.1 Root development

By modifying main root elongation and lateral root production, SLs control the overall architecture of the root system (Maqbool *et al.*, 2022). In *Arabidopsis thaliana*, SL-deficient mutants (e.g. *max*) produce more lateral roots and have petite primary roots (Sun *et al.*, 2019). The synthetic SL analog GR24 has been shown to influence meristem size and root elongation, with effects dependent on functional MORE AXILLARY GROWTH 2 (MAX2) signaling. Auxin transport and concentration gradients, crucial for root patterning, are modulated by SLs (Carbonnel *et al.*, 2020).

4.2.2 Root-hair elongation

SLs enhance root-hair elongation under normal growth conditions, particularly via MAX2-dependent signaling pathways (Seo *et al.*, 2023). Exogenous SL analogs promote root-hair growth in SL biosynthesis mutants (e.g. *max3*, *max4*), but not in SL-insensitive mutants (*max2*), highlighting the specificity of the response (Wani *et al.*, 2021).

4.2.3 Shoot branching and apical dominance

SLs act similarly to the branching preventing signal, a key factor in apical dominance (Song *et al.*, 2024). Mutations in SL biosynthetic or signaling genes (e.g. *max*, *d*, *dad*, *rms*) lead to enhanced shoot branching, a phenotype that can be reversed by GR24. Genes such as *MAX3*, *MAX4*, *MAX1*, and *MAX2* are central to SL biosynthesis and signaling in *Arabidopsis* (Wani *et al.*, 2021).

4.2.4 Seed germination

Compounds that lack the ABC-ring scaffold yet possess the ability to enhance seed germination are referred to as SL mimics (Korek and Marzec, 2023). For example, the seeds of *Striga hermonthica* exhibit a moderate response to debranones, while *para*-chlorophenoxy-debranone demonstrated enhanced activity (Kapoor *et al.*, 2024).

4.3 Reactive Oxygen Species in Plants: Roles and Regulation

4.3.1 ROS

Phototrophs keep life going by turning energy from the sun into biochemical energy. Cells produce ROS such as singlet oxygen and superoxide radical, but they are formed as unwanted by-products in cells (Edge and Truscott, 2021). These compounds have elevated reactivity and can adversely impact cellular functioning by inflicting significant damage to proteins, DNA, and lipids. Atmospheric oxygen (O_2) is a molecule characterized by minimal reactivity in its fundamental state (Napolitano *et al.*, 2022). Despite this, biological interactions, electron transport chains, ultraviolet-B radiation, and ionizing radiation may allow oxygen to go beyond spin limits, creating ROS. Different kinds of ROS react chemically, have different half-lives, and can get into cells in different ways (Brown *et al.*, 2020).

There are two main types of antioxidants that plants use to keep their redox balance when they are stressed: enzymatic components and non-enzymatic low-molecular-weight molecules. ROS levels are lowered by enzymes (Abdulfatah, 2022). In addition to phytohormonal regulation, several non-enzymatic antioxidants flavonoids and osmolytes such as proline play a crucial role in protecting plants against stress-induced oxidative damage (Hosseinifard *et al.*, 2022). These compounds function by directly scavenging ROS, maintaining cellular redox homeostasis, and stabilizing membranes, thereby contributing to enhanced stress tolerance and overall plant resilience (Zandi and Schnug, 2022). Both antioxidant systems work well, which shows that getting rid of ROS is important for keeping cells alive (Gómez *et al.*, 2021).

4.3.2 Singlet oxygen

Singlet oxygen ($^1O_2$) is produced when the valence electrons in triplet oxygen ($^3O_2$) change their spin. Although not classified as a free radical, it exists as a high-energy variant (Edge and Truscott, 2021). Environmental stressors such as salt, drought, and heavy metals cause stomata to close, which lowers the amount of carbon dioxide (CO_2) inside cells and raises the production of $^1O_2$ (Guo *et al.*, 2023). Under typical settings, the concentration of $^1O_2$ in cells is roughly 10^{-13} M; concentrations of 10^{-8} M may cause membrane damage, while levels of 10^{-5} M might result in cell death (Zeng *et al.*, 2025). It particularly undermines photosystem I (PSI) and PSII, jeopardizing the entire photosynthetic apparatus. $^1O_2$ is the main ROS that causes PSII activity to drop when light hits it. It can damage biomolecules such as proteins and lipids (Wang *et al.*, 2023).

Plants utilize antioxidants to mitigate the detrimental effects of $^1O_2$. Furthermore, the D1 protein of PSII acts as a $^1O_2$ scavenger (Bendou, 2025). Instead, $^1O_2$ affects how plants react to stress by increasing the activity of genes that protect them from photo-oxidative stress (van Breusegem and Mittler, 2023).

4.3.3 Superoxide radical

The superoxide radical ($O_2\bullet^-$) is a main part of ROS, and chloroplasts are where it is mostly made in the electron transport chain (Foyer and Hanke, 2022). PSI in thylakoid membranes generates $O_2\bullet^-$ during non-cyclic electron transport. Typically, O_2 interacts with cytochrome *c* oxidase to generate H_2O (Kozuleva and Ivanov, 2023). In certain instances, O_2 interacts with various elements of the electron transport chain to produce $O_2\bullet^-$ (Noodleman *et al.*, 2020). Although not a very destructive molecule in isolation, $O_2\bullet^-$ facilitates the generation of additional ROS. It can cause membrane lipid peroxidation by changing into more harmful hydroxyl radical ($OH\bullet$) and $^1O_2$, which damages cell structures significantly (Chowdhary and Songachan, 2025). Like $^1O_2$, $O_2\bullet^-$ has a half-life of about 4 µs, which means it is more reactive and can move several micrometers away from where it was created (Liu *et al.*, 2024a). Under typical cellular circumstances, the concentration of $O_2\bullet^-$ is significantly lower than that of hydrogen peroxide (H_2O_2) (about 10^{-12} M). None the less, the formation of H_2O_2 can be enhanced by the interaction of two $O_2\bullet^-$ molecules with protons (H^+) (Tian *et al.*, 2025). $O_2\bullet^-$ can easily interact with the structures of cellular Fe-S proteins, which can cause structural changes and the release of Fe^+ (Wang *et al.*, 2024). Through the Fenton reaction, free Fe^{2+} can join with H_2O_2 to make the very dangerous $OH\bullet$. In addition, the hydroperoxyl radical ($HO_2\bullet$), formed by the reaction of $O_2\bullet^-$ with H^+, is more stable and reactive than $O_2\bullet^-$ and can cross biological membranes more easily (Edge and Truscott, 2021).

4.3.4 Hydrogen peroxide

H_2O_2 is a moderately strong ROS, generated by the monovalent reduction and protonation of $O_2\bullet^-$ (Sharma *et al.*, 2021). $O_2\bullet^-$ can be changed into H_2O_2 in acidic conditions, either without the help of enzymes or, mostly, with the help of a process sped up by superoxide dismutase (Sachdev *et al.*, 2021). H_2O_2 is made by plant cells naturally and also in response to oxidative stressors such as drought, low temperatures, bright light, ultraviolet radiation, mechanical damage, and pathogen invasion (Ul Islam *et al.*, 2023). Stomatal closure, reduced CO_2 concentrations, and restricted carbon fixation enhance photorespiration by elevating ribulose-1,5-bisphosphate oxygenation, which is responsible for about 70% of the H_2O_2 generated during drought stress (Dalal, 2021).

4.3.5 Hydroxyl radical

$OH\bullet$ is the most reactive and poisonous member of the ROS family. It is made when H_2O_2 reacts with $O_2\bullet^-$, which is sped up by transition metals such as Fe^{2+} and Fe^{3+} in the Fenton reaction (Ovalle, 2022). It has a high chance of damaging living organisms by altering cell structures through processes such as lipid peroxidation, protein degradation, and membrane disintegration (Rezende *et al.*, 2022). $OH\bullet$ levels are often predicted to be at low concentrations of around 10^{-16} M, whereas the physiological limit is 10^{-13} M. $OH\bullet$ has a very short half-life (about 1 ns) and a very small diffusion range of only ~1 nm, which means it acts quickly where it is made and cannot move outside of cells or participate in intercellular signaling (Chen *et al.*, 2021). New research shows that $OH\bullet$ is not only a harmful substance but also plays a part in many important physiological processes (Di Meo and Venditti, 2020).

4.4 Reactive Oxygen Species Feedback to Strigolactone Signaling and Biosynthesis

The quantity and quality of SLs are raised in response to certain environmental problems in order to support and promote growth methods appropriate for these circumstances (Naorem *et al.*, 2023). Collectively, abiotic stress factors such as deficiency of water, extreme temperatures, heavy-metal toxicity, and nutrient deficiencies, along with the associated overrun of ROS, present a significant challenge to crop growth and, in turn, threaten global food security (Vasilachi *et al.*, 2023). SLs can help plants become more resilient to a variety of abiotic stressors (Soliman *et al.*, 2022).

4.4.1 SLs in drought stress

Water scarcity is a common abiotic stressor that impairs plant growth and has a detrimental effect on survival and yield (Seleiman *et al.*, 2021). ROS buildup and oxidative stress are caused by drought stress in plants, which interferes with the electron transport chain and damages organelles such as peroxisomes, and chloroplasts (Suresh *et al.*, 2026). In *Solanum lycopersicum*, Yu *et al.* (2022) found that down-regulation of the SL biosynthetic gene *SlCCD7* was positively connected with a decrease in SL levels during drought stress.

4.4.2 SLs in salt stress

In many parts of the world, salinization has become a serious agricultural and environmental problem. Currently, over 7% of the world's land area is affected by salinity, and projections suggest that this figure could rise to nearly 50% by the mid-21st century (Adler, 2022). The enhanced photosynthetic capability of *Brassica napus* indicates that exogenous GR24 therapy reduces the impacts of salt stress (Upadhyay, 2026). Liu *et al.* (2020) demonstrated that SL-deficient and SL-signaling mutants of *A. thaliana* exhibited heightened sensitivity to salinity during the germination and vegetative stages relative to their wild-type counterparts. Similarly, *Salvia nemorosa* exhibits increased levels of malondialdehyde and H_2O_2 in response to salinity (Rahmani and Radjabian, 2024). Salinity stress also led to a significant reduction in essential oil concentration. The exogenous application of GR24 reduced MDA and H_2O_2 accumulation, while enhancing essential oil yield (Zhang *et al.*, 2022a).

4.4.3 SLs in heat stress

Heat stress significantly limits root expansion in cold-season plants, as root growth necessitates lower temperatures and is highly delicate to increasing temperatures (Wang *et al.*, 2025). The positive impact of GR24 on crown root elongation and root cell quantity is evidenced by fluctuations in the appearance of genes linked to cell division (*PCNA*) (Alrajhi *et al.*, 2024). In *A. thaliana* mutants lacking SL, the control of germination by SL under heat stress has been demonstrated. According to Rodriguez *et al.* (2025), GR24 therapy reduced the negative impacts of heat stress on *Lupinus polyphyllus* seed germination.

4.4.4 SLs in chilling stress

Globally, low temperatures have a negative effect on crop output, which causes financial losses. According to Zhang *et al.* (2022b), the *MAX* and *RMS4* genes linked to SLs indicate that SLs play a major role in the low-temperature tolerance of these plants. Changes in growth characteristics of *Brassica rapa* under chilling stress were noted by Ljubej *et al.* (2021). The expression of enzymes linked to cold, such as mitogen-activated protein kinase (MAPK), NADPH oxidase, and antioxidant enzymes, was significantly elevated by GR24 (Liu *et al.*, 2024b). To alleviate chilling stress, GR24 treatment markedly decreased lipid peroxidation while enhancing proline accumulation, soluble sugar levels, lipoxygenase activity, and the activities of enzymes involved in the phenylpropanoid pathway (Zhang *et al.*, 2022a).

4.4.5 SLs in heavy-metal stress

Because of their potential to be harmful to human health, heavy metals are the second most important environmental issue in the world (Mitra *et al.*, 2022). Due to buildup in their organs, heavy-metal toxicity in plants causes growth and development retardation, which may be lessened by applying different phytohormones (Zhang *et al.*, 2024). Heavy-metal toxicity varies depending on the particular metal in question. Bhoi *et al.* (2021) investigated how cadmium affected *Panicum virgatum* absorption, while Li *et al.* (2025) conducted a comparative analysis of the role of SLs in wild-type and *D10* and *D17* mutants of *O. sativa* under arsenic stress.

4.5 Transcription Factors and Cross-Regulation

The master transcription factors MYC2, involved in the jasmonic acid (JA) signaling pathway, and EIN3/EILs, associated with the ethylene (ET) signaling pathway, function as regulatory switches in hormone responses, modulating various environmental signals (Pérez-Llorca *et al.*, 2023). MYC2 and EIN3 function antagonistically in the regulation of JA/ET during the development of the *Arabidopsis* apical hook. This mechanism operates on two levels (Kumar *et al.*, 2025). JA-activated MYC2 interacts with EIN3, inhibiting its transcriptional activation. MYC2 binds directly to the EBF1 promoter, enhancing its expression and facilitating EIN3 degradation (Li *et al.*, 2024). EIN3/EIL1 serve as a critical integration point for the synergistic regulation of root development and necrotrophic pathogen defense by JA and ET (Xu *et al.*, 2020). EIN3/EIL1 are normally repressed by JA-Zim domain (JAZ) proteins; however, JA application induces JAZ degradation, thereby activating EIN3/EIL1 and enhancing ET/JA gene expression synergy. MYC2 likely plays a role in the cross-regulation of abscisic acid (ABA) and JA signaling pathways, acting as a transcriptional activator within the ABA pathway and positively regulating the ABA-induced gene *RD22* (Wang *et al.*, 2025).

Abiotic stress factors such as nutrient scarcity, salinity, and heat trigger SL biosynthesis, leading to modifications in root architecture, proline accumulation, antioxidant activation, and auxin transport (Fig. 4.1) (Bhoi *et al.*, 2021). Additionally, SLs participate in hormonal crosstalk with phytohormones such as ABA and ethylene to fine-tune stress rejoinders.

4.5.1 Heat and cold stress

One of the main environmental factors affecting root development, particularly in plants in cool climates, is temperature stress (Calleja-Cabrera *et al.*, 2020). This is because the roots of such plants require relatively low temperatures for optimal growth and are highly sensitive to increases in ambient temperature. Novel phytohormone-based approaches are required to enhance plant resilience under temperature stress (Ali *et al.*, 2024). In this context, exogenous application of GR24 can strengthen stress tolerance by promoting antioxidant defense and photosynthesis, regulating hormone balance and increasing heat-shock proteins (Ahsan *et al.*, 2022). SLs are thought to interact with other hormones to form an integrated defense mechanism against heat stress (Al-Zahrani *et al.*, 2022).

Low-temperature stress causes yield losses and thus economic losses, especially in regions with intensive agricultural production (Junaid and Gokce, 2024). Just like drought and extreme heat, cold stress severely limits plant growth and crop productivity. Therefore, the development of low-temperature-tolerant varieties is currently among the priority targets of global agricultural research (Raza *et al.*, 2023). Recent studies have revealed that SLs play a protective role in plants against cold stress. In particular, mutations in genes involved in SL signaling and biosynthesis caused damage to the photosynthetic structure in model plants such as *A. thaliana* and *Pisum sativum*, indicating that SLs are important regulators of cold-stress tolerance (Yang *et al.*, 2023a).

While cold stress leads to adverse effects such as membrane damage, protein degradation,

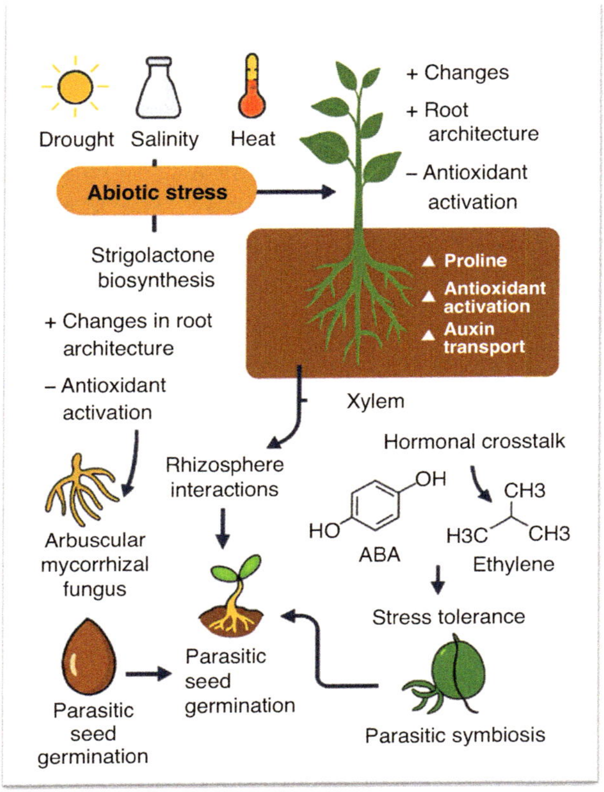

Fig. 4.1. Schematic illustration summarizing the role of strigolactones in plant responses to abiotic stress. ABA, abscisic acid.

and ROS accumulation in plants, SLs (especially GR24) significantly enhance plant cold tolerance by promoting photosynthesis, increasing antioxidant defense and activating stress-related genes (Qi *et al.*, 2024).

4.6 Biotic Stress and Pathogen Resistance

Plants have developed complex defense systems based on pathogen recognition and regulation of defense responses against biotic stresses caused by harmful organisms (Iqbal *et al.*, 2021). These organisms cause various disease symptoms such as leaf spots, wilt, and deformity, while pests damage plants by direct feeding or by carrying viruses. Plants develop complex responses to these stressors at the molecular, morphological, and biochemical levels (Dixit *et al.*, 2024). Plants also play critical roles in rhizosphere interactions such as parasitic plant germination, mycorrhizal hyphae branching, and rhizobial symbiosis (Bao *et al.*, 2022). SLs not only regulate plant growth and development but also play a role in plant defense against pathogens such as *Rhodococcus fascians*, *Pectobacterium carotovorum* and *Botrytis cinerea*; this defense effect has been suggested by regulation of SL biosynthesis genes and GR24 application. However, it has also been reported that SLs are not effective against all pathogens and in some cases do not contribute to the immune response (Hammers *et al.*, 2022).

4.7 Developmental Processes Influenced by SL–ROS Interactions

The development and growth dynamics of plants are determined by interactions between genetic traits and environmental factors (Napier *et al.*, 2023). Plants have the ability to show flexibility by reorganizing their developmental processes and metabolism to cope with changes in the environment, for example by SL–ROS interactions (Smith and Zeeman, 2020). Plant hormones are involved in providing this flexibility. To date, nine basic plant hormone types have been identified, including SLs (Nambara and van Wees, 2021). SLs are produced mostly in the roots and transported to different parts of the plant via the xylem (Fig. 4.2) (Barbier *et al.*, 2023). The interactions involved are especially intricate in the regulation of branching, with SLs acting together with auxins to affect the growth of lateral buds (Yuan *et al.*, 2023).

4.8 Conclusion

The dynamic and bidirectional interaction between SLs and ROS forms a central regulatory network in plant expansion and environmental stress variation. As a result of SL applications, an increase in the activity of antioxidant enzymes is observed. It has also been shown that oxidative stress is alleviated through the expression of SL genes. However, ROS also affect SL biosynthesis and signaling pathways by acting as signaling molecules, thus contributing to the regulation of hormonal responses based on oxidative cues. The SL–ROS interaction forms an even more integrated signaling network through crosstalk with other hormones. In the future, it will be important to identify in

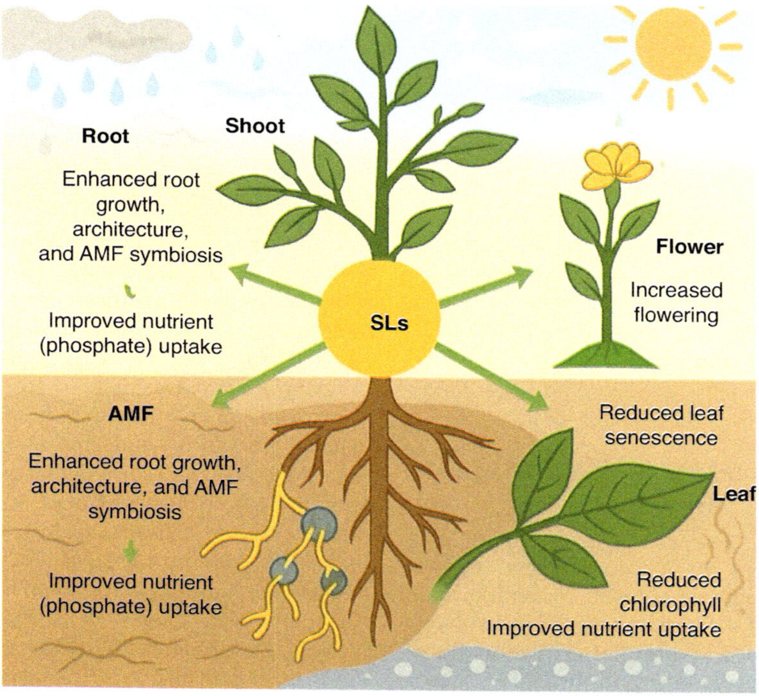

Fig. 4.2. Schematic representation summarizing the effects of strigolactones on plant development. AMF, arbuscular mycorrhizal fungi; SL, strigolactone.

more detail the key molecular nodes regulating SL–ROS interaction, such as redox-responsive transcription factors, RBOH (respiratory burst oxidase homolog) enzymes, and SL biosynthesis genes. Exogenous application of SL analogs (e.g. GR24) and biotechnological approaches offer promising strategies to enhance plant resilience to environmental stresses. Future studies should elucidate interspecific differences, study SL–ROS interactions associated with biotic stresses, and integrate omics technologies to map the regulation of these networks. This may contribute to the development of plants that are more resistant to stress conditions.

References

Abdulfatah, H.F. (2022) Non-enzymatic antioxidants in stressed plants: A review. *Journal of University of Anbar for Pure Science* 16, 25–37.

Adler, R.W. (2022) Climate change adaptation and agricultural and forestry law. In: Verschuuren, J. (ed.) *Research Handbook on Climate Change Adaptation Law*. Edward Elgar Publishing, pp. 214–249.

Ahsan, M., Zulfiqar, H., Farooq, M.A., Ali, S., Tufail, A. *et al.* (2022) Strigolactone (GR24) application positively regulates photosynthetic attributes, stress-related metabolites and antioxidant enzymatic activities of ornamental sunflower (*Helianthus annuus* cv. Vincent's Choice) under salinity stress. *Agriculture* 13(1), 50. DOI: 10.3390/agriculture13010050.

Ali, J., Mukarram, M., Ojo, J., Dawam, N., Riyazuddin, R. *et al.* (2024) Harnessing phytohormones: Advancing plant growth and defence strategies for sustainable agriculture. *Physiologia Plantarum* 176(3), e14307. DOI: 10.1111/ppl.14307.

Alrajhi, A., Alharbi, S., Beecham, S. and Alotaibi, F. (2024) Regulation of root growth and elongation in wheat. *Frontiers in Plant Science* 15, 1397337. DOI: 10.3389/fpls.2024.1397337.

Al-Zahrani, H.S., Alharby, H.F. and Fahad, S. (2022) Antioxidative defense system, hormones, and metabolite accumulation in different plant parts of two contrasting rice cultivars as influenced by plant growth regulators under heat stress. *Frontiers in Plant Science* 13, 911846. DOI: 10.3389/fpls.2022.911846.

Bansal, A., Kaushik, S. and Kukreti, S. (2022) Non-canonical DNA structures: Diversity and disease association. *Frontiers in Genetics* 13, 959258. DOI: 10.3389/fgene.2022.959258.

Bao, X., Zou, J., Zhang, B., Wu, L. and Yang, T. (2022) Arbuscular mycorrhizal fungi and microbes interaction in rice mycorrhizosphere. *Agronomy* 12(6), 1277. DOI: 10.3390/agronomy12061277.

Barbier, F., Fichtner, F. and Beveridge, C. (2023) The strigolactone pathway plays a crucial role in integrating metabolic and nutritional signals in plants. *Nature Plants* 9(8), 1191–1200. DOI: 10.1038/s41477-023-01453-6.

Bendou, O. (2025) Physiological and antioxidant response of wheat to water availability under elevated CO_2 and high temperatures and its impact on grain yield and quality traits. PhD thesis, University of Salamanca, Salamanca, Spain.

Bhoi, A., Yadu, B., Chandra, J. and Keshavkant, S. (2021) Contribution of strigolactone in plant physiology, hormonal interaction and abiotic stresses. *Planta* 254(2), 28. DOI: 10.1007/s00425-021-03678-1.

Brown, R.A., Stevanovic, S., Bottle, S., Wang, H., Hu, Z. *et al.* (2020) Relationship between atmospheric PM-bound reactive oxygen species, their half-lives, and regulated pollutants: Investigation and preliminary model. *Environmental Science and Technology* 54(8), 4995–5002. DOI: 10.1021/acs.est.9b06643.

Calleja-Cabrera, J., Boter, M., Oñate-Sánchez, L. and Pernas, M. (2020) Root growth adaptation to climate change in crops. *Frontiers in Plant Science* 11, 544. DOI: 10.3389/fpls.2020.00544.

Carbonnel, S., Das, D., Varshney, K., Kolodziej, M.C., Villaécija-Aguilar, J.A. *et al.* (2020) The karrikin signaling regulator SMAX1 controls *Lotus japonicus* root and root hair development by suppressing ethylene biosynthesis. *Proceedings of the National Academy of Sciences USA* 117(35), 21757–21765. DOI: 10.1073/pnas.2006111117.

Chen, G.-T.E., Wang, J.Y., Votta, C., Braguy, J., Jamil, M. *et al.* (2023) Disruption of the rice 4-DEOXYOROBANCHOL HYDROXYLASE unravels specific functions of canonical strigolactones.

Proceedings of the National Academy of Sciences USA 120(42), e2306263120. DOI: 10.1073/pnas.2306263120.

Chen, K.Y., Jenkins, E., Körbel, M., Ponjavic, A., Lippert, A.H. *et al.* (2021) Trapping or slowing the diffusion of T cell receptors at close contacts initiates T cell signaling. *Proceedings of the National Academy of Sciences USA* 118(39), e2024250118. DOI: 10.1073/pnas.2024250118.

Chowdhary, N.A. and Songachan, L. (2025) Plant peptides involved in ROS signalling and biotic and abiotic stress responses. *International Journal of Peptide Research and Therapeutics* 31, 53. DOI: 10.1007/s10989-025-10711-4.

Dalal, V. (2021) Modulation of photosynthesis and other proteins during water-stress. *Molecular Biology Reports* 48(4), 3681–3693. DOI: 10.1007/s11033-021-06329-6.

Dembitsky, V.M. (2023) Biological activity and structural diversity of steroids containing aromatic rings, phosphate groups, or halogen atoms. *Molecules* 28(14), 5549. DOI: 10.3390/molecules28145549.

Di Meo, S. and Venditti, P. (2020) Evolution of the knowledge of free radicals and other oxidants. *Oxidative Medicine and Cellular Longevity* 9829176. DOI: 10.1155/2020/9829176.

Dixit, S., Sivalingam, P.N., Baskaran, R.M., Senthil-Kumar, M. and Ghosh, P.K. (2024) Plant responses to concurrent abiotic and biotic stress: Unravelling physiological and morphological mechanisms. *Plant Physiology Reports* 29, 6–17. DOI: 10.1007/s40502-023-00766-0.

Dun, E.A., Brewer, P.B., Gillam, E.M. and Beveridge, C.A. (2023) Strigolactones and shoot branching: What is the real hormone and how does it work. *Plant and Cell Physiology* 64(9), 967–983. DOI: 10.1093/pcp/pcad088.

Edge, R. and Truscott, T.G. (2021) The reactive oxygen species singlet oxygen, hydroxy radicals, and the superoxide radical anion – Examples of their roles in biology and medicine. *Oxygen* 1(2), 77–95. DOI: 10.3390/oxygen1020009.

El Sabagh, A., Islam, M.S., Hossain, A., Iqbal, M.A., Mubeen, M. *et al.* (2022) Phytohormones as growth regulators during abiotic stress tolerance in plants. *Frontiers in Agronomy* 4, 765068. DOI: 10.3389/fagro.2022.765068.

Foyer, C.H. and Hanke, G. (2022) ROS production and signalling in chloroplasts: Cornerstones and evolving concepts. *Plant Journal* 111(3), 642–661. DOI: 10.1111/tpj.15856.

Fujita, M. and Hasanuzzaman, M. (2022) Approaches to enhancing antioxidant defense in plants. *Antioxidants* 11(5), 925. DOI: 10.3390/antiox11050925.

Gómez, X., Sanon, S., Zambrano, K., Asquel, S., Bassantes, M. *et al.* (2021) Key points for the development of antioxidant cocktails to prevent cellular stress and damage caused by reactive oxygen species (ROS) during manned space missions. *Npj Microgravity* 7(1), 35. DOI: 10.1038/s41526-021-00162-8.

Guo, Z., Gao, Y., Yuan, X., Yuan, M., Huang, L. *et al.* (2023) Effects of heavy metals on stomata in plants: A review. *International Journal of Molecular Sciences* 24(11), 9302. DOI: 10.3390/ijms24119302.

Hammers, D.E., Donahue, D.L., Tucker, Z.D., Ashfeld, B.L., ploplis, V.A. *et al.* (2022) Streptolysin S targets the sodium-bicarbonate cotransporter NBCn1 to induce inflammation and cytotoxicity in human keratinocytes during Group A Streptococcal infection. *Frontiers in Cellular and Infection Microbiology* 12, 1002230. DOI: 10.3389/fcimb.2022.1002230.

Homma, M., Wakabayashi, T., Moriwaki, Y., Shiotani, N., Shigeta, T. *et al.* (2023) Insights into stereoselective ring formation in canonical strigolactone: Discovery of a dirigent domain-containing enzyme catalyzing orobanchol synthesis. *Proceedings of the National Academy of Sciences USA* 121(26), e2313683121. DOI: 10.1073/pnas.2313683121.

Hosseinifard, M., Stefaniak, S., Ghorbani Javid, M., Soltani, E., Wojtyla, Ł. *et al.* (2022) Contribution of exogenous proline to abiotic stresses tolerance in plants: A review. *International Journal of Molecular Sciences* 23(9), 5186. DOI: 10.3390/ijms23095186.

Imaduwage, I. and Hewadikaram, M. (2024) Predicted roles of long non-coding RNAs in abiotic stress tolerance responses of plants. *Molecular Horticulture* 4(1), 20. DOI: 10.1186/s43897-024-00094-3.

Iqbal, Z., Iqbal, M.S., Hashem, A., Abd Allah, E.F. and Ansari, M.I. (2021) Plant defense responses to biotic stress and its interplay with fluctuating dark/light conditions. *Frontiers in Plant Science* 12, 631810. DOI: 10.3389/fpls.2021.631810.

Jain, S., Kumar, R., Gunsola, D., Živković, S., Vasić, T. *et al.* (2025) Strigolactones and control of parasitic weeds. In: Aftab, T. and Wani, K.I. (eds) *Emerging Plant Hormones*. Scrivener Publishing, Strigolactones, pp. 261–279.

Junaid, M. and Gokce, A. (2024) Global agricultural losses and their causes. *Bulletin of Biological and Allied Sciences Research* 2024(1), 66. DOI: 10.54112/bbasr.v2024i1.66.

Kapoor, R.T., Alam, P., Chen, Y. and Ahmad, P. (2024) Strigolactones in plants: From development to abiotic stress management. *Journal of Plant Growth Regulation* 43, 903–919. DOI: 10.1007/s00344-023-11148-z.

Khorobrykh, S., Havurinne, V., Mattila, H. and Tyystjärvi, E. (2020) Oxygen and ROS in photosynthesis. *Plants* 9(1), 91. DOI: 10.3390/plants9010091.

Kohli, P.S., Maurya, K., Thakur, J.K., Bhosale, R. and Giri, J. (2022) Significance of root hairs in developing stress-resilient plants for sustainable crop production. *Plant, Cell & Environment* 45(3), 677–694. DOI: 10.1111/pce.14237.

Korek, M. and Marzec, M. (2023) Strigolactones and abscisic acid interactions affect plant development and response to abiotic stresses. *BMC Plant Biology* 23(1), 314. DOI: 10.1186/s12870-023-04332-6.

Kozuleva, M.A. and Ivanov, B.N. (2023) Superoxide anion radical generation in photosynthetic electron transport chain. *Biochemistry (Moscow)* 88(8), 1045–1060. DOI: 10.1134/S0006297923080011.

Kumar, A., Partap, M. and Warghat, A.R. (2025) Jasmonic acid: A versatile phytohormone regulating growth, physiology, and biochemical responses. *Journal of Plant Growth Regulation* 44, 131–154. DOI: 10.1007/s00344-024-11376-x.

Lee, Z., Lim, J.A., Harikrishna, J.A., Islam, T., Abd Rahim, M.H. *et al.* (2024) Regulation of plant responses to temperature stress: A key factor in food security and for mitigating effects of climate change. *International Journal of Plant Production* 18, 141–159. DOI: 10.1007/s42106-024-00282-7.

Li, S., Lu, S., Wang, J., Chen, Z., Zhang, Y. *et al.* (2023) Responses of physiological, morphological and anatomical traits to abiotic stress in woody plants. *Forests* 14(9), 1784. DOI: 10.3390/f14091784.

Li, X., Xie, Z., Qin, T., Zhan, C., Jin, L. *et al.* (2025) The SLR1-OsMADS23-D14 module mediates the crosstalk between strigolactone and gibberellin signaling to control rice tillering. *New Phytologist* 246(5), 2137–2154. DOI: 10.1111/nph.20331.

Li, Z., Zhang, X., Zhao, Y., Li, Y., Zhang, G. *et al.* (2018) Enhancing auxin accumulation in maize root tips improves root growth and dwarfs plant height. *Plant Biotechnology Journal* 16(1), 86–99. DOI: 10.1111/pbi.12751.

Li, Z., Huang, Y., Shen, Z., Wu, M., Huang, M. *et al.* (2024) Advances in functional studies of plant MYC transcription factors. *Theoretical and Applied Genetics* 137(8), 195. DOI: 10.1007/s00122-024-04697-8.

Liu, J., Zhu, L., Wang, B., Wang, H., Khan, I. *et al.* (2021) BnA1.CER4 and BnC1.CER4 are redundantly involved in branched primary alcohols in the cuticle wax of *Brassica napus*. *Theoretical and Applied Genetics* 134(9), 3051–3067. DOI: 10.1007/s00122-021-03879-y.

Liu, A., Huang, Z., Du, X., Duvva, N., Du, Y. *et al.* (2024a) Biodegradable ruthenium-rhenium complexes containing nanoamplifiers: Triggering ROS-induced CO release for synergistic cancer treatment. *Advanced Science* 11(35), 2403795. DOI: 10.1002/advs.202403795.

Liu, J., Bao, Y., Liu, S., Zhu, L., Xu, X. *et al.* (2024b) Physiological and transcriptomic analyses reveal mechanisms of exogenous strigolactones to regulate cold tolerance in litchi fruit. *Postharvest Biology and Technology* 210, 112764. DOI: 10.1016/j.postharvbio.2024.112764.

Liu, X., Hu, Q., Yan, J., Sun, K., Liang, Y. *et al.* (2020) ζ-Carotene isomerase suppresses tillering in rice through the coordinated biosynthesis of strigolactone and abscisic acid. *Molecular Plant* 13, 1784–1801. DOI: 10.1016/j.molp.2020.10.001.

Ljubej, V., Radojčić Redovniković, I., Salopek-Sondi, B., Smolko, A., Roje, S. *et al.* (2021) Chilling and freezing temperature stress differently influence glucosinolates content in *Brassica oleracea* var. acephala. *Plants* 10, 1305. DOI: 10.3390/plants10071305.

Mansoor, S., Mir, M.A., Karunathilake, E., Rasool, A., Ştefănescu, D.M. *et al.* (2024) Strigolactones as promising biomolecule for oxidative stress management: A comprehensive review. *Plant Physiology and Biochemistry* 206, 108282. DOI: 10.1016/j.plaphy.2023.108282.

Maqbool, S., Hassan, M.A., Xia, X., York, L.M., Rasheed, A. *et al.* (2022) Root system architecture in cereals: Progress, challenges and perspective. *Plant Journal* 110, 23–42. DOI: 10.1111/tpj.15669.

Mareri, L., Parrotta, L. and Cai, G. (2022) Environmental stress and plants. *International Journal of Molecular Sciences* 23(10), 5416. DOI: 10.3390/ijms23105416.

Mashiguchi, K., Seto, Y. and Yamaguchi, S. (2021) Strigolactone biosynthesis, transport and perception. *Plant Journal* 105, 335–350. DOI: 10.1111/tpj.15059.

Mashiguchi, K., Seto, Y., Onozuka, Y., Suzuki, S., Takemoto, K. *et al.* (2022) A carlactonoic acid methyltransferase that contributes to the inhibition of shoot branching in *Arabidopsis*. In: *Proceedings of the National Academy of Sciences USA* (Vol. 119). DOI: 10.1073/pnas.2111565119.

Mehmood, I., Wani, K.I. and Aftab, T. (2025) Nature, structural diversity, biosynthetic pathway, and strigolactone transport in plants. In: Aftab, T. and Wani, K.I. (eds) *Strigolactones: Emerging Plant Hormones*. Scrivener Publishing, Beverly, Massachusetts, pp. 13–34.
Mitra, S., Chakraborty, A.J., Tareq, A.M., Emran, T.B., Nainu, F. *et al.* (2022) Impact of heavy metals on the environment and human health: Novel therapeutic insights to counter the toxicity. *Journal of King Saud University—Science* 34, 101865. DOI: 10.1016/j.jksus.2022.101865.
Mittler, R., Zandalinas, S.I., Fichman, Y. and van Breusegem, F. (2022) Reactive oxygen species signalling in plant stress responses. *Nature Reviews Molecular Cell Biology* 23, 663–679. DOI: 10.1038/s41580-022-00499-2.
Mizutori, Y., Nishiyama, K. and Seto, Y. (2025) Inhibition of shoot branching in *Arabidopsis* by the artificially biosynthesized canonical strigolactone. *Bioscience, Biotechnology, and Biochemistry* 89, 406–412. DOI: 10.1093/bbb/zbae202.
Mohapatra, T. (2021) Characterization of contrasting rice (*Oryza sativa* L.) genotypes reveals the Pi-efficient schema for phosphate starvation tolerance. DOI: 10.1186/s12870-021-03015-4.
Nambara, E. and van Wees, S. (2021) Plant hormone functions and interactions in biological systems. *Plant Journal* 105(2), 287–289. DOI: 10.1111/tpj.15151.
Naorem, A., Jayaraman, S., Dang, Y.P., Dalal, R.C., Sinha, N.K. *et al.* (2023) Soil constraints in an arid environment—Challenges, prospects, and implications. *Agronomy* 13, 220. DOI: 10.3390/agronomy13010220.
Napier, J.D., Heckman, R.W. and Juenger, T.E. (2023) Gene-by-environment interactions in plants: Molecular mechanisms, environmental drivers, and adaptive plasticity. *Plant Cell* 35, 109–124. DOI: 10.1093/plcell/koac322.
Napolitano, G., Fasciolo, G. and Venditti, P. (2022) The ambiguous aspects of oxygen. *Oxygen* 2, 382–409. DOI: 10.3390/oxygen2030027.
Nelson, D.C. (2021) The mechanism of host-induced germination in root parasitic plants. *Plant Physiology* 185, 1353–1373. DOI: 10.1093/plphys/kiab043.
Noodleman, L., Du, W.-G.H., McRee, D., Chen, Y., Goh, T. *et al.* (2020) Coupled transport of electrons and protons in a bacterial cytochrome c oxidase—DFT calculated properties compared to structures and spectroscopies. *Physical Chemistry Chemical Physics* 22, 26652–26668. DOI: 10.1039/d0cp04848h.
Ovalle, R. (2022) A history of the Fenton reactions (Fenton chemistry for beginners). In: Ahmad, R. (ed.) *Reactive Oxygen Species*. Intech Open, London. Available at: www.intechopen.com/chapters/80183 (accessed 17 February 2026).
Pérez-Escalante, E., Alatorre-Santamaría, S., Castañeda-Ovando, A., Salazar-Pereda, V., Bautista-Ávila, M. *et al.* (2022) Human milk oligosaccharides as bioactive compounds in infant formula: Recent advances and trends in synthetic methods. *Critical Reviews in Food Science and Nutrition* 62, 181–214. DOI: 10.1080/10408398.2020.1813683.
Pérez-Llorca, M., Pollmann, S. and Müller, M. (2023) Ethylene and jasmonates signaling network mediating secondary metabolites under abiotic stress. *International Journal of Molecular Sciences* 24, 5990. DOI: 10.3390/ijms24065990.
Qi, J., Mao, Y., Cui, J., Lu, X., Xu, J. *et al.* (2024) The role of strigolactones in resistance to environmental stress in plants. *Physiologia Plantarum* 176, e14419. DOI: 10.1111/ppl.14419.
Rahmani, N. and Radjabian, T. (2024) Integrative effects of phytohormones in the phenolic acids production in *Salvia verticillata* L. under multi-walled carbon nanotubes and methyl jasmonate elicitation. *BMC Plant Biology* 24, 56. DOI: 10.1186/s12870-023-04719-5.
Raza, A., Wang, D., Zou, X. and Prakash, C.S. (2023) Developing temperature-resilient plants: A matter of present and future concern for sustainable agriculture. *Agronomy* 13(4), 1006. DOI: 10.3390/agronomy13041006.
Rezende, L.G., Tasso, T.T., Candido, P.H. and Baptista, M.S. (2022) Assessing photosensitized membrane damage: Available tools and comprehensive mechanisms. *Photochemistry and Photobiology* 98, 572–590. DOI: 10.1111/php.13582.
Rodriguez, J.P., Herrera, J.D., Bascuñán, L. and Ostria, E. (2025) Strigolactones' role in heat and saline stress tolerance in horticultural and field crops. In: Aftab, T. and Wani, K.I. (eds) *Strigolactones: Emerging Plant Hormones*. Scrivener Publishing, Beverly, Massachusetts, pp. 109–135.
Sachdev, S., Ansari, S.A., Ansari, M.I., Fujita, M. and Hasanuzzaman, M. (2021) Abiotic stress and reactive oxygen species: Generation, signaling, and defense mechanisms. *Antioxidants* 10, 277. DOI: 10.3390/antiox10020277.

Seleiman, M.F., Al-Suhaibani, N., Ali, N., Akmal, M., Alotaibi, M. *et al.* (2021) Drought stress impacts on plants and different approaches to alleviate its adverse effects. *Plants* 10, 259. DOI: 10.3390/plants10020259.
Seo, P.J., Lee, H.G., Choi, H.-Y., Lee, S. and Park, C.-M. (2023) Complexity of SMAX1 signaling during seedling establishment. *Trends in Plant Science* 28, 902–912. DOI: 10.1016/j.tplants.2023.03.014.
Sharma, P., Jha, A.B., Dubey, R.S. and Pessarakli, M. (2021) Reactive oxygen species generation, hazards, and defense mechanisms in plants under environmental (abiotic and biotic) stress conditions. In: Pessarakli, M. (ed.) *Handbook of Plant and Crop Physiology*. CRC Press, Bocan Raton, Florida, pp. 617–658.
Sigalas, P.P., Buchner, P., Thomas, S.G., Jamois, F., Arkoun, M. *et al.* (2023) Nutritional and tissue-specific regulation of cytochrome P450 CYP711A MAX1 homologues and strigolactone biosynthesis in wheat. *Journal of Experimental Botany* 74, 1890–1910. DOI: 10.1093/jxb/erad008.
Smith, A.M. and Zeeman, S.C. (2020) Starch: A flexible, adaptable carbon store coupled to plant growth. *Annual Review of Plant Biology* 71, 217–245. DOI: 10.1146/annurev-arplant-050718-100241.
Soliman, S., Wang, Y., Han, Z., Pervaiz, T. and El-Kereamy, A. (2022) Strigolactones in plants and their interaction with the ecological microbiome in response to abiotic stress. *Plants* 11, 3499. DOI: 10.3390/plants11243499.
Song, X., Gu, X., Chen, S., Qi, Z., Yu, J. *et al.* (2024) Far-red light inhibits lateral bud growth mainly through enhancing apical dominance independently of strigolactone synthesis in tomato. *Plant, Cell and Environment* 47, 429–441. DOI: 10.1111/pce.14758.
Sun, H., Xu, F., Guo, X., Wu, D., Zhang, X. *et al.* (2019) A strigolactone signal inhibits secondary lateral root development in rice. *Frontiers in Plant Science* 10, 1527. DOI: 10.3389/fpls.2019.01527.
Sun, T., Zhu, Q., Wei, Z., Owens, L.A., Fish, T. *et al.* (2021) Multi-strategy engineering greatly enhances provitamin a carotenoid accumulation and stability in Arabidopsis seeds. *Abiotech* 2, 191–214. DOI: 10.1007/s42994-021-00046-1.
Suresh, R., Mal, S., Panchal, S. and Karuppasamy, R. (2026) Mechanisms of antioxidant and prooxidant interactions: Redox dynamics in plants. In: Kasinathan, K. and Muthukumaran, A. (eds) *Recent Advances in Oxidative Stress Associated Chronic Diseases Volume 1: A Review of the Health Benefits and Risks of the Substance*. Springer Nature, Singapore, pp. 177–198.
Sze, H., Palanivelu, R., Harper, J.F. and Johnson, M.A. (2021) Holistic insights from pollen omics: Co-opting stress-responsive genes and ER-mediated proteostasis for male fertility. *Plant Physiology* 187, 2361–2380. DOI: 10.1093/plphys/kiab463.
Tian, Q., Jing, L., Wang, W., Ye, X., Chai, X. *et al.* (2025) Hydrogen peroxide electrosynthesis via selective oxygen reduction reactions through interfacial reaction microenvironment engineering. *Advanced Materials* 37, 2414490. DOI: 10.1002/adma.202414490.
Ul Islam, S.N., Asgher, M. and Khan, N.A. (2023) Hydrogen peroxide and its role in abiotic stress tolerance in plants. In: Fatma, M., Sehar, Z. and Khan, N.A. (eds) *Gasotransmitters Signaling in Plant Abiotic Stress: Gasotransmitters in Adaptation of Plants to Abiotic Stress*. Springer Nature, Cham, Switzerland, pp. 167–195. DOI: 10.1007/978-3-031-30858-1_9.
Upadhyay, S.K. (2026) Strigolactones targeting plant–microbe dialogues from roots to soil: Ulocking pathways for sustainable agriculture. *Physiologia Plantarum* 178(1), e70787. DOI: 10.1111/ppl.70787.
van Breusegem, F. and Mittler, R. (eds) (2023) *Oxidative Stress Responses in Plants*. Academic Press, London.
Vasilachi, I.C., Stoleru, V. and Gavrilescu, M. (2023) Analysis of heavy metal impacts on cereal crop growth and development in contaminated soils. *Agriculture* 13, 1983. DOI: 10.3390/agriculture13101983.
Wang, K., Yu, J. and Zhou, J. (2025) Hormonal regulation of stress tolerance in vegetable crops. In: Ahammed, G. and Zhou, J. (eds) *Growth Regulation and Quality Improvement of Vegetable Crops: Physiological and Molecular Features*. Springer Nature, Singapore.
Wang, L., Yang, T., Pan, Y., Shi, L., Jin, Y. *et al.* (2023) The metabolism of reactive oxygen species and their effects on lipid biosynthesis of microalgae. *International Journal of Molecular Sciences* 24, 11041. DOI: 10.3390/ijms241311041.
Wang, L., Chen, F., Zhang, C., Cheng, Y.Y., Bu, W. *et al.* (2024) Reactive oxygen species and neurodegenerative diseases: Insights into nanozyme therapeutics. *MedMat* 1(2), 55–73. DOI: 10.1097/mm9.0000000000000009.
Wang, Y., Li, H., Li, Y., Zheng, A. and Dang, W. (2025) Current research status and development trends in soil moisture-heat coupled migration mechanisms: A review. *Plant and Soil* 515, 1107–1133. DOI: 10.1007/s11104-025-07668-0.

Wani, K.I., Zehra, A., Choudhary, S., Naeem, M., Khan, M.M.A. *et al.* (2021) Mechanistic insights into strigolactone biosynthesis, signaling, and regulation during plant growth and development. *Journal of Plant Growth Regulation* 40, 1836–1852. DOI: 10.1007/s00344-020-10234-w.

Wu, F., Gao, Y., Yang, W., Sui, N. and Zhu, J. (2022) Biological functions of strigolactones and their crosstalk with other phytohormones. *Frontiers in Plant Science* 13, 821563. DOI: 10.3389/fpls.2022.821563.

Xu, P., Zhao, P.-X., Cai, X.-T., Mao, J.-L., Miao, Z.-Q. *et al.* (2020) Integration of jasmonic acid and ethylene into auxin signaling in root development. *Frontiers in Plant Science* 11, 271. DOI: 10.3389/fpls.2020.00271.

Yan, Y., Zhao, N., Tang, H., Gong, B. and Shi, Q. (2020) Shoot branching regulation and signaling. *Plant Growth Regulation* 92, 131–140. DOI: 10.1007/s10725-020-00640-1.

Yang, J., Wang, P., Li, X., Zhou, D., Cai, X. *et al.* (2023a) Regulation of cold resistance by the polyamine biosynthetic gene SlSPDS2 via modulating the antioxidant enzyme system and osmotic regulatory substances in *Solanum lycopersicum*. *Environmental and Experimental Botany* 216, 105531. DOI: 10.1016/j.envexpbot.2023.105531.

Yang, Y., Abuauf, H., Song, S., Wang, J.Y., Alagoz, Y. *et al.* (2023b) The *Arabidopsis* D27-LIKE1 is a cis/cis/trans-β-carotene isomerase that contributes to strigolactone biosynthesis and negatively impacts ABA level. *Plant Journal* 113, 986–1003. DOI: 10.1111/tpj.16095.

Yu, Y., Xu, J., Wang, C., Pang, Y., Li, L. *et al.* (2022) Genome-wide analysis of the strigolactone biosynthetic and signaling genes in grapevine and their response to salt and drought stresses. *PeerJ* 10, e13551. DOI: 10.7717/peerj.13551.

Yuan, Y., Khourchi, S., Li, S., Du, Y. and Delaplace, P. (2023) Unlocking the multifaceted mechanisms of bud outgrowth: Advances in understanding shoot branching. *Plants* 12, 3628. DOI: 10.3390/plants12203628.

Zandi, P. and Schnug, E. (2022) Reactive oxygen species, antioxidant responses and implications from a microbial modulation perspective. *Biology* 11, 155. DOI: 10.3390/biology11020155.

Zeng, R., Zhou, R., Zhen, L., Lan, J., Li, Z. *et al.* (2025) Tumor-targeted nanosystem with hypoxia inducible factor 1α inhibition for synergistic chemo-photodynamic therapy against hypoxic tumor. *Colloids and Surfaces B: Biointerfaces* 248, 114456. DOI: 10.1016/j.colsurfb.2024.114456.

Zhang, X.-H., Ma, C., Zhang, L., Su, M., Wang, J. *et al.* (2022a) GR24-mediated enhancement of salt tolerance and roles of H2O2 and Ca2+ in regulating this enhancement in cucumber. *Journal of Plant Physiology* 270, 153640. DOI: 10.1016/j.jplph.2022.153640.

Zhang, Y., Zeng, D., Liu, Y. and Zhu, W. (2022b) SlSPS, a sucrose phosphate synthase gene, mediates plant growth and thermotolerance in tomato. *Horticulturae* 8, 491. DOI: 10.3390/horticulturae8060491.

Zhang, Y., Yang, C., Liu, S., Xie, Z. and Chang, H. (2024) Phytohormones-mediated strategies for mitigation of heavy metals toxicity in plants focused on sustainable production. *Plant Cell Reports* 43, 99. DOI: 10.1007/s00299-024-03189-9.

5 Reactive Nitrogen Species and Strigolactone Interaction in Plants

Narkhede Gopal Wasudeo[1], G. Harish Kumar[1], Hirakant V. Kalpande[2], K.S. Baig[3] and Shabir H. Wani[4]*

[1]Department of Genetics and Plant Breeding, School of Agriculture, SR University, Warangal, Telangana, India; [2]Department of Agricultural Botany, Vasantrao Naik Marathwada Krishi Vidyapeeth, Parbhani, Maharashtra, India; [3]Vasantrao Naik Marathwada Krishi Vidyapeeth, Parbhani, Maharashtra, India; [4]National Seed Project Laboratory, Sher-e-Kashmir University of Agricultural Sciences and Technology of Kashmir, Shalimar, Srinagar, J&K, India

Abstract

Embarking on the complexity of interactions between reactive nitrogen species (RNS) and strigolactones (SLs) in plants, this chapter discusses the growth and development of plants under stress. These RNS include nitric oxide and peroxynitrite, which might be important in signaling pathways that alter SL biosynthesis, as well as the overall activities of SLs to help plants cope with the ever-changing environment and communications from roots to shoots. The chapter investigates how RNS modulate the production and the signaling of SL and vice versa, explaining how the two link to enhance plant growth and adaptability. It reviews the potential of RNS and SLs for crop development, while analyzing the most suitable RNS and SL interactions for optimal sustainable agriculture practices to make these practices effective.

Keywords: agriculture, crop improvement, plant growth, reactive nitrogen species, signaling pathways, stress response, strigolactones

5.1 Introduction

Reactive nitrogen species (RNS) and strigolactones (SLs) are signal mediators in plant growth and stress responses. Nitric oxide (NO) and peroxynitrite ($ONOO^-$) are the representative members of RNS that regulate redox outcomes and signaling. SLs are a type of plant hormone that regulate various traits, such as shoot branching, root growth, and interaction with symbiotic fungi. Current research evidence indicates crosstalk between RNS and SLs, influencing plant responses to different environmental stresses and situations. RNS affect SL biosynthetic and signaling regulation, which in turn regulate shoot and root architecture (Berger *et al.*, 2020; Galatro *et al.*, 2020; Wani *et al.*, 2021). Understanding how these are linked can be utilized to understand better how crops can be made more resilient to adverse abiotic factors, such as drought and nutrient deficiency, and hence increase overall productivity (Petřivalský and Luhová, 2020; Staszek and Gniazdowska, 2020).

*Corresponding author: shabirhwani@skuastkashmir.ac.in

DOI: 10.1079/9781836990598.0005

5.1.1 Overview of RNS

RNS are a series of highly reactive nitrogen-based molecules implicated in various aspects of plant physiology, including development and responses to stress. These molecules include NO, $ONOO^-$, nitrogen dioxide (NO_2), nitroxyl, and *S*-nitrosothiols. RNS are mainly produced as by-products of cellular metabolism, especially in the mitochondria, chloroplasts, and peroxisomes, and they participate in redox signaling (Wu *et al.*, 2022; Selwal *et al.*, 2023).

RNS are produced by both enzymatic and non-enzymatic pathways. The main enzymatic sources are as follows (Ciacka *et al.*, 2020; Kapoor *et al.*, 2024):

- *Nitrate reductase*: nitrate is converted to nitrite and then to NO by nitrate reductase.
- *Nitric oxide synthase (NOS)*: although no NOS enzyme has been found in plants, it is assumed that there must be some NOS-like activity involved in NO generation.
- *Xanthine oxidase and peroxidases*: these participate in the production of NO and $ONOO^-$ under conditions of oxidative stress.

RNS are essential in plant signaling and metabolism, influencing several physiological processes:

- *Growth and development*: NO regulates seed germination, root development, and flowering. It interacts with other hormones such as auxins, cytokinins, and gibberellins to coordinate growth.
- *Stress responses*: RNS mediate responses to abiotic (drought, salinity, heavy metals) and biotic (pathogen attack) stresses by activating defense pathways and antioxidant mechanisms.
- *Post-translational modifications*: NO modifies proteins through *S*-nitrosylation, affecting their function and stability. This modification regulates enzyme activities, gene expression, and signaling cascades.
- *Interaction with phytohormones*: RNS crosstalk with auxins, ethylene, abscisic acid (ABA), and SLs, influencing growth and stress adaptation (Wang *et al.*, 2024).

Nitrosative stress occurs after the excessive buildup of RNS, causing damage to cellular structures and macromolecules, including proteins, lipids, and nucleic acids. High concentrations of RNS disturb cellular homeostasis, causing oxidative stress and programmed cell death. Understanding the dynamics of RNS and their regulatory controls in plants may offer a workable pathway to develop stress tolerance, growth regulation, and agricultural productivity in difficult environmental circumstances (Petřivalský and Luhová, 2020; Staszek and Gniazdowska, 2020).

5.1.2 Overview of SLs

SLs were identified initially as germination-timing stimulants for plants considered to be parasites such as members of the genus *Striga*. They have since been recognized as key regulators of plant architecture and as effective signaling molecules under environmental conditions. These hormones are a class of carotenoid-derived plant hormones that have been found to control and regulate several physiological processes encompassing shoot and root development and stress responses, as well as plant–microbe interactions (Bürger and Chory, 2020; Mashiguchi *et al.*, 2021).

Biosynthesis of SLs occurs from β-carotene in the roots of plants by a series of enzyme-catalyzed reactions. DWARF27 (D27) converts all-*trans*-β-carotene to 9-*cis*-β-carotene; carotenoid cleavage dioxygenases 7 and 8 (CCD7 and CCD8, also known as MAX3 and MAX4) then cleave 9-*cis*-β-carotene to form carlactone, a precursor of SLs. Finally, MORE AXILLARY GROWTH1 (MAX1) converts carlactone into bioactive SLs.

After biosynthesis, SLs are transported from the roots to the shoots through the xylem, where they can regulate plant architecture and interact with other signaling agents (Bürger and Chory, 2020; Mashiguchi *et al.*, 2021).

SLs have a number of functions in plants.

- *Suppression of axillary buds for shoot branching*: by degrading the BRANCHED1 (BRC1) transcriptional regulator, SLs prevent the initiation of axillary bud outgrowth, thereby stopping unnecessary outgrowth of branches.

- *Root architectural mass modifications*: SLs facilitate primary root elongation, lateral root initiation, as well as root-hair developmental proliferation to enhance nutrient uptake.
- *Responses to drought and nutritional deprivation*: SLs make it easier for plants to adapt to phosphate and nitrogen starvation by causing a transformation in root architecture to increase the symbiotic interaction of mycorrhizal fungi with the root.
- *Seed germination and parasitic plant relations*: SLs promote germination in root-parasitic plants such as *Striga* and *Orobanche*, which may affect growth positively or negatively.
- *Interaction with other hormones*: SLs cross-talk with auxins, cytokinins, gibberellins, and ABA to fine-tune plant responses for growth and in response to changes in the environment (Wang *et al.*, 2013; Yoneyama *et al.*, 2013; Zheng *et al.*, 2016; Saeed *et al.*, 2017; van Dingenen *et al.*, 2023).

SLs have potential applications in agriculture, including enhancing crop resilience to nutrient deficiency and drought, controlling shoot architecture for higher yield efficiency, and managing parasitic weed infestations through targeted SL manipulation (de Saint Germain *et al.*, 2016; Ito *et al.*, 2017; Waters *et al.*, 2017).

5.2 Biosynthesis and Signaling Pathways

As described above, bioactive SLs are synthesized from β-carotene in roots via enzymes such as D27, CCD7, CCD8, and MAX1. These hormones are transported to the shoots, where they regulate plant architecture (Soliman *et al.*, 2022; Trasoletti *et al.*, 2022; Wu *et al.*, 2022). In signaling, SLs bind to the receptor DWARF14 (D14), triggering interaction with the F-box protein MORE AXILLARY GROWTH2 (MAX2). This leads to degradation of suppressor proteins such as BRANCHED1 (BRC1), inhibiting shoot branching (Table 5.1). SLs also influence root growth, stress responses, and symbiosis with fungi, making them crucial regulators of plant development and adaptation (Korek and Marzec, 2023; Al-Babili, 2024).

5.2.1 Biosynthesis of RNS

RNS are generated during various metabolic processes, primarily through the enzymatic activity of NOS, which converts L-arginine into NO. NO can further react with superoxide (O_2^-) to form $ONOO^-$, a potent RNS. Other RNS, such as NO_2 and dinitrogen trioxide (N_2O_3), are formed through oxidation or reaction of NO with oxygen or other radicals. These species are involved in signaling, but excessive accumulation leads to nitrosative stress, damaging proteins, lipids, and DNA (Table 5.1).

5.2.2 Biosynthesis of SLs

SLs are a class of phytohormones derived from carotenoids and are synthesized mainly in plant roots, with some lesser amounts synthesized in shoots. The biosynthesis of SLs follows a well-characterized enzymatic route initiated from β-carotene, an important carotenoid, and ends with different bioactive SLs causing regulation of plant growth and environment interaction (Table 5.2). SLs are synthesized from all-*trans*-β-carotene, a C40 carotenoid that undergoes isomerization and further cleavage to give rise to SLs in various intermediates and final forms (Yoneyama, 2020; Mashiguchi *et al.*, 2021; Yoda *et al.*, 2021; Chen *et al.*, 2022).

The downstream products of carlactone modification differ between species. For example, in *Arabidopsis*, MAX1 converts carlactone to carlactonoic acid, which may be further methylated to methyl carlactonoate, while in rice, multiple MAX1 homologs lead to the formation of 5-deoxystrigol and orobanchol, two major bioactive SLs.

These compounds are collectively referred to as canonical SLs (containing the ABC-ring structure) and non-canonical SLs (lacking parts of the ring structure but still active) (Yoneyama *et al.*, 2018; Wakabayashi *et al.*, 2019; Homma *et al.*, 2024).

Table 5.1. Biosynthesis of reactive nitrogen species. Data from Arasimowicz and Floryszak-Wieczorek (2007); Santana *et al.* (2017); León and Costa-Broseta (2020); Zhou *et al.* (2021); Khan *et al.* (2023).

Pathway	Key enzymes/ mechanisms	Products	Location	Function
Enzymatic pathways				
Nitrate reduction	Nitrate reductase	Nitrite (NO_2^-), nitric oxide (NO)	Cytoplasm, chloroplasts	Converts nitrate to NO via nitrite reduction
Nitric oxide synthase (NOS)-like activity	NOS-like	NO	Cytoplasm	Produces NO for signaling and stress responses
Xanthine oxidase pathway	Xanthine oxidase	NO, superoxide (O_2^-)	Peroxisomes	Generates NO under stress conditions
Peroxidase pathway	Peroxidases	NO	Cell wall, vacuole	Involved in redox homeostasis and defense
Non-enzymatic pathways				
Nitrite reduction	Acidic or reducing conditions	NO	Apoplast, cytoplasm	Generates NO under hypoxia or stress conditions
Chemical interactions	Reaction with reactive oxygen species	Peroxynitrite ($ONOO^-$)	Cytoplasm, mitochondria	Induces oxidative/ nitrosative stress

Table 5.2. Enzymatic steps in strigolactone biosynthesis. Data from Yoneyama (2020); Mashiguchi *et al.* (2021); Wu and Li (2021); Yoda *et al.* (2021); Seto (2023).

Step	Enzyme	Function	Product
1	DWARF27 (D27)	Isomerizes all-*trans*-β-carotene to 9-*cis*-β-carotene	9-*cis*-β-carotene
2	carotenoid cleavage dioxygenase 7 (CCD7)	Cleaves 9-*cis*-β-carotene at the 9–10 double bond	9-*cis*-β-apo-10′-carotenal
3	carotenoid cleavage dioxygenase 8 (CCD8)	Converts 9-*cis*-β-apo-10′-carotenal into carlactone	Carlactone
4	MORE AXILLARY GROWTH1 (MAX1) (a cytochrome P450 monooxygenase)	Oxidizes carlactone into various active SLs	Carlactonoic acid, 5-deoxystrigol, or orobanchol

SL production is tightly regulated by internal and external cues. These include: (i) nutrient status: low phosphate (Pi) or nitrogen (N) availability enhances SL biosynthesis, promoting mycorrhizal symbiosis and root adaptations; (ii) hormonal crosstalk: auxins positively regulate SL biosynthesis by upregulating genes such as *D27*, *CCD7*, and *CCD8*; and (iii) environmental factors: drought, salinity, and light conditions can influence SL production.

After synthesis, SLs are transported from the roots to the shoots via the xylem to suppress shoot branching. Some SLs are exuded into the rhizosphere to stimulate seed germination of parasitic plants (e.g. *Striga*) and promote colonization by arbuscular mycorrhizal fungi (AMF) (Foo *et al.*, 2013; Yoneyama *et al.*, 2020; Marro *et al.*, 2021; Yuan *et al.*, 2023).

5.2.3 Signaling pathways of RNS and SLs

The signaling pathways of RNS and SLs play crucial roles in regulating plant growth, development, and stress responses (Tables 5.3 and 5.4). While they function through distinct mechanisms, both can interact and influence each other to coordinate physiological processes (Oláh *et al.*, 2020; Liu *et al.*, 2022).

5.3 Physiological Roles of Reactive Nitrogen Species and Strigolactones in Plants

The physiological roles of RNS (including NO) and SLs in plant signaling, seed germination, root development, stomatal closure, and stress responses are very critical. They are highly important for modulating gene expression, enzyme activity, and interacting with reactive oxygen species (ROS) during defense responses (Oláh *et al.*, 2020; Zhang *et al.*, 2023; Khan *et al.*, 2023). SLs act in shoot branching, root architecture, and mycorrhizal symbiosis (Tables 5.5 and 5.6), and facilitate responses against nutrient deficiencies, particularly phosphate. RNS and SLs, therefore, are both important factors in adaptation of the plant to environmental stress and in the promotion of plant growth and development through the regulation of critical physiological and molecular pathways (Boyno *et al.*, 2023; Wang *et al.*, 2024).

Table 5.3. Reactive nitrogen species (RNS) signaling pathways in plants. Data from Gupta *et al.* (2020); Hancock (2020); Yu *et al.* (2020); Chakraborty (2021); Borrowman *et al.* (2023).

Component	Role/function	Mechanism of action	Physiological impact
Nitric oxide (NO)	Central signaling molecule in RNS pathway	Modifies proteins via *S*-nitrosylation and interacts with ROS	Regulates growth, stress responses, and hormone signaling
S-Nitrosylation	Post-translational modification of proteins	Adds NO group to cysteine residues	Alters protein function, enzyme activity, and signal transduction
Tyrosine nitration	Modification of tyrosine residues in proteins	Nitration reduces enzyme activity or alters signaling proteins	Impacts defense responses and metabolic regulation
Reactive oxygen species (ROS)	Interact with RNS to form peroxynitrite ($ONOO^-$)	Combined oxidative/ nitrosative stress signaling	Coordinates redox homeostasis and stress adaptation
Phytohormone crosstalk	Interaction with auxin, ABA, gibberellins, and strigolactones	Modulates hormone biosynthesis, transport, and signaling	Fine-tunes developmental and stress responses
Transcription factors	Targeted by RNS through redox modifications or indirect signaling	Affect gene expression of defense and stress-responsive genes	Enhance tolerance to abiotic and biotic stresses
Calcium (Ca^{2+}) signaling	Interacts with NO signaling	NO modulates Ca^{2+}-dependent pathways and vice versa	Influences stomatal closure, root growth, and defense
Protein kinases	Activated downstream of NO signaling	Modulate phosphorylation cascades	Regulate cell signaling and adaptive responses

Table 5.4. Strigolactone (SL) signaling pathway in plants. Data from Yoneyama *et al.* (2013); Yao *et al.* (2017); Bürger and Chory (2020); Qu *et al.* (2020); Marro *et al.* (2021); Wu *et al.* (2022).

Component	Role/function	Mechanism of action	Physiological impact
SLs	Hormones regulating plant growth and environmental responses	Perceived by receptor D14; initiate signaling cascade	Control shoot branching, root development, and symbiosis
DWARF14 (D14)	SL receptor protein (α/β-hydrolase)	Binds and hydrolyzes SLs, and undergoes conformational change to activate MAX2	Initiates SL signal transduction
MAX2 (F-box protein)	Part of SCF E3 ubiquitin ligase complex	Interacts with activated D14, promotes degradation of repressors such as D53/SMXLs	Enables downstream gene expression
DWARF53 (D53)/SMXLs	Negative regulators of SL signaling	Targeted for ubiquitination and degradation via SCF^{MAX2} complex	Suppression of these repressors leads to gene activation
BRANCHED1 (BRC1)	Transcription factor regulated by SLs	Downregulated in active SL signaling to inhibit axillary bud growth	Suppresses shoot branching
Auxin (indole-3-acetic acid) crosstalk	Interacts with SL pathway	Auxin promotes SL biosynthesis; SLs regulate auxin transport	Coordinated control of apical dominance and root development
Root architecture modulation	SLs influence lateral root and root-hair formation	Signal through D14–MAX2 to alter expression of root development genes	Improved nutrient and water uptake
Stress and nutrient response	SL levels increase under phosphate and nitrogen deficiency	Enhances symbiotic signaling and root adaptation	Promotes arbuscular mycorrhizal fungal colonization and stress resilience

5.4 Reactive Nitrogen Species and Strigolactones in Plant Stress Responses

NO, the most representative member of RNS, is a crucial player in plant defense mechanisms, as it regulates antioxidant systems, expression of stress-responsive genes, and signaling pathways during abiotic and biotic stresses. SLs promote plant adaptation to stress through regulation of shoot and root architecture, thereby enhancing drought and salinity tolerance and improving nutrient uptake, especially under phosphate deficiency. The interplay between RNS and SLs is important, as RNS can also modulate SL biosynthesis and signaling, thereby fine-tuning stress responses. Thus, RNS and SLs converge to activate the plant's protective machinery for its growth and survival under unfavorable conditions (Trasoletti *et al.*, 2022; Korek and Marzec, 2023).

5.4.1 RNS in abiotic stress responses

RNS, particularly NO, play an essential role in signaling mediators in plant responses to a variety of abiotic stresses such as drought, salinity, extremes of temperature, heavy metals, and nutrient deficiencies. Although RNS can be toxic at elevated concentrations, in a controlled range

Table 5.5. Roles of reactive nitrogen species (RNS) in plant growth and development. Data from Mangal *et al.* (2023); Khator *et al.* (2024).

Plant process	Role of RNS
Seed germination	Promote dormancy release and enhance germination via signaling pathways
Root development	Regulate root elongation, lateral root formation, and root-hair growth
Shoot development	Modulate shoot meristem activity, and influence cell division and expansion
Stomatal movement	Mediate stomatal closure in response to stress by interacting with reactive oxygen species
Flowering	Influence floral transition and development through hormonal crosstalk
Senescence	Participate in age-related signaling and programmed cell death
Abiotic stress tolerance	Enhance resistance to drought, salinity, and heat by regulating stress genes
Biotic stress response	Act as a defense signal against pathogens, activating immune responses
Nitrogen assimilation	Regulate nitrate uptake and assimilation pathways
Signal transduction	Function as a secondary messenger in various hormonal and stress pathways

they are vital for signaling and defensive mechanisms (Alvi *et al.*, 2022; Khan *et al.*, 2024).

5.4.2 SLs in abiotic stress responses

SLs play an important regulatory role in the abiotic stress response. Under conditions of drought, salinity, and nutrient deficiency (particularly phosphate), SLs act to change the plant architecture by inhibiting shoot branching while promoting root growth for better water and nutrient absorption. They increase stress tolerance through the regulation of antioxidant enzymes, hormonal crosstalk (with ABA and auxin), and the expression of stress-responsive genes. SLs also enhance symbiotic relationships with AMF, thereby enhancing nutrient acquisition. SLs also act to maintain growth and productivity under harsh environmental conditions, thus playing a vital role as regulators in the adaptation of plants to abiotic stresses (Haider *et al.*, 2018; Mostofa *et al.*, 2018).

Table 5.6. Roles of strigolactones (SLs) in plant growth and development. Data from Nisa *et al.* (2023); Su *et al.* (2023); Fichtner *et al.* (2024); Visentin *et al.* (2024).

Plant process	Role of SLs
Shoot branching	Inhibit excessive shoot branching by suppressing axillary bud growth
Root architecture	Promote root-hair elongation and lateral root formation
Seed germination	Stimulate germination in parasitic plants (e.g. *Striga*)
Leaf senescence	Accelerate leaf aging under nutrient-deficient conditions
Mycorrhizal symbiosis	Induce hyphal branching in arbuscular mycorrhizal fungi
Nutrient uptake	Enhance phosphate and nitrogen acquisition via root–microbe interaction
Stress response	Improve tolerance to abiotic stresses such as drought and salinity
Stem elongation	Contribute to internode elongation and plant height regulation

5.4.3 RNS–SL interactions in stress adaptation

RNS, particularly NO, have been shown to have an important and interrelated role with SLs in determining stress adaptation in plants. Under different abiotic stresses such as drought, salinity, and nutrient deficiency, RNS serve as signaling molecules to trigger antioxidant defenses, gene expression, and interplay with other hormonal pathways. SLs, being plant hormones, are important for maintaining root architecture, restricting shoot branching, and optimizing water and nutrient uptake. There is evidence suggesting that RNS act as modulators or regulators of SL biosynthesis and signaling in stress responses mediated by SLs (López-Ráez *et al.*, 2015; Visentin *et al.*, 2016). Altogether, this interaction serves to promote plant resilience through joint physiological and molecular adjustments such as stomatal opening, redox homeostasis, and activation of stress-responsive genes. The interplay between RNS and SLs further calibrates the fine-tuning of plant responses to environmental agitation, promoting survival and growth in adverse conditions. An understanding of this interaction provides the basis upon which crop stress tolerance can be enhanced through breeding and biotechnological interventions (Liu *et al.*, 2015; Visentin *et al.*, 2020) (Tables 5.7 and 5.8).

5.5 Molecular Mechanisms of Reactive Nitrogen Species and Strigolactone Interaction

Signal transduction is part of the pool of complex interactions between RNS and SLs. This is because RNS, especially NO, can induce SL biosynthesis by influencing the expression of genes involved in SL biosynthesis (i.e. *D27*, *MAX1*, and *MAX3*). In addition, RNS modify SL signaling elements with redox-based post-translational adjustments, such as *S*-nitrosylation. SLs may diminish or increase RNS levels by exerting their regulatory effects on NO-generating enzymes. This interface refines the stress response, differential gene expression, and hormonal balancing during the abiotic stress response. Interactions between RNS and SLs also coordinate antioxidant defense, root development, and adaptive growth, thereby increasing environmental challenge resilience in plants (Oláh *et al.*, 2020; Liu *et al.*, 2022).

5.5.1 NO-mediated regulation of SL biosynthesis and signaling in plants

NO is an important member of RNS and primarily takes part in the regulation of

Table 5.7. Interaction between reactive nitrogen species (RNS) and strigolactones (SLs) in plant physiology.

Aspect	RNS	SLs	Interaction	Reference
Growth regulation	Regulate cell division, elongation, and differentiation	Inhibit shoot branching and promote root growth	RNS modulate SL biosynthesis and signaling, affecting overall growth	Sigalas *et al.* (2024)
Root development	Influence lateral and adventitious root formation	Regulate root-hair elongation and lateral root formation	Synergistic effect on improving the root system under stress	Antika *et al.* (2024)
Stress response	Involved in abiotic stress tolerance (e.g. drought, salinity)	Enhance stress tolerance by modulating hormonal pathways	RNS and SLs co-regulate stress-responsive genes and antioxidant activity	Visentin *et al.* (2024)
Nutrient uptake	Facilitate nitrate uptake and assimilation	Enhance phosphate acquisition via mycorrhizal association	RNS signaling can enhance SL-mediated nutrient signaling under deficiencies	Mansoor *et al.* (2024)
Hormonal crosstalk	Interact with auxin, abscisic acid (ABA), and ethylene	Interact with auxin, ABA, and cytokinin	RNS may influence SL activity through redox modifications of hormonal crosstalk	Marro *et al.* (2021)
Gene expression regulation	Modulate transcription factors and signaling proteins	Control expression of genes linked to development and stress	RNS can modify SL-responsive genes via *S*-nitrosylation and other redox changes	Li *et al.* (2021)

Table 5.8. Interaction between reactive nitrogen species (RNS) and strigolactones (SLs) in stress adaptation in plants. Data from Liu *et al.* (2022); Guo *et al.* (2023); Lu *et al.* (2024); Qi *et al.* (2024); Mansoor *et al.* (2024); Sigalas *et al.* (2024).

Stress type	Role of RNS	Role of SLs	RNS–SL interaction	Physiological outcome
Drought stress	Induce stomatal closure and enhance antioxidant activity	Promote deeper rooting and modulate stomatal conductance	RNS modulate SL biosynthesis; SLs enhance nitric oxide (NO) signaling for efficient stress response	Reduced water loss, improved water uptake, better drought tolerance
Salinity stress	Regulate ion transport (Na^+/K^+ balance) and reduce oxidative stress	Modify root architecture and enhance salt tolerance	RNS influence SL signaling; SLs enhance antioxidant gene expression via NO mediation	Improved ion homeostasis and enhanced salinity tolerance
Nutrient deficiency	Regulate nitrate signaling and assimilation	Enhance phosphate uptake via mycorrhizal interaction	NO promotes SL production under low phosphate conditions; SLs amplify NO signals for nutrient uptake	Improved nitrogen and phosphorus acquisition, optimized root structure
Heat stress	Activate heat-shock proteins and antioxidants	Regulate heat-responsive genes and hormonal crosstalk	RNS and SLs synergistically regulate expression of stress genes	Increased thermotolerance and cellular stability
Cold stress	Modulate cold-responsive pathways and enhance reactive oxygen species (ROS) scavenging	Regulate expression of cold-tolerance genes	SL signaling influenced by NO-mediated redox regulation	Improved membrane integrity and cold acclimation
Oxidative stress	Direct scavenging of ROS, and upregulate antioxidant enzymes	Indirectly enhance antioxidant responses	RNS and SLs cooperatively enhance antioxidant systems via signaling pathways	Minimized cellular damage from ROS
Hormonal crosstalk	Interact with abscisic acid (ABA), auxin, and ethylene in stress signaling	Coordinate with ABA, auxin, and cytokinins in stress response	RNS modulate SL function via redox changes; SLs amplify or modulate RNS signaling	Fine-tuned stress signaling and adaptive development
Gene regulation	Regulate transcription factors (e.g. WRKY, MYB); protein nitrosylation	Control expression of SL-responsive genes	NO modifies SL-associated transcriptional activity (e.g. via *S*-nitrosylation)	Coordinated activation of stress-tolerance genes

growth, development, and stress responses in plants. Recent evidence indicates the participation of NO in regulating SL biosynthesis and signaling in many molecular mechanisms, especially during stress (Oláh *et al.*, 2020, 2021; Liu *et al.*, 2022).

Regulation of SL biosynthesis by NO

GENE EXPRESSION MODULATION. NO influences the expression of key genes involved in SL biosynthesis, such as: (i) *D27*, involved in isomerization of all-*trans*-β-carotene; (ii) *CCD7* (*MAX3*) and *CCD8* (*MAX4*), involved in cleavage of carotenoids; and (iii) *MAX1*, a cytochrome P450 enzyme that converts intermediates to active SLs. NO can up- or down-regulate these genes depending on developmental cues or environmental stress (Oláh *et al.*, 2020, 2021).

TRANSCRIPTION FACTOR ACTIVATION. NO can activate transcription factors (e.g. MYB, WRKY) that bind to SL biosynthetic gene promoters, enhancing or repressing their transcription during nutrient deficiency or abiotic stress (Pande *et al.*, 2022).

REDOX REGULATION OF ENZYME ACTIVITY. NO can post-translationally modify biosynthetic enzymes through *S*-nitrosylation or tyrosine nitration, altering their stability or activity and affecting SL levels (Astier and Lindermayr, 2012).

Regulation of SL signaling by NO

RECEPTOR AND SIGNAL TRANSDUCTION MODULATION. SL signaling involves perception by the D14 receptor, interaction with MAX2 (an F-box protein), and degradation of repressors such as D53. NO may modulate: (i) protein–protein interactions within this pathway; (ii) the stability of signaling components via *S*-nitrosylation; and (iii) expression of *MAX2* and *D14*, thus altering sensitivity to SLs (Astier and Lindermayr, 2012; Lau *et al.*, 2021; Korek and Marzec, 2023; Wang *et al.*, 2024).

HORMONAL CROSSTALK AND NO INTEGRATION. NO influences the balance between SLs and other hormones such as auxin and ABA. For example, NO can enhance auxin transport, which feeds back into SL biosynthesis regulation, and during drought, NO and SLs work together to modulate ABA levels and promote stomatal closure (Mostofa *et al.*, 2018; Lau *et al.*, 2021; Wang *et al.*, 2024).

Stress-responsive integration

Under nutrient deficiency, particularly phosphate starvation, NO levels rise, which in turn boosts SL biosynthesis to promote mycorrhizal interactions for phosphate uptake. Under abiotic stress (e.g. drought, salinity), NO enhances SL signaling to regulate root architecture and stress-responsive gene expression. Together, they help maintain redox balance and developmental plasticity (Marzec and Melzer, 2018; Mostofa *et al.*, 2018; Oláh *et al.*, 2020).

5.5.2 SL-mediated regulation of RNS production

SLs are classes of plant hormones that control plant architecture and symbiotic interactions and also participate in plant responses to abiotic stresses. SLs have recently been shown to fine-tune RNS, predominantly NO, which is a signaling molecule important for plant growth and development and stress adaptation. Regulation of the production of RNS by SLs involves modulation of NO-generating enzymes, gene expression, and redox homeostasis (Oláh *et al.*, 2020, 2021; Liu *et al.*, 2022).

Regulation of NO biosynthesis enzymes by SLs

SLs can have an influence on the expression and activity of nitrate reductase, which is an enzyme important in regulating NO biosynthesis via the nitrate-to-nitrite pathway. Nitrate reductase activity, and thus NO production, is upregulated by SL treatment or SL-inducing conditions (e.g. phosphate starvation). Although very little is known about the NOS-like enzymes in plants, SLs may possibly exert indirect influence on NOS-like activity by modifying calcium signal transduction and/or redox states, which invariably affects these pathways (Galatro *et al.*, 2020; Oláh *et al.*, 2020; Marro *et al.*, 2021; Liu *et al.*, 2022).

SL-induced NO accumulation under stress

SL application or overexpression of SL biosynthetic genes has been shown to increase endogenous NO levels in roots and shoots, particularly under conditions of drought, salinity, and utrient deficiency (e.g. phosphate). This increase in NO contributes to stress tolerance by regulating antioxidant systems, ion homeostasis, and root growth patterns (Oláh *et al.*, 2020; Liu *et al.*, 2022).

Transcriptional control of NO-related genes

SLs modulate the expression of genes related to NO biosynthesis (e.g. nitrate reductase genes) and NO signaling components, such as *S*-nitrosoglutathione reductase and non-symbiotic hemoglobins, which regulate NO levels. Through transcriptional control, SLs help maintain optimal NO concentrations, preventing toxic over-accumulation while ensuring effective signaling (Oláh *et al.*, 2020; Liu *et al.*, 2022; Sigalas *et al.*, 2024).

SL–NO crosstalk in stress signaling

SLs increase the activities of antioxidant enzymes (e.g. SOD, CAT, and APX), some through the pathway of NO signaling, and ultimately activated due to SL-induced NO accumulation. In conjunction with ABA and NO signaling pathways, SLs also assist in mediating stomatal closure by NO, especially during drought for conserving water. SL-induced rises in NO cause the production and elongation of lateral roots to enhance resource acquisition under stress (Niu *et al.*, 2017; Oláh *et al.*, 2020; Omoarelojie *et al.*, 2020).

Hormonal crosstalk involving RNS and SLs

SLs interact with other phytohormones (i.e. auxin, ABA, ethylene), which are known to regulate or be regulated by RNS. SLs enhance auxin flow in roots, and auxin is a known regulator of NO production. SL-mediated ABA responses, such as stomatal control, involve NO as a downstream signal (Niu *et al.*, 2017; Oláh *et al.*, 2020).

5.5.3 Transcriptional and post-translational regulation

Transcriptional regulation in plants

Transcriptional regulation is the control of gene expression at the DNA to mRNA level. It determines when, where, and how much of a gene is transcribed, as described in Table 5.9.

Post-transcriptional regulation in plants

Post-transcriptional regulation in plants involves control of gene expression after mRNA synthesis but before translation. The various aspects included in this type of regulation are mRNA splicing, editing, transport, stability, and translation efficiency. The most prominent members of these regulatory molecules are microRNAs (miRNAs) and small interfering RNAs (siRNAs), which mainly guide mRNA degradation or inhibition of translation. RNA-binding proteins influence the disposition of the mRNA (Bajczyk *et al.*, 2023; Zhan and Meyers, 2023). This is a mechanism whereby rapid, dynamic developmental cues or environmental stresses can control the amount of a particular protein synthesized (Table 5.10). Significantly, it is crucial for the responses of flowering, hormone signals, and stress adaptations (Szweykowska-Kulinska and Jarmolowski, 2018; Wu *et al.*, 2020).

5.6 Ecological and Agricultural Implications

RNS and SL are linked in terms of two very significant aspects, one ecological and the other agricultural. For example, ecologically, crosstalk co-action between RNS and SLs improves stress adaptability, allowing plants to survive in conditions of scarcity, salinity, and lack of nutrients; this then comes in handy for ecosystem stability. For agriculture, RNS–SL co-action increases root architecture, uptake of nutrients, and stress endurance allowing reduced fertilizers and irrigation input. It also affects yield and resilience under climate-changing conditions (Soliman *et al.*, 2022; Quaggiotti *et al.*, 2024). In addition, SL-level management could be important with respect to parasite weed germination and sustainable weed management. Thus, exploiting them together has shown great promise for low-budget options in developing steps for green-crop management and yield improvement in sustainable agriculture (Sun *et al.*, 2021; Liu *et al.*, 2022; Soliman *et al.*, 2022).

5.6.1 RNS and SLs in plant–microbe interactions

Plant–microbe interactions impact nutrient intake, stress tolerance, and promotion of growth. RNS and SLs are one group of signaling agents that take part in these interactions. RNS and SLs regulate the interactions either into

Table 5.9. Mechanisms of transcriptional regulation in plants.

Regulatory component	Description	Examples	Role in plant physiology	Reference
Transcription factors (TFs)	Proteins that bind specific DNA sequences to activate or repress gene transcription	MYB, bZIP, WRKY, NAC, AP2/ERF	Control growth, development, hormone signaling, and stress responses	Romani and Moreno (2021)
cis-Regulatory elements	Specific DNA sequences in promoters where TFs bind	TATA box, CAAT box, ABRE, DRE, AuxRE	Determine spatial and temporal expression of genes	Lloyd and Lister (2022)
Epigenetic modifications	Heritable changes in gene expression without altering the DNA sequence	DNA methylation, histone acetylation/ methylation	Regulate gene silencing or activation; chromatin remodeling	Mendieta *et al.* (2023)
Chromatin remodeling complexes	Protein complexes that reposition or modify nucleosomes	SWI/SNF, CHD	Control accessibility of DNA to transcription machinery	Saeed *et al.* (2022)
Non-coding RNAs	Small RNAs that regulate gene expression post-transcriptionally or transcriptionally	microRNAs, small interfering RNAs, long non-coding RNAs	Silence or fine-tune expression of target genes	Tao *et al.* (2023)
Environmental signals	External stimuli influencing transcriptional regulation	Light, drought, salinity, temperature	Trigger stress-responsive TFs and gene activation	Saeed *et al.* (2022)
Hormonal regulation	Hormones modulating TF activity or gene expression	Abscisic acid, auxin, gibberellin	Mediate developmental processes and adaptive responses to stress	Zemlyanskaya *et al.* (2021)
Signal transduction cascades	Pathways that link receptor activation to transcriptional response	Mitogen-activated protein kinase (MAPK) cascade, calcium signaling	Activate TFs through phosphorylation or secondary messengers	Xie and Duan (2023)

beneficial symbiosis or effective antagonisms against pathogens, as described in Table 5.11.

5.6.2 RNS and SLs in rhizosphere signaling

The rhizosphere is the microenvironment surrounding practically all plant roots; the bulk of space in soil is occupied by microorganisms such as bacteria and fungi, and is dominated by chemical signaling processes. Plant roots release different endogenous chemical signals into the rhizosphere soil, which are specific signaling molecules and play significant roles in interacting with soil microbes (Table 5.12) (Mozes and Meijler, 2020; Kee *et al.*, 2023). For instance, RNS and SLs are the major players in the pattern formation of the relationship between plant roots and their microbial symbionts by regulating nutrient exchanges and specifying structures directed toward roots (Waters *et al.*, 2017; Aquino *et al.*, 2021; Kumar *et al.*, 2022).

5.6.3 Potential applications in agriculture

The interplay of RNS, primarily NO, and SLs constitutes a potent regulatory network to stimulate growth, nutrient efficiency, and stress tolerance in plants. Understanding and/or applying this crosstalk in agriculture can stimulate ways for sustainable crop production, reduced input of

Table 5.10. Mechanism of post-transcriptional regulation in plants. Data from Wang *et al.* (2020); Wu *et al.* (2020); Hao *et al.* (2021); Yu *et al.* (2021); Kong *et al.* (2022).

Mechanism	Description	Key components/examples	Biological significance
mRNA splicing	Removal of introns and joining of exons to form mature mRNA	Spliceosome, SR proteins	Generates mRNA diversity (alternative splicing)
mRNA editing	Alteration of nucleotide sequences after transcription	Cytidine-to-uridine editing in chloroplast/mitochondria	Corrects coding errors, regulates organellar gene expression
mRNA transport	Movement of processed mRNA from nucleus to cytoplasm	Nuclear-pore complexes, RNA-binding proteins	Ensures proper spatial and temporal protein synthesis
mRNA stability/ degradation	Regulation of mRNA lifespan and decay pathways	Exonucleases, deadenylation complexes	Controls transcript availability and prevents unwanted gene expression
RNA silencing (miRNA/ siRNA)	Small RNAs guide mRNA cleavage or repress translation	miRNAs, siRNAs, RISC	Fine-tunes gene expression, crucial for development and stress response
Translational control	Regulation of initiation or efficiency of translation	Upstream open reading frames, initiation factors, miRNAs	Rapid protein production adjustment under environmental changes
RNA-binding proteins	Proteins that bind to mRNAs to influence their processing or translation	GRPs, PUF proteins	Affect mRNA splicing, localization, translation, and stability

GRP, glycine-rich RNA-binding protein; miRNA, microRNA; siRNA, small interfering RNA; RISC, RNA-induced silencing complex

agrochemicals, and climate-resilient farming systems, as described in Table 5.13.

5.7 Future Perspectives

The complex interactions between RNS and SLs are known to play a significant role in regulating plant development, responses to stress, and plant–microbe interaction. The increasing challenges of climate change, soil degradation, nutrient depletion, and population increase have continued forcing agriculture to look into creating opportunities for sustainable crop production and resource-efficient agriculture through exploring and manipulating RNS–SL networks.

5.7.1 Crop improvement for stress tolerance

Future research on RNS–SL interactions may pave the way for developing climate-resilient crops. In terms of drought and salinity tolerance, joint regulation of antioxidant systems by RNS and SLs enhances tolerance to oxidative and osmotic stress. Future breeding and biotechnological strategies could focus on manipulating SL biosynthesis and NO signaling pathways to improve stress endurance.

For increased nutrient-use efficiency, RNS–SL interactions can be harnessed to optimize root architecture and microbial symbiosis, especially under nitrogen and phosphate deficiency.

Table 5.11. Role of reactive nitrogen species (RNS) and strigolactones (SLs) in plant–microbe interactions. Data from Aliche *et al.* (2020); Mozes and Meijler (2020); Aquino *et al.* (2021); Prandi *et al.* (2021); Kumar *et al.* (2022).

Aspect	RNS	SLs
Role in symbiosis	Involved in early signaling of legume–rhizobium and arbuscular mycorrhizal fungi (AMF) interactions	Induce hyphal branching in AMF; promote colonization under low phosphate conditions
Key microbial partners	*Rhizobium*, AMF, endophytic bacteria	AMF, rhizobacteria, parasitic plants (*Striga*, *Orobanche*)
Mechanism of action	Nitric oxide (NO) acts as a signaling molecule, modulating gene expression and redox state	Exuded into rhizosphere to attract beneficial microbes
Effect on root colonization	Facilitate nodule and arbuscule formation	Enhance AMF colonization and root–microbe signaling
Defense regulation	Trigger hypersensitive response, reactive oxygen species burst, and pathogenesis-related gene activation	Modulate defense pathways; role in pathogen resistance under stress
Gene regulation	Modulate nodulation and defense-related gene expression	Regulate SL biosynthetic and signaling genes (*MAX*, *D27*)
Crosstalk	Regulate SL biosynthesis; interact with hormone pathways (auxin, abscisic acid)	Influence NO production and redox homeostasis
Environmental trigger	Activated by biotic stress, nutrient status, and microbial presence	Secreted in response to phosphate deficiency and microbial cues
Agricultural significance	Improve nitrogen fixation, stress tolerance, and plant defense	Enhance nutrient uptake, support mycorrhization, and control weeds

Engineering crops with enhanced SL exudation and RNS responsiveness could reduce fertilizer dependency (Oláh *et al.*, 2020).

5.7.2 Sustainable rhizosphere management

Understanding the RNS–SL network offers potential for rhizosphere engineering.

Microbiome modulation

Future applications may involve targeted manipulation of SLs and NO levels to attract beneficial microbes (e.g. plant-growth-promoting rhizobacteria, AMF) and suppress pathogens, while precision agriculture could use biosensors to monitor and adjust RNS/SL levels in real time.

Biocontrol of parasitic weeds

Modifying SL profiles or disrupting SL perception pathways may help in developing *Striga*-resistant varieties, while combined use of SL analogs and RNS-regulating compounds could offer integrated weed management solutions (Chesterfield *et al.*, 2020; Thiruvengadam *et al.*, 2024).

5.7.3 Genetic and molecular tools

Advanced genomic and gene-editing technologies such as CRISPR/Cas (clustered regularly interspaced short palindromic repeats/CRISPR-associated) can be used to: (i) modify key RNS/SL biosynthesis and signaling genes (e.g. *CCD7*, *D27*, *MAX2*, nitrate reductase); (ii) develop

Table 5.12. Role of reactive nitrogen species (RNS) and strigolactones (SLs) in rhizosphere signaling. Data from Arellano-Saab *et al.* (2023); Gu *et al.* (2023); Kee *et al.* (2023); Puglia (2024).

Aspect	RNS	SLs
Source	Produced in roots in response to symbiotic signals and stress	Secreted from roots, especially under nutrient (phosphate/ nitrogen) deficiency
Primary signaling molecule	Nitric oxide (NO)	E.g. 5-deoxystrigol
Role in microbe attraction	Modulate signaling pathways that facilitate microbial colonization	Stimulate hyphal branching in arbuscular mycorrhizal fungal (AMF); attracts beneficial microbes
Impact on symbiosis	Involved in *Rhizobium*-induced nodule formation and AMF colonization	Promote AMF colonization under phosphate starvation
Regulation of root exudates	Influence secretion patterns of organic acids and flavonoids	Act as a major exudate modifying microbial behavior
Effect on root architecture	Work with hormones to shape root-hair growth and lateral roots	Inhibit lateral root formation; promote primary root elongation
Interaction with microbiome	Shape microbial community through redox signaling	Influence microbial diversity and abundance in the rhizosphere
Role in pathogen response	Activate defense genes and reactive oxygen species burst; contributes to hypersensitive response	Modulate defense indirectly; can also stimulate parasitic weeds
Crosstalk with each other	Regulate SL biosynthesis genes (e.g. *D27*, *MAX1*)	Modulate NO production via nitrate reductase regulation
Agricultural significance	Enhance nodulation, stress resistance, and microbial recruitment	Boost mycorrhization and nutrient uptake, and support weed management

tissue-specific or stress-inducible promoters to fine-tune signaling in roots, shoots, or reproductive tissues; and (iii) elucidate novel regulators and transcription factors governing RNS–SL crosstalk (Jia *et al.*, 2019).

5.7.4 Biostimulant and agronomic innovations

Bioformulations based on SL analogs or NO donors may be developed to stimulate root growth, enhance nutrient uptake, and improve stress adaptation, while integration of RNS/SL-based compounds with organic farming and low-input systems could revolutionize sustainable crop practices. Foliar or root-applied treatments using biostimulants that activate SL or RNS pathways may serve as cost-effective tools for smallholder farmers.

5.7.5 Integrated systems approach

Future research should adopt systems biology, combining transcriptomics, metabolomics, and phenotyping to understand RNS–SL interactions under field conditions. Artificial intelligence and machine learning could help model how RNS–SL networks behave under varying environments, informing precision agriculture decisions (Sun *et al.*, 2023).

5.8 Conclusion

RNS and SLs establish a complex interaction in plant biology responsible for many physiological, developmental, and environmental responses. RNS such as NO act as versatile cellular signals, while SLs rank among the foremost regulatory

Table 5.13. Applications of reactive nitrogen species (RNS) and strigolactone (SL) interaction in agriculture. Data from Oláh *et al.* (2021); Kyozuka *et al.* (2022); Boyno *et al.* (2023); Gu *et al.* (2023); Kee *et al.* (2023); Priya *et al.* (2024).

Application area	Role of RNS–SL interaction	Agricultural benefits
Nutrient-use efficiency	Enhances root development and microbial symbioses (e.g. arbuscular mycorrhizal fungi, rhizobia)	Improved nitrogen and phosphorus uptake; reduced fertilizer dependence
Drought and salinity tolerance	Coordinates stress-responsive gene expression and antioxidant defense	Increased plant resilience under abiotic stress
Root architecture optimization	Modulates lateral root formation, root elongation, and root-hair development	Better soil exploration and resource acquisition
Symbiotic interaction enhancement	Promotes effective nodulation and arbuscule formation via combined signaling	Enhanced biological nitrogen fixation and mycorrhizal colonization
Microbiome engineering	Shapes rhizosphere microbial communities through root exudation and redox signaling	Supports beneficial microbes and suppresses pathogens
Weed management	SL levels can be modulated to prevent germination of parasitic weeds such as *Striga*	Environmentally friendly control of parasitic plants
Plant growth regulation	Integrated signaling affects hormonal balance and developmental pathways	Balanced growth under suboptimal conditions
Sustainable agriculture	Reduces need for agrochemicals through improved natural nutrient cycling and defense	Cost-effective, ecofriendly farming practices
Crop yield improvement	Enhances stress tolerance, nutrient uptake, and symbiosis-driven growth	Higher productivity under stress-prone or marginal soils
Biostimulant development	Insights into RNS–SL pathways help in designing bio-based growth enhancers	Development of targeted, natural growth-promoting products

factors governing the architecture of plants and their activity in rhizosphere communication. Their *trans*-actions are significantly responsible for root development, absorptive capacity, symbiosis with microorganisms, and adaptation to both biotic and abiotic strains. On the one hand, RNS can modulate SL biosynthesis and signaling pathways, while on the other, SLs may affect RNS in the same manner. Such effects form a crosstalk of hormone and redox signaling between the two.

Molecular mechanisms involved in the modulation of RNS–SL interactions could yield enormous potential benefits while improving plant performance in resource-limited, stress-prone environments. Future research, relying on the combination of genetic, biochemical, and agronomic schemes, will engage this interaction toward sustainable agricultural practice generation, disaster-relocated crop rehabilitation in various countries, and optimization of functions of the rhizosphere. The ultimate aim of decoding the nexus between RNS and SLs will be to increase crop productivity toward an ideal environment against global agricultural challenges.

References

Al-Babili, S. (2024) A dirigent of the ring for strigolactone stereochemistry. *Proceedings of the National Academy of Sciences USA* 121(34), e2410953121. DOI: 10.1073/pnas.2410953121.

Aliche, E.B., Screpanti, C., de Mesmaeker, A., Munnik, T. and Bouwmeester, H.J. (2020) Science and application of strigolactones. *New Phytologist* 227(4), 1001–1011. DOI: 10.1111/nph.16489.

Alvi, A.F., Sehar, Z., Fatma, M., Masood, A. and Khan, N.A. (2022) Strigolactone: An emerging growth regulator for developing resilience in plants. *Plants* 11(19), 2604. DOI: 10.3390/plants11192604.

Antika, G., Cinar, Z.Ö., Dönmez, S., Seçen, E., Özbil, M. *et al.* (2024) Effects of strigolactones on NLRP3 activation, nitrosative stress, and antioxidant mox phenotype: *In vitro* and in silico evidence. *ACS Bio & Med Chem Au* 4(3), 131–136. DOI: 10.1021/acsbiomedchemau.3c00063.

Aquino, B., Bradley, J.M. and Lumba, S. (2021) On the outside looking in: Roles of endogenous and exogenous strigolactones. *Plant Journal* 105(2), 322–334. DOI: 10.1111/tpj.15087.

Arasimowicz, M. and Floryszak-Wieczorek, J. (2007) Nitric oxide as a bioactive signalling molecule in plant stress responses. *Plant Science* 172(5), 876–887. DOI: 10.1016/j.plantsci.2007.02.005.

Arellano-Saab, A., Skarina, T., Xu, Z., McErlean, C.S.P., Savchenko, A. *et al.* (2023) Structural analysis of a hormone-bound striga strigolactone receptor. *Nature Plants* 9(6), 883–888. DOI: 10.1038/s41477-023-01423-y.

Astier, J. and Lindermayr, C. (2012) Nitric oxide-dependent posttranslational modification in plants: An update. *International Journal of Molecular Sciences* 13(11), 15193–15208. DOI: 10.3390/ijms131115193.

Bajczyk, M., Jarmolowski, A., Jozwiak, M., Pacak, A., Pietrykowska, H. *et al.* (2023) Recent insights into plant mirna biogenesis: Multiple layers of mirna level regulation. *Plants* 12(2), 342. DOI: 10.3390/plants12020342.

Berger, A., Boscari, A., Horta Araújo, N., Maucourt, M., Hanchi, M. *et al.* (2020) Plant nitrate reductases regulate nitric oxide production and nitrogen-fixing metabolism during the *Medicago truncatula–Sinorhizobium meliloti* symbiosis. *Frontiers in Plant Science* 11, 1313. DOI: 10.3389/fpls.2020.01313.

Borrowman, S., Kapuganti, J.G. and Loake, G.J. (2023) Expanding roles for s-nitrosylation in the regulation of plant immunity. *Free Radical Biology & Medicine* 194, 357–368. DOI: 10.1016/j.freeradbiomed.2022.12.009.

Boyno, G., Rezaee Danesh, Y., Demir, S., Teniz, N., Mulet, J.M. *et al.* (2023) The complex interplay between arbuscular mycorrhizal fungi and strigolactone: Mechanisms, sinergies, applications and future directions. *International Journal of Molecular Sciences* 24(23), 16774. DOI: 10.3390/ijms242316774.

Bürger, M. and Chory, J. (2020) In-silico analysis of the strigolactone ligand-receptor system. *Plant Direct* 4(9), e00263. DOI: 10.1002/pld3.263.

Chakraborty, N. (2021) Salicylic acid and nitric oxide cross-talks to improve innate immunity and plant vigor in tomato against fusarium oxysporum stress. *Plant Cell Reports* 40(8), 1415–1427. DOI: 10.1007/s00299-021-02729-x.

Chen, G.T.E., Wang, J.Y., Jamil, M., Braguy, J. and Al-Babili, S. (2022) 9-cis-β-apo-10′-carotenal is the precursor of strigolactones in planta. *Planta* 256(5), 88. DOI: 10.1007/s00425-022-03999-9.

Chesterfield, R.J., Vickers, C.E. and Beveridge, C.A. (2020) Translation of strigolactones from plant hormone to agriculture: Achievements, future perspectives, and challenges. *Trends in Plant Science* 25(11), 1087–1106. DOI: 10.1016/j.tplants.2020.06.005.

Ciacka, K., Krasuska, U., Staszek, P., Wal, A., Zak, J. *et al.* (2020) Effect of nitrogen reactive compounds on aging in seed. *Frontiers in Plant Science* 11, 1011. DOI: 10.3389/fpls.2020.01011.

de Saint Germain, A., Clavé, G., Badet-Denisot, M.-A., Pillot, J.-P., Cornu, D. *et al.* (2016) An histidine covalent receptor and butenolide complex mediates strigolactone perception. *Nature Chemical Biology* 12(10), 787–794. DOI: 10.1038/nchembio.2147.

Fichtner, F., Humphreys, J.L., Barbier, F.F., Feil, R., Westhoff, P. *et al.* (2024) Strigolactone signalling inhibits trehalose 6-phosphate signalling independently of BRC1 to suppress shoot branching. *New Phytologist* 244(3), 900–913. DOI: 10.1111/nph.20072.

Foo, E., Yoneyama, K., Hugill, C.J., Quittenden, L.J. and Reid, J.B. (2013) Strigolactones and the regulation of pea symbioses in response to nitrate and phosphate deficiency. *Molecular Plant* 6(1), 76–87. DOI: 10.1093/mp/sss115.

Galatro, A., Ramos-Artuso, F., Luquet, M., Buet, A. and Simontacchi, M. (2020) An update on nitric oxide production and role under phosphorus scarcity in plants. *Frontiers in Plant Science* 11, 413. DOI: 10.3389/fpls.2020.00413.

Gu, P., Tao, W., Tao, J., Sun, H., Hu, R. *et al.* (2023) The D14-SDEL1-SPX4 cascade integrates the strigolactone and phosphate signalling networks in rice. *New Phytologist* 239(2), 673–686. DOI: 10.1111/nph.18963.

Guo, S., Wei, X., Ma, B., Ma, Y., Yu, Z. *et al.* (2023) Foliar application of strigolactones improves the desiccation tolerance, grain yield and water use efficiency in dryland wheat through modulation of non-hydraulic root signals and antioxidant defense. *Stress Biology* 3(1), 54. DOI: 10.1007/s44154-023-00127-9.

Gupta, K.J., Kolbert, Z., Durner, J., Lindermayr, C., Corpas, F.J. *et al.* (2020) Regulating the regulator: Nitric oxide control of post-translational modifications. *New Phytologist* 227(5), 1319–1325. DOI: 10.1111/nph.16622.

Haider, I., Andreo-Jimenez, B., Bruno, M., Bimbo, A., Floková, K. *et al.* (2018) The interaction of strigolactones with abscisic acid during the drought response in rice. *Journal of Experimental Botany* 69(9), 2403–2414. DOI: 10.1093/jxb/ery089.

Hancock, J.T. (2020) Nitric oxide signaling in plants. *Plants* 9(11), 1550. DOI: 10.3390/plants9111550.

Hao, W., Liu, G., Wang, W., Shen, W., Zhao, Y. *et al.* (2021) RNA editing and its roles in plant organelles. *Frontiers in Genetics* 12, 757109. DOI: 10.3389/fgene.2021.757109.

Homma, M., Wakabayashi, T., Moriwaki, Y., Shiotani, N., Shigeta, T. *et al.* (2024) Insights into stereoselective ring formation in canonical strigolactone: Identification of a dirigent domain-containing enzyme catalyzing orobanchol synthesis. *Proceedings of the National Academy of Sciences USA* 121(26), e2313683121. DOI: 10.1073/pnas.2313683121.

Ito, S., Yamagami, D., Umehara, M., Hanada, A., Yoshida, S. *et al.* (2017) Regulation of strigolactone biosynthesis by gibberellin signaling. *Plant Physiology* 174(2), 1250–1259. DOI: 10.1104/pp.17.00301.

Jia, Z., Giehl, R.F.H., Meyer, R.C., Altmann, T. and von Wirén, N. (2019) Natural variation of BSK3 tunes brassinosteroid signaling to regulate root foraging under low nitrogen. *Nature Communications* 10(1), 2378. DOI: 10.1038/s41467-019-10331-9.

Kapoor, R.T., Alam, P., Chen, Y. and Ahmad, P. (2024) Strigolactones in plants: From development to abiotic stress management. *Journal of Plant Growth Regulation* 43(3), 903–919. DOI: 10.1007/s00344-023-11148-z.

Kee, Y.J., Ogawa, S., Ichihashi, Y., Shirasu, K. and Yoshida, S. (2023) Strigolactones in rhizosphere communication: Multiple molecules with diverse functions. *Plant & Cell Physiology* 64(9), 955–966. DOI: 10.1093/pcp/pcad055.

Khan, M., Ali, S., Al Azzawi, T.N.I. and Yun, B.W. (2023) Nitric oxide acts as a key signaling molecule in plant development under stressful conditions. *International Journal of Molecular Sciences* 24(5), 4782. DOI: 10.3390/ijms24054782.

Khan, M.K., Pandey, A., Hamurcu, M., Vyhnánek, T., Zargar, S.M. *et al.* (2024) Exploring strigolactones for inducing abiotic stress tolerance in plants. *Czech Journal of Genetics and Plant Breeding* 60(2), 55–69. DOI: 10.17221/88/2023-CJGPB.

Khator, K., Parihar, S., Jasik, J. and Shekhawat, G.S. (2024) Nitric oxide in plants: An insight on redox activity and responses toward abiotic stress signaling. *Plant Signaling & Behavior* 19(1), 2298053. DOI: 10.1080/15592324.2023.2298053.

Kong, X., Yang, M., Le, B.H., He, W. and Hou, Y. (2022) The master role of siRNAs in plant immunity. *Molecular Plant Pathology* 23(10), 1565–1574. DOI: 10.1111/mpp.13250.

Korek, M. and Marzec, M. (2023) Strigolactones and abscisic acid interactions affect plant development and response to abiotic stresses. *BMC Plant Biology* 23(1), 314. DOI: 10.1186/s12870-023-04332-6.

Kumar, S., Joshi, A. and Shukla, R.K. (2022) Strigolactone signalling and plant-microbe communications. In: Horwitz, B.A. and Mukherjee, P.K. (eds) *Microbial Cross-Talk in the Rhizosphere*. Springer Nature, Singapore, pp. 25–57. DOI: 10.1007/978-981-16-9507-0_3.

Kyozuka, J., Nomura, T. and Shimamura, M. (2022) Origins and evolution of the dual functions of strigolactones as rhizosphere signaling molecules and plant hormones. *Current Opinion in Plant Biology* 65, 102154. DOI: 10.1016/j.pbi.2021.102154.

Lau, S.E., Hamdan, M.F., Pua, T.L., Saidi, N.B. and Tan, B.C. (2021) Plant nitric oxide signaling under drought stress. *Plants* 10(2), 360. DOI: 10.3390/plants10020360.

León, J. and Costa-Broseta, Á. (2020) Present knowledge and controversies, deficiencies, and misconceptions on nitric oxide synthesis, sensing, and signaling in plants. *Plant, Cell & Environment* 43(1), 1–15. DOI: 10.1111/pce.13617.

Li, S., Li, Y., Chen, L., Zhang, C., Wang, F. *et al.* (2021) Strigolactone mimic 2-nitrodebranone is highly active in *Arabidopsis* growth and development. *Plant Journal* 107(1), 67–76. DOI: 10.1111/tpj.15274.

Liu, H., Li, C., Yan, M., Zhao, Z., Huang, P. *et al.* (2022) Strigolactone is involved in nitric oxide-enhanced the salt resistance in tomato seedlings. *Journal of Plant Research* 135(2), 337–350. DOI: 10.1007/s10265-022-01371-2.

Liu, J., He, H., Vitali, M., Visentin, I., Charnikhova, T. *et al.* (2015) Osmotic stress represses strigolactone biosynthesis in *Lotus japonicus* roots: Exploring the interaction between strigolactones and ABA under abiotic stress. *Planta* 241(6), 1435–1451. DOI: 10.1007/s00425-015-2266-8.

Lloyd, J.P.B. and Lister, R. (2022) Epigenome plasticity in plants. *Nature Reviews. Genetics* 23(1), 55–68. DOI: 10.1038/s41576-021-00407-y.

López-Ráez, J.A., Fernández, I., García, J.M., Berrio, E., Bonfante, P. *et al.* (2015) Differential spatio-temporal expression of carotenoid cleavage dioxygenases regulates apocarotenoid fluxes during AM symbiosis. *Plant Science* 230, 59–69. DOI: 10.1111/grs.12417.

Lu, J., Shen, Y., Yang, Y., Li, X. and Li, X.E. (2024) Synthetic strigolactone (rac-GR24) alleviates the photosynthetic inhibition and oxidative damage in alfalfa (*Medicago sativa* L.) under salt stress. *Grassland Science* 70(1), 23–34. DOI: 10.1111/grs.12417.

Mangal, V., Lal, M.K., Tiwari, R.K., Altaf, M.A., Sood, S. *et al.* (2023) Molecular insights into the role of reactive oxygen, nitrogen and sulphur species in conferring salinity stress tolerance in plants. *Journal of Plant Growth Regulation* 42(2), 554–574. DOI: 10.1007/s00344-022-10591-8.

Mansoor, S., Mir, M.A., Karunathilake, E.M.B.M., Rasool, A., Ştefănescu, D.M. *et al.* (2024) Strigolactones as promising biomolecule for oxidative stress management: A comprehensive review. *Plant Physiology and Biochemistry* 206, 108282. DOI: 10.1016/j.plaphy.2023.108282.

Marro, N., Caccia, M. and López-Ráez, J.A. (2021) Are strigolactones a key in plant–parasitic nematodes interactions? An intriguing question. *Plant and Soil* 462(1–2), 591–601. DOI: 10.1007/s11104-021-04862-8.

Marzec, M. and Melzer, M. (2018) Regulation of root development and architecture by strigolactones under optimal and nutrient deficiency conditions. *International Journal of Molecular Sciences* 19(7), 1887. DOI: 10.3390/ijms19071887.

Mashiguchi, K., Seto, Y. and Yamaguchi, S. (2021) Strigolactone biosynthesis, transport and perception. *Plant Journal* 105(2), 335–350. DOI: 10.1111/tpj.15059.

Mendieta, J.P., Sangra, A., Yan, H., Minow, M.A.A. and Schmitz, R.J. (2023) Exploring plant cis-regulatory elements at single-cell resolution: Overcoming biological and computational challenges to advance plant research. *Plant Journal* 115(6), 1486–1499. DOI: 10.1111/tpj.16351.

Mostofa, M.G., Li, W., Nguyen, K.H., Fujita, M. and Tran, L.P. (2018) Strigolactones in plant adaptation to abiotic stresses: An emerging avenue of plant research. *Plant, Cell & Environment* 41(10), 2227–2243. DOI: 10.1111/pce.13364.

Mozes, C. and Meijler, M.M. (2020) Modulation of bacterial quorum sensing by strigolactones. *ACS Chemical Biology* 15(8), 2055–2059. DOI: 10.1021/acschembio.0c00509.

Nisa, Z., Wang, Y., Ali, N., Chen, C., Zhang, X. *et al.* (2023) Strigolactone signaling gene from soybean GmMAX2a enhances the drought and salt-alkaline resistance in *Arabidopsis* via regulating transcriptional profiles of stress-related genes. *Functional & Integrative Genomics* 23(3), 216. DOI: 10.1007/s10142-023-01151-8.

Niu, L., Yu, J., Liao, W., Yu, J., Zhang, M. *et al.* (2017) Calcium and calmodulin are involved in nitric oxide-induced adventitious rooting of cucumber under simulated osmotic stress. *Frontiers in Plant Science* 8, 1684. DOI: 10.3389/fpls.2017.01684.

Oláh, D., Feigl, G., Molnár, Á., Ördög, A. and Kolbert, Z. (2020) Strigolactones interact with nitric oxide in regulating root system architecture of *Arabidopsis thaliana*. *Frontiers in Plant Science* 11, 1019. DOI: 10.3389/fpls.2020.01019.

Oláh, D., Molnár, Á., Soós, V. and Kolbert, Z. (2021) Nitric oxide is associated with strigolactone and karrikin signal transduction in *Arabidopsis* roots. *Plant Signaling & Behavior* 16(3), 1868148. DOI: 10.1080/15592324.2020.1868148.

Omoarelojie, L.O., Kulkarni, M.G., Finnie, J.F., Pospíšil, T., Strnad, M. *et al.* (2020) Synthetic strigolactone (rac-GR24) alleviates the adverse effects of heat stress on seed germination and photosystem II function in lupine seedlings. *Plant Physiology and Biochemistry* 155, 965–979. DOI: 10.1016/j.plaphy.2020.07.043.

Pande, A., Mun, B.G., Rahim, W., Khan, M., Lee, D.S. *et al.* (2022) Phytohormonal regulation through protein S-nitrosylation under stress. *Frontiers in Plant Science* 13, 865542. DOI: 10.3389/fpls.2022.865542.

Petřivalský, M. and Luhová, L. (2020) Nitrated nucleotides: New players in signaling pathways of reactive nitrogen and oxygen species in plants. *Frontiers in Plant Science* 11, 598. DOI: 10.3389/fpls.2020.00598.

Prandi, C., Kapulnik, Y. and Koltai, H. (2021) Strigolactones: Phytohormones with promising biomedical applications. *European Journal of Organic Chemistry* 2021(29), 4019–4026. DOI: 10.1002/ejoc.202100204.

Priya, S., Somenahally, A.C., Obayomi, O., Gentry, T.J., Sarker, T.C. *et al.* (2024) Exploring the application of signaling compounds and soil amendments to modulate plant–microbe interactions for improved plant salinity tolerance. *Plant and Soil* 501(1–2), 267–288. DOI: 10.1007/s11104-024-06512-1.

Puglia, G.D. (2024) Reactive oxygen and nitrogen species (RONS) signalling in seed dormancy release, perception of environmental cues, and heat stress response. *Plant Growth Regulation* 103(1), 9–32. DOI: 10.1007/s10725-023-01094-x.

Qi, J., Mao, Y., Cui, J., Lu, X., Xu, J. *et al.* (2024) The role of strigolactones in resistance to environmental stress in plants. *Physiologia Plantarum* 176(4), e14419. DOI: 10.1111/ppl.14419.

Qu, B., Qin, Y. and Bai, Y. (2020) From signaling to function: How strigolactones regulate plant development. *Science China Life Sciences* 63(11), 1768–1770. DOI: 10.1007/s11427-020-1802-y.

Quaggiotti, S., Buzzicotti, L., Koch, K.E., Guan, J.C., Trevisan, S. *et al.* (2024) Strigolactone roles in maize tolerance to low nitrogen involve shifts in acquisition and partitioning of protein, sulfur, and iron. *Plant and Soil* 505(1–2), 41–64. DOI: 10.1007/s11104-024-06561-6.

Romani, F. and Moreno, J.E. (2021) Molecular mechanisms involved in functional macroevolution of plant transcription factors. *New Phytologist* 230(4), 1345–1353. DOI: 10.1111/nph.17161.

Saeed, S., Usman, B., Shim, S.-H., Khan, S.U., Nizamuddin, S. *et al.* (2022) CRISPR/Cas-mediated editing of cis-regulatory elements for crop improvement. *Plant Science* 324, 111435. DOI: 10.1016/j.plantsci.2022.111435.

Saeed, W., Naseem, S. and Ali, Z. (2017) Strigolactones biosynthesis and their role in abiotic stress resilience in plants: A critical review. *Frontiers in Plant Science* 8, 1487. DOI: 10.3389/fpls.2017.01487.

Santana, M.M., Gonzalez, J.M. and Cruz, C. (2017) Nitric oxide accumulation: The evolutionary trigger for phytopathogenesis. *Frontiers in Microbiology* 8. DOI: 10.3389/fmicb.2017.01947.

Selwal, N., Wani, A.K., Akhtar, N., Kaur, M. and Jassal, P.S. (2023) Molecular insights of strigolactone biosynthesis, signalling pathways, regulatory roles, and hormonal crosstalks in plant systems. *South African Journal of Botany* 160, 9–22. DOI: 10.1016/j.sajb.2023.06.046.

Seto, Y. (2023) Latest knowledge on strigolactone biosynthesis and perception. *Bioscience, Biotechnology, and Biochemistry* 88(1), 1–7. DOI: 10.1093/bbb/zbad150.

Sigalas, P.P., Bennett, T., Buchner, P., Thomas, S.G., Jamois, F. *et al.* (2024) At the crossroads: Strigolactones mediate changes in cytokinin synthesis and signalling in response to nitrogen limitation. *Plant Journal* 120(1), 139–158. DOI: 10.1111/tpj.16976.

Soliman, S., Wang, Y., Han, Z., Pervaiz, T. and El-Kereamy, A. (2022) Strigolactones in plants and their interaction with the ecological microbiome in response to abiotic stress. *Plants* 11(24), 3499. DOI: 10.3390/plants11243499.

Staszek, P. and Gniazdowska, A. (2020) Peroxynitrite induced signaling pathways in plant response to non-proteinogenic amino acids. *Planta* 252(1), 5. DOI: 10.1007/s00425-020-03411-4.

Su, C., Kokosza, A., Xie, X., Pěnčík, A., Zhang, Y. *et al.* (2023) Tree architecture: A strigolactone-deficient mutant reveals a connection between branching order and auxin gradient along the tree stem. *Proceedings of the National Academy of Sciences USA* 120(48), e2308587120. DOI: 10.1073/pnas.2308587120.

Sun, H., Guo, X., Qi, X., Feng, F., Xie, X. *et al.* (2021) SPL14/17 act downstream of strigolactone signalling to modulate rice root elongation in response to nitrate supply. *Plant Journal* 106(3), 649–660. DOI: 10.1111/tpj.15188.

Sun, H., Guo, X., Zhu, X., Gu, P., Zhang, W. *et al.* (2023) Strigolactone and gibberellin signaling coordinately regulate metabolic adaptations to changes in nitrogen availability in rice. *Molecular Plant* 16(3), 588–598. DOI: 10.1016/j.molp.2023.01.009.

Szweykowska-Kulinska, Z. and Jarmolowski, A. (2018) Post-transcriptional regulation of microrna accumulation and function: New insights from plants. *Molecular Plant* 11(8), 1006–1007. DOI: 10.1016/j.molp.2018.06.010.

Tao, Z., Yan, F., Hahn, M. and Ma, Z. (2023) Regulatory roles of epigenetic modifications in plant–Phytopathogen interactions. *Crop Health* 1(1), 6. DOI: 10.1007/s44297-023-00003-y.

Thiruvengadam, R., Venkidasamy, B., Easwaran, M., Chi, H.Y., Thiruvengadam, M. *et al.* (2024) Dynamic interplay of reactive oxygen and nitrogen species (ROS and RNS) in plant resilience: Unveiling the signaling pathways and metabolic responses to biotic and abiotic stresses. *Plant Cell Reports* 43(8), 198. DOI: 10.1007/s00299-024-03281-0.

Trasoletti, M., Visentin, I., Campo, E., Schubert, A. and Cardinale, F. (2022) Strigolactones as a hormonal hub for the acclimation and priming to environmental stress in plants. *Plant, Cell & Environment* 45(12), 3611–3630. DOI: 10.1111/pce.14461.

van Dingenen, J., de Keyser, A., Desmet, S., Clarysse, A., Beullens, S. *et al.* (2023) Strigolactones repress nodule development and senescence in pea. *Plant Journal* 116(1), 7–22. DOI: 10.1111/tpj.16421.

Visentin, I., Vitali, M., Ferrero, M., Zhang, Y., Ruyter-Spira, C. *et al.* (2016) Low levels of strigolactones in roots as a component of the systemic signal of drought stress in tomato. *New Phytologist* 212(4), 954–963. DOI: 10.1111/nph.14190.

Visentin, I., Pagliarani, C., Deva, E., Caracci, A., Turečková, V. *et al.* (2020) A novel strigolactone–miR156 module controls stomatal behaviour during drought recovery. *Plant, Cell & Environment* 43(7), 1613–1624. DOI: 10.1111/pce.13758.

Visentin, I., Ferigolo, L.F., Russo, G., Korwin Krukowski, P., Capezzali, C. *et al.* (2024) Strigolactones promote flowering by inducing the miR319–LA–SFT module in tomato. *Proceedings of the National Academy of Sciences USA* 121(19), e2316371121. DOI: 10.1073/pnas.2316371121.

Wakabayashi, T., Hamana, M., Mori, A., Akiyama, R., Ueno, K. *et al.* (2019) Direct conversion of carlactonoic acid to orobanchol by cytochrome P450 CYP722C in strigolactone biosynthesis. *Science Advances* 5(12), eaax9067. DOI: 10.1126/sciadv.aax9067.

Wang, J.Y., Chen, G.T.E., Braguy, J. and Al-Babili, S. (2024) Distinguishing the functions of canonical strigolactones as rhizospheric signals. *Trends in Plant Science* 29(8), 925–936. DOI: 10.1016/j.tplants.2024.02.013.

Wang, M., Zang, L., Jiao, F., Perez-Garcia, M.-D., Ogé, L. *et al.* (2020) Sugar signaling and post-transcriptional regulation in plants: An overlooked or an emerging topic. *Frontiers in Plant Science* 11, 578096. DOI: 10.3389/fpls.2020.578096.

Wang, Y., Sun, S., Zhu, W., Jia, K., Yang, H. *et al.* (2013) Strigolactone/MAX2-induced degradation of brassinosteroid transcriptional effector BES1 regulates shoot branching. *Developmental Cell* 27(6), 681–688. DOI: 10.1016/j.devcel.2013.11.010.

Wani, K.I., Zehra, A., Choudhary, S., Naeem, M., Khan, M.M.A. *et al.* (2021) Mechanistic insights into strigolactone biosynthesis, signaling, and regulation during plant growth and development. *Journal of Plant Growth Regulation* 40(5), 1836–1852. DOI: 10.1007/s00344-020-10234-w.

Waters, M.T., Gutjahr, C., Bennett, T. and Nelson, D.C. (2017) Strigolactone signaling and evolution. *Annual Review of Plant Biology* 68(1), 291–322. DOI: 10.1146/annurev-arplant-042916-040925.

Wu, S. and Li, Y. (2021) A unique sulfotransferase-involving strigolactone biosynthetic route in sorghum. *Frontiers in Plant Science* 12, 793459. DOI: 10.3389/fpls.2021.793459.

Wu, F., Gao, Y., Yang, W., Sui, N. and Zhu, J. (2022) Biological functions of strigolactones and their crosstalk with other phytohormones. *Frontiers in Plant Science* 13, 821563. DOI: 10.3389/fpls.2022.821563.

Wu, H., Li, B., Iwakawa, H., Pan, Y., Tang, X. *et al.* (2020) Plant 22-nt siRNAs mediate translational repression and stress adaptation. *Nature* 581(7806), 89–93. DOI: 10.1038/s41586-020-2231-y.

Xie, S.S. and Duan, C.G. (2023) Epigenetic regulation of plant immunity: From chromatin codes to plant disease resistance. *aBIOTECH* 4(2), 124–139. DOI: 10.1007/s42994-023-00101-z.

Yao, R., Wang, F., Ming, Z., Du, X., Chen, L. *et al.* (2017) ShHTL7 is a non-canonical receptor for strigolactones in root parasitic weeds. *Cell Research* 27(6), 838–841. DOI: 10.1038/cr.2017.3.

Yoda, A., Mori, N., Akiyama, K., Kikuchi, M., Xie, X. *et al.* (2021) Strigolactone biosynthesis catalyzed by cytochrome P450 and sulfotransferase in sorghum. *New Phytologist* 232(5), 1999–2010. DOI: 10.1111/nph.17737.

Yoneyama, K. (2020) Recent progress in the chemistry and biochemistry of strigolactones. *Journal of Pesticide Science* 45(2), 45–53. DOI: 10.1584/jpestics.D19-084.

Yoneyama, K., Xie, X., Kisugi, T., Nomura, T. and Yoneyama, K. (2013) Nitrogen and phosphorus fertilization negatively affects strigolactone production and exudation in sorghum. *Planta* 238(5), 885–894. DOI: 10.1007/s00425-013-1943-8.

Yoneyama, K., Mori, N., Sato, T., Yoda, A. and Xie, X. (2018) Conversion of carlactone to carlactonoic acid is a conserved function of MAX 1 homologs in strigolactone biosynthesis. *New Phytologist* 218(4), 1522–1533. DOI: 10.1111/nph.15055.

Yoneyama, K., Xie, X., Nomura, T. and Yoneyama, K. (2020) Do phosphate and cytokinin interact to regulate strigolactone biosynthesis or act independently. *Frontiers in Plant Science* 11, 438. DOI: 10.3389/fpls.2020.00438.

Yu, X., Sharma, B. and Gregory, B.D. (2021) The impact of epitranscriptomic marks on post-transcriptional regulation in plants. *Briefings in Functional Genomics* 20(2), 113–124. DOI: 10.1093/bfgp/elaa021.

Yu, Z., Cao, J., Zhu, S., Zhang, L., Peng, Y. *et al.* (2020) Exogenous nitric oxide enhances disease resistance by nitrosylation and inhibition of S-nitrosoglutathione reductase in peach fruit. *Frontiers in Plant Science* 11, 543. DOI: 10.3389/fpls.2020.00543.

Yuan, K., Zhang, H., Yu, C., Luo, N., Yan, J. *et al.* (2023) Low phosphorus promotes NSP1–NSP2 heterodimerization to enhance strigolactone biosynthesis and regulate shoot and root architecture in rice. *Molecular Plant* 16(11), 1811–1831. DOI: 10.1016/j.molp.2023.09.022.

Zemlyanskaya, E.V., Dolgikh, V.A., Levitsky, V.G. and Mironova, V. (2021) Transcriptional regulation in plants: Using omics data to crack the cis-regulatory code. *Current Opinion in Plant Biology* 63, 102058. DOI: 10.1016/j.pbi.2021.102058.

Zhan, J. and Meyers, B.C. (2023) Plant small RNAs: Their biogenesis, regulatory roles, and functions. *Annual Review of Plant Biology* 74(1), 21–51. DOI: 10.1146/annurev-arplant-070122-035226.

Zhang, Y., Wang, R., Wang, X., Zhao, C., Shen, H. *et al.* (2023) Nitric oxide regulates seed germination by integrating multiple signalling pathways. *International Journal of Molecular Sciences* 24(10), 9052. DOI: 10.3390/ijms24109052.

Zheng, Z., Guo, Y., Novák, O., Chen, W., Ljung, K. *et al.* (2016) Local auxin metabolism regulates environment-induced hypocotyl elongation. *Nature Plants* 2(4), 16025. DOI: 10.1038/nplants.2016.25.

Zhou, X., Joshi, S., Khare, T., Patil, S., Shang, J. *et al.* (2021) Nitric oxide, crosstalk with stress regulators and plant abiotic stress tolerance. *Plant Cell Reports* 40(8), 1395–1414. DOI: 10.1007/s00299-021-02705-5.

6 Strigolactone-mediated Regulation of Germination, Plant Establishment, and Vegetative Development

Mushfaq Ul-Nisa[1], Shahzadi Rafique[1], Syeda Noureen Fatima[1], Sajjad Hyder[1*], Amjad Shahzad Gondal[2], Fannie Isela Parra-Cota[3], Sergio de los Santos-Villalobos[4], Javed Iqbal[5] and Banzeer Ahsan Abbasi[6]

[1]Department of Botany, Faculty of Natural Sciences, Government College Women University, Sialkot, Pakistan; [2]Department of Plant Pathology, Bahauddin Zakariya University, Multan, Pakistan; [3]Campo Experimental Norman E. Borlaug, Instituto Nacional de Investigaciones Forestales, Agrícolas y Pecuarias (INIFAP), Ciudad Obregón, Sonora, Mexico; [4]Departamento de Ciencias Agronómicas y Veterinarias, Instituto Tecnológico de Sonora, Ciudad Obregón, Sonora, Mexico; [5]Department of Botany, Bacha Khan University, Charsadda, Khyber Pakhtunkhwa, Pakistan; [6]Department of Botany, Rawalpindi Women University, Satellite Town, Rawalpindi, Pakistan

Abstract

Strigolactones (SLs) are innovative master regulators of plant development, executing a pivotal function in plant germination and vegetative development. SLs play a significant role in arbuscular mycorrhizal fungi and plant interaction. Early findings suggested their function in parasitic seed germination, but they have since been identified as having a widespread impact on root and shoot architecture, leaf senescence, and resilience to stress. Despite their core function, modifications in SL pathways provide groundbreaking applications in the realm of agriculture, such as improving crop resilience, mitigating parasitic weeds, and beneficial interspecies interactions. As the investigation unfolds to explore the complicated signaling pathways, SLs are at the cutting edge of plant physiology, development, and resilient crop innovations. The current chapter delves into the versatile functions of SLs, unveiling their influential impact on plant growth and their potential to innovate modern agricultural approaches. The molecular recognition of the mechanisms by which SLs perform potential roles is essential and may provide new opportunities for future investigations in this captivating realm of plant hormones.

Keywords: abiotic stress, arbuscular mycorrhizal fungi, phytohormones, strigolactones

*Corresponding author: sajjad.hyder@gcwus.edu.pk

DOI: 10.1079/9781836990598.0006

6.1 Introduction to Strigolactones: Mysterious Regulators of Plant Growth

Strigolactones (SLs) are plant hormones that perform an essential role in regulating plant growth and maturation (Li *et al.*, 2019). SLs are associated with a class of metabolically active compounds recognized as semiochemicals that are utilized for interspecies data transfer (Zwanenburg *et al.*, 2016). SLs were extracted from rhizosphere metabolites of cotton plants by Cook in 1996 and the first identified SL was strigol; however, it took more than 40 years to identify SLs as comprising an exclusive division of phytohormones (Umehara *et al.*, 2008). Nevertheless, the exponential expansion of research endeavors with an emphasis on SLs indicates encouraging prospects for this type of signaling molecules (Zwanenburg and Blanco-Ania, 2018). Currently, over 30 SLs have been categorized, chiefly as germination stimulants for parasitic root weeds. Plants synthesize limited quantities of SLs, but their effects are intense. Root emissions are an ideal source for the identification of SLs as they are released in root exudates (Yoneyama and Brewer, 2021).

SLs are involved in regulating a broad spectrum of morphological characteristics, including shoot branching, leaf structure, primary and secondary stem growth, leaf senescence, root-hair formation, and adventitious root formation. SLs also play an integral part in the survival dynamics of plants under abiotic stresses, including drought, salinity, and phosphate starvation. Some of these functions have been verified in a wide number of species indicating that their roles are extensively sustained in angiosperm plants, but some functions appear to be distinct to few species evaluated so far (Rameau *et al.*, 2019).

Over the last few years, emerging biocharacteristics of SLs have been explored. A revolutionary finding was the detection of SLs as a branching factor for arbuscular mycorrhizal fungi (AMF) (Parniske, 2008), which is a symbiotic relationship among plants, fungi, and soil. An essential milestone in the development of AMF is stimulation of hyphal ontogenesis through branching factors. The separation and identification of branching factors is immensely complicated owing to their limited concentration. The first branching factor was identified from *Lotus japonicus*. Endogenous SLs have also been shown to play a crucial role in plant morphology, such as inhibition of shoot branching and suppression of bud development (Umehara *et al.*, 2008).

6.2 Chemical Basis of Strigolactones

Plants synthesize a variety of SLs. The versatile nature and basal concentration of SLs differ in plant species. Naturally, SLs are carotenoid-derivative substances distinguished by a sustained butenolide D-ring connected to a distinct R-configuration variable component (Jia *et al.*, 2018). In accordance with the second portion, SLs are grouped into two categories: canonical and non-canonical. Canonical SLs consist of a tricycle lactone ring (ABC-ring), whereas non-canonical SLs consist of distinct configurations that lack than ABC-ring. The common attributes of all identified SLs are that they all have D-rings in the R-configuration (2′R), and have D-rings associated with alkenyl moieties by an oxygen atom. Natural SLs are highly versatile and vulnerable to being attacked by nucleophiles owing to the distinctive chemical characteristics of the enol-ether D-ring. This enol-ether D-ring is regarded as a fundamental bioactive moiety that is mandatory for the metabolic activities of natural SLs (Wang *et al.*, 2021). A variety of structures of SLs can be demonstrated by modifications of the C-ring configuration, which contributes to further analysis and categorization of SLs, which are created by 5-deoxystrigol (Motonami *et al.*, 2013).

Non-canonical SLs consist of those that do not possess the features and molecular confirmation of canonical SLs, but there are no distinct categories, enzymes, or fundamental precursors for non-canonical SLs. Nevertheless, the enzymes associated with biosynthesis in distinct species maintain the diversity of the molecules (Guercio *et al.*, 2023). Various structures of

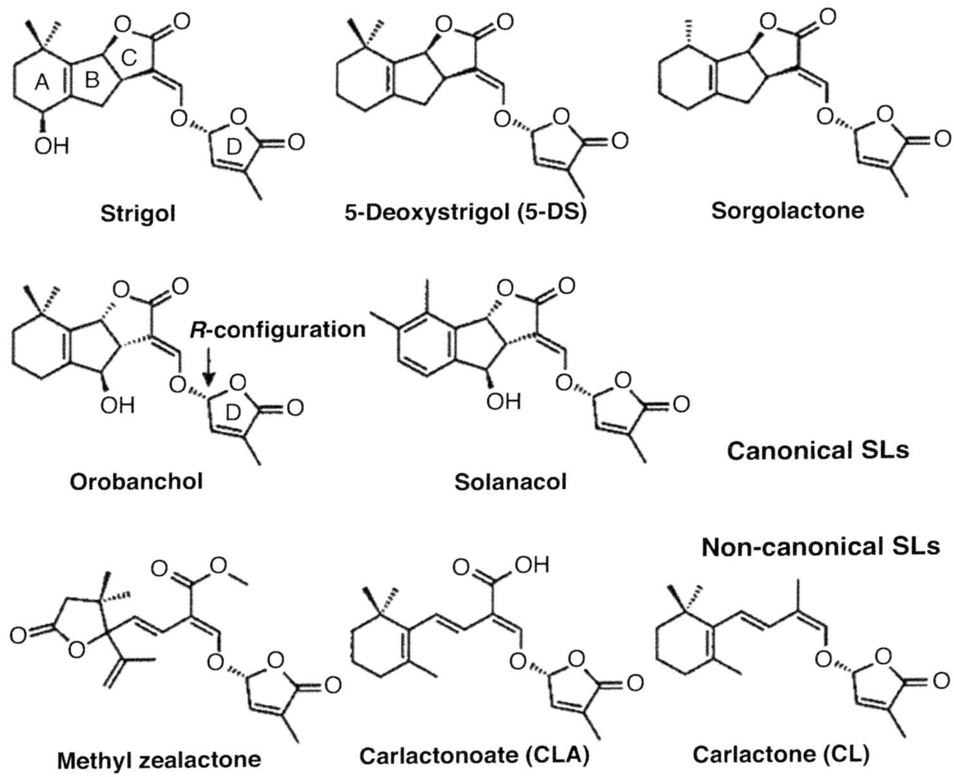

Fig. 6.1. Structure and classification of strigolactones (SLs). Canonical SLs contain a tricyclic lactone (ABC-ring) linked to a second lactone (D-ring). Non-canonical SLs have various structural units rather than the ABC-ring (Miyakawa *et al.*, 2020). Figure used with permission from Springer Nature.

canonical and non-canonical SLs are shown in Fig. 6.1.

6.3 The Molecular Path of Strigolactones: Biosynthetic Journey

SL synthesis takes place in plant roots. Intermediates of the biosynthetic pathway of SLs have been also recognized in the stem. Initially, SLs were assumed to be sesquiterpene lactones related to secondary metabolites that emerged from cytosolic C15 compound farnesyl diphosphate (de Kraker *et al.*, 1998). However, the morphological resemblance of the A-ring of SLs and the ionone rings of carotenoids contributed to the prediction that SLs might be produced from carotenoids (Bouwmeester *et al.*, 2003). Carotenoids and sesquiterpene are isoprenoids in nature and are produced from isopentenyl diphosphate (IPP). Two pathways are usually utilized by plants for IPP synthesis: the 2-C-methyl-D-erythritol 4-phosphate (MEP) pathway and the mevalonic acid (MVA) route. In the MEP route, D-glyceraldehyde 3-phosphate and pyruvate are converted to deoxy-D-xylulose 5-phosphate (DOXP) with the help of DOXP synthase. DOXP reductoisomerase converts DOXP into MEP, which is then converted to 4-(cytidine 5′-diphospho)−2-C methyl-D-erythritol (CDP-ME) by CDP-ME synthase. CDP-ME kinase converts CDP-ME into CDP-methyl-D-erythritol-2-phosphate, and then 2-C methyl-D-erythritol 2,4-cyclodiphosphate synthase modifies the isolated compound into 2-C methyl-D-erythritol 2,4-cyclodiphosphate, and reductases and dehydratases convert it to IPP (Fig. 6.2).

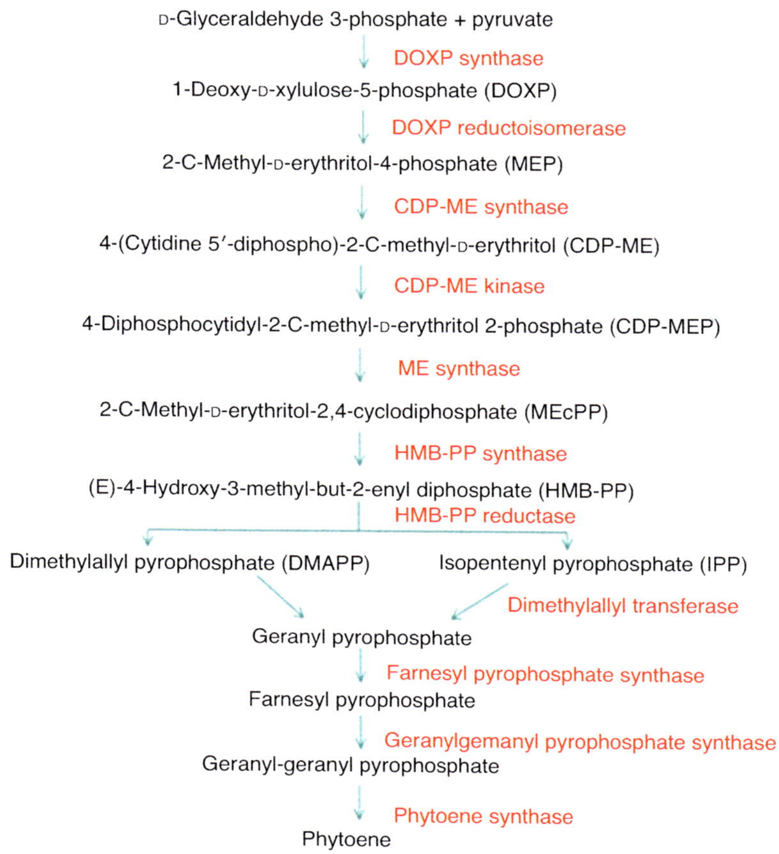

Fig. 6.2. The 2-C-methyl-D-erythritol 4-phosphate (MEP) pathway for production of isopentenyl diphosphate (IPP).

In the MVA pathway, acetyl-CoA is integrated to form 3-hydroxy-3-methylglutaryl-CoA by utilizing HMG-CoA synthase enzyme. HMG-CoA reductase then reduces 3-hydroxy-3-methylglutaryl-CoA to MVA. MVA is synthesized into MVA-phosphate and MVA-pyrophosphate. IPP is converted into DMAPP and then geranyl pyrophosphate, farnesyl pyrophosphate and finally geranyl-geranyl pyrophosphate, which is the precursor of carotenoids (Bajguz *et al*., 2020), as shown in Fig. 6.3. Phytoene synthase converts geranyl-geranyl pyrophosphate into phytoene, which then undergoes a sequence of reactions that leads to the formation of carlactone, the first strigolactone (Fig. 6.4). β-Carotene isomerase (also known as DWARF27 (D27)) converts all-*trans*-β-carotene into 9-*cis*-β-carotene. After that, carotenoid cleavage dioxygenase 7 (CCD7) and CCD8 convert 9-*cis*-β-carotene into carlactone. Carlactone is regarded as a fundamental precursor of natural SLs in plants. Consequently, cytochrome P450 mono-oxygenase oxidized carlactone in numerous canonical and non-canonical SLs.

6.4 Strigolactones as Germination Triggers

SLs are important in seed dormancy and ensure seed germination under favorable conditions. Seed dormancy is regulated by two main hormones, abscisic acid (ABA), which promotes dormancy, and gibberellins, which breaks

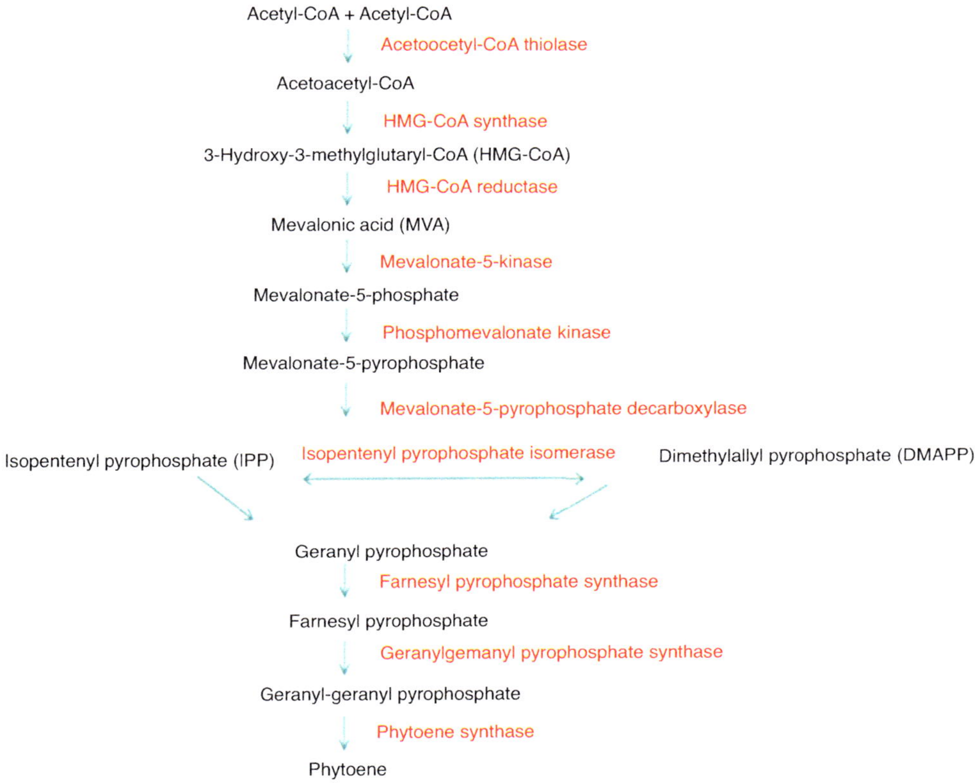

Fig. 6.3. The mevalonic acid (MVA) pathway for production of isopentenyl diphosphate (IPP).

dormancy in plants. SLs interact with both of these hormones to regulate seed dormancy. For instance, SLs downregulate the 9-*cis*-epoxy carotenoid dioxygenase (*NCED*) ABA biosynthesis gene, which inhibits ABA, and upregulate the abscisic acid 8β-hydroxylase 1 gene (*CYP707A*), which decreases ABA levels in seeds and promotes germination under stress circumstances (Al-Babili and Bouwmeester, 2015). Moreover, López-Ráez *et al.* (2017) found that SLs regulated the auxin transportation and signaling pathway to reduce seed dormancy and promote root germination by increasing plant nutrient uptake.

SLs are vital in seed germination, in both parasitic and non-parasitic germination. SLs stimulate parasitic germination of weeds such as *Striga* and *Orobanche*. SLs can be used for "suicidal germination," in which parasitic seed germination occurs without a host body, and due to nutrient deficiency, mainly phosphorus, the weeds eventually die. This mechanism has been used in the development of SL-based herbicides, as the SL analog GR24, which induces parasitic weed germination and reduces crop infestations by *Striga hermonthica* and *Orobanche cumana* (López-Ráez *et al.*, 2017; Zorrilla *et al.*, 2022). SLs also stimulate the growth of non-parasitic plants and facilitate better symbiotic relationships between plants and AMF. The fungi facilitate better nutrient acquisition including phosphorus, which enhances seedling development throughout their growth period. Under conditions of phosphate deprivation, the biosynthesis of SLs becomes more active, which in turn promotes the transcription of vital SL metabolic synthesis genes (*CCD7* and *CCD8*). The process of AMF colonization is accelerated, while seed germination improves (López-Ráez *et al.*, 2017). SLs produced in root exudate work

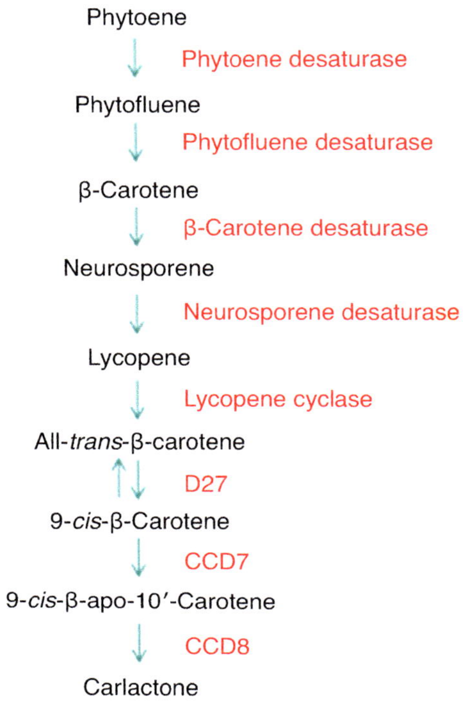

Fig. 6.4. Synthesis of carlactone, the first strigolactone. D27, DWARF 27; CCD7, carotenoid cleavage dioxygenase 7; CCD8, carotenoid cleavage dioxygenase 8.

with auxin signaling systems to enhance root structure and nutrient accessibility, which helps seeds grow better under stressful situations (Waters *et al.*, 2017).

6.5 Strigolactones and Plant Establishment and Vegetative Development: The Master Conductor of Plant Growth

6.5.1 Root architecture

A strong root system anchors plants evenly and acropetally allows efficient nutrient transportation. Furthermore, plants exhibit various modifications and different root structures as a result of environmental adaptations. Plant hormones, mainly auxin, perform a pivotal function in regulation of root development (Jiang and Feldman, 2002). SLs stimulate various plant growth attributes including the root system (Fig. 6.5). SLs have shown a significant role not only in primary, lateral, and adventitious root formation but also in elongation of root hairs. For example, external treatment with the synthetic SL GR24 promoted prolonged primary roots due to enhanced cell populations in meristematic zones in *Arabidopsis* (Ruyter-Spira *et al.*, 2011). Similarly, Rasmussen *et al.* (2012) found that exogenous application of GR24 reduced adventitious root germination in *Arabidopsis*. In root system architecture, auxin gradient functions as the primary growth regulator. For example, in *Arabidopsis*, auxin concentration gradient not only regulates cell size in the root transition and meristematic zones, and adventitious root development, but also interacts with *CCD8*, a SL-biosynthesis gene in epidermal cells, zones of elongation, and root cortical cells. SLs and auxin are regulated independently: SLs reduce adventitious root development under auxin influence and auxin promotes adventitious roots in the absence of SLs (Rasmussen *et al.*, 2012). Another study utilized *PLEIOTROPIC DRUG RESISTANCE1* (*PDR1*)-overexpressing lines in *Petunia hybrida* to show that optimization of SL transportation released feedback suppression for SL biosynthesis and promoted plant-rooting traits (Liu *et al.*, 2018). Overexpression of *PDR1*, an ABCG-class transporter associated with SL transportation to the shoot and emission in soil, even at lower auxin levels, promoted primary and lateral roots and induced root-hair elongation compared with wild-type plants with stimulated hormonal regulation of plant root development (Liu *et al.*, 2018).

6.5.2 Shoot branching

SLs significantly affect plant shoot architecture. SLs and branching inhibiting signals are both derived from carotenes and linked due to their similar activity in apical dominance (Sorefan *et al.*, 2003). In the SL-signaling process, DWARF14 (D14) is the first receptor that plays a major role in branching architecture, for instance GR24 mediated

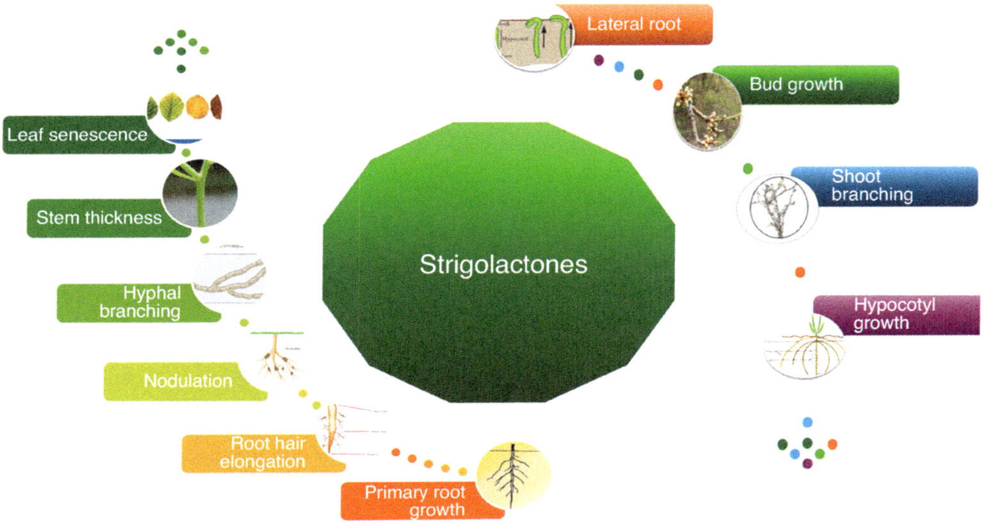

Fig. 6.5. Various pivotal roles of strigolactones in plant maturation, plant physiology, and plant growth.

degradation of SUPPRESSOR OF MAX2 1-LIKE (SMXL) 6–8 (Zhou *et al.*, 2013). *SMXL* genes lose function due to this degradation, which results in reduced shoot branching while enhancing PIN accumulation and auxin transportation in the plant stem. This demonstrated that SLs mediate shoot branching with auxin (indole-3-acetic acid (IAA)) using SMXLs (Soundappan *et al.*, 2015). Moreover, Aguilar-Martínez *et al.* (2007) and Braun *et al.* (2012) found that BRC1 is an auxiliary bud suppression factor released as a result of SMXL activity, stimulating branching. SLs degrade SMXL, which reverses this action. Similarly, BRC1 functions downstream of the SL pathway in various plants including peas, *Arabidopsis*, and rice. In another study, it was found that the BRC1 homolog TEOSINTE BRANCHED1 (TB1) functions and acts similarly in maize (Kellogg, 1997). The use of SL-deficient strains and lost-function mutants confirmed the preceding results and demonstrated that SLs inhibit shoot branching. Guan *et al.* (2012) also found that maize diameter and ear length were substantially minimized in the absence of SLs and the tassels dropped. Depending on the desired characteristic, these architectural phenotypic properties may be significant for agronomic gross primary production and can be employed in SL application to specific species based on preferred traits; for instance, wheat (*Triticum aestivum*) and rice produce more tillers, which may have a higher agronomic value, while maize favors branchless single stems, with the number of tillers associated with reduced SL production (Jamil *et al.*, 2012; Zhao *et al.*, 2020). Application of exogenous GR24 can affect the late developmental stages of crops. For instance, GR24 is applied both *in vitro* and in the field to grapevine berries. GR24 was cotreated with ABA to interrupt anthocyanin aggregation and maturation (Ferrero *et al.*, 2018). The study shows that future research into SL hormonal connections could aid in the planning of SL applications in the field. For example, synergistic effects of synthetic SLs and ABA can retard maturation, thus increasing the durability of vulnerable crop products and lowering food losses owing to post-harvest problems.

6.5.3 Leaf senescence

Leaf senescence is caused by chloroplast breakdown along with RuBisCo (ribulose-1,5-bisphosphate carboxylase/oxygenase) denaturation and chlorophyll *a*/*b*-binding protein denaturation, resulting in yellowing of leaves (Yamada and Umehara, 2015). Leaf senescence requires SL activity, as shown by experiments on SL-deficient mutants including *d27*, *d17*, and *d10* in *Oryza sativa* and *max1*, *max3*, and *max4* in *Arabidopsis thaliana* along with SL resistance mutants composed of *d3* and *d14* of *O. sativa* and *max2* and *atd14* of *A. thaliana* (Yamada *et al.*, 2014; Yamada and Umehara, 2015). Addition of GR24 to SL-deficient *Oryza sativa* mutants produced senescence restoration through changes in leaf color and chlorophyll content along with electrolyte leakage, but this did not apply to SL-resistant mutants. Experiments involving SL-related mutants showed accelerated senescence responses following GR24 treatment, indicating that SLs play a significant role in leaf aging (Yamada *et al.*, 2014). The addition of GR24 also increases chlorophyll content under drought stress. Research was conducted on rapeseed (*Brassica napus*) under salinity and wheat under drought-stress conditions that matched previous findings (Ma *et al.*, 2017; Sedaghat *et al.*, 2017). The chlorophyll *a*/*b* value of SL-deficient mutants including *d27*, *d17*, *d10*, and *max1*, *max3*, and *max4* showed reduced potential to harness light energy across different wavelengths (Dale and Causton, 1992; Yamada *et al.*, 2014; Yamada and Umehara, 2015). The variation of photosynthetic pigments as a result of the action of SLs enables leaves to capture different light energies, and reduces photosystem II (PSII) electron flow rate, the quantum yield of PSII photochemistry and non-photochemical quenching compared with wild-type levels.

6.5.4 Hyphal branching and nodulation

AMF have a symbiotic relationship with most terrestrial plants (Redecker and Raab, 2006). SLs have been shown to trigger essential hyphal growth in AMF, which leads to improved plant mycorrhization and enhanced nutrient acquisition with better resource distribution. The study of mycorrhizal hyphal branching in pea and tomato mutants has shown reduced AMF hyphal expansion when compared with wild-type plants. The mycorrhizal colonization level of wild-type plants decreased when compared with plants with normal SL-biosynthesis capacity. Hyphal spores and root infection inoculations reduced this difference between the two types of plants (Koltai *et al.*, 2010). Moreover, the development of nodules in legumes, specifically under nitrogen-deficiency conditions, depends largely on SLs. The nodulation process in peas occurs through SL- and brassinosteroid-biosynthesis genes, even when the autoregulation of nodulation system remains inactive (Koltai, 2013). Likewise, Soto *et al.* (2010) found that when alfalfa is exposed to a low GR24 concentration, there is a substantial increase in nodule numbers. GR24 treatment promotes nodulation in peas and *Lotus* plants. Thus, severe deficiency of SL in *rms4* mutant peas led to superior nodule formation compared with both wild-type and *rms1* peas (Foo and Davies, 2011; Foo *et al.*, 2013; Liu *et al.*, 2013). Similarly, in another study, the GmMAX2a protein in soybean led to higher nodule numbers in plants when the gene was overexpressed, but plants that knocked down GmMAX2a expression exhibited fewer nodules (Ahmad *et al.*, 2020). Hence, these studies confirm that SLs control nodulation development in legumes, thus promoting nitrogen-acquisition functions and crop-yield properties.

6.6 Hormonal Interplay of Strigolactones and Other Phytohormones Under Abiotic Stress Conditions

6.6.1 SLs and ABA

The relationship between ABA and SLs is crucial for systematic regulation mechanisms and adaptation of plants to environmental fluctuations (Fig. 6.6). In *Arabidopsis*, the SL-signaling-deficient mutant *max2* reduced the effect of ABA importer genes (*ABCG22/AT5G06530* and *ABCG40/AT1G15520*) under both well-irrigated and dry conditions (Ha *et al.*, 2014; Ruiz-Lozano *et al.*, 2016). SLs primarily activate ABA signaling to create a tolerance to salt stress and drought. In a SL-biosynthesis-deficient mutant of *L. japonicus*, resistance to drought was compromised, which was linked to slower stomatal closure and ascribed to ABA insensitivity (Liu *et al.*, 2015). Therefore, SLs influence stomatal closure, which depends on ABA biosynthesis, transportation, and susceptibility (Visentin *et al.*, 2020). In subsequent investigations, there was induction of high drought

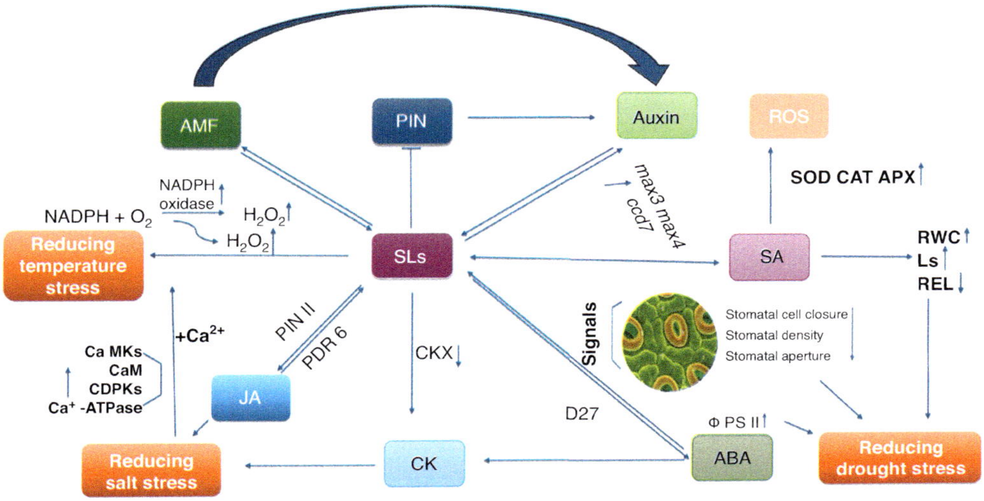

Fig. 6.6. The association between strigolactones (SLs) and other phytohormones under abiotic stress conditions. During this process, there is a synchronized crosstalk among various phytohormones and the association of numerous genes, enabling plants to maintain homeostasis under abiotic stress conditions. ABA, abscisic acid; AMF, arbuscular mycorrhizal fungi; APX, ascorbate peroxidase; CaMKs, CAM and CDPKs, calcium-binding proteins; *ccd7*, carotenoid cleavage dioxygenase 7 mutant; CK, cytokinin; CKX, CK catalyst genes; CAT, catalase; D27, DWARF 27; H_2O_2, hydrogen peroxide; JA, jasmonic acid; Ls, leaf stomatal limitation; *max3/4*, MORE AXILLARY GROWTH 3/4 mutant; NADPH, nicotinamide adenine dinucleotide phosphate; PDR6, PLEIOTROPIC DRUG RESISTANCE 6; PIN, PIN-FORMED transport protein, REL, relative electrolyte leakage; ROS, reactive oxygen species; RWC, relative water content; SA, salicyclic acid; SOD, superoxide dismutase; ΦPSII, photosystem II efficiency; ↑, increase; ↓ decrease.

tolerance in rice plants by SL-signaling-deficient mutant *d13* and SL-biosynthesis-deficient mutants *d10* and *d17*, and more ABA accumulated in mutant plants than wild-type plants. The method by which ABA correlates with SLs is unclear, but the reduced ABA-generating mutant *d27* is vulnerable to drought and is involved in the ABA signaling route (Haider *et al.*, 2018). SLs not only aid in drought tolerance but also have a favorable impact on heat and cold stress, which is correlated to ABA biosynthesis. Application of $GR24^{5DS}$ (a biologically active stereoisomer of GR24) improved the resistance of tomato to heat and cold, while ABA-deficient mutants reduced the impacts of $GR24^{5DS}$, suggesting that SL, in an ABA-dependent way, moderately enables plants to adapt to and cope with these stressors (Chi *et al.*, 2021). More study is needed to find out if SL signal upregulation could potentially be exploited to produce drought-tolerant crop varieties.

6.6.2 SLs and salicyclic acid

Several studies have shown that the simultaneous application of SL and salicyclic acid (SA) substantially damages plants compared with their individual application (Sedaghat *et al.*, 2017). Under drought stress, exogenous SL or SA reduced relative electrolyte leakage in drought-sensitive wheat cultivars; however, there was substantial inhibition of electrolyte leakage when both were applied simultaneously (Sedaghat *et al.*, 2017). Similarly, drought-sensitive wheat cultivars treated with SL or SA alone showed no statistically significant changes in relative water content or leaf stomatal limitation; however, the combined use of these plant hormones increased both of these characteristics. Additionally, exogenous treatment of GR24 and SA improves catalase (CAT), superoxide dismutase (SOD), and ascorbate peroxidase

(APX) activities in wheat crops (Sedaghat *et al.*, 2017). Correspondingly, SL-signaling *max2* mutant *Arabidopsis* plants showed a lower SA gene expression response and endogenous SA content than wild-type plants (Kusajima *et al.*, 2022). According to this study, SLs had an impact on how SA content changed in plants. Future studies should concentrate on the process by which the combined action of SLs and SA reduces stress-induced plant damage.

6.6.3 SLs and auxin

Studies indicate that auxin (IAA) positively regulates the transcription of *MAX3* and *MAX4* genes, while SLs promotes auxin feedback that aids in regulating the axillary branches (Hayward *et al.*, 2009; Omoarelojie *et al.*, 2019). Likewise, auxin directs SL synthesis and signaling in plants by stimulating translation of the SL-coding genes *CCD8* and *MAX3* (Zou *et al.*, 2006). However, SLs inhibited the auxin transporter PIN-FORMED (PIN) protein and stimulated *Arabidopsis* primary root growth while reducing lateral-root development and bud growth (Bennett *et al.*, 2006; Hayward *et al.*, 2009). Sun *et al.* (2014) found that SLs interfered with auxin pathways to regulate stem branching. Consequently, auxin and SLs can be associated with a feedback system to regulate the branching system of plants.

In another study, Zhang *et al.* (2020) found that exogenous auxin substantially elevated the gene expression of *WRKY2*, *WRKY108715*, and *WRKY56* in white clover under drought stress. Similarly, according to Wang *et al.* (2023), SL stimulates plant freezing resistance by inhibiting expression of the transcription factor WRKY41, which represses expression of *C-REPEAT BINDING FACTOR/DRE BINDING FACTOR 1* (*CBF/DREB1*). However, the association between SL and auxin that affects the *WRKY* genes under stress conditions remains elusive. Further studies should examine the interaction between SLs and auxin to reduce abiotic stress-induced plant damage.

6.6.4 SLs and cytokinins

SLs and cytokinins (CKs) have been shown to demonstrate antagonistic effects during shoot growth in plants (Dun *et al.*, 2012). Manandhar *et al.* (2018) found that elevated levels of SLs in *Zantedeschia* reduced the endogenous CK content, which inhibited the axillary bud growth. Likewise, in rice, exogenous GR24 application reduced bud growth, while CKs promoted growth (Xu *et al.*, 2015). In another study, mutant *Arabidopsis max2* plants showed inhibited expression of the CK catalyst genes, namely *CKX1*, *CKX2*, *CKX3*, and *CKX5*, compared with wild-type plants under dry conditions. Hence, this study revealed that SL signals are antagonistic to CK content.

In *Arabidopsis* and tobacco plants, the CK oxidases *CKX1*, *CKX2*, *CKX3*, and *CKX4* were overexpressed, which promoted ABA sensitivity, giving better resistance against salinity and drought stress (Nishiyama *et al.*, 2011; Li *et al.*, 2019). These findings show that SL signals antagonistic to CK oxidase genes regulate CK levels in plant cells during water-deficient conditions and stimulate ABA to regulate plant stress tolerance. Nevertheless, further studies are required to confirm the direct interlink between SLs and CKs in reducing abiotic stresses. Additionally, SL sand CKs synergistically control the lateral-root growth of plants, and during this process, these two hormones regulate biosynthetic gene expression (Zhang *et al.*, 2010; Dun *et al.*, 2012; Jiang *et al.*, 2016).

6.6.5 SLs and ethylene

SLs are associated with ethylene signaling and regulation of root-hair elongation. In ethylene-signaling-deficient mutants (*ein2* and *etr1*), SL signals did not influence root-hair growth and development (Kapulnik *et al.*, 2011). In a recent study, GR24 treatment alone showed a significant increase in root-hair elongation in *Arabidopsis* but not with an ethylene biosynthetic inhibitor (aminoethoxyvinylglycine), or even with GR24 (Lee and Yoon, 2020). This suggests that SL-mediated root-hair elongation is promoted by ethylene signals. SLs control the balance between the ethylene and IAA signaling pathways to stimulate the growth and development stages in plants, thereby regulating their biosynthesis in the root zone in response to nutrient deficiencies in soil. SLs also make associations with ethylene and stimulate IAA signaling transduction under phosphate-sufficient conditions (Koltai, 2013). Thus, ethylene associates with SLs and auxin pathways, influencing root-hair elongation. These hormones likely form a coordinated network to regulate plant development and respond to unfavorable growth circumstances.

6.7 Strigolactones as a Secret Signal for Nutrient Uptake

Plants have adopted numerous modification strategies to improve connections with their external environment. Typically, during phosphate, nitrogen, and sulfur deprivation, plants synthesize and release SLs into the rhizosphere, and consequently AMF association is established in the rhizosphere (Marro *et al.*, 2022). To enhance the concentration of phosphorus in plants, SLs stimulate the hyphal branching of AMF. In research by Boyer *et al.* (2014), GR24 was found to initiate hyphal branching of AMF. Increased biosynthesis and release of SLs during phosphorus starvation reduces shoot branching, promotes the formation of lateral roots, increases the density of root hairs, and enhances AMF association in *Oryza sativa* and *Arabidopsis* (Brewer *et al.*, 2013). Decker *et al.* (2017) investigated the impact of phosphate accessibility on germination of *Physcomitrella patens*. The results found that under phosphate deprivation, SL secretion was increased, indicating the sustained evolutionary role of SLs and their modulation under phosphate starvation in plants.

6.8 Environmental Cues and Strigolactone Signaling

6.8.1 High temperature

SLs contribute significantly to stress management by enabling plants to develop resistance to various environmental stresses and challenges (Table 6.1) (Tariq *et al.*, 2023). Global warming and unanticipated heat waves are major environmental concerns that have gained attention in recent years and have led to climate changes due to the rise in temperature, which poses a threat to tree populations and sustainable agriculture (Shafqat *et al.*, 2021). Plants rely on SL signaling and biosynthesis by using antioxidants and various mechanisms to avoid cellular degradation under extreme temperature stress (Pandey *et al.*, 2016). Signaling molecules such as SLs regulate plant growth and promote symbiotic connections with soil microbes through root exudates. Plants require specific temperature ranges either long term or short term to perform optimal physiological and metabolic functions under stress (Khan *et al.*, 2019; Bermudez *et al.*, 2021). Chi *et al.* (2021) suggested that SLs are plant regulators that enhance plant tolerance to heat stress by stimulating protein transcription and antioxidant enzyme activity. Heatwaves negatively impact plant physico-chemical and metabolic mechanisms including degradation of enzymes (Teskey *et al.*, 2015), disturbances in plant gas exchanges, excess reactive oxygen species (ROS) production (Zinta *et al.*, 2014), interruption of biomembranes (Scafaro *et al.*, 2021), dehydration (Marchin *et al.*, 2022), and imbalance of phytohormones (López *et al.*, 2022), which eventually affect plant growth and development.

SLs are phytohormones that modulate plant growth and establishment by regulating physiological, biochemical, and metabolic functions (Faizan *et al.*, 2020; El Sabagh *et al.*, 2022; Wang *et al.*, 2022). In plants, the photosynthetic efficiency of PSII and quantum yield declines under heat stress. Light is the most crucial component in determining photosynthesis in plants, and chlorophyll controls the photosynthetic activity. SLs improve photosynthetic efficiency and reduce oxidative stress (Ma *et al.*, 2017). To increase chlorophyll content, net photosynthetic rate, effective quantum yield of PSII, and photosystem photochemical efficiency, synthetic SLs can be given to plants under heat stress (Alvi *et al.*, 2022). SLs not only stimulate photosynthetic activity by increasing chlorophyll content but also promote seed germination in plants by simulating chemical CK content and balancing the ratio of gibberellins to ABA in plants (Liu *et al.*, 2022). SLs may play a role in fruit ripening and seed development, as evidenced by expression of the gene *SlCCD7* in immature green tomatoes (Vogel *et al.*, 2010). Additionally, SL-biosynthesis genes are expressed in roots and act as a germination stimulant. SLs also promote

Table 6.1. Consequences of strigolactone on biological functions of plants under abiotic stresses.

Plant	Type of stress	SL, concentration and means of application	Observed effects	Reference(s)
Zea mays (maize)	Drought	0, 10 and 20 μM GR24, foliar spray	Improved root-hair density and water-uptake efficiency, stabilized gas-exchange parameters and photosynthetic rate, enhanced chlorophyll content, improved stomatal conductance and osmotic adjustment, and elevated antioxidant enzyme activity (SOD, CAT, POD, and APX) to combat drought-induced oxidative stress	Sedaghat *et al.* (2020); Sattar *et al.* (2022)
Triticum aestivum (wheat)	Drought	10 μM GR24, foliar spray	Improved root architecture and water uptake, maintained higher photosynthetic rates, and enhanced antioxidant enzyme activity (SOD, CAT, POD, and APX) to mitigate oxidative stress	Sedaghat *et al.* (2017)
Oryza sativa (rice)	Salt stress	0.1, 0.2, 1, 5 μM GR24 with 200 mM NaCl	Increased root elongation and improved water-uptake efficiency, increased stomatal conductance, improved photosynthetic efficiency under water deficit, and increased activity of antioxidant enzymes (SOD, MAD, and POD) to reduce oxidative stress	Ling *et al.* (2020)
Arabidopsis thaliana	Drought		Promoted lateral-root formation and improved water retention, maintained chlorophyll content and photosynthetic rate under drought, modulated ABA signaling pathways to enhance drought tolerance, and enhanced the activity of ROS-scavenging enzymes to mitigate oxidative damage	Li *et al.* (2020)
Arabidopsis thaliana and *Malus domestica*	Drought, salinity, low-temperature stress	5, 10, and 20 μM GR24, MS medium	Decreased hypocotyl elongation, MdD14-His protein, and shoot branching, and increased MdD14 degradation	Yang *et al.* (2020)

Continued

Table 6.1. Continued

Plant	Type of stress	SL, concentration and means of application	Observed effects	Reference(s)
Solanum lycopersicum (tomato)	Salinity	GR24, foliar spray	Improved osmotic adjustment and reduced leaf senescence, enhanced photosynthetic efficiency and plant growth attributes under salinity, increased activity of SOD and CAT to reduce ROS accumulation	Faisal *et al.* (2023)
Glycine max (soybean)	Alkaline stress	Exogenous SL application	Improved root growth and photosynthetic rate, and decreased salt-induced damage; SLs coordinated with ABA to regulate ion homeostasis; increased activity of antioxidant enzymes to combat oxidative stress	Mi *et al.* (2023)
Cucumis sativus (cucumber)	Salinity	1, 5, 10, and 20 μM GR24	Improved salt stress tolerance by enhancing antioxidant activity (SOD, MAD, APX, and POD)	Li *et al.* (2023)
Triticum aestivum (wheat)	Cadmium stress	4 μM GR24, foliar spray	Reduced cadmium uptake, enhanced antioxidant defense and increased nitric oxide signaling; increased activity of antioxidant enzymes (SOD, CAT, and APX) to reduce oxidative damage	Shah *et al.* (2023)
Galinsoga parviflora	Cadmium stress	1 and 2 μM GR24	Cadmium content increased in shoot and decreased in root with improved phytoremediation efficacy	Yang *et al.* (2024)
Solanum lycopersicum (tomato)	Heat and cold stress	$GR24^{5DS}$	Induction of CBF1, heat-shock protein (HSP70), and improved heat and cold tolerance through enhanced antioxidant defense in ABA-mediated manner	Chi *et al.* (2021)

ABA, abscisic acid; APX, ascorbate peroxidase; CAT, catalase; POD, peroxidase; ROS, reactive oxygen species; SOD, superoxide dismutase.

branching in lower shoot sections and increase transcription of the *CCD7* and *CCD8* genes under heat stress (Alvi *et al.*, 2022).

6.8.2 Drought stress

Over the past few years, water-related stresses such as drought and salinity have increased substantially, affecting plant establishment and maturation globally. Drought stress is an abiotic factor that affects moisture levels, a major reason for global warming that is responsible for 70% of potential crop losses. Insufficient availability of water supply to the plant roots affects the electron transport pathway, contributing to oxidative stress and ROS accumulation, resulting in tissue and organ damage to the plants (Khalid *et al.*, 2022). In various studies, SLs have shown the ability to regulate drought inhibition at different plant growth and development stages (Ito *et al.*, 2022). For example, in perennial grass species, exogenous SLs inhibited tiller production and promoted the growth of lateral branches, increasing drought tolerance compared with control treatment (Zhuang *et al.*, 2017). Similarly, Sedaghat *et al.* (2017) found that SLs act as a stress signal mediator when applied to wheat plants under drought conditions, decreasing the stomatal limitation of the leaf and electrolyte leakage, and increasing relative water content. Min *et al.* (2019) observed similar findings in grapevine, where foliar application of SLs activated the antioxidant defense mechanism, improved chlorophyll content, and regulated stomatal closure under drought stress. Correspondingly, SLs significantly stimulate symbiotic relationships with AMF and nitrogen-fixing bacteria in plants to avoid nitrogen and phosphorus deficiency by adjusting root and shoot structure and stimulating hyphal branching in root-derived SLs under drought stress (Marzec, 2016; Yoneyama and Brewer, 2021; Ho-Plágaro and García-Garrido, 2022). Visentin *et al.* (2016) also found changes in physiological responses in plants affected by SL depletion in the roots, which showed increased SL levels in shoots, stimulating the stomatal sensitivity toward ABA under dry conditions. Similarly, AMF-colonized plants with high concentrations of SLs in the root area had increased water and nutrient uptake and stimulated photosynthesis to reduce the negative effects on plant germination under drought stress (Mostofa *et al.*, 2018). Additionally, Sedaghat *et al.* (2020) showed that SL application to tomatoes reduced lipid peroxidation, malondialdehyde (MDA), hydrogen peroxide (H_2O_2) content, and ROS scavenging by increasing antioxidant enzyme activity under drought conditions (Fig. 6.7). Al-Amri *et al.* (2023) and Bhoi *et al.* (2021) indicated that SLs enhance photosynthetic rates and promote the accumulation of proline and soluble sugars, facilitating osmotic adjustment in tomatoes.

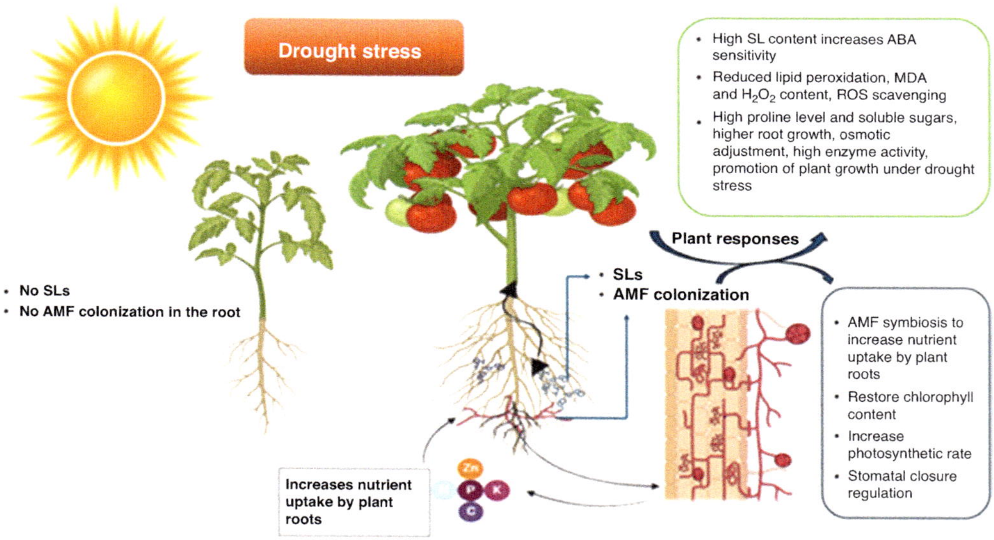

Fig. 6.7. Strigolactone (SL) synthesis and arbuscular mycorrhizal fungi (AMF) association under drought stress. SLs promote the AMF association and enhance plant efficiency. ABA, abscisic acid; MDA, malondialdehyde; ROS, reactive oxygen species.

6.8.3 Salt stress

Salinity, a key abiotic stressor, is a risk to global food security, damaging crop yields of over 800 million to 1 billion hectares in 100 nations (Duan *et al.*, 2023). It affects 23% of cultivated land and 25–30% of irrigated land, reducing plant growth parameters and crop production globally (Daba and Qureshi, 2021; Karaca *et al.*, 2022). The problem becomes even worse with the use of high-salinity irrigation water, especially in dry and semi-arid regions (Amer, 2021). Cells responded to salt stress by excluding Na^+ ions and producing osmoprotectants, reducing photosynthetic and antioxidant levels, which reduce plant growth (Balasubramaniam *et al.*, 2023). Salinity stress damages crops and vegetables such as *Triticum aestivum*, *Solanum lycopersicum*, and *Hordeum vulgare*, emphasizing the need for long-term solutions to combat this challenge and protect food production globally (Akhter *et al.*, 2021; Guo *et al.*, 2022).

SLs are a recent finding among plant growth regulators that help plants adapt to challenging conditions while also improving the signaling channel in plants (Fig. 6.8) (Lu *et al.*, 2023). They encourage hyphal branching, plant symbiotic relationships, and associations with AMF (Omoarelojie *et al.*, 2019; Alvi *et al.*, 2022). SLs impact plant growth and developmental processes, including shoot elongation, root branching, photomorphogenesis, and leaf senescence (Faizan *et al.*, 2022). Applying GR24 to salt-stressed sunflower seedlings stimulated photosynthetic activity by improving leaf chlorophyll content and antioxidant metabolism, and lowered MDA and ROS levels (Ahsan *et al.*, 2023). Ling *et al.* (2020) found that when synthetic SL GR24 was applied to rice plants, it significantly enhanced SOD and peroxidase (POD) activity while decreasing MDA levels, because the SL-signaling pathway contains enzymes such as leucine-rich repeat protein D3/MAX2, α/β hydrolase (D14) SL receptor, and Clp ATPase family, which is a protein transcriptional repressor, promoting growth and development of crops under salt stress. Similarly, in apple seedlings, exogenous GR24 enhances photosynthetic activity by improving chlorophyll levels, which increase stomatal conductance and transpiration rate, resulting in maximum plant fresh weight with a lower wilting rate under KCl stress (Zheng *et al.*, 2021). Correspondingly, GR24 treatment in wheat plants improves cell viability while inhibiting ROS damage by controlling

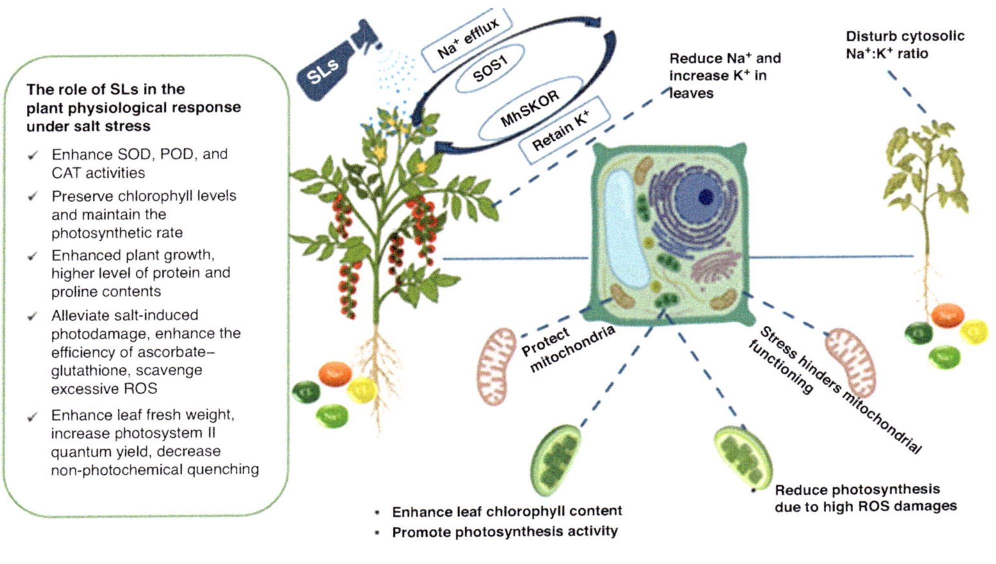

Fig. 6.8. Functions of strigolactones in plants under salinity stress.

the production of H_2O_2, superoxide (O_2^-), and MDA under salinity stress (Sedaghat *et al.*, 2017). These findings suggest that exogenous SL application improves cell viability, reduces lipid peroxidation, and protects proline content while increasing soluble sugar and protein levels, reduces electrolyte leakage by increasing the activity of CAT, POD, and SOD enzymes, contributing as a fundamental factor in plant establishment and development under salinity/alkalinity stress (Wang *et al.*, 2019). These findings have potential benefits for exploring plant resistance strategies and establishing new agricultural techniques.

6.9 Agricultural Implications of Strigolactone Research

SLs have considerable prospects in the realm of agriculture. The versatility of functions suggests the importance of utilization of SLs in horticulture and agriculture. Biosynthesis and perception of SLs are important goals for genetic engineering and plant breeding approaches for sustainable agriculture. The first practical utilization of SLs was for regulating parasitic weeds, a process known as "suicidal germination" (Kgosi *et al.*, 2012), whereby parasitic weed seeds were stimulated to germinate in the absence of a host root, leading to their death.

Given their ability to assist AMF colonization in plant roots, from an agricultural perspective SLs could be utilized as indirect biofertilizers. One of the foremost functions of AMF association with the plant is to transfer mineral nutrients, specifically phosphate. In several areas of the world, the concentration of phosphate in soil is very low, which significantly reduces plant growth. AMF could be beneficial in such areas for plant growth and health. In accordance with the fundamental role of AMF in phosphate acquisition, root exudates of phosphate-limited plants were established as more inducive to AMF in comparison with root exudates having an adequate amount of phosphate (Nagahashi and Douds, 2004).

S. hermonthica is a catastrophic root parasitic weed that affects several regions worldwide and is a major risk to global food resilience. SLs are at the forefront of the association of *Striga* and host plants, and are crucial for mitigating this catastrophic weed. The latest research indicates that the structure and extent of secretion of SLs verify the susceptibility of *Striga*. In comparison, orobanchol-like SLs are a substantially better germination activator. Consequently, modification of SL structure is an effective approach to mitigate infestation by root-penetrating parasites. Such modifications can be performed by genetic engineering or plant breeding (Jia *et al.*, 2019).

6.10 Major Obstacles and Future Perspectives: Harnessing Strigolactones for Sustainable Agriculture

SLs have been evaluated as an agroproduct for cultivated land, but have not yet been implemented as a product. For the time being, the major milestone in SL utilization is the lack of cost-effective sources of SLs. Synthetic production of natural SLs is not feasible for extensive manufacturing owing to their complex composition and stereochemistry, as their synthesis needs approximately 20 steps and it is not achievable at present to obtain SLs at an appropriate scale. Bioactive SL analogs can be prepared with maximum agronomic production and have served as key sources of SL research studies. GR24 is a commercially accessible SL analog that has the characteristics of maximum parasitic seed germination and branching suppression potential. It is consistently utilized to explore the activities of SLs. Other metabolically active SL analogs with improved configurations have been explored that have uncomplicated production and maximum yield outputs, including Nijmegen-1, which his utilized for suicidal germination of *Orobanche* and *Striga* spp. and can be prepared as an optical isomer mix in just two stages in field trials (Vurro *et al.*, 2016), although the cost is an ongoing obstacle for mass production. However, microbial engineering might create an opportunity to synthesize SLs in sufficient quantities for crop-management approaches. Remodeling of the SL synthesis pathway in microbial hosts including yeast and *Escherichia coli* may allow SL synthesis in exogenous hosts (Chesterfield *et al.*, 2020).

Additionally, implementing massive-output omics techniques using genomics, proteomics, and metabolomics may provide extensive knowledge of the various modifications occurring worldwide in plants during SL and AMF association. Such techniques can offer valuable perspectives into the regulatory pathways, biological processes, and molecular responses supporting this interactive effect. Genetic engineering approaches include gene knockout or intensified expression to explore the particular functions of crucial genes associated with SL and AMF interaction, and may assist in identifying the fundamental components and regulatory factors that trigger the beneficial outcomes and elevate the SL–AMF interaction potential (Boyno *et al.*, 2023).

6.11 Conclusion

SLs have emerged as influential bioactive hormones and master regulators of plant establishment and development, regulating core mechanisms from germination to plant morphology. To date, over 30 SLs have been recognized and characterized. It is expected further SLs will be recognized with the evolution of technology. These invisible regulators not only regulate root and shoot architecture but also promote plant growth and survival according to environmental cues. They play a crucial role in the symbiotic relationship between AMF and plants, and act as growth regulators and signaling modulators. As further findings emerge, harnessing the potential of SL molecular composition could innovate the realm of agriculture and global food production.

References

Aguilar-Martínez, J.A., Poza-Carrión, C. and Cubas, P. (2007) *Arabidopsis* BRANCHED1 acts as an integrator of branching signals within axillary buds. *Plant Cell* 19(2), 458–472. DOI: 10.1105/tpc.106.048934.

Ahmad, M.Z., Rehman, N.U., Yu, S., Zhou, Y., Haq, B.U. *et al.* (2020) GmMAX2-D14 and -KAI interaction-mediated SL and KAR signaling play essential roles in soybean root nodulation. *Plant Journal* 101(2), 334–351. DOI: 10.1111/tpj.14545.

Ahsan, M., Zulfiqar, H., Farooq, M.A., Ali, S., Tufail, A. *et al.* (2023) Strigolactone (GR24) application positively regulates photosynthetic attributes, stress-related metabolites and antioxidant enzymatic activities of ornamental sunflower (*Helianthus annuus* cv. Vincent's Choice) under salinity stress. *Agriculture* 13(1), 50. DOI: 10.3390/agriculture13010050.

Akhter, M.S., Noreen, S., Mahmood, S., Athar, H., Ashraf, M. *et al.* (2021) Influence of salinity stress on PSII in barley (*Hordeum vulgare* L.) genotypes, probed by chlorophyll-a fluorescence. *Journal of King Saud University—Science* 33(1), 101239. DOI: 10.1016/j.jksus.2020.101239.

Al-Amri, A.A., Alsubaie, Q.D., Alamri, S.A. and Siddiqui, M.H. (2023) Strigolactone analog GR24 induces seed germination and improves growth performance of different genotypes of tomato. *Journal of Plant Growth Regulation* 42(9), 5653–5666. DOI: 10.1007/s00344-023-10947-8.

Al-Babili, S. and Bouwmeester, H.J. (2015) Strigolactones, a novel carotenoid-derived plant hormone. *Annual Review of Plant Biology* 66(1), 161–186. DOI: 10.1146/annurev-arplant-043014-114759.

Alvi, A.F., Sehar, Z., Fatma, M., Masood, A. and Khan, N.A. (2022) Strigolactone: An emerging growth regulator for developing resilience in plants. *Plants* 11(19), 2604. DOI: 10.3390/plants11192604.

Amer, R. (2021) Spatial relationship between irrigation water salinity, waterlogging, and cropland degradation in the arid and semi-arid environments. *Remote Sensing* 13(6), 1047. DOI: 10.3390/rs13061047.

Bajguz, A., Chmur, M. and Gruszka, D. (2020) Comprehensive overview of the brassinosteroid biosynthesis pathways: Substrates, products, inhibitors, and connections. *Frontiers in Plant Science* 11, 1034. DOI: 10.3389/fpls.2020.01034.

Balasubramaniam, T., Shen, G., Esmaeili, N. and Zhang, H. (2023) Plants' response mechanisms to salinity stress. *Plants* 12(12), 2253. DOI: 10.3390/plants12122253.

Bennett, T., Sieberer, T., Willett, B., Booker, J., Luschnig, C. *et al.* (2006) The *Arabidopsis* MAX pathway controls shoot branching by regulating auxin transport. *Current Biology* 16(6), 553–563. DOI: 10.1016/j.cub.2006.01.058.

Bermudez, R., Stefanski, A., Montgomery, R.A. and Reich, P.B. (2021) Short- and long-term responses of photosynthetic capacity to temperature in four boreal tree species in a free-air warming and rainfall manipulation experiment. *Tree Physiology* 41(1), 89–102. DOI: 10.1093/treephys/tpaa115.

Bouwmeester, H.J., Matusova, R., Zhongkui, S. and Beale, M.H. (2003) Secondary metabolite signalling in host–parasitic plant interactions. *Current Opinion in Plant Biology* 6(4), 358–364. DOI: 10.1016/s1369-5266(03)00065-7.

Boyer, F.-D., Germain, A., Pouvreau, J.-B., Clavé, G., Pillot, J.-P. *et al.* (2014) New strigolactone analogs as plant hormones with low activities in the rhizosphere. *Molecular Plant* 7(4), 675–690. DOI: 10.1093/mp/sst163.

Boyno, G., Rezaee Danesh, Y., Demir, S., Teniz, N., Mulet, J.M. *et al.* (2023) The complex interplay between arbuscular mycorrhizal fungi and strigolactone: Mechanisms, sinergies, applications and future directions. *International Journal of Molecular Sciences* 24(23), 16774. DOI: 10.3390/ijms242316774.

Braun, N., Germain, A., Pillot, J.-P., Boutet-Mercey, S., Dalmais, M. *et al.* (2012) The pea TCP transcription factor PsBRC1 acts downstream of strigolactones to control shoot branching. *Plant Physiology* 158(1), 225–238. DOI: 10.1104/pp.111.182725.

Brewer, P.B., Koltai, H. and Beveridge, C.A. (2013) Diverse roles of strigolactones in plant development. *Molecular Plant* 6(1), 18–28. DOI: 10.1093/mp/sss130.

Chesterfield, R.J., Vickers, C.E. and Beveridge, C.A. (2020) Translation of strigolactones from plant hormone to agriculture: Achievements, future perspectives, and challenges. *Trends in Plant Science* 25(11), 1087–1106. DOI: 10.1016/j.tplants.2020.06.005.

Chi, C., Xu, X., Wang, M., Zhang, H., Fang, P. *et al.* (2021) Strigolactones positively regulate abscisic acid-dependent heat and cold tolerance in tomato. *Horticulture Research* 8(1), 237. DOI: 10.1038/s41438-021-00668-y.

Daba, A.W. and Qureshi, A.S. (2021) Review of soil salinity and sodicity challenges to crop production in the lowland irrigated areas of ethiopia and its management strategies. *Land* 10(12), 1377. DOI: 10.3390/land10121377.

Dale, M.P. and Causton, D.R. (1992) Use of the chlorophyll *a/b* ratio as a bioassay for the light environment of a plant. *Functional Ecology* 6(2), 190. DOI: 10.2307/2389754.

Decker, E.L., Alder, A., Hunn, S., Ferguson, J., Lehtonen, M.T. *et al.* (2017) Strigolactone biosynthesis is evolutionarily conserved, regulated by phosphate starvation and contributes to resistance against phytopathogenic fungi in a moss, *Physcomitrella patens*. *New Phytologist* 216(2), 455–468. DOI: 10.1111/nph.14506.

de Kraker, J.W., Franssen, M.C., de Groot, A., Konig, W.A. and Bouwmeester, H.J. (1998) (+)-Germacrene A biosynthesis: The committed step in the biosynthesis of bitter sesquiterpene lactones in chicory. *Plant Physiology* 117(4), 1381–1392. DOI: 10.1104/pp.117.4.1381.

Duan, G., Liu, M., Liang, Z., Wang, M., Yang, H. *et al.* (2023) Amendments of severe saline-sodic paddy land: Optimal combination of phosphogypsum, farmyard fertilizer, and wood peat. *Agronomy* 13(5), 1364. DOI: 10.3390/agronomy13051364.

Dun, E.A., de Saint Germain, A., Rameau, C. and Beveridge, C.A. (2012) Antagonistic action of strigolactone and cytokinin in bud outgrowth control. *Plant Physiology* 158(1), 487–498. DOI: 10.1104/pp.111.186783.

El Sabagh, A., Islam, M.S., Hossain, A., Iqbal, M.A., Mubeen, M. *et al.* (2022) Phytohormones as growth regulators during abiotic stress tolerance in plants. *Frontiers in Agronomy* 4, 765068. DOI: 10.3389/fagro.2022.765068.

Faizan, M., Cheng, S.H., Tonny, S.H. and Robab, M.I. (2022) Specific roles of strigolactones in plant physiology and remediation of heavy metals from contaminated soil. *Plant Physiology and Biochemistry* 192, 186–195. DOI: 10.1016/j.plaphy.2022.10.004.

Faisal, M., Faizan, M., Tonny, S.H., Rajput, V.D., Minkina, T. *et al.* (2023) Strigolactone-mediated mitigation of negative effects of salinity stress in *Solanum lycopersicum* through reducing the oxidative damage. *Sustainability* 15(7), 5805. DOI: 10.3390/su15075805.

Faizan, M., Faraz, A., Sami, F., Siddiqui, H., Yusuf, M. *et al.* (2020) Role of strigolactones: Signalling and crosstalk with other phytohormones. *Open Life Sciences* 15(1), 217–228. DOI: 10.1515/biol-2020-0022.

Ferrero, M., Pagliarani, C., Novák, O., Ferrandino, A., Cardinale, F. *et al.* (2018) Exogenous strigolactone interacts with abscisic acid-mediated accumulation of anthocyanins in grapevine berries. *Journal of Experimental Botany* 69(9), 2391–2401. DOI: 10.1093/jxb/ery033.

Foo, E. and Davies, N.W. (2011) Strigolactones promote nodulation in pea. *Planta* 234(5), 1073–1081. DOI: 10.1007/s00425-011-1516-7.

Foo, E., Yoneyama, K., Hugill, C.J., Quittenden, L.J. and Reid, J.B. (2013) Strigolactones and the regulation of pea symbioses in response to nitrate and phosphate deficiency. *Molecular Plant* 6(1), 76–87. DOI: 10.1093/mp/sss115.

Guan, J.C., Koch, K.E., Suzuki, M., Wu, S., Latshaw, S. *et al.* (2012) Diverse roles of strigolactone signaling in maize architecture and the uncoupling of a branching-specific subnetwork. *Plant Physiology* 160(3), 1303–1317. DOI: 10.1104/pp.112.204503.

Guercio, A.M., Palayam, M. and Shabek, N. (2023) Strigolactones: Diversity, perception, and hydrolysis. *Phytochemistry Reviews* 22(2), 339–360. DOI: 10.1007/s11101-023-09853-4.

Guo, M., Wang, X.-S., Guo, H.-D., Bai, S.-Y., Khan, A. *et al.* (2022) Tomato salt tolerance mechanisms and their potential applications for fighting salinity: A review. *Frontiers in Plant Science* 13, 949541. DOI: 10.3389/fpls.2022.949541.

Ha, C.V., Leyva-González, M.A., Osakabe, Y., Tran, U.T., Nishiyama, R. *et al.* (2014) Positive regulatory role of strigolactone in plant responses to drought and salt stress. *Proceedings of the National Academy of Sciences USA* 111(2), 851–856. DOI: 10.1073/pnas.1322135111.

Haider, I., Andreo-Jimenez, B., Bruno, M., Bimbo, A., Floková, K. *et al.* (2018) The interaction of strigolactones with abscisic acid during the drought response in rice. *Journal of Experimental Botany* 69(9), 2403–2414. DOI: 10.1093/jxb/ery089.

Hayward, A., Stirnberg, P., Beveridge, C. and Leyser, O. (2009) Interactions between auxin and strigolactone in shoot branching control. *Plant Physiology* 151(1), 400–412. DOI: 10.1104/pp.109.137646.

Ho-Plágaro, T. and García-Garrido, J.M. (2022) Molecular regulation of arbuscular mycorrhizal symbiosis. *International Journal of Molecular Sciences* 23(11), 5960. DOI: 10.3390/ijms23115960.

Ito, S., Braguy, J., Wang, J.Y., Yoda, A., Fiorilli, V. *et al.* (2022) Canonical strigolactones are not the major determinant of tillering but important rhizospheric signals in rice. *Science Advances* 8(44), eadd1278. DOI: 10.1126/sciadv.add1278.

Jamil, M., Charnikhova, T., Houshyani, B., Ast, A. and Bouwmeester, H.J. (2012) Genetic variation in strigolactone production and tillering in rice and its effect on *Striga hermonthica* infection. *Planta* 235(3), 473–484. DOI: 10.1007/s00425-011-1520-y.

Jia, K.P., Baz, L. and Al-Babili, S. (2018) From carotenoids to strigolactones. *Journal of Experimental Botany* 69(9), 2189–2204. DOI: 10.1093/jxb/erx476.

Jia, K.P., Li, C., Bouwmeester, H.J. and Al-Babili, S. (2019) Strigolactone biosynthesis and signal transduction. In: Koltai, H. and Prandi, C. (eds) *Strigolactones—Biology and Applications*. Springer Nature, Cham, Switzerland, pp. 1–45. DOI: 10.1007/978-3-030-12153-2_1.

Jiang, K. and Feldman, L.J. (2002) Root meristem establishment and maintenance: The role of auxin. *Journal of Plant Growth Regulation* 21(4), 432–440. DOI: 10.1007/s00344-002-0037-9.

Jiang, L., Matthys, C., Marquez-Garcia, B., Cuyper, C., Smet, L. *et al.* (2016) Strigolactones spatially influence lateral root development through the cytokinin signaling network. *Journal of Experimental Botany* 67(1), 379–389. DOI: 10.1093/jxb/erv478.

Kapulnik, Y., Delaux, P.-M., Resnick, N., Mayzlish-Gati, E., Wininger, S. *et al.* (2011) Strigolactones affect lateral root formation and root-hair elongation in *Arabidopsis*. *Planta* 233(1), 209–216. DOI: 10.1007/s00425-010-1310-y.

Karaca, C., Aslan, G.E., Buyuktas, D., Kurunc, A., Bastug, R. *et al.* (2022) Effects of salinity stress on drip-irrigated tomatoes grown under mediterranean-type greenhouse conditions. *Agronomy* 13(1), 36. DOI: 10.3390/agronomy13010036.

Kellogg, E.A. (1997) Plant evolution: The dominance of maize. *Current Biology* 7(7), R411–413. DOI: 10.1016/s0960-9822(06)00204-1.

Kgosi, R.L., Zwanenburg, B., Mwakaboko, A.S. and Murdoch, A.J. (2012) Strigolactone analogues induce suicidal seed germination of *Striga* spp. *Weed Research* 52(3), 197–203. DOI: 10.1111/j.1365-3180.2012.00912.x.

Khalid, M.F., Iqbal Khan, R., Jawaid, M.Z., Shafqat, W., Hussain, S. *et al.* (2022) Nanoparticles: The plant saviour under abiotic stresses. *Nanomaterials* 12(21), 3915. DOI: 10.3390/nano12213915.

Khan, N., Bano, A., Rahman, M.A., Rathinasabapathi, B. and Babar, M.A. (2019) UPLC-HRMS-based untargeted metabolic profiling reveals changes in chickpea (*Cicer arietinum*) metabolome following long-term drought stress. *Plant, Cell & Environment* 42(1), 115–132. DOI: 10.1111/pce.13195.

Koltai, H. (2013) Strigolactones activate different hormonal pathways for regulation of root development in response to phosphate growth conditions. *Annals of Botany* 112(2), 409–415. DOI: 10.1093/aob/mcs216.

Koltai, H., LekKala, S.P., Bhattacharya, C., Mayzlish-Gati, E., Resnick, N. *et al.* (2010) A tomato strigolactone-impaired mutant displays aberrant shoot morphology and plant interactions. *Journal of Experimental Botany* 61(6), 1739–1749. DOI: 10.1093/jxb/erq041.

Kusajima, M., Fujita, M., Soudthedlath, K., Nakamura, H., Yoneyama, K. *et al.* (2022) Strigolactones modulate salicylic acid-mediated disease resistance in *Arabidopsis thaliana*. *International Journal of Molecular Sciences* 23(9), 5246. DOI: 10.3390/ijms23095246.

Lee, H.Y. and Yoon, G.M. (2020) Strigolactone elevates ethylene biosynthesis in etiolated arabidopsis seedlings. *Plant Signaling & Behavior* 15(11), 1805232. DOI: 10.1080/15592324.2020.1805232.

Li, C., Lu, X., Liu, Y., Xu, J. and Yu, W. (2023) Strigolactone alleviates the adverse effects of salt stress on seed germination in cucumber by enhancing antioxidant capacity. *Antioxidants* 12(5), 1043. DOI: 10.3390/antiox12051043.

Liu, J., He, H., Vitali, M., Visentin, I., Charnikhova, T. *et al.* (2015) Osmotic stress represses strigolactone biosynthesis in lotus japonicus roots: Exploring the interaction between strigolactones and ABA under abiotic stress. *Planta* 241(6), 1435–1451. DOI: 10.1007/s00425-015-2266-8.

Li, W., Herrera-Estrella, L. and Tran, L.S.P. (2019) Do cytokinins and strigolactones crosstalk during drought adaptation? *Trends in Plant Science* 24(8), 669–672. DOI: 10.1016/j.tplants.2019.06.007.

Li, W., Gupta, A., Tian, H., Nguyen, K.H., Tran, C.D. *et al.* (2020) Different strategies of strigolactone and karrikin signals in regulating the resistance of *Arabidopsis thaliana* to water-deficit stress. *Plant Signaling & Behavior* 15(9), 1789321. DOI: 10.1080/15592324.2020.1789321.

Ling, F., Su, Q., Jiang, H., Cui, J., He, X. *et al.* (2020) Effects of strigolactone on photosynthetic and physiological characteristics in salt-stressed rice seedlings. *Scientific Reports* 10(1), 6183. DOI: 10.1038/s41598-020-63352-6.

Liu, G., Pfeifer, J., Brito Francisco, R., Emonet, A., Stirnemann, M. *et al.* (2018) Changes in the allocation of endogenous strigolactone improve plant biomass production on phosphate-poor soils. *New Phytologist* 217(2), 784–798. DOI: 10.1111/nph.14847.

Liu, H., Li, C., Yan, M., Zhao, Z., Huang, P. *et al.* (2022) Strigolactone is involved in nitric oxide-enhanced salt resistance in tomato seedlings. *Journal of Plant Research* 135(2), 337–350. DOI: 10.1007/s10265-022-01371-2.

Liu, J., Novero, M., Charnikhova, T., Ferrandino, A., Schubert, A. *et al.* (2013) Carotenoid cleavage dioxygenase 7 modulates plant growth, reproduction, senescence, and determinate nodulation in the model legume lotus japonicus. *Journal of Experimental Botany* 64(7), 1967–1981. DOI: 10.1093/jxb/ert056.

López, R., Ramírez-Valiente, J.A. and Pita, P. (2022) How plants cope with heatwaves in a drier environment. *Flora* 295, 152148. DOI: 10.1016/j.flora.2022.152148.

López-Ráez, J.A., Shirasu, K. and Foo, E. (2017) Strigolactones in plant interactions with beneficial and detrimental organisms: The yin and yang. *Trends in Plant Science* 22(6), 527–537. DOI: 10.1016/j.tplants.2017.03.011.

Lu, X., Liu, X., Xu, J., Liu, Y., Chi, Y. *et al.* (2023) Strigolactone-mediated trehalose enhances salt resistance in tomato seedlings. *Horticulturae* 9(7), 770. DOI: 10.3390/horticulturae9070770.

Ma, N., Hu, C., Wan, L., Hu, Q., Xiong, J. *et al.* (2017) Strigolactones improve plant growth, photosynthesis, and alleviate oxidative stress under salinity in rapeseed (*Brassica napus* L.) by regulating gene expression. *Frontiers in Plant Science* 8, 1671. DOI: 10.3389/fpls.2017.01671.

Manandhar, S., Funnell, K.A., Woolley, D.J. and Cooney, J.M. (2018) Interaction between strigolactone and cytokinin on axillary and adventitious bud development in Zantedeschia. *Journal of Plant Physiology and Pathology* 6(1), 1000172. DOI: 10.4172/2329-955X.1000172.

Marchin, R.M., Backes, D., Ossola, A., Leishman, M.R., Tjoelker, M.G. *et al.* (2022) Extreme heat increases stomatal conductance and drought-induced mortality risk in vulnerable plant species. *Global Change Biology* 28(3), 1133–1146. DOI: 10.1111/gcb.15976.

Marro, N., Lidoy, J., Chico, M.Á., Rial, C., García, J. *et al.* (2022) Strigolactones: New players in the nitrogen–phosphorus signalling interplay. *Plant, Cell & Environment* 45(2), 512–527. DOI: 10.1111/pce.14212.

Marzec, M. (2016) Strigolactones as part of the plant defence system. *Trends in Plant Science* 21(11), 900–903. DOI: 10.1016/j.tplants.2016.08.010.

Mi, X., Anwar, S., Chen, C., Jin, X., Yu, L. *et al.* (2023) Strigolactone enhances alkaline tolerance in soybean seeds germination by altering expression profiles of ABA biosynthetic and signaling genes. *Journal of Plant Biology* 66(4), 373–381. DOI: 10.1007/s12374-022-09357-2.

Miyakawa, T., Xu, Y. and Tanokura, M. (2020) Molecular basis of strigolactone perception in root-parasitic plants: Aiming to control its germination with strigolactone agonists/antagonists. *Cellular and Molecular Life Sciences* 77(6), 1103–1113. DOI: 10.1007/s00018-019-03318-8.

Mostofa, M.G., Li, W., Nguyen, K.H., Fujita, M. and Tran, L.P. (2018) Strigolactones in plant adaptation to abiotic stresses: An emerging avenue of plant research. *Plant, Cell & Environment* 41(10), 2227–2243. DOI: 10.1111/pce.13364.

Min, Z., Li, R., Chen, L., Zhang, Y., Li, Z. *et al.* (2019) Alleviation of drought stress in grapevine by foliar-applied strigolactones. *Plant Physiology and Biochemistry* 135, 99–110. DOI: 10.1016/j.plaphy.2018.11.037.

Motonami, N., Ueno, K., Nakashima, H., Nomura, S., Mizutani, M. *et al.* (2013) The bioconversion of 5-deoxystrigol to sorgomol by the sorghum, *Sorghum bicolor* (L.) Moench. *Phytochemistry* 93, 41–48. DOI: 10.1016/j.phytochem.2013.02.017.

Nagahashi, G. and Douds, D.D. (2004) Isolated root caps, border cells, and mucilage from host roots stimulate hyphal branching of the arbuscular mycorrhizal fungus, *Gigaspora gigantea*. *Mycological Research* 108(Pt 9), 1079–1088. DOI: 10.1017/s0953756204000693.

Nishiyama, R., Watanabe, Y., Fujita, Y., Le, D.T., Kojima, M. *et al.* (2011) Analysis of cytokinin mutants and regulation of cytokinin metabolic genes reveals important regulatory roles of cytokinins in drought, salt and abscisic acid responses, and abscisic acid biosynthesis. *Plant Cell* 23(6), 2169–2183. DOI: 10.1105/tpc.111.087395.

Omoarelojie, L.O., Kulkarni, M.G., Finnie, J.F. and van Staden, J. (2019) Strigolactones and their crosstalk with other phytohormones. *Annals of Botany* 124(5), 749–767. DOI: 10.1093/aob/mcz100.

Pandey, A., Sharma, M. and Pandey, G.K. (2016) Emerging roles of strigolactones in plant responses to stress and development. *Frontiers in Plant Science* 7, 434. DOI: 10.3389/fpls.2016.00434.

Parniske, M. (2008) Arbuscular mycorrhiza: The mother of plant root endosymbioses. *Nature Reviews Microbiology* 6(10), 763–775. DOI: 10.1038/nrmicro1987.

Rameau, C., Goormachtig, S., Cardinale, F., Bennett, T. and Cubas, P. (2019) Strigolactones as plant hormones. In: Koltai, H. and Prandi, C. (eds) *Strigolactones—Biology and Applications*. Springer Nature, Cham, Switzerland, pp. 47–87. DOI: 10.1007/978-3-030-12153-2_2.

Rasmussen, A., Mason, M.G., Cuyper, C., Brewer, P.B., Herold, S. *et al.* (2012) Strigolactones suppress adventitious rooting in Arabidopsis and pea. *Plant Physiology* 158(4), 1976–1987. DOI: 10.1104/pp.111.187104.

Redecker, D. and Raab, P. (2006) Phylogeny of the Glomeromycota (arbuscular mycorrhizal fungi): Recent developments and new gene markers. *Mycologia* 98(6), 885–895. DOI: 10.1080/15572536.2006.11832618.

Ruiz-Lozano, J.M., Aroca, R., Zamarreño, Á.M., Molina, S., Andreo-Jiménez, B. *et al.* (2016) Arbuscular mycorrhizal symbiosis induces strigolactone biosynthesis under drought and improves drought tolerance in lettuce and tomato. *Plant, Cell & Environment* 39(2), 441–452. DOI: 10.1111/pce.12631.

Ruyter-Spira, C., Kohlen, W., Charnikhova, T., Zeijl, A., Bezouwen, L. *et al.* (2011) Physiological effects of the synthetic strigolactone analog GR24 on root system architecture in *Arabidopsis*: Another below-ground role for strigolactones? *Plant Physiology* 155(2), 721–734. DOI: 10.1104/pp.110.166645.

Sattar, A., Ul-Allah, S., Ijaz, M., Sher, A., Butt, M. *et al.* (2022) Exogenous application of strigolactone alleviates drought stress in maize seedlings by regulating the physiological and antioxidants defense mechanisms. *Cereal Research Communications* 50(2), 263–272. DOI: 10.1007/s42976-021-00171-z.

Scafaro, A.P., Fan, Y., Posch, B.C., Garcia, A., Coast, O. *et al.* (2021) Responses of leaf respiration to heatwaves. *Plant, Cell & Environment* 44(7), 2090–2101. DOI: 10.1111/pce.14018.

Sedaghat, M., Tahmasebi-Sarvestani, Z., Emam, Y. and Mokhtassi-Bidgoli, A. (2017) Physiological and antioxidant responses of winter wheat cultivars to strigolactone and salicylic acid in drought. *Plant Physiology and Biochemistry* 119, 59–69. DOI: 10.1016/j.plaphy.2017.08.015.

Sedaghat, M., Sarvestani, Z.T., Emam, Y., Bidgoli, A.M. and Sorooshzadeh, A. (2020) Foliar-applied GR24 and salicylic acid enhanced wheat drought tolerance. *Russian Journal of Plant Physiology* 67(4), 733–739. DOI: 10.1134/S1021443720040159.

Shafqat, W., Naqvi, S.A., Maqbool, R., Haider, M.S., Jaskani, M.J. *et al.* (2021) Climate change and citrus. In: Kahn, M.S. and Khan, I.A. (eds) *Citrus—Research, Development and Biotechnology*. Intech Open, London. Available at: www.intechopen.com/chapters/74767 (accessed 29 January 2026).

Shah, T., Khan, Z., Asad, M., Imran, A., Khan Niazi, M.B. *et al.* (2023) Alleviation of cadmium toxicity in wheat by strigolactone: Regulating cadmium uptake, nitric oxide signaling, and genes encoding antioxidant defense system. *Plant Physiology and Biochemistry* 202, 107916. DOI: 10.1016/j.plaphy.2023.107916.

Sorefan, K., Booker, J., Haurogné, K., Goussot, M., Bainbridge, K. *et al.* (2003) MAX4 and RMS1 are orthologous dioxygenase-like genes that regulate shoot branching in *Arabidopsis* and pea. *Genes & Development* 17(12), 1469–1474. DOI: 10.1101/gad.256603.

Soto, M.J., Fernández-Aparicio, M., Castellanos-Morales, V., García-Garrido, J.M., Ocampo, J.A. *et al.* (2010) First indications for the involvement of strigolactones on nodule formation in alfalfa (*Medicago sativa*). *Soil Biology and Biochemistry* 42(2), 383–385. DOI: 10.1016/j.soilbio.2009.11.007.

Soundappan, I., Bennett, T., Morffy, N., Liang, Y., Stanga, J.P. *et al.* (2015) SMAX1-LIKE/D53 family members enable distinct MAX2-dependent responses to strigolactones and karrikins in *Arabidopsis*. *Plant Cell* 27(11), 3143–3159. DOI: 10.1105/tpc.15.00562.

Sun, H., Tao, J., Liu, S., Huang, S., Chen, S. *et al.* (2014) Strigolactones are involved in phosphate- and nitrate-deficiency-induced root development and auxin transport in rice. *Journal of Experimental Botany* 65(22), 6735–6746. DOI: 10.1093/jxb/eru029.

Tariq, A., Ullah, I., Sardans, J., Graciano, C., Mussarat, S. *et al.* (2023) Strigolactones can be a potential tool to fight environmental stresses in arid lands. *Environmental Research* 229, 115966. DOI: 10.1016/j.envres.2023.115966.

Teskey, R., Wertin, T., Bauweraerts, I., Ameye, M., McGuire, M.A. *et al.* (2015) Responses of tree species to heat waves and extreme heat events. *Plant, Cell & Environment* 38(9), 1699–1712. DOI: 10.1111/pce.12417.

Umehara, M., Hanada, A., Yoshida, S., Akiyama, K., Arite, T. *et al.* (2008) Inhibition of shoot branching by new terpenoid plant hormones. *Nature* 455(7210), 195–200. DOI: 10.1038/nature07272.

Visentin, I., Vitali, M., Ferrero, M., Zhang, Y., Ruyter-Spira, C. *et al.* (2016) Low levels of strigolactones in roots as a component of the systemic signal of drought stress in tomato. *New Phytologist* 212, 954–963. DOI: 10.1111/nph.14190.

Visentin, I., Pagliarani, C., Deva, E., Caracci, A., Turečková, V. *et al.* (2020) A novel strigolactone-miR156 module controls stomatal behaviour during drought recovery. *Plant, Cell & Environment* 43(7), 1613–1624. DOI: 10.1111/pce.13758.

Vogel, J.T., Walter, M.H., Giavalisco, P., Lytovchenko, A., Kohlen, W. *et al.* (2010) SlCCD7 controls strigolactone biosynthesis, shoot branching and mycorrhiza-induced apocarotenoid formation in tomato. *Plant Journal* 61(2), 300–311. DOI: 10.1111/j.1365-313X.2009.04056.x.

Vurro, M., Prandi, C. and Baroccio, F. (2016) Strigolactones: How far is their commercial use for agricultural purposes? *Pest Management Science* 72(11), 2026–2034. DOI: 10.1002/ps.4254.

Wang, D.-W., Yu, S.-Y., Pang, Z.-L., Ma, D.-J., Liang, L. *et al.* (2021) Discovery of a broad-spectrum fluorogenic agonist for strigolactone receptors through a computational approach. *Journal of Agricultural and Food Chemistry* 69(36), 10486–10495. DOI: 10.1021/acs.jafc.1c03471.

Wang, J.Y., Braguy, J., Chen, G.T.E., Jamil, M., Balakrishna, A. *et al.* (2022) Perspectives on the metabolism of strigolactone rhizospheric signals. *Frontiers in Plant Science* 13, 1062107. DOI: 10.3389/fpls.2022.1062107.

Wang, M., Dai, W., Du, J., Ming, R., Dahro, B. *et al.* (2019) ERF109 of trifoliate orange (*Poncirus trifoliata* (L.) Raf.) contributes to cold tolerance by directly regulating expression of *Prx1* involved in antioxidative process. *Plant Biotechnology Journal* 17(7), 1316–1332. DOI: 10.1111/pbi.13056.

Wang, X., Li, Z., Shi, Y., Liu, Z., Zhang, X. *et al.* (2023) Strigolactones promote plant freezing tolerance by releasing the WRKY41-mediated inhibition of CBF/DREB1 expression. *EMBO Journal* 42(19), e112999. DOI: 10.15252/embj.2022112999.

Waters, M.T., Gutjahr, C., Bennett, T. and Nelson, D.C. (2017) Strigolactone signaling and evolution. *Annual Review of Plant Biology* 68(1), 291–322. DOI: 10.1146/annurev-arplant-042916-040925.

Yamada, Y. and Umehara, M. (2015) Possible roles of strigolactones during leaf senescence. *Plants* 4(3), 664–677. DOI: 10.3390/plants4030664.

Yang, Y., Zheng, Q., Zhou, K., Xiao, Y., Huang, C. *et al.* (2024) Effects of exogenous strigolactone on the cadmium accumulation in *Galinsoga parviflora* Cav. *Chemistry and Ecology* 40(3), 292–304. DOI: 10.1080/02757540.2024.2305339.

Yang, Y.Y., Ren, Y.R., Zheng, P.F., Zhao, L.L., You, C.X. *et al.* (2020) Cloning and functional identification of a strigolactone receptor gene MdD14 in apple. *Plant Cell, Tissue and Organ Culture* 140(1), 197–208. DOI: 10.1007/s11240-019-01722-3.

Yamada, Y., Furusawa, S., Nagasaka, S., Shimomura, K., Yamaguchi, S. *et al.* (2014) Strigolactone signaling regulates rice leaf senescence in response to a phosphate deficiency. *Planta* 240(2), 399–408. DOI: 10.1007/s00425-014-2096-0.

Xu, J., Zha, M., Li, Y., Ding, Y., Chen, L. *et al.* (2015) The interaction between nitrogen availability and auxin, cytokinin, and strigolactone in the control of shoot branching in rice (*Oryza sativa* L.). *Plant Cell Reports* 34(9), 1647–1662. DOI: 10.1007/s00299-015-1815-8.

Yoneyama, K. and Brewer, P.B. (2021) Strigolactones, how are they synthesized to regulate plant growth and development. *Current Opinion in Plant Biology* 63, 102072. DOI: 10.1016/j.pbi.2021.102072.

Zhang, S., Li, G., Fang, J., Chen, W., Jiang, H. *et al.* (2010) The interactions among DWARF10, auxin and cytokinin underlie lateral bud outgrowth in rice. *Journal of Integrative Plant Biology* 52(7), 626–638. DOI: 10.1111/j.1744-7909.2010.00960.x.

Zhang, Y., Li, Y., Hassan, M.J., Li, Z. and Peng, Y. (2020) Indole-3-acetic acid improves drought tolerance of white clover via activating auxin, abscisic acid and jasmonic acid related genes and inhibiting senescence genes. *BMC Plant Biology* 20(1), 150. DOI: 10.1186/s12870-020-02354-y.

Zhao, B., Wu, T.T., Ma, S.S., Jiang, D.J., Bie, X.M. *et al.* (2020) TaD27-B gene controls the tiller number in hexaploid wheat. *Plant Biotechnology Journal* 18(2), 513–525. DOI: 10.1111/pbi.13220.

Zheng, X., Li, Y., Xi, X., Ma, C., Sun, Z. *et al.* (2021) Exogenous strigolactones alleviate KCl stress by regulating photosynthesis, ROS migration and ion transport in *Malus hupehensis* Rehd. *Plant Physiology and Biochemistry* 159, 113–122. DOI: 10.1016/j.plaphy.2020.12.015.

Zhou, F., Lin, Q., Zhu, L., Ren, Y., Zhou, K. *et al.* (2013) D14–SCF^{D3}-dependent degradation of D53 regulates strigolactone signalling. *Nature* 504(7480), 406–410. DOI: 10.1038/nature12878.

Zhuang, L., Wang, J. and Huang, B. (2017) Drought inhibition of tillering in *Festuca arundinacea* associated with axillary bud development and strigolactone signaling. *Environmental and Experimental Botany* 142, 15–23. DOI: 10.1016/j.envexpbot.2017.07.017.

Zinta, G., AbdElgawad, H., Domagalska, M.A., Vergauwen, L., Knapen, D. *et al.* (2014) Physiological, biochemical, and genome-wide transcriptional analysis reveals that elevated co2 mitigates the impact of combined heat wave and drought stress in *Arabidopsis thaliana* at multiple organizational levels. *Global Change Biology* 20(12), 3670–3685. DOI: 10.1111/gcb.12626.

Zorrilla, J.G., Rial, C., Varela, R.M., Molinillo, J.M.G. and Macías, F.A. (2022) Strategies for the synthesis of canonical, non-canonical and analogues of strigolactones, and evaluation of their parasitic weed germination activity. *Phytochemistry Reviews* 21(5), 1627–1659. DOI: 10.1007/s11101-022-09801-8.

Zwanenburg, B. and Blanco-Ania, D. (2018) Strigolactones: New plant hormones in the spotlight. *Journal of Experimental Botany* 69(9), 2205–2218. DOI: 10.1093/jxb/erx487.

Zou, J., Zhang, S., Zhang, W., Li, G., Chen, Z. *et al.* (2006) The rice HIGH-TILLERING DWARF1 encoding an ortholog of *Arabidopsis* MAX3 is required for negative regulation of the outgrowth of axillary buds. *Plant Journal* 48(5), 687–698. DOI: 10.1111/j.1365-313X.2006.02916.x.

Zwanenburg, B., Pospíšil, T. and Ćavar Zeljković, S. (2016) Strigolactones: New plant hormones in action. *Planta* 243(6), 1311–1326. DOI: 10.1007/s00425-015-2455-5.

7 Regulatory Role of Strigolactones in Photosynthesis and Respiration

Shikha Arora, Tajammul Husain, Shivangi Chaudhary, Om Sajeevan Patel, Kewat Sanjay Kumar and Sheo Mohan Prasad*

Ranjan Plant Physiology and Biochemistry Laboratory, Department of Botany, University of Allahabad, Prayagraj, India

Abstract

Strigolactones (SLs) are a newly discovered class of plant growth regulators or plant hormones that contribute to the growth and development of plants. These phytohormones were initially recognized as stimulating seed germination in parasitic plants; however, later research discovered that they regulate plant architecture by controlling shoot and root growth under continuously changing environmental conditions, including light availability, nutrient levels, and temperature variations. This chapter discusses the function of SLs in controlling the physiological processes of photosynthesis and respiration. It also includes the mechanisms involved in processes that may facilitate future research in the enhancement of plant resilience under adverse conditions.

Keywords: abiotic stress, chlorophyll, photosynthesis, phytohormones, strigolactones

7.1 Introduction

Several abiotic and biotic factors impede the growth of plants, resulting in inferior crop quality and reduced agricultural productivity. Plant secondary metabolites are the chemical signals that the plants synthesize to regulate physiological processes under stress conditions (Yoneyama and Bennett, 2024). These chemical signals help communication with adjacent plants, microorganisms, and pathogens and play a critical role in developing resistance against stress factors (Pang *et al.*, 2021).

Strigolactones (SLs) are phytohormones derived from carotenoid catabolism and have recently gained attention as plant growth regulators. They were originally identified as chemical signals released by host plants that act as inducers of seed germination in root-parasitic plants, particularly members of the genera *Striga* and *Orobanche* (Al-Babili and Bouwmeester, 2015). Subsequently, it was found that the function of SLs is influential as signaling molecules in communication with different fungi and bacteria that influence plant development (Marzec and Muszynska, 2015). These biologically active substances are commonly referred to as non-traditional plant growth regulators, highlighting their distinction from other plant hormonal actions. The primary site of SL production is the root system, where they are synthesized and consequently released into the rhizosphere (Arora *et al.*, 2024). Among thousands of structurally different SLs that have been identified across plant groups, only 30 of these compounds have been thoroughly

*Corresponding author: profsmprasad@gmail.com

DOI: 10.1079/9781836990598.0007

characterized. As well as acting as chemical signals, SLs also control the plant architecture by controlling root and shoot growth (Arora *et al.*, 2024). Additionally, SLs govern the plant's survival and fitness by determining the thickness of the stem, growth of lateral roots, and density of the root hairs, and also by suppressing the growth of axillary buds and delaying the process of leaf senescence (Al-Babili and Bouwmeester, 2015). Furthermore, SLs are crucial players in developing resilience of terrestrial plants against various stressors by governing the hyphal branching in symbiosis with arbuscular mycorrhizal fungi (AMF) (Pandey *et al.*, 2016). AMF develop a network of mycelia with plant roots, thereby accelerating growth by providing an easy passage for the roots to reach a larger soil surface area (Bowles *et al.*, 2016). AMF are potential contributors against a wide range of abiotic stressors, such as high temperatures, cold stress, alkali stress, nutrient stress, and salinity (Begum *et al.*, 2019). The ongoing research on SLs and their multi-dimensional roles continues to unveil the complex nature of plant hormonal regulation and its application in agriculture and ecological interactions. A comprehensive knowledge of the functional aspects of SLs could facilitate innovative approaches for enhancing crop productivity in diverse climatic conditions. This chapter focuses on the role of SLs in the regulation of the plant physiological processes of photosynthesis and respiration.

7.2 Role of Strigolactones in Growth and Development of Plants

SLs are a newly discovered carotenoid derivative, obtained from the root exudate and known to boost seed germination related to root parasites such as *Striga* (Yoneyama *et al.*, 2008). They are key signaling molecules that promote plant interactions with microbes in the rhizosphere and help in the mycorrhizal symbiosis between Glomeromycota fungi and plants. These fungal communications result in the formation of arbuscular mycorrhiza with the roots of terrestrial plants. These AMF colonize the roots of nearly 80% of terrestrial plants (Brundrett, 2009). Under stressful conditions, AMF stimulate the rate of photosynthesis and gaseous exchange, which helps plants flourish in harsh environments (Birhane *et al.*, 2012). SL signaling helps in establishing this symbiosis by synthesizing a short-chain oligomer of chitin that activates the signaling pathway linked with symbiosis (López-Ráez *et al.*, 2017). As a result of the exogenous application of the synthetic SL analog GR24 to fungal mycelium, various changes, such as growth and hyphal branching, enhancement of the rate of respiration, and increases in the production of ATP and NADPH, have been observed (Rozpądek *et al.*, 2018). Plants regulate the formation and exudation of SLs on the basis of nutrient accessibility, especially phosphate (Yoneyama, 2019). In mycotrophic plants, a shortage of phosphate results in an increase of SL production, whereas an excessive amount of phosphate suppresses it (Yoneyama, 2019). Plants intensify SL exudation to encourage AMF symbiosis, which helps them acquire inorganic nutrients, especially phosphate (Boyno *et al.*, 2023). A deficiency of nitrogen also enhances the synthesis of SLs and increases their exudation (Marzec *et al.*, 2013; Yoneyama *et al.*, 2013). As well as their role in the roots, SLs also help in internode elongation. The internodes of SL-deficient mutants in rice, canola, pea, *Arabidopsis*, and tomato exhibit a short internode phenotype compared with the wild type, indicating that SLs play a crucial role in internodal elongation (Rehman *et al.*, 2021). The rice *htd1/d17* mutants, deficient in the synthesis of SLs, exhibit a reduction in height in comparison with the wild type (Zou *et al.*, 2006). It has also been shown that the level of gibberellic acids is affected by SLs, and SLs play a role in GA signaling in pea plants (Rehman *et al.*, 2021). However, SLs and gibberellic acids regulate the elongation of the internode independently in pea plants (de Saint Germain *et al.*, 2013).

The role of SLs is crucial in photosynthesis and senescence of leaves (Fig. 7.1). Senescence of the leaf occurs with the degradation of chloroplasts along with the enzyme RuBisCO (ribulose-1,5-bisphosphate carboxylase/oxygenase) and denaturation of the chlorophyll *a*/*b*-binding protein, resulting in the leaf changing color (Yamada and Umehara, 2015). The involvement of SLs in senescence in the leaves has been demonstrated by treatment with GR24 in mutants deficient in SLs, namely *d27*, *d17*, and *d10* (*Oryza*

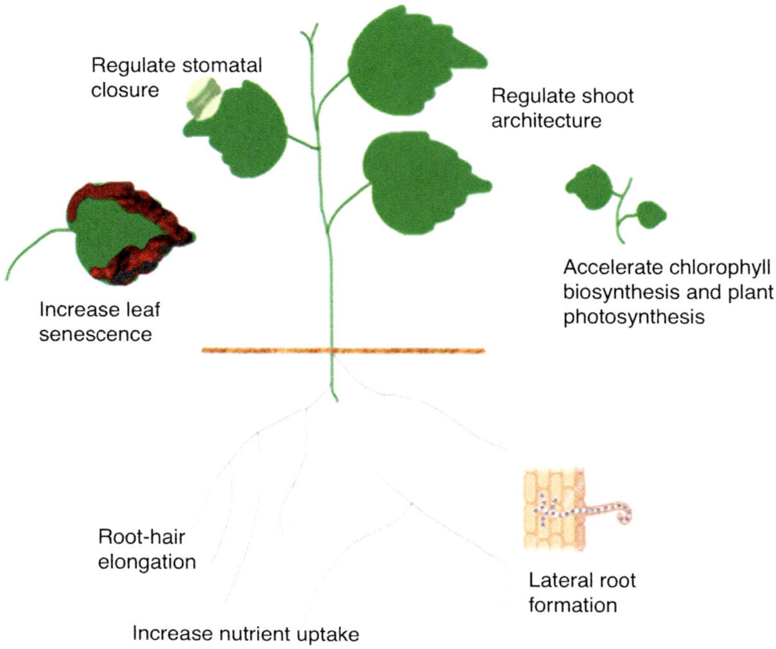

Fig. 7.1. The different roles of strigolactones in regulating plant growth and development.

sativa), and *max1*, *max3*, and *max4* (*Arabidopsis thaliana*) (Yamada *et al*., 2014). When GR24 was applied to *O. sativa* mutants deficient in SLs, the senescence activity was recovered in the form of color of the leaf, content of chlorophyll, and leakage of electrolytes, but this did not occur in mutants insensitive to SLs. SL-related mutants have been shown to both accelerate and delay senescence when GR24 was present, suggesting a functional aspect of SLs in leaf senescence (Yamada *et al*., 2014). Under the condition of drought, GR24 treatment increased the chlorophyll level and photosynthesis. Parallel findings were noted in rapeseed exposed to saline stress (Ma *et al*., 2017) and wheat under drought stress (Sedaghat *et al*., 2017). It has been demonstrated that SLs can reduce the effects of drought, which harms a variety of plants. Plants close their stomata, reduce transpiration, scavenge reactive oxygen species (ROS), minimize lipid peroxidation, control chlorophyll concentration, and increase photosynthesis to escape drought stress (Fig. 7.2) (Zhang *et al*., 2018). Application of SLs under drought conditions in wheat and *Vitis vinifera* resulted in a decrease in malondialdehyde levels, hydrogen peroxide (H_2O_2), and leakage of electrolytes, as well as better water content, membrane stability, and photosynthesis (Min *et al*., 2019). This suggests that SLs activate ROS-scavenging mechanisms (Table 7.1) (Sedaghat *et al*., 2020).

Transcriptional regulation of SL-biosynthetic genes such as *carotenoid cleavage dioxygenase 7* (*CCD7*), *CCD8*, and *MORE AXILLARY GROWTH 1* (*MAX1*) and the SL-signaling gene *MAX2* is enhanced under cold stress (Chi *et al*., 2021). In addition, it has been observed that SL treatment is capable of reversing the damaging effects of cold stress. Under conditions of cold stress, osmoregulatory substances such as proline and soluble sugars accumulate, which is the plant's adaptive response (Hussain *et al*., 2018). Therefore, the amount of such substances might alter the resistance of plants under cold stress. Soluble sugar and proline levels increase in mung bean seedlings under cold stress following treatment with GR24 (Omoarelojie *et al*., 2021). According to Zhang *et al*. (2020), in cold stress conditions, pretreatment with GR24 of

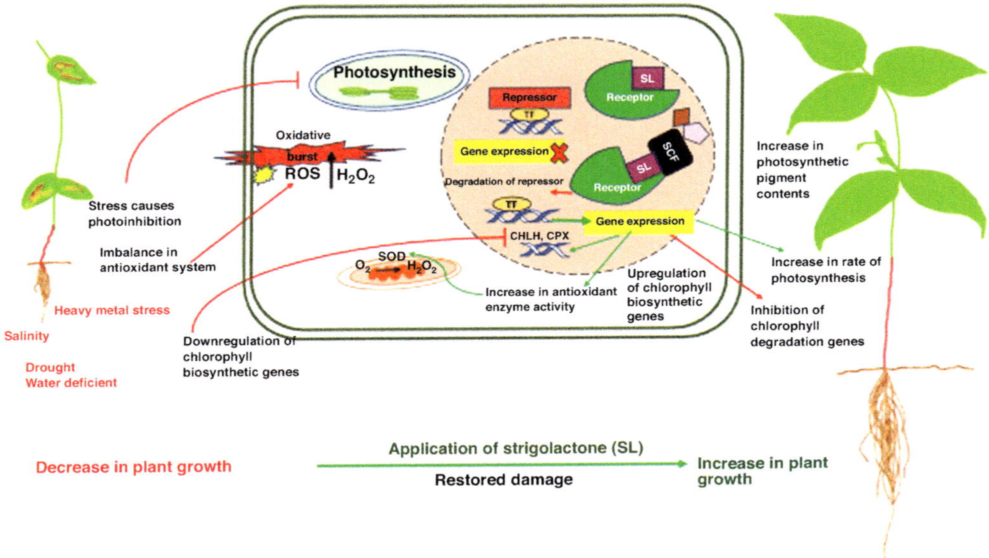

Fig. 7.2. Schematic representation of the effects of exogenous application of strigolactone on the physiology of plants under stress. CHLH, magnesium chelatase subunit H; CPX, coproporphyrinogen III oxidase; H_2O_2, hydrogen peroxide; ROS, reactive oxygen species; SCF, F-box protein–SKP1–Cullin; SOD, superoxide dismutase; TF, transcription factor.

rape leaf cells decreases the levels of superoxide (O_2^-), H_2O_2, and malondialdehyde, and reverses the cell-membrane damage caused by low-temperature stress.

7.3 Strigolactone-mediated Regulation of Photosynthesis in Plants

SLs modulate several plant physiological processes under stressful situations, such as the regulation of photosynthesis and enhancement in plant tolerance against numerous environmental constraints. Ling *et al.* (2020) documented that the seedlings of Jinongda rice grown under NaCl stress (200 mM) showed a rise in photosynthetic process and other physiological attributes when treated with synthetic strigolactone (GR24). The negative impact of salt stress on the growth along with various physiological processes (i.e. photosynthesis) was less pronounced in the GR24-treated seedlings in comparison with their respective control (Ling *et al.*, 2020). The responses of oilseed rape variety Zhongshuang 11 exposed to NaCl (0, 100, and 200 mM) and treated with GR24 (0.18 μM) were noted by growing replications in a controlled growth environment for 7 days (Ma *et al.*, 2017). Thereafter, various physiological and biochemical attributes were analyzed and it was found that with the increase in salt concentration, the amount of chlorophyll and gaseous exchange parameters (net rate of photosynthesis, stomatal conductance, intercellular concentration of CO_2, and rate of transpiration) decreased dramatically, while GR24 treatment partially counteracted these effects (Ma *et al.*, 2017). Furthermore, it was noticed that the application of GR24 also enhanced the quantum yield of photosystem II (PSII) and reduced the non-photochemical quenching (Ma *et al.*, 2017). Similarly, Zhang *et al.* (2022) reported that plants exposed to NaCl with GR24 pretreatment showed elevated levels of chlorophyll and an increase in photosynthetic efficiency compared with non-GR24-pretreated plants exposed to NaCl. In another study, the application of varied GR24 concentrations on foliar parts of *Artemisia annua* was analyzed (Wani *et al.*,

Table 7.1. Regulatory role of strigolactones (SLs) on the physiology of plants under stress.

Physiological process	Plant species	Physiological effect	Reference
Root development	*Arabidopsis thaliana*	GR24 actively intervenes in initial root extension and positively monitors many meristematic cells	Ruyter-Spira *et al.* (2011)
Shoot branching	*Oryza sativa*, *Pisum sativum*, *A. thaliana*	Increase expression of branching inhibiting signal (BIS)	Kretzschmar *et al.* (2012)
Leaf senescence	*O. sativa*	Reduced expression of the RuBisCO and chlorophyll *a*/*b*-binding protein precursor genes	Yamada *et al.* (2014)
Stomatal closure	*Solanum lycopersicum*	SLs triggered Slow Anion Channel-Associated1 (SLAC1)	Visentin *et al.* (2020)
Chlorophyll synthesis	*Vitis vinifera*	Chlorophyll restoration by exogenous application of GR24	Min *et al.* (2019)
Seed germination	*Lupinus angustifolius*	High germination indices, amplified proline content, and reduced lipid peroxidation were all seen in seeds primed with GR24; increase seed germination	Omoarelojie *et al.* (2020)
Nodulation	*P. sativum*	Endogenous SLs positively regulate nodulation in the pea plant	Foo and Davies (2011)
Bud outgrowth	*P. sativum*, *A. thaliana*	SLs act downstream of auxin to regulate bud outgrowth	Brewer *et al.* (2009)
Hyphal branching	*Gigaspora rosea*	Upregulation of the genes involved in hyphal development and mitochondrial metabolism, as well as stimulation of the mitotic activity of fungi	Besserer *et al.* (2008)
Transpiration rate	*Malus hupehensis*	SLs accelerate transpiration rate, and increase the activities of POD and CAT, and sodium and magnesium content	Zheng *et al.* (2021)
Photosynthesis	*A. thaliana*, *P. sativum*	Reduced rate of photosynthesis	Cooper *et al.* (2018)
Chlorophyll and photosynthesis	*Zea mays*	Increased chlorophyll, carotenoids, photosynthetic rate, transpiration and stomatal conductance	Iftikhar *et al.* (2023)
Cell-cycle dynamics	*Festuca arundinacea*	Upregulation of genes involved in cell-cycle functions, cell number, and crown root elongation, increased expression of *D3* and *D14* genes, and downregulation of auxin transporter *PIN1*, *TIR1*, *PIN2*, and *PIN5* genes	Zhang *et al.* (2018)
Photosynthesis	*S. lycopersicum*	Increased root growth and higher photosynthesis rate	Ruiz-Lozano *et al.* (2016)
Proline and sugar	*S. lycopersicum*	Higher proline and soluble sugar content	Sedaghat *et al.* (2020)
Chlorophyll and photosynthesis	*Hordeum vulgare*	Increase in chlorophyll and plant photosynthesis	Qiu *et al.* (2021)

CAT, catalase; *D3*, *DWARF3; D14*, DWARF14; GR24, synthetic analog of SL; *PIN*, *PIN-FORMED*; POD, peroxidase; RuBisCO, ribulose-1,5-bisphosphate carboxylase/oxygenase; *TIR1*, *TRANSPORT INHIBITOR RESPONSE 1*.

2023). Compared with control, the supplementation of GR24 at 4 μM concentration improved overall growth performance, rate of photosynthesis, and other physiological characters. Earlier studies indicated that the plant biomass and internal CO_2 levels, stomatal conductance and total chlorophyll concentration, chlorophyll fluorescence, and overall photosynthesis increased following application of GR24 (Wani *et al.*, 2023). Additionally, a rise in the activity of several enzymes involved in photosynthesis, such as RuBisCO and carbonic anhydrase, was noted (Wani *et al.*, 2023). Under harsh environmental conditions, the level of ROS is found to be enhanced (Husain *et al.*, 2024). Similarly, abrupt production of ROS was observed under drought conditions, which disturbed the functioning of PSII and ultimately reduced the efficiency of photosynthesis and annual yield (Cui *et al.*, 2019). Roig-Oliver *et al.* (2021) observed that, under drought conditions, the elevated level of leaf water relations has the potential to elevate the net photosynthetic rate in wheat. The exogenous supplementation of SLs improves and positively regulates the photosynthetic efficiency under drought stress (Song *et al.*, 2023). Similarly, it was noted that exogenous application of SLs on *Pennisetum purpureum* grown under water scarcity significantly recovered the adverse effect of drought. It has been shown that SLs significantly mitigate the detrimental effect of drought (Li *et al.*, 2022). The application of SLs enhanced the rate of photosynthesis, transpiration, and stomatal conductance in *P. purpureum* leaves, thereby indicating a prominent recovery in the reduction of maximum photochemical efficiency of PSII and the overall performance index (Li *et al.*, 2022). The ability of the plant to utilize light energy across various wavelengths is indicated by the chlorophyll *a*/*b* ratio (Alvi *et al.*, 2022), which was diminished in mutants deficient in SLs (*d27*, *d17*, *d10*, *max1*, *max3*, and *max4*) (Alvi *et al.*, 2022). Under low light, GR24 showed a positive impact on gaseous exchange parameters and chlorophyll fluorescence, thereby increasing photosynthetic efficiency in tall fescue (*Festuca arundinacea*) (Gong *et al.*, 2024). It can be inferred from the above findings that SLs can positively regulate the rate of photosynthesis by modulating the chlorophyll concentration and improving light energy capture efficiency in leaves.

7.4 Strigolactone-mediated Regulation of Respiration in Plants

Respiration serves as a fundamental aspect of plant life, playing vital roles in carbon and energy metabolism. The energy captured in carbohydrate molecules during photosynthesis is subsequently released through the cellular oxidation of these carbohydrates into carbon dioxide (CO_2), water, and energy. During the process of respiration, various organic food substances, referred to as respiratory substrates, undergo oxidation. These substrates include carbohydrates, fats, and proteins. The metabolic pathways of the Embden–Meyerhof pathway in the cytosol, the citric-acid cycle in the mitochondrial matrix, and the electron transport chain lead to the release of energy (i.e. adenosine triphosphate (ATP)) from carbohydrates. SLs are plant hormones that indirectly affect respiration through a variety of mechanisms. SLs can influence respiration by modifying photosynthesis, antioxidant capacity, and stress responses, which in turn can affect energy production and consumption (Table 7.1) (Saeed *et al.*, 2017; Soliman *et al.*, 2022). Liu *et al.* (2024) noted that through regulating the amount of antioxidant enzymes and their activity, SL treatment preserved the antioxidant equilibrium, prevented weight loss and alterations in color, and progressively diminished the respiratory activity in freshly cut apples. Huang *et al.* (2021) showed that in strawberries that were subjected to different doses of exogenous SL (0.0, 0.5, 1, 2, and 4 μM), 1 μM SL significantly lowered the rate of weight loss and respiratory intensity. Therefore, exogenous SL effectively diminished respiratory intensity and prevented weight loss in strawberries while in storage. Key factors in postharvest fruit-quality decline are based on deterioration and senescence. The application of the SL analog GR24 (200 μM) showed an increase in respiration by 24%, resulting in an increase in organoleptic quality in sweet oranges postharvest (Ma *et al.*, 2022).

SLs are found in root exudates and are recognized for initiating the seed germination of parasitic plants and mediate AMF to establish symbiosis with root systems, thereby maintaining mutualistic relationships (Sabbagh, 2011). These plant hormones influence the metabolism of AMF by triggering their mitochondrial respiration (Kowalczyk and Hrynkiewicz, 2018). The rapid and strong activation of mitochondrial function in fungi in response to SLs constitutes the most notable impact of SLs on AMF. This mitochondrial response was noted in *Gigaspora rosea* and *Glomus intraradices*, two distinct AMF; this phenomenon was also linked to a sudden rise in mitochondrial respiration and density of mitochondria, as demonstrated in *Glomus intraradices* (Besserer *et al.*, 2006). In a separate study, the researchers revealed respiratory activation occurring within minutes in *Gigaspora rosea* following the application of GR24, which quickly elevated its ATP content, NADH concentration, and NADH dehydrogenase activity (Besserer *et al.*, 2008). In another study, the rate of respiration was presented with relation to the diameter of a fungal (*Mucor* sp.) colony, and it was observed that the colony diameter decreased in a concentration-dependent manner. Doses of GR24 of 50–1000 nM have shown inhibition in the growth of mycelia, whereas lower doses (up to 10 nM) did not negatively affect mycelial growth: at a lower concentration, GR24 functions to promote the mutualistic interaction, while at a higher concentration, it has potential impacts in plant defense (Rozpądek *et al.*, 2018).

7.5 Molecular Mechanism of Strigolactone-mediated Photosynthesis and Respiration

Biosynthesis of SLs starts with the formation of its precursor, carlactone, in the plastids, which involves two carotenoid cleavage dioxygenases, CCD7 and CCD8 (Arora *et al.*, 2024). Carlactone forms carlactonic acid in the cytoplasm via MAX1 or MAX1-type mono-oxygenases (Arora *et al.*, 2024). Carlactonic acid is converted to 5-deoxystrigol, which is the main precursor of other SLs (Marzec, 2016). SLs bind to their receptor (e.g. AtD14 receptor in *Arabidopsis*), which causes hydrolysis of SLs along with a conformational change in the receptor. The latter then binds to an F-box protein (e.g. MAX2 in *Arabidopsis*)–SKP1–Cullin (SCF) complex (Fig. 7.2). Thereafter, the modified receptor comprising of SLs bound to the SCF complex binds to the repressor (D53). The repressor is degraded via the ubiquitin–proteasomal pathway, which allows transcription factors to bind for gene expression (Marzec, 2016).

SLs are involved in various functions in plant growth, including photomorphogenesis, modulating photosynthesis and respiration. Light-related growth responses are positively modulated by a slight rise in the level of SLs via the F-box protein MAX2 (Tsuchiya *et al.*, 2010). Application of GR24 in *Arabidopsis* results in light-related gene expression, which is repressed in mutants deficient in SLs (Mashiguchi *et al.*, 2009).

Chlorophyll constitutes the essential foundation for the process of photosynthesis, and variations in chlorophyll levels are crucial for assessing plant growth and resistance to drought conditions. Earlier research has indicated that SLs augment chlorophyll levels under drought stress across a range of crops (Table 7.1), such as alfalfa, wheat, and maize (Song *et al.*, 2023). Song *et al.* (2023) studied the function of SLs in ameliorating drought stress in wheat seedlings. They reported that treatment with SLs in the leaves resulted in the downregulation of chlorophyll degradation genes, while simultaneously positively regulating the genes involved in chlorophyll biosynthesis (Fig. 7.2). These observations imply that SLs may augment the level of chlorophyll by promoting the gene expression associated with its synthesis while inhibiting the expression of those linked to its breakdown during drought stress (Fig. 7.2). Furthermore, they also noted that application of SLs upregulated genes involved in light-harvesting, chlorophyll *a*/*b*-binding proteins, and electron transport. Conversely, expression levels of genes implicated in chlorophyll degradation and photorespiration were observed to be downregulated following SL priming. SLs also interact with other phytohormones and regulate many physiological processes in plants. Cytokinins and SLs upregulate photosynthesis in plants by modulating the expression of genes involved in photosynthesis (Brenner and Schmülling, 2012). A hybridization approach was used to generate cDNA libraries for genes differentially expressed in maize pathogen *Sporisorium reilianum* exposed to GR24 (Sabbagh *et al.*, 2012). Cell respiration increased within 1 h following the addition of GR24, but decreased after 5 and 8 h. Cells treated with GR24 were used for the construction of a cDNA library comprising 1440 clones. The cDNA expressed sequence tags (ESTs) were placed on macro-array and hybridized with ^{32}P-labeled cDNA probes, derived from mRNA isolated from the yeast cultures that were either treated three times GR24 (100 nM) or not treated. Around 670 ESTs were recognized as differentially expressed genes across three time points in response to GR24. A group of 36 candidate genes was examined using quantitative reverse transcriptase-polymerase chain reaction (qRT-PCR), which showed induction of the genes at 1 h, thereby confirming the hybridization data. The induced genes primarily influenced catalytic functions (45%), such as cell respiration (27%) and cell signaling (26%).

7.6 Conclusion and Future Perspectives

This chapter has discussed the regulatory functions of SLs in modulating a range of physiological processes that facilitate the growth of plants. SL application as plant hormones could potentially alleviate the detrimental effects of stressors on plant health. Consequently, this could prove beneficial in promoting global agricultural productivity, even amidst fluctuating environmental conditions. Despite previous investigations providing substantial understanding of the physiological functions of SLs, ongoing research is essential to enhance our understanding of the pathways related to the biosynthesis and interactions of SLs with other plant hormones. The mechanisms through which SLs regulate these processes are yet to be fully elucidated.

References

Al-Babili, S. and Bouwmeester, H.J. (2015) Strigolactones, a novel carotenoid-derived plant hormone. *Annual Review of Plant Biology* 66, 161–186. DOI: 10.1146/annurev-arplant-043014-114759.

Alvi, A.F., Sehar, Z., Fatma, M., Masood, A. and Khan, N.A. (2022) Strigolactone: An emerging growth regulator for developing resilience in plants. *Plants* 11(19), 2604. DOI: 10.3390/plants11192604.

Arora, S., Husain, T., Pandey, A. and Prasad, S.M. (2024) Cross-talk of strigolactones with auxin and cytokinin. In: Bashri, G., Hayat, S. and Bajguz, A. (eds) *Strigolactones: Synthesis, Application and Role in Plants*. Academic Press, London, pp. 89–102. DOI: 10.1016/B978-0-443-13521-7.00006-3.

Begum, N., Qin, C., Ahanger, M.A., Raza, S., Khan, M.I. *et al*. (2019) Role of arbuscular mycorrhizal fungi in plant growth regulation: Implications in abiotic stress tolerance. *Frontiers in Plant Science* 10, 1068. DOI: 10.3389/fpls.2019.01068.

Besserer, A., Puech-Pagès, V., Kiefer, P., Gomez-Roldan, V., Jauneau, A. *et al*. (2006) Strigolactones stimulate arbuscular mycorrhizal fungi by activating mitochondria. *PLOS Biology* 4(7), e226. DOI: 10.1371/journal.pbio.0040226.

Besserer, A., Bécard, G., Jauneau, A., Roux, C. and Séjalon-Delmas, N. (2008) GR24, a synthetic analog of strigolactones, stimulates the mitosis and growth of the arbuscular mycorrhizal fungus *Gigaspora rosea* by boosting its energy metabolism. *Plant Physiology* 148(1), 402–413. DOI: 10.1104/pp.108.121400.

Birhane, E., Sterck, F.J., Fetene, M., Bongers, F. and Kuyper, T.W. (2012) Arbuscular mycorrhizal fungi enhance photosynthesis, water use efficiency, and growth of frankincense seedlings under pulsed water availability conditions. *Oecologia* 169(4), 895–904. DOI: 10.1007/s00442-012-2258-3.

Bowles, T.M., Barrios-Masias, F.H., Carlisle, E.A., Cavagnaro, T.R. and Jackson, L.E. (2016) Effects of arbuscular mycorrhizae on tomato yield, nutrient uptake, water relations, and soil carbon dynamics under deficit irrigation in field conditions. *Science of the Total Environment* 566–567, 1223–1234. DOI: 10.1016/j.scitotenv.2016.05.178.

Boyno, G., Rezaee Danesh, Y., Demir, S., Teniz, N., Mulet, J.M. *et al*. (2023) The complex interplay between arbuscular mycorrhizal fungi and strigolactone: Mechanisms, sinergies, applications and future directions. *International Journal of Molecular Sciences* 24(23), 16774. DOI: 10.3390/ijms242316774.

Brenner, W.G. and Schmülling, T. (2012) Transcript profiling of cytokinin action in *Arabidopsis* roots and shoots discovers largely similar but also organ-specific responses. *BMC Plant Biology* 12(1), 112. DOI: 10.1186/1471-2229-12-112.

Brewer, P.B., Dun, E.A., Ferguson, B.J., Rameau, C. and Beveridge, C.A. (2009) Strigolactone acts downstream of auxin to regulate bud outgrowth in pea and *Arabidopsis*. *Plant Physiology* 150(1), 482–493. DOI: 10.1104/pp.108.134783.

Brundrett, M.C. (2009) Mycorrhizal associations and other means of nutrition of vascular plants: Understanding the global diversity of host plants by resolving conflicting information and developing reliable means of diagnosis. *Plant and Soil* 320(1–2), 37–77. DOI: 10.1007/s11104-008-9877-9.

Chi, C., Xu, X., Wang, M., Zhang, H., Fang, P. *et al.* (2021) Strigolactones positively regulate abscisic acid-dependent heat and cold tolerance in tomato. *Horticulture Research* 8(1), 237. DOI: 10.1038/s41438-021-00668-y.

Cooper, J.W., Hu, Y., Beyyoudh, L., Yildiz Dasgan, H., Kunert, K. *et al.* (2018) Strigolactones positively regulate chilling tolerance in pea and in *Arabidopsis*. *Plant, Cell & Environment* 41(6), 1298–1310. DOI: 10.1111/pce.13147.

Cui, G., Zhao, Y., Zhang, J., Chao, M., Xie, K. *et al.* (2019) Proteomic analysis of the similarities and differences of soil drought and polyethylene glycol stress responses in wheat (*Triticum aestivum* L.). *Plant Molecular Biology* 100(4–5), 391–410. DOI: 10.1007/s11103-019-00866-2.

de Saint Germain, A., Bonhomme, S., Boyer, F.-D. and Rameau, C. (2013) Novel insights into strigolactone distribution and signalling. *Current Opinion in Plant Biology* 16(5), 583–589. DOI: 10.1016/j.pbi.2013.06.007.

Foo, E. and Davies, N.W. (2011) Strigolactones promote nodulation in pea. *Planta* 234(5), 1073–1081. DOI: 10.1007/s00425-011-1516-7.

Gong, J., Wang, R., Liu, B., Zhu, T., Li, H. *et al.* (2024) Regulatory mechanism of strigolactone in tall fescue to low-light stress. *Plant Physiology and Biochemistry* 215, 109054. DOI: 10.1016/j.plaphy.2024.109054.

Huang, D., Wang, Y., Zhang, D., Dong, Y., Meng, Q. *et al.* (2021) Strigolactone maintains strawberry quality by regulating phenylpropanoid, NO, and H2S metabolism during storage. *Postharvest Biology and Technology* 178, 111546. DOI: 10.1016/j.postharvbio.2021.111546.

Husain, T., Prasad, S.M. and Singh, V.P. (2024) Ethylene and hydrogen sulfide regulate hexavalent chromium toxicity in two pulse crops: Implication on growth, photosynthetic activity, oxidative stress and ascorbate glutathione cycle. *Plant Physiology and Biochemistry* 216, 109170. DOI: 10.1016/j.plaphy.2024.109170.

Hussain, H.A., Hussain, S., Khaliq, A., Ashraf, U., Anjum, S.A. *et al.* (2018) Chilling and drought stresses in crop plants: Implications, cross talk, and potential management opportunities. *Frontiers in Plant Science* 9, 393. DOI: 10.3389/fpls.2018.00393.

Iftikhar, I., Shahbaz, M. and Wahid, M.A. (2023) Potential role of foliage applied strigolactone (GR24) on photosynthetic pigments, gas exchange attributes, mineral nutrients and yield components of *Zea mays* (L.) under saline regimes. *Gesunde Pflanzen* 75, 577–591. DOI: 10.1007/s10343-022-00720-4.

Kowalczyk, A. and Hrynkiewicz, K. (2018) Strigolactones as mediators between fungi and plants. *Acta Mycologica* 53(2), 1–8. DOI: 10.5586/am.1110.

Kretzschmar, T., Kohlen, W., Sasse, J., Borghi, L., Schlegel, M. *et al.* (2012) A petunia ABC protein controls strigolactone-dependent symbiotic signalling and branching. *Nature* 483(7389), 341–344. DOI: 10.1038/nature10873.

Li, Y., Li, S., Feng, Q., Zhang, J., Han, X. *et al.* (2022) Effects of exogenous strigolactone on the physiological and ecological characteristics of *Pennisetum purpureum* Schum. seedlings under drought stress. *BMC Plant Biology* 22(1), 578. DOI: 10.1186/s12870-022-03978-y.

Ling, F., Su, Q., Jiang, H., Cui, J., He, X. *et al.* (2020) Effects of strigolactone on photosynthetic and physiological characteristics in salt-stressed rice seedlings. *Scientific Reports* 10(1), 6183. DOI: 10.1038/s41598-020-63352-6.

Liu, Z., Zhang, X., Zhu, S. and Huang, D. (2024) Strigolactone preserves fresh-cut apple quality during shelf life. *Agriculture* 14(9), 1588. DOI: 10.3390/agriculture14091588.

López-Ráez, J.A., Shirasu, K. and Foo, E. (2017) Strigolactones in plant interactions with beneficial and detrimental organisms: The yin and yang. *Trends in Plant Science* 22(6), 527–537. DOI: 10.1016/j.tplants.2017.03.011.

Ma, N., Hu, C., Wan, L., Hu, Q., Xiong, J. *et al.* (2017) Strigolactones improve plant growth, photosynthesis, and alleviate oxidative stress under salinity in rapeseed (*Brassica napus* L.) by regulating gene expression. *Frontiers in Plant Science* 8, 1671. DOI: 10.3389/fpls.2017.01671.

Ma, Q., Lin, X., Zhan, M., Chen, Z., Wang, H. *et al.* (2022) Effect of an exogenous strigolactone GR24 on the antioxidant capacity and quality deterioration in postharvest sweet orange fruit stored at ambient temperature. *International Journal of Food Science & Technology* 57(1), 619–630. DOI: 10.1111/ijfs.15415.

Marzec, M. (2016) Strigolactones as part of the plant defence system. *Trends in Plant Science* 21(11), 900–903. DOI: 10.1016/j.tplants.2016.08.010.

Marzec, M. and Muszynska, A. (2015) *In silico* analysis of the genes encoding proteins that are involved in the biosynthesis of the RMS/MAX/D pathway revealed new roles of strigolactones in plants. *International Journal of Molecular Sciences* 16(4), 6757–6782. DOI: 10.3390/ijms16046757.

Marzec, M., Muszynska, A. and Gruszka, D. (2013) The role of strigolactones in nutrient-stress responses in plants. *International Journal of Molecular Sciences* 14(5), 9286–9304. DOI: 10.3390/ijms14059286.

Mashiguchi, K., Sasaki, E., Shimada, Y., Nagae, M., Ueno, K. *et al.* (2009) Feedback-regulation of strigolactone biosynthetic genes and strigolactone-regulated genes in Arabidopsis. *Bioscience, Biotechnology, and Biochemistry* 73(11), 2460–2465. DOI: 10.1271/bbb.90443.

Min, Z., Li, R., Chen, L., Zhang, Y., Li, Z. *et al.* (2019) Alleviation of drought stress in grapevine by foliar-applied strigolactones. *Plant Physiology and Biochemistry* 135, 99–110. DOI: 10.1016/j.plaphy.2018.11.037.

Omoarelojie, L.O., Kulkarni, M.G., Finnie, J.F., Pospíšil, T., Strnad, M. *et al.* (2020) Synthetic strigolactone (rac-GR24) alleviates the adverse effects of heat stress on seed germination and photosystem II function in lupine seedlings. *Plant Physiology and Biochemistry* 155, 965–979. DOI: 10.1016/j.plaphy.2020.07.043.

Omoarelojie, L.O., Kulkarni, M.G., Finnie, J.F. and van Staden, J. (2021) Strigolactone analog (rac-GR24) enhances chilling tolerance in mung bean seedlings. *South African Journal of Botany* 140, 173–181. DOI: 10.1016/j.sajb.2021.03.044.

Pandey, A., Sharma, M. and Pandey, G.K. (2016) Emerging roles of strigolactones in plant responses to stress and development. *Frontiers in Plant Science* 7, 434. DOI: 10.3389/fpls.2016.00434.

Pang, Z., Chen, J., Wang, T., Gao, C., Li, Z. *et al.* (2021) Linking plant secondary metabolites and plant microbiomes: A review. *Frontiers in Plant Science* 12, 621276. DOI: 10.3389/fpls.2021.621276.

Qiu, C.W., Zhang, C., Wang, N.H., Mao, W. and Wu, F. (2021) Strigolactone GR24 improves cadmium tolerance by regulating cadmium uptake, nitric oxide signaling and antioxidant metabolism in barley (*Hordeum vulgare* L.). *Environmental Pollution* 273, 116486. DOI: 10.1016/j.envpol.2021.116486.

Rehman, N.U., Li, X., Zeng, P., Guo, S., Jan, S. *et al.* (2021) Harmony but not uniformity: Role of strigolactone in plants. *Biomolecules* 11(11), 1616. DOI: 10.3390/biom11111616.

Roig-Oliver, M., Fullana-Pericàs, M., Bota, J. and Flexas, J. (2021) Adjustments in photosynthesis and leaf water relations are related to changes in cell wall composition in *Hordeum vulgare* and *Triticum aestivum* subjected to water deficit stress. *Plant Science* 311, 111015. DOI: 10.1016/j.plantsci.2021.111015.

Rozpądek, P., Domka, A.M., Nosek, M., Ważny, R., Jędrzejczyk, R.J. *et al.* (2018) The role of strigolactone in the cross-talk between *Arabidopsis thaliana* and the endophytic fungus Mucor s. *Frontiers in Microbiology* 9, 441. DOI: 10.3389/fmicb.2018.00441.

Ruiz-Lozano, J.M., Aroca, R., Zamarreño, Á.M., Molina, S., Andreo-Jiménez, B. *et al.* (2016) Arbuscular mycorrhizal symbiosis induces strigolactone biosynthesis under drought and improves drought tolerance in lettuce and tomato. *Plant, Cell & Environment* 39(2), 441–452. DOI: 10.1111/pce.12631.

Ruyter-Spira, C., Kohlen, W., Charnikhova, T., van Zeijl, A., van Bezouwen, L. *et al.* (2011) Physiological effects of the synthetic strigolactone analog GR24 on root system architecture in *Arabidopsis*: Another belowground role for strigolactones. *Plant Physiology* 155(2), 721–734. DOI: 10.1104/pp.110.166645.

Sabbagh, S.K. (2011) Effect of GR24, a synthetic analogue of strigolactones, on gene expression of solopathogenic strain of *Sporisorium reilianum*. *African Journal of Biotechnology* 10(70), 15739–15743. DOI: 10.5897/AJB11.393.

Sabbagh, S., Mazaheri, M., Panjehkeh, N. and Salari, M. (2012) Transcriptomic analysis of *Sporisorium reilianum* in response to the strigolactone analogue GR24. *Phytopathologia Mediterranea* 51(2), 283–291.

Saeed, W., Naseem, S. and Ali, Z. (2017) Strigolactones biosynthesis and their role in abiotic stress resilience in plants: A critical review. *Frontiers in Plant Science* 8, 1487. DOI: 10.3389/fpls.2017.01487.

Sedaghat, M., Tahmasebi-Sarvestani, Z., Emam, Y. and Mokhtassi-Bidgoli, A. (2017) Physiological and antioxidant responses of winter wheat cultivars to strigolactone and salicylic acid in drought. *Plant Physiology and Biochemistry* 119, 59–69. DOI: 10.1016/j.plaphy.2017.08.015.

Sedaghat, M., Sarvestani, Z.T., Emam, Y., Bidgoli, A.M. and Sorooshzadeh, A. (2020) Foliar-applied GR24 and salicylic acid enhanced wheat drought tolerance. *Russian Journal of Plant Physiology* 67(4), 733–739. DOI: 10.1134/S1021443720040159.

Soliman, S., Wang, Y., Han, Z., Pervaiz, T. and El-Kereamy, A. (2022) Strigolactones in plants and their interaction with the ecological microbiome in response to abiotic stress. *Plants* 11(24), 3499. DOI: 10.3390/plants11243499.

Song, M., Hu, N., Zhou, S., Xie, S., Yang, J. *et al.* (2023) Physiological and RNA-Seq analyses on exogenous strigolactones alleviating drought by improving antioxidation and photosynthesis in wheat (*Triticum aestivum* L.). *Antioxidants* 12(10), 1884. DOI: 10.3390/antiox12101884.

Tsuchiya, Y., Vidaurre, D., Toh, S., Hanada, A., Nambara, E. *et al.* (2010) A small-molecule screen identifies new functions for the plant hormone strigolactone. *Nature Chemical Biology* 6(10), 741–749. DOI: 10.1038/nchembio.435.

Visentin, I., Pagliarani, C., Deva, E., Caracci, A., Turečková, V. *et al.* (2020) A novel strigolactone–miR156 module controls stomatal behaviour during drought recovery. *Plant, Cell & Environment* 43(7), 1613–1624. DOI: 10.1111/pce.13758.

Wani, K.I., Zehra, A., Choudhary, S., Naeem, M. and Khan, M.M.A. (2023) Exogenous strigolactone (GR24) positively regulates growth, photosynthesis, and improves glandular trichome attributes for enhanced artemisinin production in *Artemisia annua*. *Journal of Plant Growth Regulation* 42(8), 4606–4615. DOI: 10.1007/s00344-022-10654-w.

Yamada, Y. and Umehara, M. (2015) Possible roles of strigolactones during leaf senescence. *Plants* 4(3), 664–677. DOI: 10.3390/plants4030664.

Yamada, Y., Furusawa, S., Nagasaka, S., Shimomura, K., Yamaguchi, S. *et al.* (2014) Strigolactone signaling regulates rice leaf senescence in response to a phosphate deficiency. *Planta* 240(2), 399–408. DOI: 10.1007/s00425-014-2096-0.

Yoneyama, K. (2019) How do strigolactones ameliorate nutrient deficiencies in plants? *Cold Spring Harbor Perspectives in Biology* 11(8), a034686. DOI: 10.1101/cshperspect.a034686.

Yoneyama, K. and Bennett, T. (2024) Whispers in the dark: Signals regulating underground plant-plant interactions. *Current Opinion in Plant Biology* 77, 102456. DOI: 10.1016/j.pbi.2023.102456.

Yoneyama, K., Xie, X., Sekimoto, H., Takeuchi, Y., Ogasawara, S. *et al.* (2008) Strigolactones, host recognition signals for root parasitic plants and arbuscular mycorrhizal fungi, from Fabaceae plants. *New Phytologist* 179(2), 484–494. DOI: 10.1111/j.1469-8137.2008.02462.x.

Yoneyama, K., Xie, X., Kisugi, T., Nomura, T. and Yoneyama, K. (2013) Nitrogen and phosphorus fertilization negatively affects strigolactone production and exudation in sorghum. *Planta* 238(5), 885–894. DOI: 10.1007/s00425-013-1943-8.

Zhang, X., Zhang, L., Sun, Y., Zheng, S., Wang, J. *et al.* (2020) Hydrogen peroxide is involved in strigolactone induced low temperature stress tolerance in rape seedlings (*Brassica rapa* L.). *Plant Physiology and Biochemistry* 157, 402–415. DOI: 10.1016/j.plaphy.2020.11.006.

Zhang, X., Zhang, L., Ma, C., Su, M., Wang, J. *et al.* (2022) Exogenous strigolactones alleviate the photosynthetic inhibition and oxidative damage of cucumber seedlings under salt stress. *Scientia Horticulturae* 297, 110962. DOI: 10.1016/j.scienta.2022.110962.

Zhang, Y., Lv, S. and Wang, G. (2018) Strigolactones are common regulators in induction of stomatal closure in planta. *Plant Signaling & Behavior* 13(3), e1444322. DOI: 10.1080/15592324.2018.1444322.

Zheng, X., Li, Y., Xi, X., Ma, C., Sun, Z. *et al.* (2021) Exogenous strigolactones alleviate KCl stress by regulating photosynthesis, ROS migration and ion transport in Malus hupehensis Rehd. *Plant Physiology and Biochemistry* 159, 113–122. DOI: 10.1016/j.plaphy.2020.12.015.

Zou, J., Zhang, S., Zhang, W., Li, G., Chen, Z. *et al.* (2006) The rice HIGH-TILLERING DWARF1 encoding an ortholog of Arabidopsis MAX3 is required for negative regulation of the outgrowth of axillary buds. *Plant Journal* 48(5), 687–698. DOI: 10.1111/j.1365-313X.2006.02916.x.

8 Diverse Functional Roles of Strigolactones in Postharvest Biology

Rakhshinda Sadiq[1], Nadia Sharif[1], Mohammad Mehdizadeh[2] and Zainul Abideen[3,4*]

[1]Department of Biotechnology, Women University Mardan, Khyber Pakhtunkhwa, Pakistan; [2]Department of Agronomy and Plant Breeding, Faculty of Agriculture and Natural Resources, University of Mohaghegh Ardabili, Ardabil, Iran; [3]Dr. Muhammad Ajmal Khan Institute of Sustainable Halophyte Utilization, University of Karachi, Karachi, Pakistan; [4]Sharjah Seed Bank and Herbarium, Environment and Protected Areas Authority, Sharjah, United Arab Emirates

Abstract

The world population is growing with accelerating edible production demands, which leads to massive food waste. Degradation of fresh food is accelerated by a number of variables, including quick ripening, physiological alterations, and susceptibility to microbial infection. Fruits and vegetables are a major area of concern because of their high perishability, which makes them especially sensitive and responsible for about half of all food waste. This chapter introduces the emerging field of phytohormone therapies in postharvest management, focusing specifically on the diverse functional roles of strigolactones (SLs), a critical recently characterized class of plant hormones. The primary focus is a deep dive into the protective mechanisms resulting from the exogenous application of SLs and the potent synthetic analog GR24. These mechanisms are multi-faceted, extending the lifespan of produce by: (i) modulating water balance through the control of respiration and transpiration rates; (ii) slowing the visible progression of decay (senescence and yellowing); and (iii) strengthening the internal antioxidant defense system. This enzymatic reinforcement, which boosts enzymes such as superoxide dismutase and catalase, is crucial for neutralizing harmful reactive oxygen species, thus maintaining cell integrity and nutritional quality during prolonged storage.

Keywords: antioxidant defense system, disease management, fruit diseases, reactive oxygen species (ROS), respiration, senescence

8.1 Introduction

Global food demand has increased dramatically in recent years, mostly due to the world's population growth (Godfray *et al.*, 2010; Ehrlich and Harte, 2015; Raak *et al.*, 2017; Ema *et al.*, 2024). However, despite this growing need, more than 30% of the agricultural products grown globally is lost each year throughout the whole supply chain. Interestingly, the fruit and vegetable industry accounts for more than half of these losses (Poveda, 2020; Lau *et al.*, 2021; Ema *et al.*, 2024). Enhancing food security and advancing sustainable food systems require

*Corresponding author: zuabideen@uok.edu.pk

DOI: 10.1079/9781836990598.0008

focused efforts to decrease postharvest fresh fruit and vegetable losses, which is urgently needed (Papargyropoulou *et al.*, 2014; Raak *et al.*, 2017; Munir *et al.*, 2025). Numerous factors, such as risk of infection with pathogenic organisms and vulnerability to physiological changes that accelerate senescence, are the main causes of the highly perishable nature of fruits and vegetables (Asghari and Aghdam, 2010; Aghdam and Bodbodak, 2013; Aghdam *et al.*, 2023; Saud *et al.*, 2023). To enhance the quality control of produce across the supply chain and minimize losses, a variety of postharvest technologies have been created and implemented (Ma *et al.*, 2019; Bílková *et al.*, 2020; Ferreira *et al.*, 2024; Patil *et al.*, 2024; Zhang, 2024).

Cold storage and the use of artificial chemicals are two of the most popular postharvest procedures for extending the shelf life of fruits and vegetables (Hawkins, 1922; Kader, 2002; Kader and Rolle, 2004; Riva *et al.*, 2020). However, due to increasing concerns about the possible negative effects on health and the environment, the use of synthetic chemicals is gradually being phased out (Bose *et al.*, 2019, Bose *et al.*, 2021). Chilling injury, internal browning, and shriveling are examples of physiological disorders that frequently arise during prolonged storage periods and are not entirely prevented by cold storage alone (González-Aguilar *et al.*, 2000; Zhang *et al.*, 2015a; Youryon *et al.*, 2019; Rasouli *et al.*, 2019; Fawole *et al.*, 2020).

Therefore, to protect and preserve the nutritional value of fresh fruits and vegetables along the supply chain, it is essential to develop cutting-edge technologies that may be successfully incorporated into food-packaging systems. Recently, supplemental plant hormone therapies have emerged as a practical natural postharvest technique for maintaining the nutritional value and lifespan of fruits and vegetables while they are being stored (Han *et al.*, 2003; Ghorbani and Pakkish, 2014; Gao *et al.*, 2015, 2016; Liu *et al.*, 2016a; Aghdam *et al.*, 2019, 2020; Baswal *et al.*, 2020; Hanif *et al.*, 2020; Adhikary *et al.*, 2021; Islam *et al.*, 2022; Gutiérrez-Villamil *et al.*, 2023). Naturally occurring organic substances called plant hormones, or phytohormones, are essential for managing the progression of plants, coordinating physiological functions, and improving plant reactions to both biological and environmental stressors (Creelman and Mullet, 1997; Peian *et al.*, 2021; Munir *et al.*, 2025). The capacity of external plant hormone treatments to promote the production of bioactive chemicals, postpone ripening and senescence, increase tolerance to a range of physiological diseases during postharvest storage, and prolong shelf life is what makes them effective (Ghorbani and Pakkish, 2014; Gao *et al.*, 2015; Gao *et al.*, 2016; Atia *et al.*, 2018; Aghdam *et al.*, 2019; Aghdam *et al.*, 2020; Baswal *et al.*, 2020; Adhikary *et al.*, 2021; Hu *et al.*, 2022; Islam *et al.*, 2022; Massolo *et al.*, 2022; Gutiérrez-Villamil *et al.*, 2023). Phytohormones are also essential for improving resistance to infections (Su *et al.*, 2017; Song *et al.*, 2022). *Citrus* (Siboza *et al.*, 2014; Tavallali and Zareiyan, 2018), pomegranates (Sayyari *et al.*, 2009, 2011a, b), papaya (Hanif *et al.*, 2020), avocado (Glowacz *et al.*, 2017), peach (Jin *et al.*, 2009), pineapple, cucumber (Zhang *et al.*, 2015b; Sangprayoon *et al.*, 2019), mango (González-Aguilar *et al.*, 2000; Zaharah and Singh, 2012), kiwifruit (Wang *et al.*, 2020), and bananas (Li *et al.*, 2018) are just a few of the fresh horticultural produce that has been shown to benefit from hormone treatment.

The potential of external phytohormone treatments in maintaining food quality is highlighted by these findings. Minimizing food wastage and losses, lowering dependence on artificial chemical preservatives, and encouraging the food industry to adopt more sustainable approaches are just a few benefits of such treatments. Understanding the effects, causes, and real-world uses of these bioactive compounds in postharvest management is therefore attracting increasing attention (Gomez-Roldan *et al.*, 2008; Umehara *et al.*, 2008; Zwanenburg and Pospíšil, 2013; Aliche *et al.*, 2020). Strigolactones (SLs) are a recently discovered class of phytohormones. In addition to controlling growth of plants, SLs interact effectively with other important phytohormones such as abscisic acid (ABA) and cytokinins (CKs). Toh *et al.* (2012) highlighted the importance of the reactions of SLs to environmental stress by showing that they help reduce heat stress by modifying ABA levels. In contrast, exogenous CK treatment inhibits SL development in rice root tissues (Yoneyama *et al.*, 2020), indicating an active and reciprocal hormonal interplay. These interactions demonstrate how SLs use

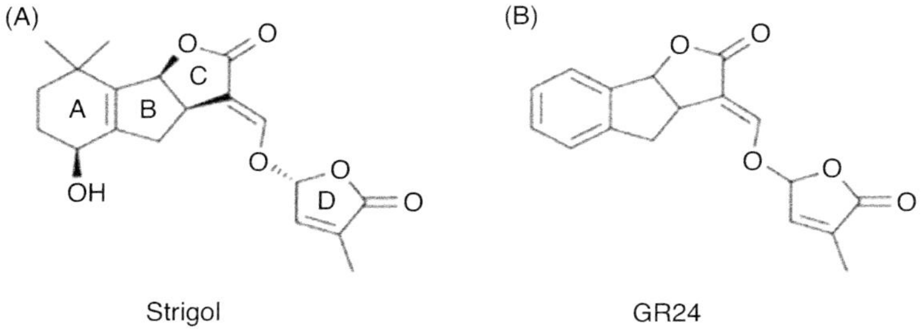

Fig. 8.1. Structure of the naturally occurring strigol (A) and its synthetic analogue GR24 (B) (Dell'Oste *et al.*, 2021). Figure used under license CC-BY-4.0.

intricate hormonal networks to coordinate plant growth and environmental adaption. Although research studies regarding external plant-based hormones for postharvest quality preservation are expanding, little is known about the synthesis, use, and mechanisms of recently discovered hormones such as SLs. This chapter examines the physicochemical characteristics, biological roles, and practical uses of SLs in conserving quality following harvest of perishable crops.

8.2 Strigolactone Hormones in Plants: Structure and Synthesis

Carotenoid-derived hormones called SLs are identified by their four-ring structure, which consists of an ABC tricyclic core joined to a fourth unit called the D-ring (Fig. 8.1) (Dell'Oste *et al.*, 2021). Research on SLs is still limited in comparison with more established plant hormones, despite increased interest in their potential uses in sustainable agriculture (Dell'Oste *et al.*, 2021; Kapoor *et al.*, 2024). SLs are generally divided into two groups: canonical and non-canonical. They are identified by whether the entire ABC-ring structure is present, but the D-ring is the same for both types (Wang and Bouwmeester, 2018; Dell'Oste *et al.*, 2021; Wu *et al.*, 2022; Kapoor *et al.*, 2024). Little information is available regarding the biological functions of SLs in connection to their structural variety, despite advancements in structural characterization. In order to investigate the possible agricultural uses of SLs, current understanding of their signaling and functional roles has depended mostly on synthetic analogs, such as GR3, GR5, GR7, GR24, and Nijmegen-1a. GR24 is the most used synthetic version of SLs, noted for its high bioactivity and effectiveness in mimicking natural SL functions (Dell'Oste *et al.*, 2021; Wu *et al.*, 2022; Li *et al.*, 2023; Selwal *et al.*, 2023). β-Carotene is the primary source of carlactone, a crucial intermediate that is necessary for the natural production of SLs (Fig. 8.2).

Three essential enzymes: DWARF27 (D27), carotenoid cleavage dioxygenase 7 (CCD7), and carotenoid cleavage dioxygenase 8 (CCD8) are involved in the strigolactone production pathway (Fig. 8.2). A crucial step in guiding carotenoid metabolism toward strigolactone formation is the isomerization of all-*trans*-β-carotene into 9-*cis*-β-carotene, which is catalyzed by D27 at the start of the process. Afterwards, CCD7 cleaves the intermediate 9-*cis*-β-carotene to yield 9-*cis*-apo-10′-carotenal. CCD8 subsequently transforms this molecule into (*Z*)-(*R*)-carlactone, a key precursor in the biosynthesis of SL. The cytochrome P450 mono-oxygenase MORE AXILLARY GROWTH 1 (MAX1) or its analogous enzymes catalyze further oxidation of the resultant carlactone, forming a wide variety of SLs. Prior to being transferred to other plant tissues where they carry out different physiological tasks, these SLs mostly gather in the roots, which also act as their primary storage location. There is currently little information available on the biosynthesis of SLs in fruits. Interestingly, depending on the stage of growth, tissue type, and environmental conditions, a single plant species can produce several forms of SLs, frequently in different quantities (Visentin *et al.*, 2016; Dell'Oste *et al.*, 2021;

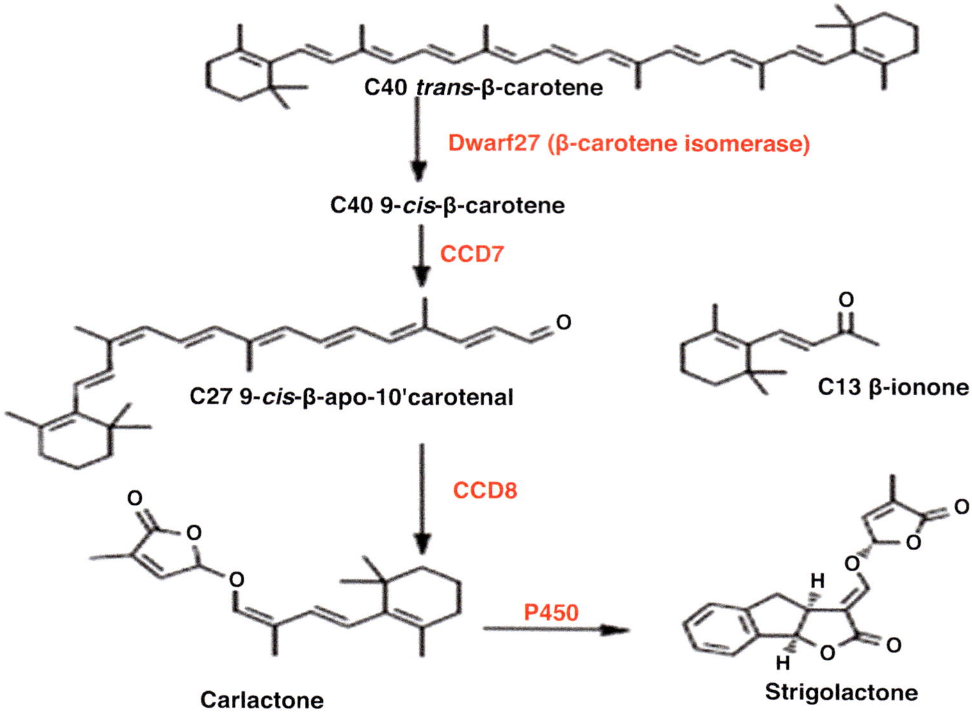

Fig. 8.2. Biosynthesis of strigolactones in plants (Dell'Oste *et al.*, 2021). CCD7, carotenoid cleavage dioxygenase 7; CCD8, carotenoid cleavage dioxygenase 8. Figure used under license CC-BY-4.0.

Wu *et al.*, 2022; Kapoor *et al.*, 2024; Nomura *et al.*, 2024).

8.3 Biological Functions of Strigolactones

8.3.1 Practical applications in agriculture

SLs have strong biological activity, which makes them of great interest for use in agriculture. However, synthesis on a kilogram scale is not feasible due to the complicated structure of naturally occurring SLs, which presents a significant obstacle to large-scale production. Many SL mimics and analogs, including Nijmegen-1, GR5, GR7, and GR24, have been created in order to address this. These substances are just as bioactive as their natural counterparts while having far simpler structures. Their synthesis can easily be increased up to multi-gram amounts and is much simpler, frequently including only a few stages. Because of this, the production costs of these analogs are significantly less than those of research-grade natural SLs. Although SLs have shown great promise in agriculture, Vurro *et al.* (2016) may have been a little too optimistic in their assessment.

SL analogs show promise in the management of pathogenic weeds in food crops (Dawood *et al.*, 2026). Notably, Nijmegen-1 has demonstrated promising outcomes in controlling tobacco infestations of *Orobanche ramosa*. Its strong effectiveness as a sprouting stimulant is highlighted by its low application rate, which is particularly excellent at about 8g ha^{-1}. To obtain the correct equilibrium between bioactivity, soil stability, toxicity, and phytotoxicity, more studies are necessary to optimize its structure.

Furthermore, for real-world field applications, the creation of an ideal formulation and assessment of crucial agricultural factors are still necessary.

One of the main obstacles to the commercial use of SL analogs is the legal registration procedure, which can be expensive and complicated. The function of SLs such as growth factors for arbuscular mycorrhizal fungi (AMF) is of special practical significance. By increasing the amount of both nitrogen and phosphate in the soil, these fungi improve plant nutrient uptake and function as an environmentally friendly biofertilizer. Few SLs have been used in practice to promote AMF. The creation of simpler SL analogs with notable activity as branching factors is still absent, despite the fact that important insights into the structural needs for SL-induced hyphal branching in AMF have been described (Akiyama *et al.*, 2010). Currently, no structurally simplified SLs have proven effective enough to be used practically to encourage AMF symbiosis.

Therefore, further research is needed in this area. SLs can be thought of as potential regulators of plant growth because they are essential in controlling the architecture of plants, including the branching of shoots and the development of roots. Their versatility underscores their potential for wider agricultural use beyond their present applications. SLs can function sequentially or synergistically with other important plant hormones, including as auxins, CKs, gibberellins, and ABA, in a process known as hormonal crosstalk. Although the importance of SLs in these interactions is widely acknowledged, little is known about the exact chemical mechanisms causing this crosstalk. Current knowledge of these intricate relationships has been enriched by recent reviews (Koltai, 2011; Cheng *et al.*, 2013; Umehara *et al.*, 2015). There is little question that SLs will be crucial to future plant growth regulation tactics, although it is still too early to put this knowledge into practice.

8.3.2 SLs as emerging regulators in postharvest preservation

Exogenous application of SLs has shown considerable potential for extending shelf life and enhancing quality management of fresh produce during postharvest management. Among these, GR24—a chemically synthesized SL analog—is the most widely used due to its high bioactivity and the inherent variability of SLs found in plants naturally.

Foliar spraying is a popular treatment technique for leafy vegetables because it is rapid and efficient while reducing excessive moisture exposure, to which these plants are especially vulnerable. In contrast, dipping is recommended for fruits as it enables deeper penetration of the substance into the fruit tissues and more consistent absorption (Huang *et al.*, 2021; Ma *et al.*, 2022; Li *et al.*, 2023, 2024). Despite these encouraging results, further research is required to optimize GR24 concentrations and crop-specific delivery methods. To encourage its wider use in the industry, research on how it interacts with other hormones produced by plants as well as assessments of its effectiveness and cost-efficiency in commercial postharvest settings are essential.

8.3.3 SLs mitigate respiration, transpiration, and postharvest quality loss

The postharvest condition and shelf life of fruits and vegetables are greatly impacted by two important physiological processes: respiration and transpiration. According to recent research, SLs can be applied exogenously to plants, such as fruits and vegetables, to alter their respiration and transpiration rates in a variety of environmental settings. Treatment with GR24 at concentrations of between 0.001 and 10 μM, for example, dramatically increased water content, stomatal conductance, intercellular carbon dioxide concentration, transpiration rate, and net photosynthetic rate, reducing the negative effects of drought and salt stress (Ling *et al.*, 2020; Sedaghat *et al.*, 2020; Wang *et al.*, 2021). However, in newly harvested strawberry fruit kept at 0°C for 10 days, 1 μM GR24 maintained a higher water content while dramatically suppressing weight loss and raising the respiration rate (Huang *et al.*, 2021). In postharvest fruits, increased respiration and transpiration can cause unfavorable outcomes such as water loss, fruit softening, and nutrient and flavor degradation. A controlled rise in respiration, however, can be advantageous, as it provides the energy and metabolites required for wound healing, improved resistance to disease, and the production of critical beneficial characteristics such as color, taste, and smell. It has been demonstrated that moderately dehydrating fresh citrus fruits after harvest, which is a traditional pre-storage treatment in China, improves fruit firmness, encourages the accumulation of hydrogen peroxide (H_2O_2), raises ascorbate peroxidase (APX) activity, and strengthens the fruit's resistance to infections by pathogens (Yun *et al.*, 2015).

Externally administered SLs have been demonstrated to affect the catabolism of naturally occurring acids and the metabolism of carbohydrates in plant cells (Wang *et al.*, 2021). For instance, when 10 μM SL was applied topically to healthy 'Hamlin' sweet orange (*Citrus sinensis* L. Osbeck) trees, the sucrose and starch contents of the mature leaves increased by 26% and 166%, respectively, in comparison with the untreated controls. It is interesting to note that the same treatment considerably increased the accumulation of starch in the roots of trees impacted by huanglongbing (citrus greening disease) while decreasing the amount of starch in the leaves (Zheng *et al.*, 2018). These results imply that SLs might be involved in controlling the distribution and storage of carbohydrates in response to the health condition of plants. It has been demonstrated that foliar treatment with GR24 increases the amount of proline and soluble solid content, which improves drought tolerance in winter wheat (Sedaghat *et al.*, 2020). Similarly, administration of 1 μM GR24 markedly increased the amounts of fructose, sucrose, and soluble sugar in tobacco axillary buds (Fan *et al.*, 2021). These results confirm the possible role of GR24 in maintaining the quality of fruits after harvest by indicating that it has a positive impact on internal as well as external quality aspects.

8.3.4 SLs improve antioxidant systems to combat ROS accumulation in postharvest fruits

SLs have been demonstrated to increase the antioxidant system's activity in plants. One of the main causes of fruit senescence is an excessive buildup of reactive oxygen species (ROS) (Tian *et al.*, 2018). The most stable ROS among these is H_2O_2, which can seriously harm cells by oxidizing proteins, lipids, and DNA, possibly resulting in cell death (García-Caparrós *et al.*, 2021). Exogenous SL treatment has been shown in numerous studies to decrease ROS generation, especially H_2O_2, which in turn reduces lipid peroxidation in membranes and electrolyte leakage, and increases plant resistance to a variety of abiotic stressors. For example, Zhang *et al.* (2020) found that external application of GR24 reduced comparative electrolytic conductivity in rapeseed seedlings during low-temperature stress and prevented the generation of H_2O_2 and superoxide (O_2^-). Similarly, by reducing leaching of electrolytes, malondialdehyde (MDA), and H_2O_2 levels, SLs applied topically enhanced drought tolerance (Sedaghat *et al.*, 2020; Bidabadi and Sharifi, 2021).

Zulfiqar *et al.* (2021) showed in an *in vitro* investigation that treatment with 0.01–320 mg l^{-1} GR24 considerably mitigated the negative effects of salt stress in sunflower callus tissues by reducing H_2O_2 and the amount of MDA by 65.78% and 30.76%, respectively. Together, these results imply that SLs protect plants by enhancing antioxidant defense systems and regulating oxidative stress reactions.

It has been demonstrated that applying GR24 to plants can lessen oxidative stress by lowering the buildup of dangerous ROS. For example, Lu *et al.* (2019) found that GR24 treatment reduced the levels of H_2O_2 and MDA in tomato cultivars under low-light stress, which decreased photosystem damage brought on by ROS and excess light energy. In a related study, Qiu *et al.* (2021) discovered that 1 μM GR24 efficiently protected plant tissue from oxidative damage by considerably reducing H_2O_2 and MDA buildup in barley caused by 10 mM cadmium stress. Despite these encouraging results, more research is necessary to fully understand the precise function of GR24 in controlling postharvest fruit senescence brought on by high ROS. Advanced endogenous defense systems have been developed by plants to shield their cells from ROS damage. These defenses consist of both non-enzymatic antioxidants, such as ascorbic acid (AsA), glutathione (GSH), phenolic acids, flavonoids, and carotenoids, and enzymatic antioxidants, such as catalase (CAT), superoxide dismutase (SOD), peroxidase (POD), APX, and glutathione reductase (GR) (Hasanuzzaman *et al.*, 2020). Among these, CAT directly contributes to ROS detoxification by catalyzing the conversion of H_2O_2 into oxygen and water.

The AsA–GSH cycle is a vital part of the antioxidant defense system and is principally in charge of neutralizing H_2O_2 (Pandey *et al.*, 2015). In this cycle, dehydroascorbic acid (DHA) is produced as a by-product of APX reducing H_2O_2 to water using AsA as an electron donor. GSH eventually oxidizes to glutathione disulfide (GSSG), which is then used in a reaction with DHA to convert it back to AsA. In order to preserve the redox balance, GR subsequently converts GSSG back to GSH (García-Caparrós *et al.*, 2021). The exogenous administration of GR24 considerably improved the antioxidant defense in barley plants under cadmium stress by raising the pools of AsA and GSH, and enhancing the activity of APX, GR, and the enzyme dehydroascorbic acid reductase (Qiu *et al.*, 2021). According to these results, GR24 might support the AsA–GSH cycle and the plant's general antioxidant defenses against oxidative stress.

GR24 treatment dramatically increased the enzyme activity of SOD, CAT, and POD in salt-stressed sunflower plants by 38.1%, 67.87%, and 124.01%, respectively (Zulfiqar *et al.*, 2021). Similarly, GR24 treatment increased the level of expression of genes linked to antioxidant enzymes in cold-sensitive rape seedlings (Zhang *et al.*, 2020). Treatment with 1 μM GR24 successfully inhibited the increase in POD activity and the decrease in CAT and SOD activities in strawberries kept at 0°C. Additionally, it maintained lower levels of GSH and lowered AsA, which improved the fruit's ability to scavenge ROS (Huang *et al.*, 2021). Similar results were found in a study on "Newhall" navel oranges, where GR24 administration resulted in increased GSH and AsA levels, as well as increased activity of APX and GR in the AsA–GSH cycle. Furthermore, over the course of the storage period, GR24 increased the fruit pulp's total phenolic content and CAT activity. According to these findings, GR24 reduces oxidative damage caused by H_2O_2 in postharvest orange fruits by fortifying the antioxidant defense system (Ma *et al.*, 2022).

Treating loquat fruit with SL has been demonstrated to successfully reduce chilling damage during cold storage by modifying several physical and biochemical processes. In particular, under low-temperature stress, it postpones the buildup of lignin, a critical component linked to tissue stiffening and quality degradation. Furthermore, SL improves the performance of both enzymatic and non-enzymatic antioxidant systems, assisting in the maintenance of increased levels of antioxidant activity. By lowering oxidative damage brought on by ROS, this enhanced antioxidant defense maintains fruit quality and increases postharvest shelf life (Cao *et al.*, 2025). To sum up, these results jointly demonstrate how important SLs—in particular GR24—are for strengthening antioxidant defense systems and reducing oxidative stress in plants and postharvest fruits. By controlling ROS levels and promoting the AsA–GSH cycle, SLs greatly enhance stress tolerance and prolong shelf life.

8.3.5 SLs affect fruit senescence process

Studies have demonstrated the critical roles that SLs play in a variety of biological processes, including leaf senescence (Yamada *et al.*, 2014; Ueda and Kusaba, 2015; Tian *et al.*, 2018; Hu *et al.*, 2021; Takahashi *et al.*, 2021; Xu *et al.*, 2021), stress responses (Omoarelojie *et al.*, 2019), and plant growth and development. Their function in controlling senescence in other plant organs has, however, received little attention. In horticultural crops, fruits and vegetables are vital production organs, so it is crucial to fully understand how SLs affect their senescence.

According to Ma *et al.* (2022), the firmness and soluble sugar content of orange fruits were considerably increased after being treated with 200 μM GR24. The treatment also increased antioxidant activity, which postponed senescence. The plant hormone ABA is the primary regulator of ripening in citrus fruits, which are typical non-climacteric fruits (Romero *et al.*, 2019). ABA and natural SLs are both terpenoids generated from carotenoids. Interestingly, it has been demonstrated that exogenous SLs prevent the production of endogenous ABA in several species (Omoarelojie *et al.*, 2019). This points to a possible connection between SLs and ABA in controlling the senescence of citrus fruits, although more investigation is required to validate this theory.

Similar effects of SLs on other horticultural crops have been shown in other investigations. For example, Wang *et al.* (2025) discovered that adding 1 μM GR24 enhanced the nutritional value and flavor of tomato fruits that were still developing. The treatment altered sugar levels and composition, improved fragrance qualities, and decreased the overall amount of organic acid. Similarly, exogenous GR24 treatment successfully postponed postharvest yellowing and deterioration in celery, according to Li *et al.* (2024). In addition to preserving taste ingredients during storage, it significantly increased antioxidant activity, including total phenols, vitamin C, and DPPH (2,2-diphenyl-1-picrylhydrazyl)) radical scavenging ability, and slowed the breakdown of chlorophyll.

8.3.6 Postharvest quality preservation using exogenous plant hormones: Limitations and possibilities for the future

Although exogenous plant hormones are essential for postharvest preservation, their use presents safety risks, especially with regard to toxicity and bioaccumulation. Studies on gibberellic acid and CKs have shown that high quantities may be harmful to both human health and the environment. Residue limits have been set by regulatory agencies to reduce these dangers, but worldwide compliance is hampered by uneven enforcement and a lack of analytical capacity. Because of this, striking a balance between safety and efficacy is still quite difficult (Wang *et al.*, 2011; Helmy *et al.*, 2015; Liu *et al.*, 2016b; Koo *et al.*, 2020; Le *et al.*, 2020).

Furthermore, new plant hormones such as SLs are gaining attention due to their possible uses in medicine, for example in the management of cancer and inflammation. However, the safe concentration limits for these applications are still unclear. This emphasizes how urgently thorough study is required to guarantee the safe and efficient application of SLs in biomedical and agricultural settings. The responsible use of these substances requires, in particular, the completion of thorough toxicological evaluations and the establishment of precise regulatory standards (Dell'Oste *et al.*, 2021).

8.4 Conclusion

Exogenous plant hormones such as SL analogs can be used to improve the quality and lifespan of fruits and vegetables through a variety of physiological processes, making them essential for postharvest preservation. However, worries about toxicity at high doses emphasize the necessity of stringent control and secure application procedures. For their reliable and sustainable usage, cutting-edge technologies, precise regulations, and cross-sector cooperation will be necessary. For commercial use, environmentally sustainable and economical distribution methods should be the focus of future studies.

References

Adhikary, T., Gill, P.S., Jawandha, S.K., Bhardwaj, R.D. and Anurag, R.K. (2021) Browning and quality management of pear fruit by salicylic acid treatment during low temperature storage. *Journal of the Science of Food and Agriculture* 101(3), 853–862. DOI: 10.1002/jsfa.10692.

Aghdam, M.S. and Bodbodak, S. (2013) Physiological and biochemical mechanisms regulating chilling tolerance in fruits and vegetables under postharvest salicylates and jasmonates treatments. *Scientia Horticulturae* 156, 73–85. DOI: 10.1016/j.scienta.2013.03.028.

Aghdam, M.S., Luo, Z., Jannatizadeh, A., Sheikh-Assadi, M., Sharafi, Y. *et al.* (2019) Employing exogenous melatonin applying confers chilling tolerance in tomato fruits by upregulating *ZAT2/6/12* giving rise to promoting endogenous polyamines, proline, and nitric oxide accumulation by triggering arginine pathway activity. *Food Chemistry* 275, 549–556. DOI: 10.1016/j.foodchem.2018.09.157.

Aghdam, M.S., Luo, Z., Li, L., Jannatizadeh, A., Fard, J.R. *et al.* (2020) Melatonin treatment maintains nutraceutical properties of pomegranate fruits during cold storage. *Food Chemistry* 303, 125385. DOI: 10.1016/j.foodchem.2019.125385.

Aghdam, M.S., Mukherjee, S., Flores, F.B., Arnao, M.B., Luo, Z. *et al.* (2023) Functions of melatonin during postharvest of horticultural crops. *Plant and Cell Physiology* 63(12), 1764–1786. DOI: 10.1093/pcp/pcab175.

Akiyama, K., Ogasawara, S., Ito, S. and Hayashi, H. (2010) Structural requirements of strigolactones for hyphal branching in AM fungi. *Plant and Cell Physiology* 51(7), 1104–1117. DOI: 10.1093/pcp/pcq058.

Aliche, E.B., Screpanti, C., de Mesmaeker, A., Munnik, T. and Bouwmeester, H.J. (2020) Science and application of strigolactones. *New Phytologist* 227(4), 1001–1011. DOI: 10.1111/nph.16489.

Asghari, M. and Aghdam, M.S. (2010) Impact of salicylic acid on post-harvest physiology of horticultural crops. *Trends in Food Science & Technology* 21(10), 502–509. DOI: 10.1016/j.tifs.2010.07.009.

Atia, A., Abdelkarim, D., Younis, M. and Alhamdan, A. (2018) Effects of calcium chloride and salicylic acid postharvest treatments on the quality of Khalal Barhi dates at different ripening levels during cold storage. *Journal of Food Measurement and Characterization* 12, 1156–1166. DOI: 10.25165/j.ijabe.20201303.4983.

Baswal, A.K., Dhaliwal, H.S., Singh, Z., Mahajan, B.V.C. and Gill, K.S. (2020) Postharvest application of methyl jasmonate, 1-methylcyclopropene and salicylic acid extends the cold storage life and maintain the quality of 'Kinnow' mandarin (*Citrus nobilis* L. × *C. deliciosa* L.) fruit. *Postharvest Biology and Technology* 161, 111064. DOI: 10.1016/j.postharvbio.2019.111064.

Bidabadi, S.S. and Sharifi, P. (2021) Strigolactone and methyl jasmonate-induced antioxidant defense and the composition alterations of different active compounds in *Dracocephalum kotschyi* Boiss under drought stress. *Journal of Plant Growth Regulation* 40(2), 878–889. DOI: 10.1007/s00344-020-10157-6.

Bose, S.K., Howlader, P., Wang, W. and Yin, H. (2021) Oligosaccharide is a promising natural preservative for improving postharvest preservation of fruit: A review. *Food Chemistry* 341, 128178. DOI: 10.1016/j.foodchem.2020.128178.

Bose, S.K., Howlader, P., Jia, X., Wang, W. and Yin, H. (2019) Alginate oligosaccharide postharvest treatment preserve fruit quality and increase storage life via Abscisic acid signaling in strawberry. *Food Chemistry* 283, 665–674. DOI: 10.1016/j.foodchem.2019.01.060.

Bílková, A., Baďurová, K., Svobodová, P., Vávra, R., Jakubec, P. *et al.* (2020) Content of major phenolic compounds in apples: Benefits of ultra-low oxygen conditions in long-term storage. *Journal of Food Composition and Analysis* 92, 103587. DOI: 10.1016/j.jfca.2020.103587.

Cao, S., Qiao, L., Wang, X., Huang, T., Wu, C. *et al.* (2025) Discussion on postharvest quality, strategies of storage and transportation techniques of blueberries: A comprehensive review. *Food Quality and Safety* 9, fyaf038. DOI: 10.1093/fqsafe/fyaf038.

Cheng, X., Ruyter-Spira, C. and Bouwmeester, H. (2013) The interaction between strigolactones and other plant hormones in the regulation of plant development. *Frontiers in Plant Science* 4, 199. DOI: 10.3389/fpls.2013.00199.

Creelman, R.A. and Mullet, J.E. (1997) Biosynthesis and action of jasmonates in plants. *Annual Review of Plant Physiology and Plant Molecular Biology* 48, 355–381. DOI: 10.1146/annurev.arplant.48.1.355.

Dawood, M.F.A., Eissa, M.A., Mir, M.I., Muhammad, M., Abideen, Z. *et al.* (2026) Strigolactones for boosting plant stress tolerance. In: Latef, A.A.H.A. (ed.) *Using Stimulants to Improve Plant Health from Stress to Strength*. Elsevier (in press).

Dell'Oste, V., Spyrakis, F. and Prandi, C. (2021) Strigolactones, from plants to human health: Achievements and challenges. *Molecules* 26(15), 4579. DOI: 10.3390/molecules26154579.

Ehrlich, P.R. and Harte, J. (2015) Food security requires a new revolution. *International Journal of Environmental Studies* 72(6), 908–920. DOI: 10.1080/00207233.2015.1067468.

Ema, N.R., Mithu, M.A.H. and Sayem, A. (2024) Exploring driving factors in employing waste reduction tools to alleviate the global food security and sustainability. *Heliyon* 10(7), e28192. DOI: 10.1016/j.heliyon.2024.e28192.

Fan, W., Pan, Z., Wang, Y., Yang, S., Zeng, R. *et al.* (2021) Effects of strigolactone on the growth, antioxidant enzyme activities and sugar contents of axillary buds in tobacco. *Plant Physiology Journal* 57, 910. DOI: 10.1016/j.pmatsci.2025.101526.

Fawole, O.A., Riva, S.C. and Opara, U.L. (2020) Efficacy of edible coatings in alleviating shrivel and maintaining quality of Japanese plum (*Prunus salicina* Lindl.) during export and shelf life conditions. *Agronomy* 10(7), 1023. DOI: 10.3390/agronomy10071023.

Ferreira, C., Ribeiro, C. and Nunes, F.M. (2024) Effect of storage conditions on phenolic composition, vitamin C and antioxidant activity of 'Golden Delicious' and 'Red Delicious' apples. *Postharvest Biology and Technology* 210, 112754. DOI: 10.1016/j.postharvbio.2023.112754.

Gao, H., Kang, L., Liu, Q., Cheng, N., Wang, B. *et al.* (2015) Effect of 24-epibrassinolide treatment on the metabolism of eggplant fruits in relation to development of pulp browning under chilling stress. *Journal of Food Science and Technology* 52(6), 3394–3401. DOI: 10.1007/s13197-014-1402-y.

Gao, H., Zhang, Z., Lv, X., Cheng, N., Peng, B. *et al.* (2016) Effect of 24-epibrassinolide on chilling injury of peach fruit in relation to phenolic and proline metabolisms. *Postharvest Biology and Technology* 111, 390–397. DOI: 10.1016/j.postharvbio.2015.07.031.

García-Caparrós, P., de Filippis, L., Gul, A., Hasanuzzaman, M., Ozturk, M. *et al.* (2021) Oxidative stress and antioxidant metabolism under adverse environmental conditions: A review. *Botanical Review* 87(4), 421–466. DOI: 10.1007/s12229-020-09231-1.

Ghorbani, B. and Pakkish, Z. (2014) Brassinosteroid enhances cold stress tolerance of Washington Navel orange (*Citrus sinensis* L.) fruit by regulating antioxidant enzymes during storage. *Agriculturae Conspectus Scientificus* 79(2), 109–114.

Glowacz, M., Bill, M., Tinyane, P.P. and Sivakumar, D. (2017) Maintaining postharvest quality of cold stored 'Hass' avocados by altering the fatty acids content and composition with the use of natural volatile compounds—methyl jasmonate and methyl salicylate. *Journal of the Science of Food and Agriculture* 97(15), 5186–5193. DOI: 10.1002/jsfa.8400.

Godfray, H.C.J., Beddington, J.R., Crute, I.R., Haddad, L., Lawrence, D. *et al.* (2010) Food security: The challenge of feeding 9 billion people. *Science* 327(5967), 812–818. DOI: 10.1126/science.1185383.

Gomez-Roldan, V., Fermas, S., Brewer, P.B., Puech-Pagès, V., Dun, E.A. *et al.* (2008) Strigolactone inhibition of shoot branching. *Nature* 455(7210), 189–194. DOI: 10.1038/nature07271.

González-Aguilar, G.A., Fortiz, J., Cruz, R., Baez, R. and Wang, C.Y. (2000) Methyl jasmonate reduces chilling injury and maintains postharvest quality of mango fruit. *Journal of Agricultural and Food Chemistry* 48(2), 515–519. DOI: 10.1021/jf9902806.

Gutiérrez-Villamil, D.A., Balaguera-López, H.E. and Álvarez-Herrera, J.G. (2023) Brassinosteroids improve postharvest quality, antioxidant compounds, and reduce chilling injury in 'Arrayana' mandarin fruits under cold storage. *Horticulturae* 9(6), 622. DOI: 10.3390/horticulturae9060622.

Han, T., Wang, Y., Li, L. and Ge, X. (2003) Effect of exogenous salicylic acid on post harvest physiology of peaches. *Acta Horticulturae* 628(628), 583–589. DOI: 10.17660/ActaHortic.2003.628.74.

Hanif, A., Ahmad, S., Shahzad, S., Liaquat, M. and Anwar, R. (2020) Postharvest application of salicylic acid reduced decay and enhanced storage life of papaya fruit during cold storage. *Journal of Food Measurement and Characterization* 14(6), 3078–3088. DOI: 10.1007/s11694-020-00555-5.

Hasanuzzaman, M., Bhuyan, M.H.M.B., Zulfiqar, F., Raza, A., Mohsin, S.M. *et al.* (2020) Reactive oxygen species and antioxidant defense in plants under abiotic stress: Revisiting the crucial role of a universal defense regulator. *Antioxidants* 9(8), 681. DOI: 10.3390/antiox9080681.

Hawkins, L.A. (1922) The effect of low-temperature storage and freezing on fruits and vegetables. *American Journal of Botany* 9(10), 551–556. DOI: 10.2307/2435289.

Helmy, G., Ahmed, S. and Mahrous, S. (2015) Evaluation of synthetic plant growth regulators residues in fruits and vegetables and health risk assessment in Giza, Egypt. *Journal of Soil Sciences and Agricultural Engineering* 6(9), 1075–1089. DOI: 10.21608/jssae.2015.42828.

Hu, Q., Ding, F., Li, M., Zhang, X., Zhang, S. *et al.* (2021) Strigolactone and ethylene inhibitor suppressing dark-induced leaf senescence in perennial ryegrass involving transcriptional downregulation of chlorophyll degradation. *Journal of the American Society for Horticultural Science* 146(2), 79–86. DOI: 10.21273/JASHS04933-20.

Hu, S., Ma, Y., Xie, B., Hou, Y., Jia, Z. *et al.* (2022) 24-epibrassinolide improves chilling tolerance by regulating PpCBF5-mediated membrane lipid metabolism in peach fruit. *Postharvest Biology and Technology* 186, 111844. DOI: 10.1016/j.postharvbio.2022.111844.

Huang, D., Wang, Y., Zhang, D., Dong, Y., Meng, Q. *et al.* (2021) Strigolactone maintains strawberry quality by regulating phenylpropanoid, NO, and H_2S metabolism during storage. *Postharvest Biology and Technology* 178, 111546. DOI: 10.1016/j.postharvbio.2021.111546.

Islam, M., Ali, S., Nawaz, A., Naz, S., Ejaz, S. *et al.* (2022) Postharvest 24-epibrassinolide treatment alleviates pomegranate fruit chilling injury by regulating proline metabolism and antioxidant activities. *Postharvest Biology and Technology* 188, 111906. DOI: 10.1016/j.postharvbio.2022.111906.

Jin, P., Wang, K.U., Shang, H.T., Tong, J.M. and Zheng, Y.H. (2009) Low-temperature conditioning combined with methyl jasmonate treatment reduces chilling injury of peach fruit. *Journal of the Science of Food and Agriculture* 89(10), 1690–1696. DOI: 10.1002/jsfa.3642.

Kader, A.A. (2002) *Postharvest Technology of Horticultural Crops*, 3rd edn. University of California Agriculture and Natural Resources, Davis, California.

Kader, A.A. and Rolle, R.S. (2004) *The Role of Post-Harvest Management in Assuring the Quality and Safety of Horticultural Produce*, FAO Agricultural Services Bulletin, Vol. 152. Food and Agriculture Organization, Rome.

Kapoor, R.T., Alam, P., Chen, Y. and Ahmad, P. (2024) Strigolactones in plants: From development to abiotic stress management. *Journal of Plant Growth Regulation* 43(3), 903–919. DOI: 10.1007/s00344-023-11148-z.

Koltai, H. (2011) Strigolactones are regulators of root development. *New Phytologist* 190(3), 545–549. DOI: 10.1111/j.1469-8137.2011.03678.x.

Koo, Y.M., Heo, A.Y. and Choi, H.W. (2020) Salicylic acid as a safe plant protector and growth regulator. *Plant Pathology Journal* 36(1), 1–10. DOI: 10.5423/PPJ.RW.12.2019.0295.

Lau, K.Q., Sabran, M.R. and Shafie, S.R. (2021) Utilization of vegetable and fruit by products as functional ingredient and food. *Frontiers in Nutrition* 8, 661693. DOI: 10.3389/fnut.2021.661693.

Le, V.N., Nguyen, Q.T., Nguyen, T.D., Nguyen, N.T., Janda, T. *et al.* (2020) The potential health risks and environmental pollution associated with the application of plant growth regulators in vegetable production in several suburban areas of Hanoi, Vietnam. *Biologia Futura* 71(3), 323–331. DOI: 10.1007/s42977-020-00041-5.

Li, T., Yun, Z., Wu, Q., Zhang, Z., Liu, S. *et al.* (2018) Proteomic profiling of 24-epibrassinolide-induced chilling tolerance in harvested banana fruit. *Journal of Proteomics* 187, 1–12. DOI: 10.1016/j.jprot.2018.05.011.

Li, C., Lu, X., Liu, Y., Xu, J. and Yu, W. (2023) Strigolactone alleviates the adverse effects of salt stress on seed germination in cucumber by enhancing antioxidant capacity. *Antioxidants* 12, 1043. DOI: 10.3390/antiox12051043.

Li, M., Li, X., Du, J., Li, W., He, R. *et al.* (2024) Effects of exogenous application of the strigolactone GR24 on quality and flavor components during postharvest storage of celery. *Postharvest Biology and Technology* 212(50), 112900. DOI: 10.1016/j.postharvbio.2024.112900.

Liu, Q., Xi, Z., Gao, J., Meng, Y., Lin, S. *et al.* (2016a) Effects of exogenous 24-epibrassinolide to control grey mould and maintain postharvest quality of table grapes. *International Journal of Food Science & Technology* 51(5), 1236–1243. DOI: 10.1111/ijfs.13066.

Liu, S., Huang, X., He, H., Jin, Q. and Zhu, G. (2016b) Evaluation of selected plant growth regulators and fungicide residues in fruits for dietary risk assessment. *Human and Ecological Risk Assessment* 22(6), 1386–1395. DOI: 10.1080/10807039.2016.1185357.

Ling, F.L., Su, Q.W., Jiang, H., Cui, J.J., He, X.L. *et al.* (2020) Effects of strigolactone on photosynthetic and physiological characteristics in salt-stressed rice seedlings. *Scientific Reports* 10(1), 6183. DOI: 10.1038/s41598-020-63352-6.

Lu, T., Yu, H.J., Li, Q., Chai, L. and Jiang, W.J. (2019) Improving plant growth and alleviating photosynthetic inhibition and oxidative stress from low-light stress with exogenous GR24 in tomato (*Solanum lycopersicum* L.) seedlings. *Frontiers in Plant Science* 10. DOI: 10.3389/fpls.2019.00490.

Ma, Q., Lin, X., Zhan, M., Chen, Z., Wang, H. *et al.* (2022) Effect of an exogenous strigolactone GR24 on the antioxidant capacity and quality deterioration in postharvest sweet orange fruit stored at ambient temperature. *International Journal of Food Science & Technology* 57(1), 619–630. DOI: 10.1111/ijfs.15415.

Massolo, J.F., Sánchez, R., Zaro, M.J., Concellón, A. and Vicente, A.R. (2022) Low-dose prestorage 24-epibrassinolide spray enhances postharvest chilling tolerance in zucchini squash (*Cucurbita pepo* L.) by eliciting peroxidase and phenolic antioxidants. *Journal of Food Processing and Preservation* 46(5), e16576. DOI: 10.1111/jfpp.16576.

Ma, Y., Ban, Q., Shi, J., Dong, T., Jiang, C.-Z. *et al.* (2019) 1-Methylcyclopropene (1-MCP), storage time, and shelf life and temperature affect phenolic compounds and antioxidant activity of 'Jonagold' apple. *Postharvest Biology and Technology* 150, 71–79. DOI: 10.1016/j.postharvbio.2018.12.015.

Munir, N., Sarwar, Z., Hanif, M., Siddiqui, Z.S., Siddique, K.H.M. *et al.* (2025) Enhancing crop resilience and agricultural sustainability: Synergistic interactions between phytohormones and nanoparticles. *Environmental Sustainability* 8(3), 441–456. DOI: 10.1007/s42398-025-00382-8.

Nomura, T., Seto, Y. and Kyozuka, J. (2024) Unveiling the complexity of strigolactones: Exploring structural diversity, biosynthesis pathways, and signaling mechanisms. *Journal of Experimental Botany* 75(4), 1134–1147. DOI: 10.1093/jxb/erad412.

Omoarelojie, L.O., Kulkarni, M.G., Finnie, J.F. and van Staden, J. (2019) Strigolactones and their crosstalk with other phytohormones. *Annals of Botany* 124(5), 749–767. DOI: 10.1093/aob/mcz100.

Pandey, P., Singh, J., Achary, V.M.M. and Reddy, M.K. (2015) Redox homeostasis via gene families of ascorbate–glutathione pathway. *Frontiers in Environmental Science* 3, 25. DOI: 10.3389/fenvs.2015.00025.

Papargyropoulou, E., Lozano, R., Steinberger, J.K., Wright, N. and Ujang, Z. bin (2014) The food waste hierarchy as a framework for the management of food surplus and food waste. *Journal of Cleaner Production* 76, 106–115. DOI: 10.1016/j.jclepro.2014.04.020.

Patil, M., Sharma, S., Sridhar, K., Anurag, R.K., Grover, K. *et al.* (2024) Effect of postharvest treatments and storage temperature on the physiological, nutritional, and shelf-life of broccoli (*Brassica oleracea*) microgreens. *Scientia Horticulturae* 327, 112805. DOI: 10.1016/j.scienta.2023.112805.

Peian, Z., Haifeng, J., Peijie, G., Sadeghnezhad, E., Qianqian, P. *et al.* (2021) Chitosan induces jasmonic acid production leading to resistance of ripened fruit against *Botrytis cinerea* infection. *Food Chemistry* 337, 127772. DOI: 10.1016/j.foodchem.2020.127772.

Poveda, J. (2020) Use of plant-defense hormones against pathogen-diseases of postharvest fresh produce. *Physiological and Molecular Plant Pathology* 111, 101521. DOI: 10.1016/j.pmpp.2020.101521.

Qiu, C.-W., Zhang, C., Wang, N.-H., Mao, W. and Wu, F. (2021) Strigolactone GR24 improves cadmium tolerance by regulating cadmium uptake, nitric oxide signaling and antioxidant metabolism in barley (*Hordeum vulgare* L.). *Environmental Pollution* 273, 116486. DOI: 10.1016/j.envpol.2021.116486.

Raak, N., Symmank, C., Zahn, S., Aschemann-Witzel, J. and Rohm, H. (2017) Processing- and product-related causes for food waste and implications for the food supply chain. *Waste Management* 61, 461–472. DOI: 10.1016/j.wasman.2016.12.027.

Rasouli, M., Koushesh Saba, M. and Ramezanian, A. (2019) Inhibitory effect of salicylic acid and aloe vera gel edible coating on microbial load and chilling injury of orange fruit. *Scientia Horticulturae* 247, 27–34. DOI: 10.1016/j.scienta.2018.12.004.

Riva, S.C., Opara, U.O. and Fawole, O.A. (2020) Recent developments on postharvest application of edible coatings on stone fruit: a review. *Scientia Horticulturae* 262, 109074. DOI: 10.1016/j.scienta.2019.109074.

Romero, P., Lafuente, M.T. and Rodrigo, M.J. (2019) A sweet orange mutant impaired in carotenoid biosynthesis and reduced ABA levels results in altered molecular responses along peel ripening. *Scientific Reports* 9(1), 9813. DOI: 10.1038/s41598-019-46365-8.

Sangprayoon, P., Supapvanich, S., Youryon, P., Wongs-Aree, C. and Boonyaritthongchai, P. (2019) Efficiency of salicylic acid or methyl jasmonate immersions on internal browning alleviation and physicochemical quality of Queen pineapple cv. “Sawi” fruit during cold storage. *Journal of Food Biochemistry* 43(12), e13059. DOI: 10.1111/jfbc.13059.

Saud, S., Jiang, Z., Chen, S. and Fahad, S. (2023) Interaction of melatonin on post-harvest physiology and quality of horticultural crops. *Scientia Horticulturae* 321, 112286. DOI: 10.1016/j.scienta.2023.112286.

Sayyari, M., Babalar, M., Kalantari, S., Serrano, M. and Valero, D. (2009) Effect of salicylic acid treatment on reducing chilling injury in stored pomegranates. *Postharvest Biology and Technology* 53, 152–154. DOI: 10.1016/j.postharvbio.2009.03.005.

Sayyari, M., Babalar, M., Kalantari, S., Martínez-Romero, D., Guillén, F. *et al.* (2011a) Vapour treatments with methyl salicylate or methyl jasmonate alleviated chilling injury and enhanced antioxidant potential during postharvest storage of pomegranates. *Food Chemistry* 124(3), 964–970. DOI: 10.1016/j.foodchem.2010.07.036.

Sayyari, M., Castillo, S., Valero, D., Díaz-Mula, H.M. and Serrano, M. (2011b) Acetyl salicylic acid alleviates chilling injury and maintains nutritive and bioactive compounds and antioxidant activity during postharvest storage of pomegranates. *Postharvest Biology and Technology* 60(2), 136–142. DOI: 10.1016/j.postharvbio.2010.12.012.

Sedaghat, M., Sarvestani, Z.T., Emam, Y., Bidgoli, A.M. and Sorooshzadeh, A. (2020) Foliar-applied GR24 and salicylic acid enhanced wheat drought tolerance. *Russian Journal of Plant Physiology* 67(4), 733–739. DOI: 10.1134/S1021443720040159.

Selwal, N., Wani, A.K., Akhtar, N., Kaur, M. and Jassal, P.S. (2023) Molecular insights of strigolactone biosynthesis, signalling pathways, regulatory roles, and hormonal crosstalks in plant systems. *South African Journal of Botany* 160(9), 9–22. DOI: 10.1016/j.sajb.2023.06.046.

Siboza, X.I., Bertling, I. and Odindo, A.O. (2014) Salicylic acid and methyl jasmonate improve chilling tolerance in cold-stored lemon fruit (*Citrus limon*). *Journal of Plant Physiology* 171(18), 1722–1731. DOI: 10.1016/j.jplph.2014.05.012.

Song, Y., Hu, C., Xue, Y., Gu, J., He, J. *et al.* (2022) 24-epibrassinolide enhances mango resistance to *Colletotrichum gloeosporioides* via activating multiple defense response. *Scientia Horticulturae* 303, 111249. DOI: 10.1016/j.scienta.2022.111249.

Su, Y., Xia, S., Wang, R. and Xiao, L. (2017) Phytohormonal quantification based on biological principles. In: Li, J., Li, C. and Smith, S.M. (eds) *Hormone Metabolism and Signaling in Plants*. Academic Press, Cambridge, Massachusetts, pp. 431–470. DOI: 10.1016/B978-0-12-811562-6.00013-X.

Takahashi, I., Jiang, K. and Asami, T. (2021) Counteractive effects of sugar and strigolactone on leaf senescence of rice in darkness. *Agronomy* 11(6), 1044. DOI: 10.3390/agronomy11061044.

Tavallali, V. and Zareiyan, F. (2018) Maintenance of physicochemical qualities of lime during cold storage using vacuum infiltration with salicylic acid. *Journal of Food Measurement and Characterization* 12(4), 2955–2963. DOI: 10.1007/s11694-018-9911-4.

Tian, M., Jiang, K., Takahashi, I. and Li, G. (2018) Strigolactone-induced senescence of a bamboo leaf in the dark is alleviated by exogenous sugar. *Journal of Pesticide Science* 43(3), 173–179. DOI: 10.1584/jpestics.D18-003.

Toh, S., Kamiya, Y., Kawakami, N., Nambara, E., McCourt, P. *et al.* (2012) Thermoinhibition uncovers a role for strigolactones in *Arabidopsis* seed germination. *Plant and Cell Physiology* 53(1), 107–117. DOI: 10.1093/pcp/pcr176.

Ueda, H. and Kusaba, M. (2015) Strigolactone regulates leaf senescence in concert with ethylene in *Arabidopsis*. *Plant Physiology* 169(1), 138–147. DOI: 10.1104/pp.15.00325.

Umehara, M., Hanada, A., Yoshida, S., Akiyama, K., Arite, T. *et al.* (2008) Inhibition of shoot branching by new terpenoid plant hormones. *Nature* 455(7210), 195–200. DOI: 10.1038/nature07272.

Umehara, M., Cao, M., Akiyama, K., Akatsu, T., Seto, Y. *et al.* (2015) Structural requirements of strigolactones for shoot branching inhibition in rice and *Arabidopsis*. *Plant Cell Physiology* 56(6), 1059–1072. DOI: 10.1093/pcp/pcv028.

Visentin, I., Vitali, M., Ferrero, M., Zhang, Y., Ruyter-Spira, C. *et al.* (2016) Low levels of strigolactones in roots as a component of the systemic signal of drought stress in tomato. *New Phytologist* 212(4), 954–963. DOI: 10.1111/nph.14190.

Vurro, M., Prandi, C. and Baroccio, F. (2016) Strigolactones: How far is their commercial use for agricultural purposes? *Pest Management Science* 72(11), 2026–2034. DOI: 10.1002/ps.4254.

Wang, K.S., Lu, C.Y. and Chang, S.H. (2011) Evaluation of acute toxicity and teratogenic effects of plant growth regulators by *Daphnia magna* embryo assay. *Journal of Hazardous Materials* 190(1–3), 520–528. DOI: 10.1016/j.jhazmat.2011.03.068.

Wang, Y. and Bouwmeester, H.J. (2018) Structural diversity in the strigolactones. *Journal of Experimental Botany* 69(9), 2219–2230. DOI: 10.1093/jxb/ery091.

Wang, J., Song, L., Gong, X., Xu, J. and Li, M. (2020) Functions of jasmonic acid in plant regulation and response to abiotic stress. *International Journal of Molecular Sciences* 21, 1446. DOI: 10.1016/j.lwt.2019.108833.

Wang, J., Jin, L., Zhu, W., Jin, N., Sun, Q. *et al.* (2025) Effects of exogenous strigolactones on tomato-fruit quality. *Postharvest Biology and Technology* 224, 113461. DOI: 10.1016/j.postharvbio.2025.113461.

Wang, W.-N., Min, Z., Wu, J.-R., Liu, B.-C., Xu, X.-L. *et al.* (2021) Physiological and transcriptomic analysis of Cabernet sauvignon (*Vitis vinifera* L.) reveals the alleviating effect of exogenous strigolactones on the response of grapevine to drought stress. *Plant Physiology and Biochemistry* 167, 400–409. DOI: 10.1016/j.plaphy.2021.08.010.

Wu, F., Gao, Y., Yang, W., Sui, N. and Zhu, J. (2022) Biological functions of strigolactones and their crosstalk with other phytohormones. *Frontiers in Plant Science* 13, 821563. DOI: 10.3389/fpls.2022.821563.

Xu, X., Jibran, R., Wang, Y., Dong, L., Flokova, K. *et al.* (2021) Strigolactones regulate sepal senescence in *Arabidopsis*. *Journal of Experimental Botany* 72(15), 5462–5477. DOI: 10.1093/jxb/erab199.

Yamada, Y., Furusawa, S., Nagasaka, S., Shimomura, K., Yamaguchi, S. *et al.* (2014) Strigolactone signaling regulates rice leaf senescence in response to a phosphate deficiency. *Planta* 240(2), 399–408. DOI: 10.1007/s00425-014-2096-0.

Yoneyama, K., Xie, X.N., Nomura, T. and Yoneyama, K. (2020) Do phosphate and cytokinin interact to regulate strigolactone biosynthesis or act independently? *Frontiers in Plant Science* 11, 438. DOI: 10.3389/fpls.2020.00438.

Youryon, P., Supapvanich, S. and Wongs-Aree, C. (2019) Internal browning alleviation of Queen pineapple cv. 'Sawi' under cold storage using salicylic acid or abscisic acid peduncle infiltration. *Journal of Horticultural Science and Biotechnology* 94(6), 744–752. DOI: 10.1080/14620316.2019.1621684.

Yun, Z., Zhu, F., Liu, P., Zeng, Y.L., Xu, J. *et al.* (2015) Sweating treatment enhances citrus fruit disease resistance by inducing the accumulation of amino acids and salicylic acid-induced resistance pathway. *Physiologia Plantarum* 155(2), 109–125. DOI: 10.1111/ppl.12340.

Zaharah, S.S. and Singh, Z. (2012) Role of brassinosteroids in mango fruit ripening. In: *Proceedings of the XXVIII International Horticultural Congress on Science and Horticulture for People (IHC2010): International Symposium on Postharvest Technology in the Global Market, Leuven, Belgium*, pp. 929–935. DOI: 10.17660/ActaHortic.2012.934.124.

Zhang, Y. (2024) Post-harvest cold shock treatment enhanced antioxidant capacity to reduce chilling injury and improves the shelf life of guava (*Psidium guajava* L.). *Frontiers in Sustainable Food Systems* 8, 1297056. DOI: 10.3389/fsufs.2024.1297056.

Zhang, Q., Liu, Y., He, C. and Zhu, S. (2015a) Postharvest exogenous application of abscisic acid reduces internal browning in pineapple. *Journal of Agricultural and Food Chemistry* 63(22), 5313–5320. DOI: 10.1021/jf506279x.

Zhang, Y., Zhang, M. and Yang, H. (2015b) Postharvest chitosan-*g*-salicylic acid application alleviates chilling injury and preserves cucumber fruit quality during cold storage. *Food Chemistry* 174, 558–563. DOI: 10.1016/j.foodchem.2014.11.106.

Zhang, X.H., Zhang, L., Sun, Y.P., Zheng, S., Wang, J. *et al.* (2020) Hydrogen peroxide is involved in strigolactone induced low temperature stress tolerance in rape seedlings (*Brassica rapa* L.). *Plant Physiology and Biochemistry* 157, 402–415. DOI: 10.1016/j.plaphy.2020.11.006.

Zheng, Y., Kumar, N., Gonzalez, P. and Etxeberria, E. (2018) Strigolactones restore vegetative and reproductive developments in huanglongbing (HLB) affected, greenhouse-grown citrus trees by modulating carbohydrate distribution. *Scientia Horticulturae* 237, 89–95. DOI: 10.1016/j.scienta.2018.04.017.

Zulfiqar, H., Shahbaz, M., Ahsan, M., Nafees, M., Nadeem, H. *et al.* (2021) Strigolactone (GR24) induced salinity tolerance in sunflower (*Helianthus annuus* L.) by ameliorating morpho-physiological and biochemical attributes under *in vitro* conditions. *Journal of Plant Growth Regulation* 40(5), 2079–2091. DOI: 10.1007/s00344-020-10256-4.

Zwanenburg, B. and Pospíšil, T. (2013) Structure and activity of strigolactones: New plant hormones with a rich future. *Molecular Plant* 6(1), 38–62. DOI: 10.1093/mp/sss141.

9 Evolution of Strigolactone as an Oxidative Stress Mitigator in Plants

Muhammad Ilyas, Zunaira Malik, Zaib-un-Nisa* and Naila Ali
Institute of Molecular Biology and Biotechnology, University of Lahore, Lahore, Pakistan

Abstract

Strigolactones (SLs), a class of carotenoid-derived phytohormones, have long been recognized for their regulatory roles in plant development, shoot branching, and symbiotic associations. However, emerging evidence has illuminated their significant function in mitigating oxidative stress, an adaptive role that appears to have evolved in response to environmental challenges. Oxidative stress, caused primarily by the overproduction of reactive oxygen species (ROS), disrupts cellular homeostasis and threatens plant survival. The evolution of SLs as key signaling molecules capable of modulating redox balance marks a pivotal development in plant defense mechanisms. This chapter explores how biotechnology has been instrumental in unraveling the evolutionary and functional dimensions of SLs in oxidative stress mitigation. Biotechnological advances such as clustered regularly interspaced short palindromic repeats/CRISPR-associated protein 9 (CRISPR/Cas9) gene editing, transcriptomics, and metabolomics have enabled the identification and functional analysis of key genes involved in SL biosynthesis (*D27*, *CCD7*, *CCD8*, *MAX1*) and signaling (*D14*, *MAX2*). Through transgenic and mutant studies, researchers have demonstrated that SLs regulate the expression of ROS-scavenging enzymes including superoxide dismutase, catalase, and peroxidases, thereby enhancing cellular antioxidant capacity. In addition, high-throughput omics approaches have provided insights into the crosstalk between SLs and other stress-related hormones such as abscisic acid, jasmonic acid, and salicylic acid. Furthermore, synthetic biology and marker-assisted selection are being utilized to engineer plants with enhanced SL pathways for improved oxidative stress tolerance. These strategies allow the development of crop varieties better suited to withstanding environmental stressors linked to climate change. By elucidating the evolutionary trajectory and regulatory potential of SLs, biotechnology not only deepens our understanding of plant adaptation but also offers tangible applications in sustainable agriculture. This discussion underscores the central role of biotechnology in decoding and directing the adaptive functions of SLs, revealing them as critical modulators of oxidative stress resilience in modern plants.

Keywords: antioxidant defense system, climate change resilience, CRISPR/Cas9, hormonal crosstalk, oxidative stress, reactive oxygen species, strigolactones, sustainable agriculture

*Corresponding author: zaib.nisa@imbb.uol.edu.pk

 Strigolactone in Plants (eds M. Waseem *et al.*)
DOI: 10.1079/9781836990598.0009

9.1 Introduction

9.1.1 Overview of oxidative stress in plants

Oxidative stress in plants arises when the production of reactive oxygen species (ROS) surpasses the plant's antioxidant defense capacity, leading to cellular damage. ROS, including superoxide anions (O_2^-), hydrogen peroxide (H_2O_2), hydroxyl radicals (•OH), and singlet oxygen ($^1O_2$), are generated as by-products of metabolic processes in organelles such as chloroplasts, mitochondria, and peroxisomes (Hasanuzzaman *et al.*, 2020). Under normal conditions, ROS function as signaling molecules regulating growth and stress responses. However, environmental stresses such as drought, salinity, heat, cold, flooding, and heavy-metal exposure can disrupt this balance, causing oxidative stress (Sachdev *et al.*, 2021). For instance, drought stress impairs photosynthesis, leading to excess light energy and increased ROS production in chloroplasts, which results in lipid peroxidation and membrane damage (Kumar, 2024). Salinity stress induces ROS accumulation, affecting proteins, lipids, and nucleic acids, and disrupting cellular homeostasis (Zhou *et al.*, 2024). Heavy-metal toxicity, such as cadmium exposure, significantly elevates H_2O_2 and malondialdehyde levels, clear indicators of oxidative damage (Parwez *et al.*, 2022).

Plants counteract oxidative stress through enzymatic antioxidants such as superoxide dismutase (SOD), catalase (CAT), ascorbate peroxidase (APX), and glutathione reductase, which work to neutralize ROS (Hasanuzzaman *et al.*, 2020). Non-enzymatic antioxidants, including ascorbic acid, glutathione, carotenoids, and flavonoids, also play crucial roles in scavenging ROS and maintaining redox balance (Rao *et al.*, 2025). Recent studies highlight the dual role of ROS as damaging agents and essential signaling molecules in stress responses, where regulated ROS levels activate defense genes and trigger acclimation processes (Rao *et al.*, 2025). As a result, insights into ROS signaling and antioxidant defense mechanisms are driving the development of stress-resilient crops through advanced breeding, genetic engineering, and agronomic interventions (Haakim *et al.*, 2025).

9.1.2 Importance of stress-mitigating phytohormones

Stress-mitigating phytohormones are essential regulators that enable plants to perceive, respond to, and survive under adverse environmental conditions, including drought, salinity, heat, cold, and heavy-metal stress. Among these, abscisic acid (ABA) is a key hormone that accumulates rapidly during drought and salt stress, leading to stomatal closure, activation of stress-responsive genes, and enhancement of antioxidant enzyme activities to limit oxidative damage (Sah *et al.*, 2016). Salicylic acid (SA) contributes to defense against biotic stress and also enhances tolerance to abiotic stresses by stimulating ROS-scavenging systems and stabilizing cellular membranes under heat and salinity stress (Rai, 2023). Jasmonic acid (JA) and its derivatives regulate defense against wounding and pathogens and play a crucial role in cold and osmotic stress tolerance by modulating gene expression and secondary metabolite accumulation (Rehman *et al.*, 2023).

Ethylene, although traditionally associated with senescence and fruit ripening, also contributes to tolerance against flooding, heavy metals, and salinity by modulating root architecture and activating detoxification pathways (Wang *et al.*, 2021). Cytokinins delay stress-induced senescence and maintain chlorophyll content and photosynthetic capacity, especially under drought and salinity (Mostofa *et al.*, 2021). Auxins help maintain root growth under stress and assist in modulating other hormonal pathways through synergistic and antagonistic interactions (Rai, 2023). Similarly, brassinosteroids have been shown to improve tolerance to multiple stresses by enhancing antioxidant defenses, stabilizing proteins and membranes, and modulating stress-related gene expression (Zulfiqar and Ashraf, 2021). Gibberellins, although growth-promoting under normal conditions, are often downregulated during stress to conserve energy, with controlled application helping in recovery after stress episodes (Ahanger *et al.*, 2023).

Recent studies underscore the importance of hormonal crosstalk, where phytohormones act in complex networks rather than in isolation. For instance, ABA and JA often act synergistically during drought, while SA and JA may antagonize each other in response to pathogens (Pandey *et al.*, 2023). This cross-regulation enables plants to fine-tune their responses to specific environmental cues. Exogenous application of hormones such as ABA, SA, brassinosteroids, and cytokinins has been reported to significantly enhance plant tolerance to abiotic stress in crops such as rice (*Oryza sativa*), wheat (*Triticum aestivum*), maize (*Zea mays*), and tomato (*Solanum lycopersicum*) (Manzoor *et al.*, 2023). Therefore, understanding and harnessing stress-mitigating phytohormones is vital for developing climate-resilient crops through breeding, genetic engineering, and agronomic interventions.

9.1.3 Introduction to strigolactones

Strigolactones (SLs) are a class of carotenoid-derived phytohormones that play vital roles in regulating plant development and responses to environmental stress. Originally discovered as germination stimulants for parasitic plants such as *Striga* and *Orobanche*, SLs are now recognized as multi-functional signaling molecules influencing shoot branching, root development, leaf senescence, and symbiotic interactions with arbuscular mycorrhizal fungi (AMF) (Hossain *et al.*, 2021). Structurally, SLs consist of a tricyclic lactone (ABC-ring) connected via an enol-ether bridge to a butenolide (D-ring), and they exist in both canonical and non-canonical forms (Mehmood *et al.*, 2025b).

In plant architecture, SLs act as inhibitors of axillary bud outgrowth, maintaining apical dominance and optimizing resource allocation under limited nutrient conditions (Liu *et al.*, 2024). SLs also modulate root system architecture, promoting primary root elongation while suppressing lateral root formation, especially under phosphorus and nitrogen deficiency, thus enhancing nutrient foraging (Lucido *et al.*, 2024). Moreover, SLs facilitate symbiotic associations with AMF, which help plants improve nutrient uptake and stress tolerance (Naseer *et al.*, 2024).

Recent research highlights the role of SLs in abiotic stress tolerance, such as drought, salinity, and temperature extremes. Under water deficit, SL biosynthesis genes are upregulated, suggesting that SLs contribute to stomatal closure, antioxidant defense, and stress-responsive gene expression (Sharma *et al.*, 2024). For instance, application of synthetic SL analogs such as GR24 improved drought resistance in rice and *Arabidopsis* by modulating ABA signaling pathways and reducing ROS accumulation (Khalid *et al.*, 2024). Furthermore, SL-deficient mutants often display hypersensitivity to environmental stress, confirming their protective function. SLs also play an emerging role in biotic stress interactions, modulating immune responses and potentially enhancing resistance against pathogens through hormone crosstalk with JA and SA pathways (Wang *et al.*, 2022). Their dual function in regulating both growth and stress responses makes SLs promising targets for crop improvement strategies, particularly in nutrient-poor and arid regions.

9.2 Strigolactones: Biosynthesis and Chemical Nature

9.2.1 Key biosynthetic enzymes: D27, CCD7, CCD8, and MAX1

SLs are a group of plant-derived terpenoid lactones that act as phytohormones and rhizospheric signaling molecules, playing essential roles in plant growth regulation, stress responses, and interactions with symbiotic and parasitic organisms. The biosynthesis of SLs is a multistep process that begins in plastids and involves a series of key enzymes: DWARF27 (D27), carotenoid cleavage dioxygenase 7 (CCD7), carotenoid cleavage dioxygenase 8 (CCD8), and MORE AXILLARY GROWTH 1 (MAX1), all of which are evolutionarily conserved across land plants (Zhang *et al.*, 2022). The enzyme D27 catalyzes the isomerization of all-*trans*-β-carotene to 9-*cis*-β-carotene, which is the precursor for downstream reactions (Fig. 9.1) (Gulyás *et al.*, 2022). Following this, CCD7 cleaves 9-*cis*-β-carotene at the 9–10 double bond to produce 9-*cis*-β-apo-10′-carotenal. This intermediate is then converted by CCD8 through oxidative cleavage and molecular rearrangement into carlactone (CL), a core SL precursor molecule (Votta *et al.*, 2023).

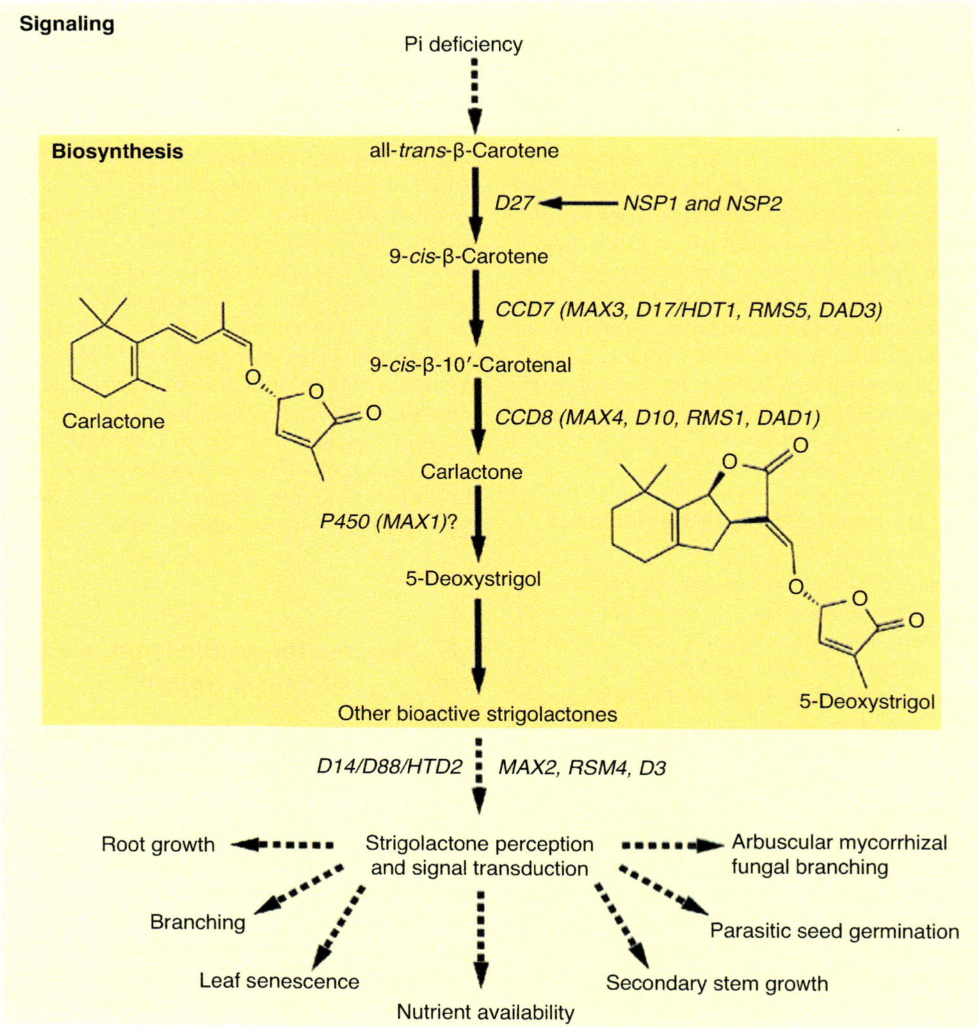

Fig. 9.1. Strigolactone biosynthetic and signaling pathways. The biosynthesis involves two carotenoid cleavage dioxygenases, CCD7 (also known as MAX3, RMS5, D17/HTD1, DAD3) and CCD8 (also known as MAX4, RMS1, D10, DAD1), one cytochrome P450 monooxygenase (MAX1), and one novel iron-containing protein (D27). In *Medicago*, SL biosynthesis also requires GRAS-type transcription factors NODULATION SIGNALING PATHWAY1 (NSP1) and NSP2. SL signaling involves MAX2/RMS4/D3, an F-box leucine-rich protein and DWARF14 (D14)/DAD2/D88/HTD2, a member of the a/b-fold hydrolase superfamily (Czarnecki *et al.*, 2013). Pi, phosphate; Figure used under license CC-BY-4.0.

CL is structurally similar to SLs and has been shown to exhibit weak hormonal activity, indicating its role as a biosynthetic hub. The final transformation of CL into bioactive SLs is catalyzed by MAX1, a cytochrome P450 mono-oxygenase. In *Arabidopsis thaliana*, MAX1 oxidizes CL into carlactonoic acid while in rice, multiple MAX1 homologs convert CL into a variety of canonical SLs such as 4-deoxyorobanchol and orobanchol, which play roles in shoot branching, mycorrhizal colonization, and suppression of parasitic weed germination (Chen *et al.*, 2023). These downstream modifications vary by species and explain the diversity of SL molecules identified across different plant taxa.

Structurally, SLs are characterized by a conserved butenolide D-ring connected via an enol-ether bridge to a variable tricyclic lactone (ABC-ring) or open-chain structures in non-canonical SLs. This structural variability underlies the functional diversity of SLs, allowing them to act both as internal hormones and external signals (Alvi *et al.*, 2022). For instance, tomato mutants deficient in CCD8 (*slccd8*) show increased shoot branching, reduced root development, and impaired symbiosis with AMF phenotypes that are reversed by application of synthetic SLs such as GR24 (Hashem and Khalil, 2023). Similarly, in barley (*Hordeum vulgare*), loss-of-function mutations in MAX1 reduce the production of orobanchol and compromise drought tolerance and root elongation, confirming the pivotal role of MAX1 in stress resilience (Marzec *et al.*, 2020).

Recent advancements in clustered regularly interspaced short palindromic repeats/CRISPR-associated protein 9 (CRISPR/Cas9) gene editing have enabled precise functional studies of these biosynthetic genes. For example, targeted knockout of *D27* in rice leads to severe reduction in SL levels, increased tillering, and reduced phosphate uptake, highlighting its upstream role in the pathway (Usman *et al.*, 2020). Moreover, expression of these biosynthetic genes is tightly regulated by nutrient availability, especially phosphate deficiency, which acts as a primary trigger for SL biosynthesis upregulation across species. This coordination between nutrient signaling and SL production further emphasizes their integrative function in plant adaptive responses (Wang *et al.*, 2021).

9.2.2 Transport and perception mechanisms

The transport and perception of SLs are critical for their function as both endogenous hormones and exogenous signaling molecules in the rhizosphere. After their biosynthesis in plant roots, SLs or their precursors such as CL and methyl carlactonoate are transported to aerial tissues through the xylem, where they influence shoot architecture, particularly by suppressing axillary bud outgrowth (Mehmood *et al.*, 2025b). Transport is mediated by ATP-binding cassette (ABC) transporters, especially PLEIOTROPIC DRUG RESISTANCE 1 (PDR1) in *Petunia hybrida*, and its orthologs in *Arabidopsis* and rice (Banasiak and Jasiński, 2022). These transporters are localized in root hypodermal cells and direct SLs both acropetally (upward to the shoots) and into the soil as rhizospheric signals to attract AMF (Demurtas *et al.*, 2023).

Perception of SLs occurs through the α/b-hydrolase receptor DWARF14 (D14), which hydrolyzes SL molecules and undergoes a conformational change that allows interaction with the F-box protein MORE AXILLARY GROWTH 2 (MAX2) (Korek and Marzec, 2024). This interaction leads to the formation of the SCF^{MAX2} complex, which ubiquitinates and degrades SUPPRESSOR OF MAX2-LIKE (SMXL) proteins, repressors of SL signaling, thereby activating downstream responses (Marzec and Korek, 2024). For example, in *A. thaliana*, degradation of SMXL6, -7, and -8 is essential for shoot branching inhibition and proper root development under nutrient-limited conditions (Nakandala *et al.*, 2020).

Recent studies highlight that SL signaling and transport are also regulated by environmental cues, such as phosphate starvation, which enhances SL exudation and internal redistribution (Bai *et al.*, 2025). Interestingly, the D14 receptor has been shown to recognize both natural and synthetic SL analogs such as GR24, providing a powerful tool for studying SL perception mechanisms across species (Mashiguchi *et al.*, 2021). Disruption of SL transport or perception pathways results in phenotypes such as bushy shoots, reduced root colonization by AMF, and increased susceptibility to parasitic weeds, emphasizing the role of SLs in plant development and environmental adaptation (Ozturk *et al.*, 2023).

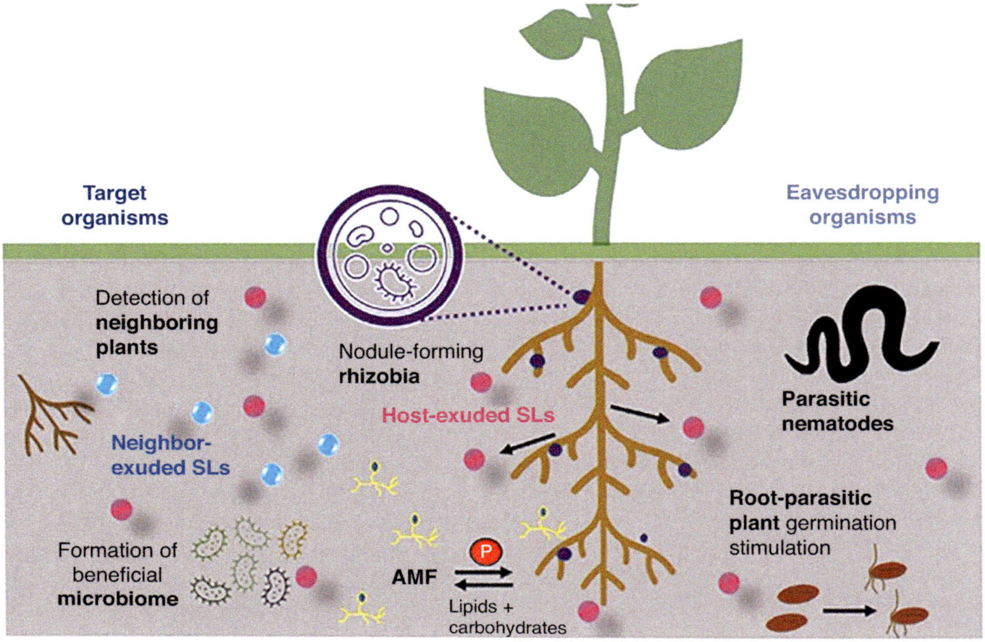

Fig. 9.2. The roles of strigolactones (SLs) as rhizosphere signals. SLs act as rhizosphere signals to detect neighboring plants, induce the hyphal branching of arbuscular mycorrhizal fungi (AMF), shape the microbial community, and induce the germination of parasitic plants (Clark and Bennett, 2024). Figure used under license CC-BY-4.0.

9.3 Evolutionary Origin and Diversification of Strigolactones

9.3.1 SL function in early land plants and non-vascular plants

SLs are ancient plant signaling molecules with origins tracing back to the early stages of land plant evolution. Recent phylogenomic and comparative studies suggest that the biosynthesis of SLs originated in the common ancestor of bryophytes and vascular plants, highlighting their early role in terrestrial adaptation (Kodama *et al.*, 2022). In non-vascular plants such as liverworts (*Marchantia polymorpha*) and mosses (*Physcomitrella patens*), SL biosynthetic genes like CCD7, CCD8, and MAX1 are conserved and functionally expressed, although their signaling mechanisms vary significantly from those in vascular plants (Jibran *et al.*, 2024).

The function of SLs in these early-diverging plant lineages is associated primarily with the establishment of beneficial relationships with AMF (Fig. 9.2). These symbiotic associations were crucial for enhancing nutrient uptake, especially phosphorus, in nutrient-poor soils during the colonization of land by plants (Khaliq *et al.*, 2022). For example, *M. polymorpha* secretes SL-like compounds into the rhizosphere, promoting AMF hyphal branching, a key step in establishing symbiosis (Kee *et al.*, 2023).

Interestingly, while SL biosynthesis is conserved, the canonical SL receptor D14 is absent in non-vascular plants, which instead utilize KAI2 homologs to perceive karrikin-like and SL-related molecules, pointing to an evolutionary divergence in signaling (Aquino *et al.*, 2021).

Experiments using the synthetic SL analog GR24 in *P. patens* have shown effects on protonemal branching and rhizoid elongation, despite the lack of D14, indicating that perception of SL-like signals occurs through alternative pathways, potentially involving ancient KAI2-like proteins (Blázquez *et al.*, 2020). This supports the idea that SLs originally evolved as external cues for environmental interactions, such as symbiosis and competition, before their recruitment as internal developmental hormones in vascular plants (Helmich, 2025).

Furthermore, evolutionary analyses reveal that diversification of SL biosynthetic and signaling genes coincided with the emergence of vascular tissues and root systems, allowing SLs to coordinate complex developmental processes such as shoot branching, root architecture modulation, and stress responses in higher plants (Sharma *et al.*, 2024). The transition from a purely symbiotic role in non-vascular plants to a hormonal regulatory function in vascular plants reflects the adaptive evolution of SL pathways to suit increasingly complex plant body plans and ecological niches (Abel and Naumann, 2024). Thus, SLs serve as a model for understanding how ancient chemical signals are co-opted and modified during evolution to meet new functional demands (Wang *et al.*, 2022).

9.3.2 Phylogenetic insights and evolutionary transitions

Phylogenetic insights into SL evolution reveal a gradual and modular assembly of the SL biosynthesis and signaling pathways across plant lineages. The SL biosynthesis genes, including D27, CCD7, CCD8, and MAX1, are conserved among land plants, suggesting that SL production originated before the divergence of bryophytes and vascular plants (Walker *et al.*, 2019). However, the full signaling machinery, particularly the SL-specific receptor D14, is absent in most non-vascular plants, pointing to a major evolutionary transition where SLs shifted from rhizospheric signals to internal hormones in vascular plants (Josse, 2022). For instance, while liverworts such as *Marchantia paleacea* synthesize SLs such as bryosymbiol, they lack the D14 receptor and rely on KAI2-like proteins, which originally evolved to detect smoke-derived karrikins but may have also sensed ancestral SLs (Bythell-Douglas *et al.*, 2017). This suggests an evolutionary sequence in which KAI2 served as a proto-receptor for diverse molecules before the recruitment and specialization of D14 in seed plants, aligning with the increasing complexity of plant body plans and ecological interactions (Kamran *et al.*, 2024).

In mosses such as *P. patens*, SL biosynthesis genes are functional and responsive to nutrient signals, but the absence of downstream canonical SL signaling proteins further supports a stepwise evolution (Stanga, 2022). Phylogenomic studies highlight that the D14–SCFMAX2–SMXL signaling cascade was likely assembled in vascular plants through gene duplication and neofunctionalization events (Taulera, 2022). These evolutionary transitions, particularly the origin of hormonal roles for SLs, paralleled major land plant innovations such as vascular tissues and branching architectures (Blázquez *et al.*, 2020). In flowering plants such as *A. thaliana*, SLs now regulate shoot branching, root development, and responses to biotic stress, demonstrating the extensive diversification of SL function following their phylogenetic emergence (Sharma *et al.*, 2024).

9.4 Mechanisms of Oxidative Stress and Role of Strigolactones

9.4.1 ROS generation and antioxidant defenses

Oxidative stress arises when the production of ROS surpasses the capacity of a biological system's antioxidant defenses, leading to potential damage to proteins, lipids, and nucleic acids. ROS, including superoxide anions (O_2^-), hydrogen peroxide (H_2O_2), and hydroxyl radicals (•OH), are by-products of cellular metabolism, particularly in organelles such as mitochondria, chloroplasts, and peroxisomes (Sachdev *et al.*, 2021). Under environmental stresses such as drought, salinity, and high light, ROS levels increase drastically, triggering oxidative damage if not adequately neutralized. Plants have evolved complex enzymatic antioxidants such as SOD, CAT, and APX, which play pivotal roles in converting harmful ROS into less toxic molecules (Hasanuzzaman *et al.*, 2020).

Non-enzymatic antioxidants, such as ascorbic acid, glutathione, flavonoids, and carotenoids, also contribute to cellular redox balance by directly scavenging ROS (Bajaj *et al.*,

2024). For example, under high-temperature stress, rice plants showed enhanced expression of *SOD* and *APX* genes, improving tolerance by maintaining cellular ROS at manageable levels (Rao *et al.*, 2025). In *A. thaliana*, mitochondrial alternative oxidase limits excessive ROS generation during respiratory stress, showing how alternative pathways help mitigate oxidative stress at the metabolic level (Garmash, 2023). Additionally, transcription factors (TFs) such as the NAC and WRKY families are known to regulate antioxidant gene expression in response to ROS signaling, acting as a feedback loop for cellular defense (Li *et al.*, 2024).

ROS also function as signaling molecules, modulating growth, defense, and programmed cell death, highlighting a dual role in both damage and signaling (Wang *et al.*, 2023). The intricate balance between ROS production and scavenging is thus essential for plant survival under stress. Recent advancements in redox proteomics and ROS imaging have deepened our understanding of these mechanisms and offer insights for engineering stress-resilient crops (Hasanuzzaman *et al.*, 2020).

9.4.2 SL-induced activation of enzymatic (SOD, CAT, POD) and non-enzymatic antioxidants

SLs have emerged as key regulators in plant responses to abiotic stress, particularly through their role in modulating oxidative stress by activating antioxidant defense systems. Recent studies show that SLs can stimulate both enzymatic antioxidants, such as SOD, CAT, and peroxidase (POD), and non-enzymatic antioxidants such as ascorbic acid and glutathione, thereby enhancing a plant's ability to detoxify ROS (Zheng *et al.*, 2023). In rice, application of the synthetic SL analog GR24 significantly increased the activity of SOD, CAT, and POD under drought and salinity stress, reducing lipid peroxidation and enhancing stress tolerance (Ahsan *et al.*, 2025).

In tomato, exogenous GR24 enhanced ascorbate and glutathione content, along with the expression of antioxidant-related genes, under high salinity conditions, suggesting SLs play a role in activating redox homeostasis (Alsubaie *et al.*, 2024). Similarly, in wheat, SL treatment improved tolerance to water stress by boosting the antioxidant enzyme system and lowering H_2O_2 accumulation. These findings suggest that SLs help maintain cellular redox balance by not only upregulating detoxifying enzymes but also enhancing the levels of scavenging molecules (Aouz *et al.*, 2023).

SLs also play a significant role in enhancing plant stress tolerance by activating non-enzymatic antioxidant defense systems. Non-enzymatic antioxidants such as ascorbic acid, glutathione, tocopherols, and flavonoids are crucial in mitigating oxidative stress caused by environmental stressors such as drought, salinity, and heavy metals (Saroy and Garg, 2021). Novel studies have shown that SL treatment leads to an increased accumulation of these antioxidants, thereby enhancing the plant's oxidative stress tolerance. For instance, in *A. thaliana* and tomato plants, exogenous application of GR24 significantly boosted ascorbic acid and glutathione levels under salinity stress, helping maintain redox homeostasis (Hashem and Khalil, 2023). Similarly, SLs enhanced flavonoid and carotenoid content in rice and wheat seedlings exposed to drought conditions, indicating their role in ROS scavenging (Mostofa *et al.*, 2021). These responses suggest that SLs act upstream in signaling pathways that regulate antioxidant biosynthesis, possibly through modulation of TFs and stress-responsive genes (Rehaman *et al.*, 2025). Moreover, SL-mediated antioxidant activation appears to be tightly linked with hormonal crosstalk, especially involving ABA and auxin (Bhoi *et al.*, 2021).

Furthermore, SL signaling mutants often exhibit elevated ROS accumulation and reduced antioxidant enzyme activities under stress, confirming the intrinsic link between SL pathways and oxidative stress responses (Mostofa *et al.*, 2021). The involvement of SLs in modulating antioxidant pathways is increasingly seen as a crucial aspect of plant adaptive strategies under adverse environmental conditions, offering potential for improving crop resilience through targeted hormonal regulation (Hossain *et al.*, 2022).

9.5 Strigolactones in Abiotic Stress Mitigation

9.5.1 Evidence from drought, salinity, and temperature stress studies

SLs play a critical role in mitigating abiotic stresses, including drought, salinity, and temperature extremes, by regulating plant growth, antioxidant defense systems, and hormonal signaling pathways. Under drought stress, synthetic SL analogs such as GR24 have been shown to improve water-use efficiency, root architecture, and photosynthetic performance in wheat, leading to increased drought resilience (Moosavi *et al.*, 2025). Similarly, in maize, SL application elevated antioxidant enzyme activity and maintained chlorophyll content under water-deficient conditions (Seleiman *et al.*, 2023).

In salinity-stress scenarios, tomato plants treated with SLs exhibited reduced sodium (Na^+) accumulation, improved osmotic balance, and enhanced expression of salt-tolerance genes, with trehalose and nitric oxide identified as key downstream signals (Guo *et al.*, 2022). In rice, SLs improved ionic homeostasis and upregulated the antioxidant defense network, resulting in lower ROS accumulation and membrane damage under salt stress (Sharma *et al.*, 2024).

Regarding temperature stress, SLs have been found to confer thermotolerance by modulating ABA signaling and enhancing the expression of heat-shock proteins in *A. thaliana* (Pandey *et al.*, 2023). In tomato, GR24 application increased proline levels and boosted the activity of SOD and CAT enzymes under both heat and cold stress (Feng *et al.*, 2024). Additionally, studies on alfalfa and lupine demonstrated that SL analogs promoted germination and sustained photosynthetic efficiency under high-temperature conditions (Oyebamiji *et al.*, 2024). Collectively, these findings underscore the multi-faceted role of SLs in abiotic stress mitigation, making them promising targets for crop improvement in the face of climate change.

9.5.2 Role of SLs in maintaining cellular redox homeostasis

SLs play a pivotal role in maintaining cellular redox homeostasis by modulating ROS levels and enhancing antioxidant defense mechanisms in plants (Fig. 9.3). SLs regulate the balance between ROS production and scavenging, thus preventing oxidative damage under both normal and stress conditions (Khalid *et al.*, 2024). For example, in rice, application of GR24 increased the activities of key enzymatic antioxidants such as SOD, CAT, and POD, which effectively reduced H_2O_2 accumulation during salinity stress (Ahsan *et al.*, 2022). Similarly, in tomato, SL treatment enhanced non-enzymatic antioxidants such as ascorbate and glutathione, contributing to improved ROS detoxification and cellular redox balance under drought stress (Zhou *et al.*, 2023). Mutants deficient in SL biosynthesis or signaling commonly exhibit higher ROS levels and impaired antioxidant responses, highlighting the essential role of SL pathways in redox regulation (Ahanger *et al.*, 2023).

Furthermore, SLs interact with other hormones such as ABA to fine-tune redox signaling networks, promoting stress adaptation (Wawrzyńska and Sirko, 2024). In *A. thaliana*, SL-mediated modulation of redox homeostasis involves upregulation of ROS-scavenging genes and stabilization of mitochondrial function, which are critical for maintaining cellular energy balance under stress (Raza *et al.*, 2023). These insights reveal that SLs function as integral regulators of redox homeostasis, enhancing

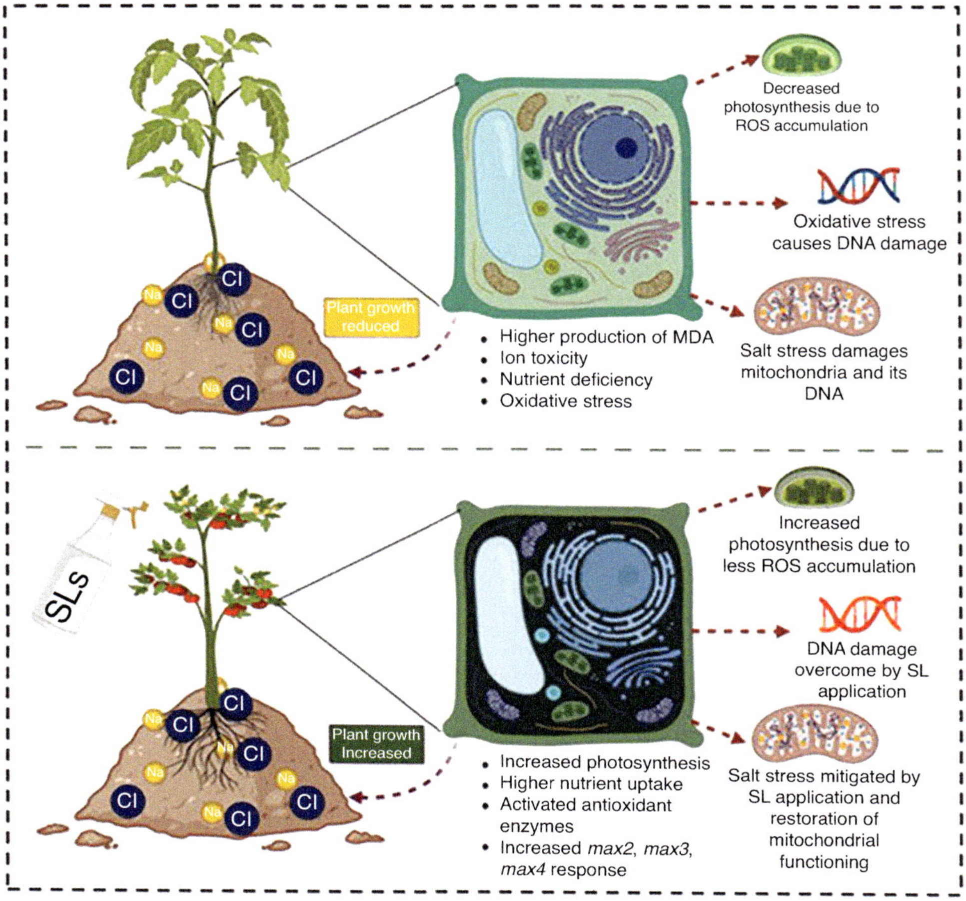

Fig. 9.3. The role of strigolactones (SLs) in improving antioxidant defense mechanisms in tomato plants. The primary stress effects are osmotic and ionic imbalances, which affect the ROS generator and antioxidant systems. This effect may be positive or negative, as indicated and discussed (Faisal *et al.*, 2023). MDA, malondialdehyde; ROS, reactive oxygen species. Figure used under license CC-BY-4.0.

plant resilience by orchestrating antioxidant defenses and signaling pathways.

9.6 Molecular Mechanisms and Gene Regulation

9.6.1 SL-responsive TFs under oxidative stress

SLs influence molecular mechanisms underlying plant responses to oxidative stress primarily through the regulation of SL-responsive TFs. Recent studies have identified several key TF families, including NAC, WRKY, and bZIP (basic-leucine zipper), which are modulated by SL signaling to orchestrate antioxidant defenses and stress adaptation. For instance, in *A. thaliana*, SL signaling activates the NAC TF *ANAC013* gene, which regulates mitochondrial retrograde signaling to enhance ROS detoxification during oxidative stress (Li *et al.*, 2024). Similarly, SLs induce expression of WRKY TFs such as *WRKY40* and *WRKY53* that mediate transcriptional activation of antioxidant genes such as *CAT* and *SOD*, improving cellular redox balance under drought and salinity (Khan *et al.*, 2022). In rice, the bZIP TF *OsbZIP23* is upregulated by SL treatment and positively regulates ABA-responsive antioxidant pathways, integrating hormonal crosstalk during oxidative stress (Kaya, 2025). Furthermore, SL-deficient mutants exhibit reduced expression of these TFs and compromised antioxidant enzyme activities, highlighting their functional importance (Salvi *et al.*, 2021). Evidence also points to the role of SLs in modulating microRNA expression that targets oxidative stress-related TFs, fine-tuning gene-regulation networks (Manzoor *et al.*, 2023). Together, these findings reveal that SL-responsive TFs act as molecular hubs connecting SL signaling to gene networks that mitigate oxidative damage, thus enhancing plant stress resilience.

9.6.2 Impact on gene expression networks for stress adaptation

SLs significantly impact gene expression networks that regulate plant adaptation to abiotic stresses, including drought, salinity, and oxidative stress. SL signaling modulates the transcription of stress-responsive genes involved in antioxidant defense, osmoprotection, and hormonal crosstalk. For example, in *A. thaliana*, SL treatment upregulates genes encoding antioxidant enzymes such as *SOD*, *CAT*, and *APX*, which mitigate ROS damage under drought conditions (Ikram *et al.*, 2025). In rice, SLs influence expression of genes involved in osmolyte biosynthesis and ion transporters, helping maintain cellular homeostasis under salinity stress (Singh *et al.*, 2022). Transcriptomic analyses in tomato have revealed that SLs activate key TFs such as the NAC and WRKY families, which orchestrate broad stress-adaptation pathways by regulating downstream gene networks (Fernandes and Ghag, 2022).

Moreover, SL-deficient mutants exhibit altered expression of ABA biosynthesis genes and stress-responsive microRNAs, indicating the role of SLs in integrating hormonal and epigenetic regulation during stress responses (Khan *et al.*, 2023). Recent research also highlights SL-mediated modulation of mitogen-activated protein kinase signaling cascades, further fine-tuning stress gene expression and enhancing plant resilience. Communally, these results demonstrate that SLs act as crucial regulators of gene expression networks, coordinating multifaceted molecular responses essential for plant stress tolerance.

9.7 Applications in Stress-resilient Crop Development

9.7.1 Use of SL analogs and transgenic approaches

SLs and their synthetic analogs have gained considerable attention for developing stress-resilient crops through both exogenous application and transgenic approaches. Application of SL analogs such as GR24 have been used to enhance drought and salinity tolerance in important crops such as wheat, maize, and tomato by improving antioxidant enzyme activities and regulating stomatal closure, which reduces water loss (Raza *et al.*, 2023). Transgenic overexpression of SL biosynthesis genes such as *D27* and *MAX1* in rice has led to enhanced root architecture and increased resistance to abiotic stresses by improving nutrient uptake and hormonal balance (Selwal *et al.*, 2023). Similarly, CRISPR/Cas9-mediated editing of SL signaling components in tomato improved salt tolerance by modulating ion homeostasis and stress-responsive gene expression (Kumar *et al.*, 2023).

Transgenic *Arabidopsis* lines overexpressing SL-responsive TFs exhibited enhanced ROS scavenging and improved survival under heat and drought stresses (Shi *et al.*, 2023). Continued research on SL analog optimization and gene-editing tools promises to accelerate the development of climate-resilient crop varieties, contributing to global food security under changing environmental conditions.

9.7.2 Potential for future breeding strategies

SLs hold significant potential for future breeding strategies aimed at enhancing crop resilience to environmental stresses. With growing climate instability, breeding programs are increasingly focused on integrating SL-related traits to improve drought tolerance, nutrient uptake, and stress adaptability. For example, natural variation in SL biosynthesis genes such as *D27* and *MAX1* has been exploited in rice to select genotypes with improved root architecture and drought tolerance (Mukherjee and Aftab, 2023). Marker-assisted selection using SL-pathway gene markers offers a promising tool for breeding stress-tolerant varieties in crops such as sorghum and maize, which exhibit differential SL production under stress (Atta *et al.*, 2023).

Furthermore, genomic-selection approaches incorporating SL-responsive gene networks are being tested in wheat and barley to predict stress-resilient genotypes based on SL-associated traits (Mangal *et al.*, 2024). Advances in CRISPR/Cas9 technology have enabled precise editing of SL-pathway genes, allowing custom development of cultivars with enhanced abiotic stress responses without compromising yield (Kumar *et al.*, 2023). Additionally, SL-related traits are being included in breeding platforms for sustainable agriculture, particularly for legumes such as chickpea and cowpea, which benefit from SL-enhanced symbiosis and nitrogen fixation under stress (Benitez-Alfonso *et al.*, 2023). Integration of omics technologies such as transcriptomics and metabolomics with SL signaling pathways also offers a systems-level approach to uncover new SL-related alleles for breeding (Roychowdhury *et al.*, 2023). These advances underscore the growing importance of SL biology in breeding pipelines for the development of future-ready, stress-resilient crops.

9.8 Conclusion and Future Perspectives

9.8.1 Summary of findings

Recent research highlights the pivotal role of SLs in plant development, stress adaptation, and cellular homeostasis, making them promising targets for crop improvement. Evolutionary studies suggest that SLs originated as rhizospheric signaling molecules in early land plants and were later recruited for internal hormonal functions, with evidence from liverworts and mosses supporting their ancient signaling roles (Woudenberg *et al.*, 2022). SLs regulate redox balance by activating antioxidant defenses, including enzymatic (SOD, CAT, POD) and non-enzymatic (ascorbate, glutathione) components, enhancing resilience against oxidative stress (Raja *et al.*, 2024). Under drought, salinity, and temperature extremes, SLs modulate gene networks involved in ABA signaling, ion homeostasis, and ROS detoxification, significantly improving plant tolerance (Pamanick *et al.*, 2022). SL-responsive TFs such as NAC, WRKY, and bZIP are crucial in linking SL signaling to stress-responsive gene expression (Thilakarathne *et al.*, 2025). Synthetic SL analogs such as GR24 and genetic modifications of SL biosynthesis genes have been used to enhance root growth and stress resilience in crops such as rice, tomato, and wheat (Mehmood *et al.*, 2025a). Moreover, breeding strategies now incorporate SL-related markers and gene-editing approaches to develop climate-resilient varieties, especially in legumes and cereals (Pixley *et al.*, 2023). These findings collectively establish SLs as central regulators in plant stress physiology and sustainable crop development.

9.8.2 Gaps in knowledge and research opportunities

Despite significant advances in understanding SLs, several critical gaps in knowledge remain, presenting exciting opportunities for future research. One major gap lies in the detailed elucidation of SL biosynthesis and transport mechanisms in non-model and crop species, especially under complex field conditions. While core biosynthetic genes such as *D27*, *CCD7*, and *MAX1* have been identified in model plants, their regulation and isoform-specific roles under stress remain poorly understood in cereals and legumes. Additionally, the full diversity of SL molecules and their structure–function relationships are still not completely characterized, particularly across different plant lineages (Hanako *et al.*, 2023). The precise crosstalk between SLs and other hormones such as ethylene, JA, and cytokinins during stress responses also requires deeper investigation (Samanta and Roychoudhury, 2025).

Another knowledge gap is the role of SLs in biotic stress responses and plant–microbe interactions beyond mycorrhizal symbiosis. The potential for engineering SL pathways in crops using gene-editing tools such as CRISPR remains underexplored, especially regarding field-level performance and ecological safety (Gajardo *et al.*, 2023). Furthermore, while SL analogs such as GR24 show promise in stress mitigation, their long-term effects, metabolic fate, and interaction with native SLs *in planta* are largely unknown (Naz *et al.*, 2025). High-throughput phenotyping and integrative omics approaches could be utilized to link SL signaling to phenotype under real environmental conditions (Yang *et al.*, 2020). Opportunities also exist in breeding for SL-associated traits using genome-wide association studies and systems biology to predict genotype-by-environment interactions (Bernousi *et al.*, 2024). Lastly, the role of SLs in seed dormancy, longevity, and post-harvest physiology remains an emerging but understudied area with agricultural relevance (Elhamouly *et al.*, 2022).

References

Abel, S. and Naumann, C. (2024) Evolution of phosphate scouting in the terrestrial biosphere. *Philosophical Transactions of the Royal Society B: Biological Sciences* 379(1914), 20230355. DOI: 10.1098/rstb.2023.0355.

Ahanger, M.A., Bhat, J.A., Ahmad, P. and John, R. (eds) (2023) *Improving Stress Resilience in Plants: Physiological and Biochemical Basis and Utilization in Breeding*. Elsevier, London.

Ahsan, M., Zulfiqar, H., Farooq, M.A., Ali, S., Tufail, A. *et al.* (2022) Strigolactone (GR24) application positively regulates photosynthetic attributes, stress-related metabolites and antioxidant enzymatic activities of ornamental sunflower (*Helianthus annuus* cv. Vincent's choice) under salinity stress. *Agriculture* 13(1), 50. DOI: 10.3390/agriculture13010050.

Ahsan, M., Abdullah, M., Radicetti, E., Elsadek, M.F., Bakhsh, A. *et al.* (2025) Cadmium toxicity affects morpho-physiological and biochemical attributes in *Gladiolus grandiflorus* L.: The mitigating roles of exogenous strigolactone (GR24). *Water, Air, & Soil Pollution* 236(7), 450. DOI: 10.1007/s11270-025-08108-6.

Alsubaie, Q.D., Al-Amri, A.A., Siddiqui, M.H. and Alamri, S. (2024) Strigolactone and nitric oxide collaborate synergistically to boost tomato seedling resilience to arsenic toxicity via modulating physiology and antioxidant system. *Plant Physiology and Biochemistry* 207, 108412. DOI: 10.1016/j.plaphy.2024.108412.

Alvi, A.F., Sehar, Z., Fatma, M., Masood, A. and Khan, N.A. (2022) Strigolactone: An emerging growth regulator for developing resilience in plants. *Plants* 11(19), 2604. DOI: 10.3390/plants11192604.

Aouz, A., Khan, I., Chattha, M.B., Ahmad, S., Ali, M. *et al.* (2023) Silicon induces heat and salinity tolerance in wheat by increasing antioxidant activities, photosynthetic activity, nutrient homeostasis, and osmo-protectant synthesis. *Plants* 12(14), 2606. DOI: 10.3390/plants12142606.

Aquino, B., Bradley, J.M. and Lumba, S. (2021) On the outside looking in: Roles of endogenous and exogenous strigolactones. *Plant Journal* 105(2), 322–334. DOI: 10.1111/tpj.15087.

Atta, K., Mondal, S., Gorai, S., Singh, A.P., Kumari, A. *et al.* (2023) Impacts of salinity stress on crop plants: Improving salt tolerance through genetic and molecular dissection. *Frontiers in Plant Science* 14, 1241736. DOI: 10.3389/fpls.2023.1241736.

Bai, L., Ju, C., Han, C., Yu, Z., Wang, M.Y. *et al.* (2025) The interaction of nutrient uptake with biotic and abiotic stresses in plants. *Journal of Integrative Plant Biology* 67(3), 455–487. DOI: 10.1111/jipb.13827.

Bajaj, S., Singh, S. and Sharma, P. (2024) Role of antioxidants in neutralizing oxidative stress. In: Keservani, R.K., Kesharwani, R.K., Emerald, M. and Sharma, A.K. (eds) *Nutraceutical Fruits and Foods for Neurodegenerative Disorders*. Academic Press, London, pp. 353–378. DOI: 10.1016/B978-0-443-18951-7.00020-7.

Banasiak, J. and Jasiński, M. (2022) ATP-binding cassette transporters in nonmodel plants. *New Phytologist* 233(4), 1597–1612. DOI: 10.1111/nph.17779.

Benitez-Alfonso, Y., Soanes, B.K., Zimba, S., Sinanaj, B., German, L. *et al.* (2023) Enhancing climate change resilience in agricultural crops. *Current Biology* 33(23), R1246–R1261. DOI: 10.1016/j.cub.2023.10.028.

Bernousi, H., Alipour, H., Abdi, I., Jafarzadeh, J. and Rabieyan, E. (2024) Morpho-colorimetric diversity and genome-wide association study of wheat spike architecture based on image processing under well-watered and rain-fed conditions. *Journal of Plant Growth Regulation* 44, 850–867. DOI: 10.1007/s00344-024-11492-8.

Bhoi, A., Yadu, B., Chandra, J. and Keshavkant, S. (2021) Contribution of strigolactone in plant physiology, hormonal interaction and abiotic stresses. *Planta* 254(2), 28. DOI: 10.1007/s00425-021-03678-1.

Blázquez, M.A., Nelson, D.C. and Weijers, D. (2020) Evolution of plant hormone response pathways. *Annual Review of Plant Biology* 71(1), 327–353. DOI: 10.1146/annurev-arplant-050718-100309.

Bythell-Douglas, R., Rothfels, C.J., Stevenson, D.W.D., Graham, S.W., Wong, G.K.-S. *et al.* (2017) Evolution of strigolactone receptors by gradual neo-functionalization of KAI2 paralogues. *BMC Biology* 15(1), 52. DOI: 10.1186/s12915-017-0397-z.

Chen, G.-T.E., Wang, J.Y., Votta, C., Braguy, J., Jamil, M. *et al.* (2023) Disruption of the rice 4-DEOXYOROBANCHOL HYDROXYLASE unravels specific functions of canonical strigolactones. *Proceedings of the National Academy of Sciences USA* 120(42), e2306263120. DOI: 10.1073/pnas.2306263120.

Clark, J. and Bennett, T. (2024) Cracking the enigma: Understanding strigolactone signalling in the rhizosphere. *Journal of Experimental Botany* 75(4), 1159–1173. DOI: 10.1093/jxb/erad335.

Czarnecki, O., Yang, J., Weston, D.J., Tuskan, G.A. and Chen, J.G. (2013) A dual role of strigolactones in phosphate acquisition and utilization in plants. *International Journal of Molecular Sciences* 14(4), 7681–7701. DOI: 10.3390/ijms14047681.

Demurtas, O.C., Nicolia, A. and Diretto, G. (2023) Terpenoid transport in plants: How far from the final picture. *Plants* 12(3), 634. DOI: 10.3390/plants12030634.

Elhamouly, N.A., Hewedy, O.A., Zaitoon, A., Miraples, A., Elshorbagy, O.T. *et al.* (2022) The hidden power of secondary metabolites in plant–fungi interactions and sustainable phytoremediation. *Frontiers in Plant Science* 13, 1044896. DOI: 10.3389/fpls.2022.1044896.

Faisal, M., Faizan, M., Tonny, S.H., Rajput, V.D., Minkina, T. *et al.* (2023) Strigolactone-mediated mitigation of negative effects of salinity stress in *Solanum lycopersicum* through reducing the oxidative damage. *Sustainability* 15(7), 5805. DOI: 10.3390/su15075805.

Feng, J., Zhang, J., Feng, N., Zheng, D., Khan, A. *et al.* (2024) Strigolactone alleviates NaCl stress by regulating antioxidant capacity and hormone levels in rice (*Oryza sativa* L.) seedlings. *Agriculture* 14(9), 1662. DOI: 10.3390/agriculture14091662.

Fernandes, L.B. and Ghag, S.B. (2022) Molecular insights into the jasmonate signaling and associated defense responses against wilt caused by *Fusarium oxysporum*. *Plant Physiology and Biochemistry* 174, 22–34. DOI: 10.1016/j.plaphy.2022.01.032.

Gajardo, H.A., Gómez-Espinoza, O., Boscariol Ferreira, P., Carrer, H. and Bravo, L.A. (2023) The potential of CRISPR/Cas technology to enhance crop performance on adverse soil conditions. *Plants* 12(9), 1892. DOI: 10.3390/plants12091892.

Garmash, E.V. (2023) Suppression of mitochondrial alternative oxidase can result in upregulation of the ROS scavenging network: Some possible mechanisms underlying the compensation effect. *Plant Biology* 25(1), 43–53. DOI: 10.1111/plb.13477.

Gulyás, Z., Moncsek, B., Hamow, K.A., Stráner, P., Tolnai, Z. *et al.* (2022) D27-LIKE1 carotenoid isomerase has a preference towards trans/cis and cis/cis conversions in *Arabidopsis*. *Plant Journal* 112(6), 1377–1395. DOI: 10.1111/tpj.16017.

Guo, M., Wang, X.-S., Guo, H.-D., Bai, S.-Y., Khan, A. *et al.* (2022) Tomato salt tolerance mechanisms and their potential applications for fighting salinity: A review. *Frontiers in Plant Science* 13, 949541. DOI: 10.3389/fpls.2022.949541.

Haakim, Z., Ahmad, Y., Iqbal, J., Abbasi, B.A., Ijaz, S. *et al.* (2025) Climate-proofing tomatoes: Molecular strategies for adapting to changing climate. In: Waseem, M., Prakash, C.S., Faiz, S. and Li, Z. (eds) *Omics Approaches for Tomato Yield and Quality Trait Improvement*. Springer Nature Singapore, Singapore, pp. 95–112. DOI: 10.1007/978-981-96-3890-1_4.

Hanako, S., Kaien, F. and Chikara, M. (2023) Metagenomic analyses of viruses in the orchid mycorrhizal interaction using improved assemble tools. *Methods in Molecular Biology* 2732, 67–81. DOI: 10.1007/978-1-0716-3515-5_5.

Hasanuzzaman, M., Bhuyan, M.H.M.B., Zulfiqar, F., Raza, A., Mohsin, S.M. *et al.* (2020) Reactive oxygen species and antioxidant defense in plants under abiotic stress: Revisiting the crucial role of a universal defense regulator. *Antioxidants* 9(8), 681. DOI: 10.3390/antiox9080681.

Hashem, H.A. and Khalil, R. (2023) Insight into the interaction of strigolactones, abscisic acid, and reactive oxygen species signals. In: Faizan, M., Hayat, S. and Ahmed, S.M. (eds) *Reactive Oxygen Species: Prospects in Plant Metabolism*. Springer Nature Singapore, Singapore, pp. 179–211. DOI: 10.1007/978-981-19-9794-5_11.

Helmich, R. (2025) Unveiling the strigolactone signaling pathway: From receptors to responses. In: Aftab, T. and Wani, K.I. (eds) *Strigolactones: Emerging Plant Hormones*. Scrivener Publishing, Beverly, Massachusetts, pp. 35–67. DOI: 10.1002/9781394302826.

Hossain, A., Raza, A., Maitra, S., Asaduzzaman, M., Islam, M.R. *et al.* (2021) Strigolactones: A novel carotenoid-derived phytohormone biosynthesis, transporters, signalling, and mechanisms in abiotic stress. In: Aftab, T. and Hakeem, K.R. (eds) *Plant Growth Regulators: Signalling under Stress Conditions*. Springer, Cham, Switzerland, pp. 275–303. DOI: 10.1007/978-3-030-61153-8_13.

Hossain, A., Pamanick, B., Venugopalan, V.K., Ibrahimova, U., Rahman, M.A. *et al.* (2022) Emerging roles of plant growth regulators for plants adaptation to abiotic stress-induced oxidative stress. In: Aftab, T. and Naeem, M. (eds) *Emerging Plant Growth Regulators in Agriculture*. Academic Press, UK, pp. 1–72. DOI: 10.1016/B978-0-323-91005-7.00010-2.

Ikram, M., Khalid, B., Batool, M., Ullah, M., Zitong, J. *et al.* (2025) Secondary metabolites as biostimulants in salt stressed plants: Mechanisms of oxidative defense and signal transduction. *Plant Stress* 16, 100891. DOI: 10.1016/j.stress.2025.100891.

Jibran, R., Tahir, J., Andre, C.M., Janssen, B.J., Drummond, R.S.M. *et al.* (2024) *DWARF27* and *CAROTENOID CLEAVAGE DIOXYGENASE 7* genes regulate release, germination and growth of gemma in *Marchantia polymorpha*. *Frontiers in Plant Science* 15, 1358745. DOI: 10.3389/fpls.2024.1358745.

Josse, M.R. (2022) The role of strigolactone and KAI2 signalling in seedling development. PhD thesis, University of Leeds, Leeds, UK.

Kamran, M., Melville, K.T. and Waters, M.T. (2024) Karrikin signalling: Impacts on plant development and abiotic stress tolerance. *Journal of Experimental Botany* 75(4), 1174–1186. DOI: 10.1093/jxb/erad476.

Kaya, C. (2025) Molecular crosstalk of transcription factors and hormones in drought tolerance genes in plants. *Turkish Journal of Botany* 49(3), 145–164. DOI: 10.55730/1300-008X.2852.

Kee, Y.J., Ogawa, S., Ichihashi, Y., Shirasu, K. and Yoshida, S. (2023) Strigolactones in rhizosphere communication: Multiple molecules with diverse functions. *Plant & Cell Physiology* 64(9), 955–966. DOI: 10.1093/pcp/pcad055.

Khalid, M.F., Shafqat, W., Khan, R.I., Jawaid, M.Z., Hussain, S. *et al.* (2024) Unveiling the resilience mechanism: Strigolactones as master regulators of plant responses to abiotic stresses. *Plant Stress* 12, 100490. DOI: 10.1016/j.stress.2024.100490.

Khaliq, A., Perveen, S., Alamer, K.H., Zia Ul Haq, M., Rafique, Z. *et al.* (2022) Arbuscular mycorrhizal fungi symbiosis to enhance plant–Soil interaction. *Sustainability* 14(13), 7840. DOI: 10.3390/su14137840.

Khan, A., Khan, V., Pandey, K., Sopory, S.K. and Sanan-Mishra, N. (2022) Thermo-priming mediated cellular networks for abiotic stress management in plants. *Frontiers in Plant Science* 13, 866409. DOI: 10.3389/fpls.2022.866409.

Khan, S., Sehar, Z., Albaqami, M. and Khan, N.A. (2023) Ethylene crosstalk with isoprenoid-derived signaling molecules in the context of salinity tolerance. *Environmental and Experimental Botany* 212, 105379. DOI: 10.1016/j.envexpbot.2023.105379.

Kodama, K., Rich, M.K., Yoda, A., Shimazaki, S., Xie, X. *et al.* (2022) An ancestral function of strigolactones as symbiotic rhizosphere signals. *Nature Communications* 13(1), 3974. DOI: 10.1038/s41467-022-31708-3.

Korek, M. and Marzec, M. (2024) An update on strigolactone signaling in plants. In: Bashri, G., Hayat, S. and Bajguz, A. (eds) *Strigolactones: Synthesis, Application and Role in Plants*. Academic Press, London, pp. 53–73. DOI: 10.1016/B978-0-443-13521-7.00004-X.

Kumar, G. (2024) Unraveling the dynamics of antioxidant defense in plants under drought. In: Nikalje, G.C., Shahnawaz, M., Parihar, J., Qazi, H.A. and Patil, V.N. (eds) *Plant Secondary Metabolites and Abiotic Stress*. Wiley, Hoboken, New Jersey, pp. 35–79. DOI: 10.1002/9781394186457.

Kumar, M., Prusty, M.R., Pandey, M.K., Singh, P.K., Bohra, A. *et al.* (2023) Application of CRISPR/Cas9-mediated gene editing for abiotic stress management in crop plants. *Frontiers in Plant Science* 14, 1157678. DOI: 10.3389/fpls.2023.1157678.

Li, S., Khoso, M.A., Xu, H., Zhang, C., Liu, Z. *et al.* (2024) WRKY transcription factors (TFs) as key regulators of plant resilience to environmental stresses: Current perspective. *Agronomy* 14(10), 2421. DOI: 10.3390/agronomy14102421.

Liu, Y., Chen, S., Pal, S., Yu, J., Zhou, Y. *et al.* (2024) The hormonal, metabolic, and environmental regulation of plant shoot branching. *New Crops* 1, 100028. DOI: 10.1016/j.ncrops.2024.100028.

Lucido, A., Andrade, F., Basallo, O., Eleiwa, A., Marin-Sanguino, A. *et al.* (2024) Modeling the effects of strigolactone levels on maize root system architecture. *Frontiers in Plant Science* 14, 1329556. DOI: 10.3389/fpls.2023.1329556.

Mangal, V., Verma, L.K., Singh, S.K., Saxena, K., Roy, A. *et al.* (2024) Triumphs of genomic-assisted breeding in crop improvement. *Heliyon* 10(15), e35513. DOI: 10.1016/j.heliyon.2024.e35513.

Manzoor, M.A., Xu, Y., lv, Z., Xu, J., Wang, Y. *et al.* (2023) Fruit crop abiotic stress management: A comprehensive review of plant hormones mediated responses. *Fruit Research* 3, 30. DOI: 10.48130/FruRes-2023-0030.

Marzec, M., Situmorang, A., Brewer, P.B. and Brąszewska, A. (2020) Diverse roles of MAX1 homologues in rice. *Genes* 11(11), 1348. DOI: 10.3390/genes11111348.

Marzec, M. and Korek, M. (2024) An update on strigolactone signaling in plants. In: Bashri, G., Hayat, S. and Bajguz, A. (eds) *Strigolactones: Synthesis, Application and Role in Plants*. Academic Press, London, pp. 53–73. DOI: 10.1016/B978-0-443-13521-7.00004-X.

Mashiguchi, K., Seto, Y. and Yamaguchi, S. (2021) Strigolactone biosynthesis, transport and perception. *Plant Journal* 105(2), 335–350. DOI: 10.1111/tpj.15059.

Mehmood, I., Wani, K.I. and Aftab, T. (2025a) Strigolactone interplay with other phytohormones under stressed and normal conditions. In: Aftab, T. and Wani, K.I. (eds) *Strigolactones: Emerging Plant Hormones*. Scrivener Publishing, Beverly, Massachusetts, pp. 161–191. DOI: 10.1002/9781394302826.ch8.

Mehmood, I., Wani, K.I. and Aftab, T. (2025b) Nature, structural diversity, biosynthetic pathway, and strigolactone transport in plants. In: Aftab, T. and Wani, K.I. (eds) *Strigolactones: Emerging Plant Hormones*. Scrivener Publishing, Beverly, Massachusetts, pp. 13–34. DOI: 10.1002/9781394302826.ch2.

Moosavi, M., Khorassani, R. and Tavakkol Afshari, R. (2025) Strigolactone (GR24) and abscisic acid induced drought tolerance in wheat by ameliorating nutrient uptake. *Journal of Plant Growth Regulation* 44(2), 972–987. DOI: 10.1007/s00344-024-11508-3.

Mostofa, M.G., Ha, C.V., Rahman, M.M., Nguyen, K.H., Keya, S.S. *et al.* (2021) Strigolactones modulate cellular antioxidant defense mechanisms to mitigate arsenate toxicity in rice shoots. *Antioxidants* 10(11), 1815. DOI: 10.3390/antiox10111815.

Mukherjee, S. and Aftab, T. (eds) (2023) *Strigolactones, Alkamides and Karrikins in Plants: Recent Updates and Future Prospects*. CRC Press, Boca Raton, Florida. DOI: 10.1201/9781003224525.

Nakandala, N.D.U.S., Ranaweera, L.T., Weebadde, C.K. and Sooriyapathirana, S.D.S.S. (2020) Strigolactone, a novel hormone with essential functions in planta possesses a significant value as a cancer therapeutic agent: A review. *Journal of Technology and Value Addition* 2(1), 106–153.

Naseer, M.A., Zhang, Z.Q., Mukhtar, A., Asad, M.S., Wu, H.Y. *et al.* (2024) Strigolactones: A promising tool for nutrient acquisition through arbuscular mycorrhizal fungi symbiosis and abiotic stress tolerance. *Plant Physiology and Biochemistry* 215, 109057. DOI: 10.1016/j.plaphy.2024.109057.

Naz, M., Afzal, M.R., Raza, M.A., Tariq, M., Yan, M. *et al.* (2025) The significant effects of strigolactones on plant growth and microbe interactions: A review. *Plant Growth Regulation* 105(4), 937–957. DOI: 10.1007/s10725-025-01325-3.

Oyebamiji, Y.O., Adigun, B.A., Shamsudin, N.A.A., Ikmal, A.M., Salisu, M.A. *et al.* (2024) Recent advancements in mitigating abiotic stresses in crops. *Horticulturae* 10(2), 156. DOI: 10.3390/horticulturae10020156.

Ozturk, M., Bhat, R.A., Ashraf, M., Tonelli, F. and Unal, B.T. (eds) (2023) *Phytohormones and Stress Responsive Secondary Metabolites*. Academic Press, Cambridge, Massachusetts.

Pamanick, B., Hossain, A., Venugopalan, V.K., Ibrahimova, U., Rahman, M.A. *et al.* (2022) Emerging roles of plant growth regulators for plants adaptation to abiotic stress-induced oxidative stress. In: Aftab, T. and Naeem, M. (eds) *Emerging Plant Growth Regulators in Agriculture*. Academic Press, London, pp. 1–72. DOI: 10.1016/B978-0-323-91005-7.00010-2.

Pandey, P., Tripathi, A., Dwivedi, S., Lal, K. and Jhang, T. (2023) Deciphering the mechanisms, hormonal signaling, and potential applications of endophytic microbes to mediate stress tolerance in medicinal plants. *Frontiers in Plant Science* 14, 1250020. DOI: 10.3389/fpls.2023.1250020.

Parwez, R., Naeem, M., Aftab, T., Ansari, A.A. and Gill, S.S. (2022) Heavy metal toxicity and underlying mechanisms for heavy metal tolerance in medicinal legumes. In: Naeem, M., Aftab, T., Ansari, A.A., Gill, S.S. and Macovei, A. (eds) *Hazardous and Trace Materials in Soil and Plants*. Academic Press, London, pp. 141–177. DOI: 10.1016/B978-0-323-91632-5.00024-0.

Pixley, K.V., Cairns, J.E., Lopez-Ridaura, S., Ojiewo, C.O., Dawud, M.A. *et al.* (2023) Redesigning crop varieties to win the race between climate change and food security. *Molecular Plant* 16(10), 1590–1611. DOI: 10.1016/j.molp.2023.09.003.

Rai, K.K. (2023) *The Role of Salicylic Acid and Nitric Oxide in Plant Heat Response*. Cambridge Scholars Publishing, Newcastle upon Tyne, UK.

Raja, V., Singh, K., Qadir, S.U., Singh, J. and Kim, K.H. (2024) Alleviation of cadmium-induced oxidative damage through application of zinc oxide nanoparticles and strigolactones in solanum lycopersicum l. *Environmental Science* 11(6), 2633–2654. DOI: 10.1039/D3EN00796K.

Rao, M.J., Duan, M., Zhou, C., Jiao, J., Cheng, P. *et al.* (2025) Antioxidant defense system in plants: Reactive oxygen species production, signaling, and scavenging during abiotic stress-induced oxidative damage. *Horticulturae* 11(5), 477. DOI: 10.3390/horticulturae11050477.

Raza, A., Charagh, S., Najafi-Kakavand, S., Abbas, S., Shoaib, Y. *et al.* (2023) Role of phytohormones in regulating cold stress tolerance: Physiological and molecular approaches for developing cold-smart crop plants. *Plant Stress* 8, 100152. DOI: 10.1016/j.stress.2023.100152.

Rehaman, A., Khan, S., Rawat, B., Gaira, K.S., Asgher, M. *et al.* (2025) Mechanistic insights into plant drought tolerance: A multi-level perspective. *Journal of Crop Health* 77(2), 53. DOI: 10.1007/s10343-025-01115-x.

Rehman, M., Saeed, M.S., Fan, X., Salam, A., Munir, R. *et al.* (2023) The multifaceted role of jasmonic acid in plant stress mitigation: An overview. *Plants* 12(23), 3982. DOI: 10.3390/plants12233982.

Roychowdhury, R., Das, S.P., Gupta, A., Parihar, P., Chandrasekhar, K. *et al.* (2023) Multi-omics pipeline and omics-integration approach to decipher plant's abiotic stress tolerance responses. *Genes* 14(6), 1281. DOI: 10.3390/genes14061281.

Sachdev, S., Ansari, S.A., Ansari, M.I., Fujita, M. and Hasanuzzaman, M. (2021) Abiotic stress and reactive oxygen species: Generation, signaling, and defense mechanisms. *Antioxidants* 10(2), 277. DOI: 10.3390/antiox10020277.

Sah, S.K., Reddy, K.R. and Li, J. (2016) Abscisic acid and abiotic stress tolerance in crop plants. *Frontiers in Plant Science* 7, 571. DOI: 10.3389/fpls.2016.00571.

Salvi, P., Manna, M., Kaur, H., Thakur, T., Gandass, N. *et al.* (2021) Phytohormone signaling and crosstalk in regulating drought stress response in plants. *Plant Cell Reports* 40(8), 1305–1329. DOI: 10.1007/s00299-021-02683-8.

Samanta, S. and Roychoudhury, A. (2025) Molecular crosstalk of jasmonate with major phytohormones and plant growth regulators during diverse stress responses. *Journal of Plant Growth Regulation* 44(1), 62–88. DOI: 10.1007/s00344-024-11412-w.

Saroy, K. and Garg, N. (2021) Role of non-enzymatic antioxidant defense mechanisms in imparting heavy metal, salt, water and temperature stress tolerance in plants. In: Latef, A.A.H.A. (ed.) *Organic Solutes, Oxidative Stress, and Antioxidant Enzymes Under Abiotic Stressors*. CRC Press, Boca Raton, Florida, pp. 321–338. DOI: 10.1201/9781003022879-15.

Seleiman, M.F., Ahmad, A., Alhammad, B.A. and Tola, E. (2023) Exogenous application of zinc oxide nanoparticles improved antioxidants, photosynthetic, and yield traits in salt-stressed maize. *Agronomy* 13(10), 2645. DOI: 10.3390/agronomy13102645.

Selwal, N., Wani, A.K., Akhtar, N., Kaur, M. and Jassal, P.S. (2023) Molecular insights of strigolactone biosynthesis, signalling pathways, regulatory roles, and hormonal crosstalks in plant systems. *South African Journal of Botany* 160, 9–22. DOI: 10.1016/j.sajb.2023.06.046.

Sharma, P., Jha, A.B. and Dubey, R.S. (2024) Strigolactones: Coordination with other phytohormones and enhancement of abiotic stress responses. *Environmental and Experimental Botany* 223, 105782. DOI: 10.1016/j.envexpbot.2024.105782.

Shi, L., Li, X., Fu, Y. and Li, C. (2023) Environmental stimuli and phytohormones in anthocyanin biosynthesis: A comprehensive review. *International Journal of Molecular Sciences* 24(22), 16415. DOI: 10.3390/ijms242216415.

Singh, P., Choudhary, K.K., Chaudhary, N., Gupta, S., Sahu, M. *et al.* (2022) Salt stress resilience in plants mediated through osmolyte accumulation and its crosstalk mechanism with phytohormones. *Frontiers in Plant Science* 13, 1006617. DOI: 10.3389/fpls.2022.1006617.

Stanga, J.P. (2022) Recent developments in parasitic plant biology. *CABI Reviews 2022*. DOI: 10.1079/cabireviews202217059.

Taulera, Q. (2022) Perception and responses to strigolactones in arbuscular mycorrhizal fungi. PhD thesis, Université Paul Sabatier-Toulouse III, Toulouse, France.

Thilakarathne, A.S., Liu, F. and Zou, Z. (2025) Plant signaling hormones and transcription factors: Key regulators of plant responses to growth, development, and stress. *Plants* 14(7), 1070. DOI: 10.3390/plants14071070.

Usman, B., Nawaz, G., Zhao, N., Liao, S., Liu, Y. *et al.* (2020) Precise editing of the OsPYL9 gene by RNA-guided Cas9 nuclease confers enhanced drought tolerance and grain yield in rice (*Oryza sativa* L.) by regulating circadian rhythm and abiotic stress responsive proteins. *International Journal of Molecular Sciences* 21(21), 7854. DOI: 10.3390/ijms21217854.

Votta, C., Wang, J.Y., Cavallini, N., Savorani, F., Liew, K.X. *et al.* (2023) Integration of apocarotenoid profile and expression pattern of carotenoid cleavage dioxygenases during mycorrhization in rice. *bioRxiv (Preprint)*. DOI: 10.1101/2023.02.24.529886.

Walker, C.H., Siu-Ting, K., Taylor, A., O'Connell, M.J. and Bennett, T. (2019) Strigolactone synthesis is ancestral in land plants, but canonical strigolactone signalling is a flowering plant innovation. *BMC Biology* 17(1), 70. DOI: 10.1186/s12915-019-0689-6.

Wang, B., Wang, Y., Zhang, J., Hu, C., Jiang, J. *et al.* (2023) ROS-induced lipid peroxidation modulates cell death outcome: Mechanisms behind apoptosis, autophagy, and ferroptosis. *Archives of Toxicology* 97(6), 1439–1451. DOI: 10.1007/s00204-023-03476-6.

Wang, L.-J., Gao, X. and Jia, G.-X. (2021) Stomata and ROS changes during botrytis elliptica infection in diploid and tetraploid lilium rosthornii diels. *Plant Physiology and Biochemistry* 167, 366–375. DOI: 10.1016/j.plaphy.2021.08.008.

Wang, Q., Smith, S.M. and Huang, J. (2022) Origins of strigolactone and karrikin signaling in plants. *Trends in Plant Science* 27(5), 450–459. DOI: 10.1016/j.tplants.2021.11.009.

Wang, Y., Duran, H.G.S., Haarst, J.C., Schijlen, E.G.W.M., Ruyter-Spira, C. *et al.* (2021) The role of strigolactones in P deficiency induced transcriptional changes in tomato roots. *BMC Plant Biology* 21(1), 1–21. DOI: 10.1186/s12870-021-03124-0.

Wawrzyńska, A. and Sirko, A. (2024) Sulfate availability and hormonal signaling in the coordination of plant growth and development. *International Journal of Molecular Sciences* 25(7), 3978. DOI: 10.3390/ijms25073978.

Woudenberg, S., Renema, J., Tomescu, A.M.F., Rybel, B. and Weijers, D. (2022) Deep origin and gradual evolution of transporting tissues: Perspectives from across the land plants. *Plant Physiology* 190(1), 85–99. DOI: 10.1093/plphys/kiac304.

Yang, W., Feng, H., Zhang, X., Zhang, J., Doonan, J.H. *et al.* (2020) Crop phenomics and high-throughput phenotyping: Past decades, current challenges, and future perspectives. *Molecular Plant* 13(2), 187–214. DOI: 10.1016/j.molp.2020.01.008.

Zhang, X., Lai, C., Liu, M., Xue, X., Zhang, S. *et al.* (2022) Whole genome analysis of sls pathway genes and functional characterization of DLSMXL6 in longan early somatic embryo development. *International Journal of Molecular Sciences* 23(22), 14047. DOI: 10.3390/ijms232214047.

Zheng, M., Liu, Y., Zhang, G., Yang, Z., Xu, W. *et al*. (2023) The applications and mechanisms of superoxide dismutase in medicine, food, and cosmetics. *Antioxidants* 12(9), 1675. DOI: 10.3390/antiox12091675.

Zhou, H., Cheng, L. and Wang, Z. (2023) Current understanding of unlocking the power of melatonin and other phytohormones to boost abiotic stress tolerance in solanaceous vegetable crops. *Scientia Horticulturae* 321, 112313. DOI: 10.1016/j.scienta.2023.112313.

Zhou, H., Shi, H., Yang, Y., Feng, X., Chen, X. *et al*. (2024) Insights into plant salt stress signaling and tolerance. *Journal of Genetics and Genomics* 51(1), 16–34. DOI: 10.1016/j.jgg.2023.08.007.

Zulfiqar, F. and Ashraf, M. (2021) Bioregulators: Unlocking their potential role in regulation of the plant oxidative defense system. *Plant Molecular Biology* 105(1–2), 11–41. DOI: 10.1007/s11103-020-01077-w.

10 Strigolactone-mediated Drought Stress Mitigation in Plants

Hummera Nawaz[1*], Saifullah[2], Ameer Khan[1], Amina Ameer[3] and Umer Hayat[4]

[1]Department of Botany, Division of Science and Technology, University of Education, Lahore, Pakistan; [2]Department of Botany, University of Agriculture, Faisalabad, Pakistan; [3]College of Geographical Sciences, Faculty of Geographical Science and Engineering, Henan University, Zhengzhou, China; [3]Department of Life Sciences, Khwaja Fareed University of Engineering and Information Technology, Rahim Yar Khan, Pakistan

Abstract

Abiotic stress negatively impacts agricultural production worldwide, reducing average yields of the majority of crops by up to 50%. Among the several abiotic stresses impacting agricultural productivity, drought stress is thought to be the primary cause of yield losses worldwide. Drought stress is predicted to cause a significant food scarcity by 2050 as the world's population grows. The condition may worsen as a result of expected changes in the global climate, which could increase the frequency, severity, and length of these abiotic stressors. As a result, there is an urgent need to increase our knowledge of the complex mechanisms of drought-stress tolerance and to produce advanced drought-resistant cultivars. The different abiotic stresses, including drought stress, which have a significant effect on crop productivity and quality, endanger human life and economic progress. Strigolactones (SLs), novel plant hormones generated from carotenoids, are synthesized in the roots of plants. SLs have been found in recent years to regulate plant responses to abiotic stressors. Eliminating oxidative stress generated by reactive oxygen species may boost chlorophyll content and photosynthetic rate, leading to improved plant resistance against drought stress. This chapter briefly explains the plant responses to drought stress, SLs, and their role in drought stress with special emphasis on morphological, physiological, and molecular regulation, and possible interactions with other hormones, as well as their possible role under drought-stress conditions.

Keywords: abiotic stress, climate change, osmolytes, reactive oxygen species, strigolactones

10.1 Introduction

Rising temperatures and altered precipitation patterns are linked to climate change. Globally, changes in environmental conditions lead to a decline in net primary production and damage to terrestrial ecosystems worldwide. The numerous abiotic stresses, including cold, temperature, drought, salt, and heat, brought about by harsh weather conditions, may affect the capacity of different plants to grow as well as to survive (Guo *et al.*, 2025; Igboeli *et al.*, 2025; Shakespear *et al.*, 2025; Varadharajan *et al.*, 2025). In the last few decades, human intervention has increased abiotic stress, reducing plant development, particularly reproductive organ production, and vegetative growth (Hoque *et al.*, 2021; Kumari *et al.*, 2022). With the rise in the generation of different reactive

*Corresponding author: hummera.nawaz@ue.edu.pk

DOI: 10.1079/9781836990598.0010

oxygen species (ROS), these abiotic stresses lead to severe oxidative damage, which impacts the growth and development of numerous plant species. Abiotic stresses are now considered the biggest threat worldwide, and may diminish the agricultural sector by reducing crop yields.

For instance, stress caused by low temperatures decreases rice output by 30–40%, whereas stress from high temperatures and drought lowers the yield of grains by about 9–10% (Zhang *et al.*, 2019; Choukri *et al.*, 2020; Tippani *et al.*, 2025). In addition, various abiotic stressors are predicted to lead to severe impacts on soils by 2050, while extreme salinity can result in the loss of as much as 50% of the agricultural land. In general, plants rely on their inherent ability to survive stressful situations, which includes developing diverse physiological, hormonal, and structural responses to abiotic pressures. Drought or water scarcity is the leading cause of decreased worldwide productivity of plants. It is a multi-dimensional environmental limitation that can influence molecular attributes to overall plant health and physiology (Caparas *et al.*, 2021; Rhaman *et al.*, 2021; Waadt *et al.*, 2022; Phour and Sindhu, 2023).

Drought has harmful impacts on the entire health of plants by reducing plant productivity, increasing susceptibility to pathogen attack and mortality, and making plants more vulnerable to various abiotic stresses. As a consequence, there is a high demand for improved plant strategies for controlling water stress (Ghanbary *et al.*, 2021). Plants have a complicated stress-response mechanism that requires the activation of coordinated pathways in response to stress. Plant hormones are important endogenous substances that regulate many physiological mechanisms involved in the plant's response to stress (El-Sheshtawy *et al.*, 2022). The primary plant hormones involved in abiotic stressors are ethylene, jasmonic acid (JA), brassinosteroids (BRs), abscisic acid (ABA), salicylic acid (SA), cytokinins (CKs), auxins, gibberellins (GAs), and strigolactones (SLs). ABA is a prominent plant hormone associated with coping mechanisms to stress (Fouda and Sofy, 2022). BRs, JA, SLs, SA, and ethylene are mainly produced under biotic stress and frequently serve vital functions in modulating defense responses of plants to abiotic challenges (El-Beltagi *et al.*, 2019).

Drought stress is a complicated feature characterized by modifications at the molecular, physiological, and morphological stages; plant hormones serve a protective function by causing leaf rolling, increased root:shoot ratio, transpiration, and osmolyte accumulation, reduced leaf area, stomatal closure, and the expression of stress-associated genes (Tiwari *et al.*, 2017). The most recent investigations related to the stress-associated pathway revealed significant proof for interactions among plant hormones that regulate the plant defense response (Tiwari *et al.*, 2017). Furthermore, the different chemical characteristics and relationships of secondary metabolites with a variety of plant hormones involved in defense responses, as well as their signaling mechanisms, are interconnected, allowing the formation of an efficient and sophisticated stress response (Abu-Shahba *et al.*, 2021). These environmental abnormalities can take the shape of severe temperatures, salinity, drought stress and flooding. Of these, drought stands out as a global negative factor affecting crop eco-agronomical characteristics, ultimately reducing overall yield capability. Drought has a significant impact on a wide range of cellular activities, including nutrition acquisition, photosynthesis, and signal perception, and as well as biochemical and molecular functions (Sharma *et al.*, 2020). As a result of such cellular disturbances, the plant's general physiological process suffers, including chlorophyll breakdown, decreased carbon availability, stunted development, disturbed turgor pressure, reduced membrane integrity, and chlorosis (Kapoor *et al.*, 2020).

As an adverse response paradigm, plants remodel their metabolic systems by influencing changes in the expression genes. At the cellular stage, such modifications help the plant adapt with certain biochemical and physiological properties that help the cell cope with stressful situations. Drought stress causes intracellular dehydration, which promotes secondary stressors, including osmotic and oxidative damage (Ghosh *et al.*, 2020). This causes an increase in ROS levels, which is damaging to macromolecular functioning and the integrity of membranes (Saxena *et al.*, 2020). However, multiple studies have shown that ROS also has a positive role in the cascade of stress signaling by promoting appropriate perception of signals and stimulating the expression of genes (Manna *et al.*, 2019). The cell may only generate ROS to a specific level, which is sufficient to ensure the proper execution of metabolic and cellular signaling events essential for the response to different stresses. Continuous ROS stress impairs cellular activities, influencing plant development (Miller *et al.*, 2010). Detoxifying procedures take place through the antioxidative defense system, which includes both non-enzymatic and enzymatic components (Soares *et al.*, 2019). Four ascorbate–glutathione cycle enzymes—ascorbate peroxidase, dehydroascorbate reductase, monodehydroascorbate reductase, and glutathione reductase—work in tandem to regulate the cellular redox status in addition to catalase (CAT), superoxide dismutase (SOD) and other enzymes. The non-enzymatic component contains low-molecular-weight molecules such as ascorbic acid (AsA), carotene, glutathione (GSH), polyphenols, and tocopherol (Turner, 2018). As a substitute for the reaction to drought conditions, the cell reduces its water potential (Ψ) as osmotic regulation is achieved by collecting specialized solutes. Such solutes are referred to as "compatible solutes" as their enhanced levels in the cytosol have no effect on the metabolic processes within the cell. These compatible solutes are biological molecules that retain neutrality at their natural pH and are primarily sugars, amino acids, polyhydric alcohols, quaternary amines, and others (Salvi *et al.*, 2021). Numerous other reactions including vigorous root architecture/growth, closure of stomata, and a boost in anti-stress proteins can assist plants in coping with the negative effects of drought conditions. These activities are all primarily driven through one or more plant hormone regulatory mechanism that promotes cellular flexibility and assists plants in developing a successful way to cope with severe drought (Ghosh *et al.*, 2020). Several studies have shown that the interactions of these hormones, along with hormone-associated transcriptional control, are required to begin and regulate the response for tolerance to drought stress. As a result, exact integration in the interactions of these hormones is required to ensure their proper function in plants.

SLs are a class of terpene-based carotenoid lactones generated in the roots of plants that were once considered to contribute to the development of root parasites such as *Striga*. There exist around 25 different types of natural SLs discovered in various plants to date. Numerous studies have demonstrated that SLs can control plant physiological functions during conditions of drought, thereby facilitating the development and growth of plants. Recent research has revealed that SLs serve a range of roles during drought conditions. For instance, SLs enhance interactions with water and boost gas exchange and pigment (photosynthetic) formation during conditions of drought, in addition to antioxidant functions, providing drought tolerance to seedlings of maize.

Exogenous application of the synthetic SL analog GR24 can modify the closure of stomata via ROS or ABA, influence photosynthesis and chlorophyll formation, and trigger antioxidant defense potential to reduce the detrimental impacts of drought on grapevine. Associations with other phytohormones, particularly ABA, are proposed as one of the key processes in this mechanism (Min *et al.*, 2019; Sattar *et al.*, 2022). Furthermore, the complex phytohormone reaction framework is integrated across several signaling channels to help the plant respond to drought-induced stress. Winter wheat types subjected to SA and SLs revealed increased drought tolerance, as evidenced by increased membrane stability index, larger relative water content, antioxidant enzyme functions, less electrolyte leakage, and leaf stomatal limitation. Additionally, there also exists a relationship between methyl jasmonate (MeJA) and SLs. *Dracocephalum* plants treated with MeJA and SL had higher dry and fresh weight, malondialdehyde (MDA), antioxidant capacity, and decreased electrolyte leakage, H_2O_2 levels, overall antioxidant content, and complete phenol concentration, indicating greater drought resistance (Sedaghat *et al.*, 2017; Shirani Bidabadi and Sharifi, 2021). Furthermore, during drought conditions, SL-treated and non-MeJA plants had the minimum essential oil concentration, whereas SL- and MeJA-treated plants had the greatest quantity. When both plant hormones were given together, the oil concentration was maximum, about 99.2%, during water stress, and tolerance to drought was improved. These findings imply that SL and MeJA contain links that can improve drought tolerance and optimize the yield of *Dracocephalum* essential oil. 6-Benzylaminopurine (6-BA) inhibits the SL signaling-associated gene only under dry conditions. Application of 6-BA to leaves of tall fescue dramatically reduced signs of stress, and treatment with 6-BA inhibited the drought-associated activation of SL signaling-related genes such as *FaMAX2* and *FaD14* (Rezaei Ghaleh *et al.*, 2020; Shirani Bidabadi and Sharifi, 2021). These findings indicate that SLs may work as a growth promoter in drought-stressed crops. Given its function in plant resistance to stress caused by abiotic factors, it is critical to study possible pathways for plants to use SL-induced drought-stress alleviation. MORE AXILLARY GROWTH 2 (MAX2) aids in responding to drought conditions by increasing the expression of stress-related genes such as *RD29b*, *RD29a*, and *RD22*, which enhances drought tolerance. Furthermore, the major *SMXL7/6/8* gene transcription factors can influence plant reactions toward drought conditions by changing the level of stability within the SMXL7/6/8 protein. SMXL7/6/8 may additionally attach to the specific promoter SnRK2.3 and repress transcription, resulting in responses to ABA (Lian *et al.*, 2023; Nisa *et al.*, 2023).

10.2 Plant Responses to Drought Stress

The response of the plant to drought is highly dependent on crop stage, physiological status, and drought severity (Moursi *et al.*, 2020). Drought-associated molecular, biochemical, and physiological alterations occur in plants. Particularly during water stress, plants work to retain the water potential in tissues by shutting stomata and minimizing the loss of water via transpiration. Furthermore, lower carbon dioxide (CO_2) levels and increased accumulation and production of mannitol, proline, ABA, and radical scavenging substances (e.g. α-tocopherol, GSH, ascorbate), mRNAs, and stress-related proteins result in decreased plant growth and photosynthesis (Abobatta, 2020). Drought stress disrupts the main mechanism of carbohydrate metabolism, causing partial closure of stomata at sites of carboxylation and a decrease in CO_2 availability (Hu *et al.*, 2019). Noctor *et al.* (2014) described the impact of stomatal closing on various components of photosynthesis in plants, where photosystem I (PSI) reduces oxygen to produce superoxide (O_2^-) and hydrogen peroxide (H_2O_2), which quickens the water–water cycle (Abdelaal *et al.*, 2022). While O_2^- and H_2O_2 or singlet oxygen ($^1O_2$) are produced in the chloroplast as a result of photorespiration, excessive reductions in electron transport in photosystem II (PSII) result in the production of $^1O_2$, which increases H_2O_2 production in the peroxisome (Saini *et al.*, 2018). The rate of photosynthesis is inadvertently impacted by the extra reduction of the photosynthetic electron transport chain caused by the possible generation of $^1O_2$ in PSII. Shoot respiration increases in response to stress caused by drought to regulate metabolic processes. As a result, glucose stored in citrus plants starts to deplete (Fahad *et al.*, 2017). Many phytohormones are the only mediators of avoidance, tolerance, and the deleterious impacts of water stress; thus, there are numerous current investigations being conducted on such pathways (Fig. 10.1). Plant hormones control a plant's growth, development, and drought-stress responses throughout its life (Iqbal *et al.*, 2022).

10.3 Strigolactones

SLs, a class of phytohormones derived from carotenoids, where first identified as compounds promoting the growth of seeds from weeds in the genera *Orobanche* and *Striga*. They are comparatively a recent discovery in the group of phytohormones, and researchers are now interested in learning more about their biology, function, and effects on the architecture of the plant's roots and shoots (Mishra *et al.*, 2017; Jia *et al.*, 2018). Plant roots synthesize SLs, which have been shown to be secreted into the rhizosphere. SLs and ABA are thought to have originated from a carotenoid-derived terpenoid lactone, which is considered a biosynthetic precursor of SLs (Soliman *et al.*, 2022; Samynathan *et al.*, 2024). Many enzymes, such as DWARF 27 (D27), carotenoid cleavage dioxygenase 7 (CCD7), carotenoid cleavage dioxygenase 8 (CCD8), and MORE AXYLLARY GROWTH 1 (MAX1), participate actively in the production of SLs, which are produced in the root region and subsequently transferred to the rhizospheric zone via transporters proteins (Sasse

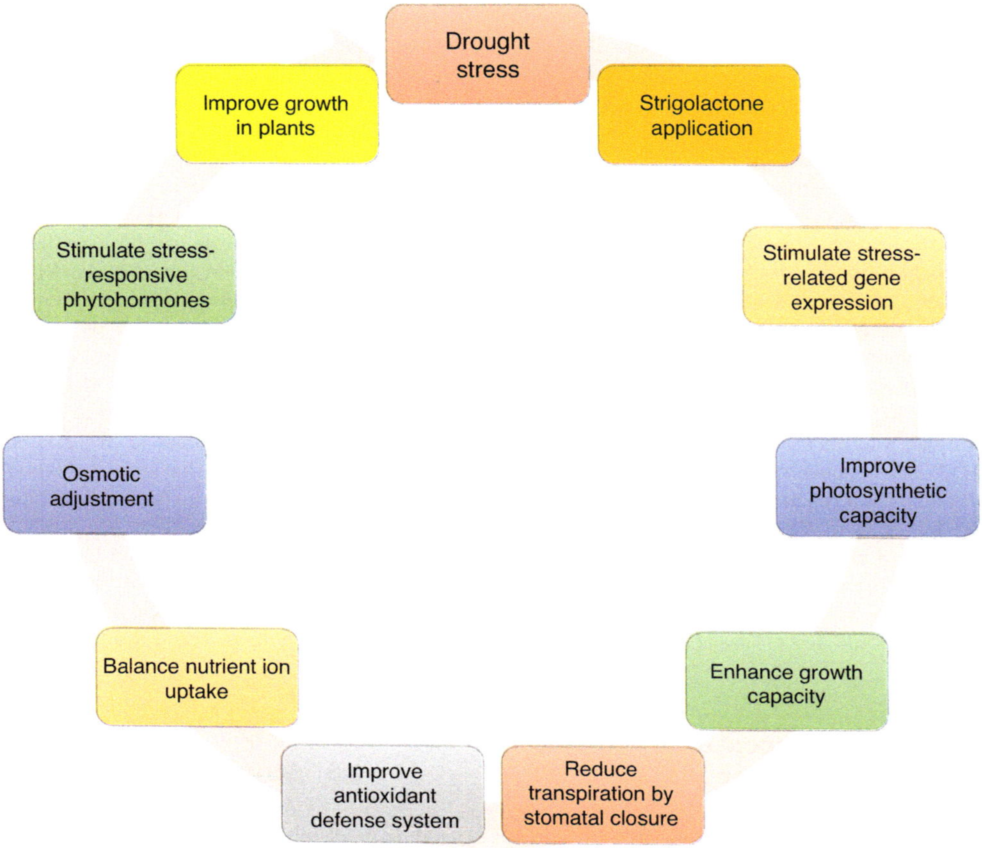

Fig. 10.1. Strigolactones mitigate the drought stress effect in plants through different changes in the plant.

et al., 2015). SLs have been shown to function in mycorrhizal infestations as a compound that signals hyphal branching and host recognition (Crosino and Genre, 2022). Until recently, their primary reported roles were as a plant-growth-enhancing compound and as signaling molecules between plants and other organisms; however, their function in reducing the negative impact of harsh circumstances such as drought, chilling, and heat is now being considered, and is currently a focus of research (Mishra *et al.*, 2017; Bhatt *et al.*, 2020; Rhaman *et al.*, 2024). Using various approaches, including liquid chromatography-tandem mass spectrometry, enzyme-specific inhibitors, and ABA-deficient mutants, a link between SLs and ABA was discovered in tomato (*Solanum lycopersicum*) plants grown in water-deficient environments. The primary reason for this association might be the same precursor involved in their biosynthesis, as both are produced from carotenoid-derived terpenoid lactones (Bhatt and Bhatt, 2020). When mutants of tomato, *Lotus japonicus*, and *Arabidopsis thaliana* with SL deficiency were subjected to drought conditions, they revealed an increased stomatal response to extracellular and intracellular concentrations of ABA throughout developmental phases and were extremely susceptible to water shortage (Cardinale *et al.*, 2018; Salvi *et al.*, 2021). Furthermore, during periods of extreme drought conditions, *Arabidopsis* showed an exponential rise in SL-associated biosynthetic gene expression, emphasizing the function of SLs in plant defense (Ha *et al.*,

2014). These results in several plants indicate that endogenous SLs perform a key function in maintaining drought tolerance by enhancing sensitivity to ABA. In contrast to these results, it was discovered that when drought stress was applied to non-mycorrhizal *L. japonicus* and tomato plants, the degree of gene expression associated with SL transport and biosynthesis was inhibited immediately as SL accumulated in the plant root systems (Chen *et al.*, 2022). As seen in plants under drought stress, these results showed a negative correlation between ABA and SLs, and externally applied SLs inhibited and negatively regulated the accumulation of ABA in *L. japonicus* plants. Because roots have been reported to be a primary site for SL production under typical conditions, the findings observed were significantly distinct. This suggests a distinct dynamic between SLs produced by shoots and those formed from roots. This was proposed as evidence of decreased SL production in the root region during drought stress, which may have facilitated ABA to accumulate selectively in the sap xylem elements. Because the plant depends mostly on signaling of ABA under water shortage conditions, this could be a systemic response to decreased root water availability (Zhou *et al.*, 2024). Additionally, the decline in the root water potential could also be a localized indicator of stress, impacting the physiology of the plant as a whole and regulating the SL flow through the root system to the other parts of the plant (Kohlen *et al.*, 2011; Mostofa *et al.*, 2018; Korek and Marzec, 2023). The control of translocation of ABA in SL-deficient mutants results in ABA-mediated SL signaling occurring, which is shown to be the most prevalent association in controlling growth events and plant responses to stress, involving intricate interactions. Overall, SLs are important for maintaining plant physiology when exposed to various environmental stressors, such as drought.

10.4 Types of Strigolactones

In 1966, the first SL was identified as strigol in the rhizosphere of *Gossypium hirsutum* L., indicating that strigol might represent a new category of plant hormones (Cook *et al.*, 1966; Rodenburg *et al.*, 2016). SLs were initially identified as potent germination stimulants for the seeds of parasitic weeds, specifically *Striga*. A bioactive chemical comparable to strigol called sorgolactone was discovered from the rhizosphere of *Sorghum*, a potential host for *Striga* in 1992. Over 30 different kinds of SLs have been identified in different plant species. Natural SLs are categorized into two classes based on their stereochemical characteristics at the B-C-ring junction: strigol and orobanchol-type (Zwanenburg *et al.*, 2016; Waters *et al.*, 2017). The β-derivatives of the C-ring of sorgomol, strigol, 5-deoxystrigol, and sorgolactone are examples of strigol class of SLs. SLs containing an α-derived C-ring, including fabacol, 7-oxorobanchol, orobanchol, solanacol, and 7-hydroxyorobanchol, are classified as the orobanchol class of SLs.

Strigol-class SLs of 5-deoxystrigol isolated from the root extract of *L. japonicus* plants have been proposed as a broad precursor of a different type of SLs. Strigol-class SLs are likely to arise from 5-deoxystrigol, while orobanchol-class SLs tend to originate from ent-2′-epi-5-deoxystrigol. SLs from the orobanchol category are particularly common in rice and pea plants (Xie *et al.*, 2013; Ćavar *et al.*, 2015). The discovery of zeapyranolactone, avenaol, zealactone, and heliolactone confirmed the existence of non-canonical SLs as a distinct family of SLs. SLs with enol-ether moieties of the D-ring but no ABC-ring are referred to as non-canonical. Heliolactone and zealactonewere discovered as non-canonical SLs present in sunflower and maize, respectively. The chemical makeup of non-canonical SLs includes methyl carlactonoate, which was initially identified in *Lotus* plants (Ueno *et al.*, 2014; Flematti *et al.*, 2016; Charnikhova *et al.*, 2017; Carlsson *et al.*, 2018; Tsukano *et al.*, 2024).

10.4.1 Synthetic SLs

The synthetic category of SLs is classed as mimics or analogs according to their function and structure. Analogs are comparable to natural SLs based on their structure, while mimics are both functionally and structurally related. The synthetic SL analog GR24 is used to improve crop production and agricultural research, including genetic focuses on breeding programs, plant–rhizosphere interactions, improvement in drought tolerance, yield improvement, and management of parasitic weeds (Mostofa *et al.*, 2018; Chesterfield *et al.*, 2020). Earlier work has described the production of nitrophenlactones and a variety of methyl phenlactonoates using carlactone and carlactonoic acid structures. GR24 is commonly used to study SL reactions, as it performs better than natural or native SLs in terms of stability. The derived SL derivatives of CISA-1 were more active than GR24 in the germination of seeds of *Orobanche*. CISA-1, a fluorescent SL analog, induces branching in arbuscular mycorrhizal fungi (AMF) while remaining non-toxic to host plant species (Goossens *et al.*, 2015; Jia *et al.*, 2016; Waters *et al.*, 2017). Synthetic SLs are similar to natural SLs and lightly promote the germination of seeds in different plant species. Synthetic SL mimics of AR36 chemicals strongly suppress pea plant branching. A set of SL derivatives of methyl phenlactonoates that have the most structural similarity to carlactonoic acid and non-canonical SLs has been described, and freshly prepared derivatives were examined in terms of regulating shoot branching, induction of seed germination, and the impact of leaf senescence across different root parasite plant species, while another study evaluated the action of MP16–MP25 analogs, which regulate shoot and rice architecture and growth (Boyer *et al.*, 2014; Jamil *et al.*, 2020).

10.5 Mechanism of Strigolactone Biosynthesis

SLs are carotenoid-originated sesquiterpene lactones with a four-ring system configuration consisting of a tricyclic ABC-ring connected to a fourth ring (D-ring) via an enol-ether linkage. Structure-wise, SLs are classified as non-canonical and canonical. Canonical SLs are composed of an ABC-ring with a butenolide (D-ring) connected to the C-ring by an enol-ether bridge. The D- and C-rings of the AB-rings should contain methyl furanone and lactone, respectively, but the number of double bonds, functional groups, and carbon atoms may differ (Bürger and Chory, 2020; Dell'Oste *et al.*, 2021; Mashiguchi *et al.*, 2021). Canonical SLs are categorized into two classes depending on their C-ring stereochemistry: orobanchol (α-derived) and strigol-type (β-derived). Non-canonical SLs lack the ABC-ring but include the biologically necessary enol-ether moiety (D-ring). Non-canonical SLs lack common precursors, classes, or core enzymes, so their biosynthesis enzymes produce a diverse range of compounds (Zorrilla *et al.*, 2022; Guercio *et al.*, 2023). However, non-canonical SLs are typically more unstable than canonical SLs resulting in their gradual breakdown during storage and purification. Additionally, compared with canonical SLs, which are broadly disseminated in various species of plants, non-canonical SLs show diverse structural variations and are specific to species (Yoneyama *et al.*, 2018; Ito *et al.*, 2022). Over 30 different SLs (both non-canonical and canonical) have been recognized as promoters for the germination of parasitic root weeds using bioassay-dependent purification process; non-canonical SLs such as avenaol, zealactone, heliolactone, and lotuslactone from maize, wild oats, sunflower, and *L. japonicus*, respectively, have been identified more recently (Kim *et al.*, 2014; Ueno *et al.*, 2014; Charnikhova *et al.*, 2017; Xie *et al.*, 2017; Yoneyama and Brewer, 2021). However, due to their greater structural diversity compared with canonical SLs, it is likely than more non-canonical SLs will be found soon, although this will require the development of efficient purification and storage techniques. The genes DECREASED API CAL DOMINANCE (DAD) in *Petunia*, MORE AXILLARY GROWTH (MAX) in *Arabidopsis*, RAMO SUS (RMS) in pea and DWARF (D) in rice are implicated in the biosynthesis and perception of SLs (Fig. 10.2).

Mutants of shoot branching have been screened, revealing that many enzymes, such as CCD7, CCD8, cytochrome P450 monooxygenases, and D27, regulate SL production. All-β-*trans*-carotene is the starting point for SL biosynthesis. The isomerase enzyme D27 converts 9-*cis*-β-carotene, which then undertakes a sequence of interactions with the CCD8 and CCD7 enzymes to generate carlactone (CL). CL was identified in *Escherichia coli* and has the potential to suppress rice tillering while also promoting germination of root-parasitic plants (Alder *et al.*, 2012; Yoneyama and Brewer, 2021). Later research demonstrated that CL is a required precursor for both forms of SL in rice and *Arabidopsis*. The cytochrome

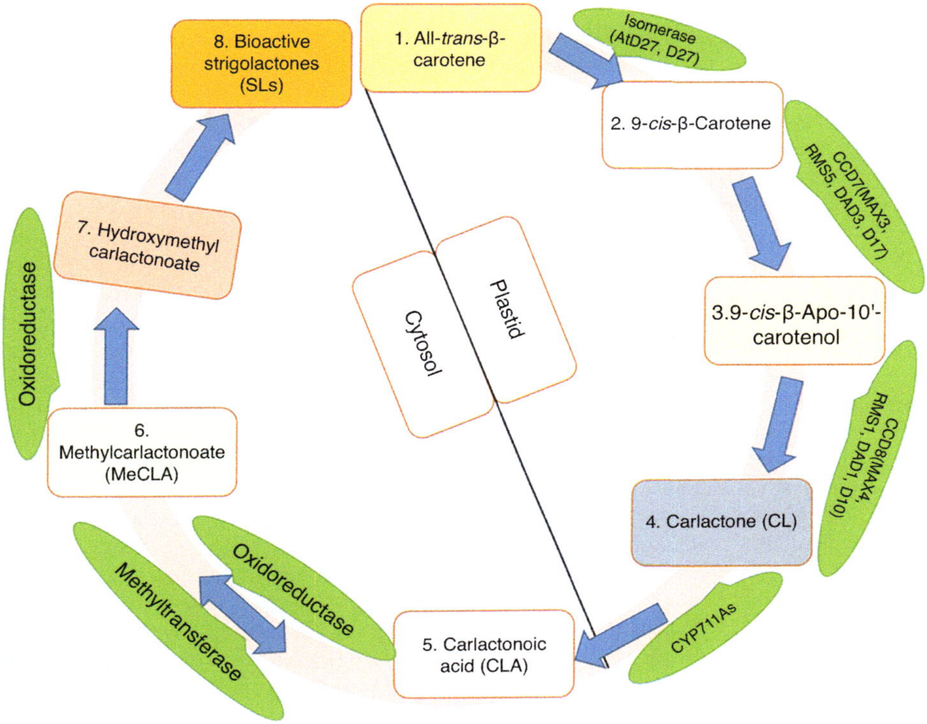

Fig. 10.2. Biosynthesis of strigolactones in the plastid and cytosol of the plant cells.

P450 MAX1 (CYP711A1) oxidizes CL to form carlactonoic acid (CLA), a critical rhizosphere signaling molecule, which is then transformed to methyl carlactonoate (MeCLA) by an unidentified methyltransferase in *Arabidopsis* (Abe *et al.*, 2014; Seto *et al.*, 2014). Other studies have used reverse genetics and transcriptomics to describe oxidoreductase (LBO), a lateral branching-causing enzyme, which encodes am iron (II)-dependent dioxygenase and 2-oxoglutarate. It was also discovered that LBO might transform MeCLA into hydroxymethyl carlactonoate (1-OH-MeCLA) due to the significant co-expression of LBO and SL biosynthesis genes. Although it varies among plants, this is a typical method for SL synthesis. For example, rice contains five homologous MAX1 genes compared with *Arabidopsis*. Os1400 (CYP711A3) and Os900 (CYP711A2) transform CL to CLA, while Os900 additionally closes the BC-ring and stereo-selectively converts CLA into 4-deoxyorobanchol (4DO), a major rice SL (Brewer *et al.*, 2016; Yoneyama *et al.*, 2018, 2020). The hydroxylation of 4DO into orobanchol is then catalyzed by Os1400. However, tomato plants cannot produce 4DO because the MAX1 homolog CYP711A21 cannot convert 4DO into orobanchol. CYP722C, a cytochrome P450 lineage distinct from MAX1, transforms CLA directly into orobanchol in cowpea without forming a 4DO intermediate. In cotton, CYP722C transforms CLA into 5-deoxystrigol (5DS). In *Sorghum*, 5DS is metabolized to sorgomol by an extra P450 (CYP728B35), which belongs to another lineage (Yoneyama *et al.*, 2018; Wakabayashi *et al.*, 2021). In contrast, canonical SLs are produced by three P450 clades, while MeCLA is considered a precursor to the formation of non-canonical SLs. For example, sunflower plants can convert MeCLA to heliolactone. However, it is difficult to determine the biological roles of non-canonical SLs because they are less stable than canonical SLs. Overall, the evidence shows that MAX1 performs three types of enzymatic function: producing CLA from CL, releasing CLA as a key signal in the rhizosphere, and converting CLA into MeCLA, a precursor of non-canonical SLs (Wakabayashi *et al.*, 2020; Yoneyama *et al.*, 2020). As a result, MAX1 is an important biosynthetic enzyme that regulates rhizosphere and endogenous signaling. Moreover, the fact that this enzyme has developed into many applications remains impressive.

10.6 Methodology for Exogenous Application of Strigolactones

Plants naturally produce SLs. However, some environmental conditions cause SL deficiency, which impacts their development and growth. Over the past few years, considerable research has been conducted on the biological and structural features of SLs, and a number of techniques have been developed for creation of synthetic SLs with exceptional bioactivity. These can be applied to plants in a variety of ways, such as foliar sprays, growth media, and solution. By altering the morphology, physiology, and molecular controls of many plant species, their exogenous use in various crops intensifies abiotic stresses, especially those brought about by drought, temperature, and salinity (Bhoi *et al.*, 2021).

10.7 Strigolactones Function in Response to Drought Conditions

The stress caused by drought stimulates the biosynthesis of SL in different plant species, and they have been shown to mitigate the harm caused by drought to plants. The surviving rate of *Arabidopsis* treated with GR24 was 100% under drought circumstances, while that of plants given water was 29%, showing that plants given SL had a strong drought resistance (Ha *et al.*, 2014). Under drought stress, spraying with GR24 notably increased the plant height, root length, and fresh weight of roots, stems, and leaves of alfalfa (Yang *et al.*, 2023). Drought stress dramatically shrunk grape epidermal cells, while exogenous GR24 kept the epidermal cells smooth (Qi *et al.*, 2024). Additionally, the relative water content (RWC) of plant leaves in drought conditions may serve as an indicator parameter of water deficit, potentially revealing the plant's drought tolerance (Badr and Brüggemann, 2020). Use of GR24 mitigated the drop in RWC and improved the relative electrolyte leakage induced by severe drought, indicating that GR24 might reduce water loss in drought environments (Qi *et al.*, 2024). Furthermore, spraying GR24 reduced the rise in MDA concentration produced by extreme drought in wheat and grape (Sedaghat *et al.*, 2017; Min *et al.*, 2019). The leaves of plants were checked for weight at regular intervals to calculate RWC and rates of water loss during dehydration testing, revealing that SL-deficient *max4* and *max3* mutants as well as SL-response *max2* mutant of *Arabidopsis* plants showed higher rates of evaporation of water than the wild type (Ha *et al.*, 2014), showing that SLs lower the transpiration rate in plants during dehydration conditions. Additionally, *smxl7*, *-6*, and *-8* mutants of *Arabidopsis* plants lacking SL signals adversely exhibited a greater leaf surface temperature, slower leaf water loss, reduced cuticular water permeability, and higher RWC under dry conditions, showing that SLs can help plants to retain water. Thus, SLs increase drought tolerance by boosting plant retention of water during periods of drought (Li *et al.*, 2020).

Understanding the exact mechanisms of SL action that reduce the harm done by abiotic stressors in plants is critical, and it is expected that SLs will be exploited in future plant production. The methods by which SLs mitigate the effects of drought stress in plants to improve plant growth are summarized in Fig. 10.3. These are discussed in more detail in the following sections.

10.8 Physiological Regulation by Strigolactones

10.8.1 Photosynthetic system regulation

Previous research has demonstrated that SLs regulate photosystems, photosynthetic

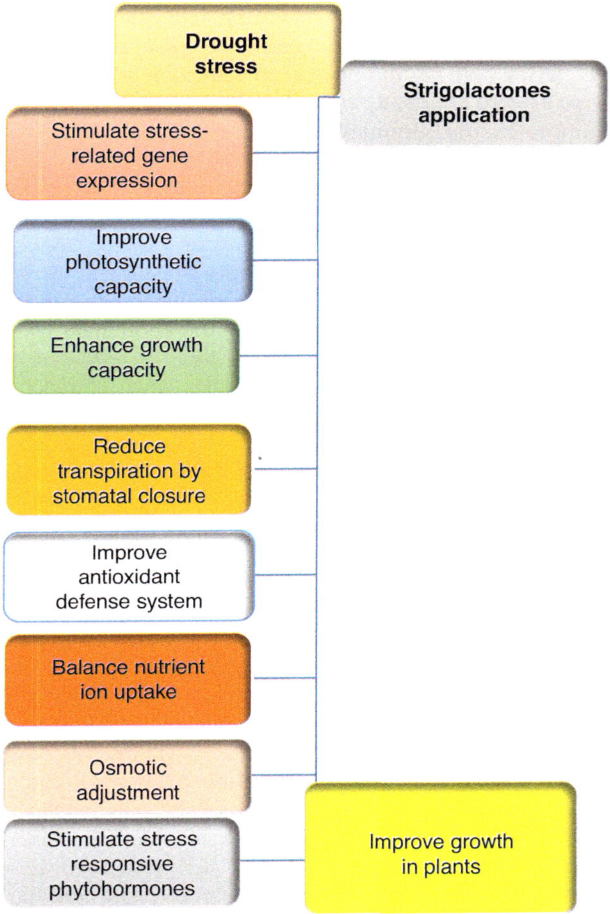

Fig. 10.3. Methods by which strigolactones mitigate the effect of drought stress in plants.

pigments, and leaf stomata, all of which contribute to stress-induced harm to the plant photosynthetic system. The *hvd14.d* mutant of barley contains a missense mutation in the gene that encodes the SL-specific HvD14 receptor and has been utilized to explore how barley responds to water shortages (Marzec *et al.*, 2020b; Qi *et al.*, 2024). In comparison with wild type, the mutant of HvD14 barley displayed reduced leaf photosynthesis, altered stomatal density, delayed stomatal closure, and disorganized chloroplast structure (Marzec *et al.*, 2020b). Furthermore, under cadmium stress, the net transpiration rate, stomatal conductance, photosynthetic rate, and chlorophyll content of barley and switch grass were severely reduced, but each of these parameters increased dramatically following GR24 treatment (Tai *et al.*, 2017; Gong *et al.*, 2024). Application of GR24 also greatly increased the net transpiration rate, stomatal conductance, photosynthetic rate, and water consumption efficiency of grape, rice, maize, and wheat leaves (Min *et al.*, 2019; Ling *et al.*, 2020). Investigations have revealed that SLs reduce the impact of stress on the photosynthetic system by controlling the levels of photosynthetic pigments. Exogenous application of GR24 can rescue seedlings of *Malus hupehensis* from saline–alkali stress by protecting chlorophyll levels and regulating the photosynthetic system (Ma *et al.*, 2022). In contrast, carotenoid levels declined dramatically under salt conditions, and priming of the achenes using GR24 boosted carotenoids in sunflower (Sarwar and Shahbaz, 2020). At the same time, GR24 applications led to a rise in the ratio of chlorophyll *a*/*b* contents (Liu *et al.*, 2022; Qi *et al.*, 2024). Likewise, salt stress drastically reduced carotenoids, overall chlorophyll level, and average chlorophyll *a*/*b* in maize, but GR24 considerably enhanced the photosynthetic pigment characteristics of the leaves, suggesting that SLs might minimize photoinhibition in plants during low-light stress by preserving the quantities and amounts of photosynthetic pigments (Liu *et al.*, 2022; Mehrabi *et al.*, 2024; Qi *et al.*, 2024). Further research has revealed that SLs have a function in quantum yield and electron transport in photosynthetic systems at the anatomic levels. GR24 enhanced chlorophyll *a* and *b* content, the photochemical activity (F_v/F_m) of PSII, the effective quantum yield PSI and PSII (Y(I) and Y(II)), and photosynthetic rate in seedlings of tomatoes exposed to low-light treatment (Lu *et al.*, 2019). Additionally, GR24 improved the electron transfer speed/rate of PSII and PSI, the quantum yield ratio of moving electron flow to Y(II), and non-photochemical quenching, which can be induced by the pool size of oxidized plastoquinone (Lu *et al.*, 2019). Furthermore, GR24 improved the maximal photochemical capacity of PSII, in addition to the efficiency and quantum yield of electron transport over the electron acceptor Q_A in the seedlings of mung bean exposed to a low-temperature environment (Omoarelojie *et al.*, 2021). These results suggest that by promoting electron transport between the two photosystems and lowering PSII and PSI photoinhibition under stress, SL can improve photosynthetic efficiency.

10.8.2 Chlorophyll synthesis

Chlorophyll is regarded as one of the key components representing the ability of plants to carry out the process of photosynthesis. Abiotic stressors such as salt and drought have been found to primarily limit photosynthesis and CO_2 diffusion by decreasing conductivity within mesophyll and stomatal cells. However, little is known about how SL application affects oxidative stress, transcriptomics, and photosynthesis in different plants grown in drought or salinity conditions. As salinity/alkalinity stress increases, *Leymus chinensis* has previously demonstrated decreases in chlorophyll *b* and carotenoid content; this could be due to either increased activity of the enzyme that breaks down chlorophyll or inhibition of chlorophyll synthesis. Furthermore, the CO_2 supply for RuBisCO (ribulose-1,5-bisphosphate carboxylase/oxygenase) may have decreased as a result of decreased stomatal conductivity, which would have prevented CO_2 assimilation (Lin *et al.*, 2017; Bhoi *et al.*, 2021). Similarly, another study demonstrated that the chlorophyll content of *Brassica napus* decreases as its salt concentration rises, but when GR24 was applied exogenously, the chlorophyll composition under salinity stress significantly improved (Ma *et al.*, 2017). Another study found that when *Vitis vinifera* was subjected to drought stress, its chlorophyll content decreased, but when GR24 was applied exogenously, the amount of chlorophyll increased (Min *et al.*, 2019). SLs have been shown to be positive regulators of plant light-harvesting complexes in a number of investigations. These comprise multi-subunit membrane-protein complexes as well as PSI and PSII components (energy transfer system). The *SlORT1* tomato mutant with SL deficiency had a low level of chlorophyll because expression of the RuBisCO genes and chlorophyll *a*/*b*-binding protein precursor was decreased; in comparison, external administration of GR24 increased expression of several of these genes, revealing that SLs perform a significant function in the control of components associated with light harvesting (Mayzlish-Gati *et al.*, 2010; Bhoi *et al.*, 2021). Much research suggests to the critical involvement of SLs in the light-harvesting complex; however, additional studies are needed to identify the connecting points among light-harvesting components and SLs. In addition, because light is the key element of SLs and carotenoid production is considered to be formed via this mechanism, it is possible that the association between light-associated pathways and SLs adopts the feedback loop required for optimal plant development and coordinated growth.

10.9 Biochemical Regulation by Strigolactones

SLs cause numerous biochemical differences in plants under drought conditions.

10.9.1 Antioxidant enzyme activity

As mentioned above, SLs may reduce oxidation in stressed plants by decreasing ROS levels while modulating the AsA–GSH cycle and enhance the function of different antioxidant enzymes. GR24 treatment significantly decreased the amounts of O_2, H_2O_2, and OH in cucumber leaves under salt stress conditions (Zhang *et al.*, 2022). Additionally, GR24 treatment dramatically reduced wheat's H_2O_2 content in drought conditions (Sedaghat *et al.*, 2021). Compared to wild-type plants, the rosette leaves of *smxl6*, *-7*, and *-8* mutants of *Arabidopsis* plants produced less H_2O_2 and O_2, and the growth inhibition level decreased when treated with paraquat and 3-amino-1,2,4-triazole, implying that the mutant plants outperformed the wild-type plants in terms of ROS detoxification (Li *et al.*, 2020). SLs then had additional effects on the antioxidant enzymes and antioxidants, mainly anthocyanins. In rice, sunflower, and *Malus hupehensis*, GR24 spraying enhanced the function of enzymes such as POD, CAT, and SOD, thereby preventing the generation of ROS (Ling *et al.*, 2020; Ma *et al.*, 2022). Non-enzymatic antioxidants such as GSH and AsA work with AsA–GSH cycle enzymes to promote the coordination of plant stress tolerance. In *Arabidopsis*, SLs also positively regulate the biosynthesis of anthocyanins, which helps to increase drought resistance by eliminating drought-induced ROS. Under dry conditions, *smxl6*, *−7*, and *−8* mutants had considerably larger anthocyanin concentrations than wild-type plants (Bartoli *et al.*, 2017; Li *et al.*, 2020; An *et al.*, 2024). When barley roots and aboveground parts were exposed to cadmium stress, GR24 effectively enhanced the function of glutathione peroxidase, dehydroascorbate reductase, monodehydroascorbate reductase, ascorbate peroxidase, and glutathione reductase. Under cadmium stress, the higher AsA/dehydroascorbate (DHA) and lower GSH/oxidized GSH (GSSG) percentage dropped dramatically, while application of GR24 significantly enhanced the DHA/AsA and GSSG/GSH concentrations in barley (Qiu *et al.*, 2021). This implies that by boosting AsA–GSH cycle enzyme activity, SLs may enhance plants' antioxidant capacity. Thus, SLs decrease ROS by boosting antioxidant enzyme activity, regulating the AsA–GSH cycle, and increasing antioxidant levels such as anthocyanin.

10.9.2 Molecular regulation of SLs

SLs are a new addition to the plant's classic growth promoters, which include ethylene, CKs, and ABA. Recent research on the characterization of shoot-branching mutants in plants identified SLs as possible phytohormones (Pandey *et al.*, 2016). SLs, like ABA, are apocarotenoids that together operate as stress-triggered phytohormones. ABA is thought to be a universal stress plant hormone that regulates abiotic stressors in addition to SLs (G.-T.E. Chen *et al.*, 2022). SL levels were reduced in tomato plants treated with abamine SG and a 9-*cis*-epoxy carotenoid dioxygenase (NCED) inhibitor, an enzyme that limits the rate of ABA biosynthesis (López *et al.*, 2022). SLs are critical to the *Arabidopsis* drought response because *max3* and *max4* mutants show stress sensitivity by delaying ABA-induced stomatal closure (Bruno *et al.*, 2017; Saeed *et al.*, 2017; Abuauf *et al.*, 2018). SL-depleted *LjCCD7* and *SlCCD7* mutants demonstrated increased stomatal conductance when ABA was administered to different plant species, confirming the function of SLs in stomatal closure (Liu *et al.*, 2015a). Abiotic stress leads mycorrhizal plants to accumulate high levels of SLs and ABA. As stress levels rise, the SL levels in the roots increase, providing resistance to unfavorable conditions and avoiding fungal colonization in the rhizosphere. A negative association between SLs and ABA has been observed in lettuce and tomato plants, where exogenous treatment with GR24 downregulated *LjNCED2* and prevented ABA accumulation (López-Ráez, 2016). Similarly, reducing the levels of SLs enhances ABA accumulation, which aids in coping with abiotic stresses in the absence of mycorrhiza. Auxins are thought to stimulate growth in peas and rice by activating SL-associated biosynthetic genes such as *CCD8* and *CCD7* (López-Ráez, 2016; Banerjee and Roychoudhury, 2018). The synthetic auxin 1-naphthalene acetic acid promotes primary roots and cortical tissues in the root zone of *Arabidopsis d27* mutants by generating positive feedback regulation of SLs, which activates *MAX4* and *MAX3* expression. It has been established that SL reduces shoot growth while promoting root networks by indirectly suppressing shoots through reduced auxin canalization (Rasmussen *et al.*, 2012; Waldie *et al.*, 2014). GR24 foliar spray has been reported to substantially increase root growth in drought conditions by improving drought resistance and water absorption (Xu *et al.*, 2023). Apple seedlings exposed to GR24 revealed enhanced expression of hormone-related genes such as ethylene and indole-3-acetic acid (IAA) (Fan *et al.*, 2022). Shim *et al.* (2018) found that overexpressing *OsNAC14* increased drought tolerance in rice, while Su *et al.* (2020) discovered that VvNAC17, a grapevine NAC transcription factor, modulates the resistance to drought in transgenic *Arabidopsis* (Shim *et al.*, 2018). The use of GR24 in crab apples during drought stress resulted in the upregulation of several ethylene/AP2 response element transcription factors, indicating that SLs perform a function in ethylene signaling, suggesting that SLs regulate complex transcriptional networks by boosting resistance to drought conditions and regulating transcription factors (Xu *et al.*, 2023). The D-group genes *MAX2* and *D14* have been shown to have a significant function in alleviating drought stress by SLs in seedlings of grape (Wang *et al.*, 2009).

10.10 Strigolactone-associated Morphological Regulation

SLs have been linked to a variety of developmental processes, including flower development, root growth, photomorphogenesis, and leaf senescence, as well as their important functions in the architecture of plant shoots. However, studies have shown that SLs hinder the initiation and growth of lateral roots, whereas they enhance senescence, height of the plants, root hairs, and growth of secondary plant bodies. Indeed, comprehensive studies of mutants from many species have indicated that SLs regulate many elements of development and growth in plants (Brewer *et al.*, 2013; Zhang *et al.*, 2018).

10.10.1 Modification of root architecture

SLs have been shown to alter the architecture of roots, specifically by suppressing the formation of the lateral branches and roots. This type of root growth and development pattern allows the roots to reach much deeper into different soil layers, which improves water and nutrient intake. This can benefit plants in stressful situations by reducing the requirements for nutrients and water, lowering the oxidative stress danger in the plant roots. SLs regulate root architecture through interactions with hormones such as CKs and IAA. The application of *rac*-GR24 (racemic mixture of GR24 enantiomers) to the root system of *A. thaliana* lowered the distribution and presence of auxin efflux carriers such as PIN-FORMED 1 (PIN1), PIN3, and PIN7 (Pandya-Kumar *et al.*, 2014; Mujahid *et al.*, 2023). This influences carrier localization and may disrupt IAA transport, which is required for lateral root formation. The same phenomenon is observed during axillary bud development, in which SLs activate endocytosis of PIN proteins, preventing transportation of IAA. Therefore, applications of *rac*-GR24 increase the IAA concentrations within axillary buds, inhibiting further bud development (Pandya-Kumar *et al.*, 2014; Jiang *et al.*, 2015; Marzec *et al.*, 2016). The interactions between IAA and SLs have varying effects on the shoot and root tissues. However, the clear effect of such relationships on subcellular location and modulation of the function of PIN proteins in the different tissues of the root system suggests that SLs are crucial to the regulation of IAA transportation. Both SLs and CKs inhibit elongation of lateral roots, and there has been speculation about a possible link and signaling crosstalk among these classes of hormones (Pandya-Kumar *et al.*, 2014; Marzec and Melzer, 2018; Mujahid *et al.*, 2023). A plausible proposal suggests that SLs may have variable regulatory roles throughout root development based on the surrounding environment, specifically the supply of phosphorus and nitrogen. In an environment with nutrient deficiency, SLs are considered stimulants, triggering increases in the length of lateral roots, although it is expected that their modulation actions will change under a controlled growing environment (Raza *et al.*, 2021; Liu *et al.*, 2023b). It is essential to recognize the flexible nature of SL activities, which are affected by the current milieu of plant development process, emphasizing the importance of developing extensively controlled experimental activities and conditions (Jiang *et al.*, 2016; Marzec and Melzer, 2018). Additionally, monocotyledonous plants with fibrous roots may respond differently to SLs than dicotyledonous plants with tap-root systems. As a result, conducting specific phenotypic assessments of SL mutants under highly monitored conditions in experiments is essential (Raza *et al.*, 2021; Sun *et al.*, 2022). These studies, which aim to elucidate SL-dependent root features, require a variety of growth platforms, including soil-based and hydroponic culture, in order to offer a complete knowledge of SL-associated regulatory processes in the development of plant roots (Marzec and Melzer, 2018; Sun *et al.*, 2022; Liu *et al.*, 2023b).

10.10.2 Modification of shoot architecture

SLs, in conjunction with D14, a signaling receptor, and SMXLs, play an important role in shoot architectural alterations. *Arabidopsis* mutants carrying the four mutations *smxl6*, *−7*, and *−8* and *max2* quadruple enhance PIN1 accumulation and auxin transportation in stems, indicating that SLs are involved in controlling shoot branching through auxin interactions, and they interact with auxins to regulate branching arrangements in shoots (Soundappan *et al.*, 2015; Bennett *et al.*, 2016; Ligerot *et al.*, 2017). In axillary buds, SMXLs inhibit BRANCHED 1 (BRC1) expression by migrating downstream during signaling pathways of SLs in *Arabidopsis* and pea, while BRC1 expression prevents the axillary development of buds (Braun *et al.*, 2012; Soundappan *et al.*, 2015). The metabolite portioning and shoot structure of the yield organ can be impacted by exogenous SL supplementation of seedlings grown in the field. For example, SL foliar treatment increased maize production by 20% in several field studies. When applied to *B. napus*, GR24 significantly increased plant organic matter within a week (Ma *et al.*, 2017). Furthermore, synthetic SL and ABA cooperate in agricultural settings to prolong the shelf life of grapevines by preventing farm products from ripening (Ferrero *et al.*, 2018).

10.10.3 Stomatal regulation

Plants have developed several kinds of coping strategies to deal with different stresses, such as minimizing the loss of water by closing their stomata, which improves tolerance to drought conditions. SLs increase the synthesis of NO and H_2O_2, and in addition trigger the SLOW ANION CHANNEL-ASSOCIATED1 (SLAC1) anion channel, which promotes the closing of stomata and increases plant resilience to environmental stresses.

SLAC1, plasma membrane protein, plays an essential role in the closing of stomata in response to calcium ions, CO_2, ABA, humidity changes, dark/light transitions, ozone, NO, and H_2O_2 (Zhang *et al.*, 2018; Bhoi *et al.*, 2021). To enhance the responses associated with SLs, guard cells need both NO and H_2O_2. Stomatal closure caused by SLs revealed a *SLAC1* mutant, indicating that SLAC1 is required for SL-associated stomatal closure. Thus, destroying the SLAC1 anion channel resulted in SL and ABA insensitivity in stomata, indicating that both pathways of signaling interact at the SLAC1 anion channel by altering the osmotic pressure that causes the stomatal movement. The coordination between SLs and ABA plays a crucial role in the integration of stress signaling and regulation of stomatal growth and function (Ha *et al.*, 2014; Lv *et al.*, 2018). Furthermore, SL biosynthetic (*max3-9*, *max1-1*, and *max4-1*) and signaling (*max2-1* and *d14-5*) *Arabidopsis* mutants have a larger amount of stomatal opening compared with wild-type plants, suggesting that SLs effectively control the closure of stomata (Lv *et al.*, 2018). Recent investigations show that SL mutants of *L. japonicus* have more extensively opened stomata and increased stomatal conductance (Liu *et al.*, 2015a). Currently, researchers have discovered that SLs enhance stomatal mobility as a systemic stress signal element in tomato. Exogenous applications of SLs have also been shown to increase matured miR156 concentrations in tomato leaves, in the absence and presence of severe drought, confirming SLs as the first identified molecular component that induces miR156 and its stress-associated effects in the drought-regulated pathway (Liu *et al.*, 2015b; Visentin *et al.*, 2020).

10.10.4 Leaf senescence

Many abiotic conditions, such as dryness and temperature, can trigger the senescence of leaves in plants. During the early stages of senescence, expression patterns of genes and cellular metabolic activities dramatically change. The breakdown of chloroplasts, which results in the decomposition of chlorophyll, and the continuous denaturation of chloroplast proteins, including chlorophyll *a*/*b*-binding protein and RuBisCO, are the most noticeable changes, as they cause the leaves to turn from green to yellow. Phytohormones such as ethylene, JA, and ABA speed up the senescence process, although CKs also inhibit it. SLs are thought to be a category of plant hormone that regulates the process of leaf senescence because SL-insensitive mutants and SL-deficient plants showed a delay in this process (Yamada *et al.*, 2014; Yamada and Umehara, 2015). In *A. thaliana* SL-deficient mutants *max4*, *max3*, and *max1* and rice (*Oryza sativa*) mutants *d10*, *d17*, and *d27*, GR24 applied exogenously accelerated the rate of leaf senescence. However, SL-insensitive *d14* and *max2* mutants of *A. thaliana* and *d3* and *d14* mutants of *O. sativa* were unaffected. Studies have reported that exogenous treatment with GR24 restored the usual process of senescence in terms of chlorophyll content, leaf color, and discharge of electrolytes in rice mutants that were SL deficient but not SL insensitive. The SL-related mutants showed a prolonged senescence phenotype, suggesting that SLs enhance the process of leaf senescence, whereas GR24 increased senescence (Yamada *et al.*, 2014; Yamada and Umehara, 2015).

10.10.5 Interaction with mycorrhizal fungi

SLs also contribute to the growth of symbiotic associations between AMF and plants. This relationship allows stressed plants to better absorb phosphorus, which may be a limiting factor under stress conditions. SLs indirectly minimize oxidative stress associated with dietary deficiencies by enhancing nutrient absorption (Foo *et al.*, 2013; Marquez-Garcia *et al.*, 2014). Research experiments have revealed that SLs critical for recognition signals from the roots to stimulate growth of hyphae of AMF, which coexist in a beneficial relationship with almost every species of land plant (Xie *et al.*, 2010). SLs possess the ability to influence the expression and regulation of stress-associated genes. They enhance the expression of such genes related to tolerance of various types of stress, particularly drought stress, including those that are involved in ROS detoxification and antioxidant defense. SLs have been related to increased plant resilience to abiotic stressors, including high light, salt, temperature, chilling, drought, and heat, which may all contribute to oxidative stress (Ha *et al.*, 2014; Ma *et al.*, 2017; Ren *et al.*, 2018; Mostofa *et al.*, 2021).

10.11 Interaction with Other Hormones

SLs interact with various other phytohormones, including JA and ABA, which have been associated with stress reactions. These hormone relationships may assist plants in regulating oxidative stress by coordinating many stress-associated processes. SLs and auxin have a reciprocal connection in diverse species, with auxin causing the regulation of SL-associated genes and interacting with developmental activities, the branching of shoots, and auxin transportation. SLs communicate with GAs to regulate branching of shoots and affect GA signaling, namely the DELLA protein (Xu *et al.*, 2015; López-Ráez, 2016; Marzec, 2017; Marzec *et al.*, 2020a; Zhang *et al.*, 2020).

10.11.1 Interactions with ABA

SLs are now recognized as key regulators of drought and salt stress tolerance, principally by facilitating ABA signaling pathways. A case study reported that *L. japonicus* mutants lacking SL production showed less tolerance to drought because of low stomata closing action, which was strongly related to insensitivity to ABA. It is important to emphasize that the effect of SLs on stomatal closure is dependent on the synthesis, sensitivity, and transport of ABA (Xu *et al.*, 2015; Visentin *et al.*, 2020). Under stress conditions, the level of ABA normally rises quickly, although the quantity of SLs varies among different species. ABA has a positive feedback loop, and orchestrates an increase in SL concentrations as well as regulating the critical signaling-related genes, thereby strengthening plant resistance and acclimatization under saline conditions. This behavior was confirmed using *Sesbania cannabina* as an experimental plant (Ren *et al.*, 2018). Furthermore, enhancing the rate of transcription of SL-associated genes from extracts obtained from the root of rice improves plant tolerance to water scarcity (Haider *et al.*, 2018).

10.11.2 Association of CKs with SLs

SL responses, concentrations, and regulation as well as expression of SL-associated genes in response to stresses such as changes in osmotic pressure can vary significantly among plant species. In some cases, both expression of SL genes and SL concentrations linked to SL biosynthesis decreased in *L. japonicus* and tomato (Liu *et al.*, 2015a; Visentin *et al.*, 2016). SLs and CKs show direct interactions throughout plant buds, performing a critical function in the expression of transcriptional activities and gene regulation such as *BRC1* in pea and *Arabidopsis*, while the *FINE CULM 1* (*FC1*) gene in *O. sativa* is strongly associated with *BRC1* (Braun *et al.*, 2012; Xu *et al.*, 2015). *BRC1* acts as an important stimulator of bud formation, affecting the ability to initiate buds in many plants. SLs and CKs have different regulatory effects on plant responses to dry environments. Mutants of *Arabidopsis* that lack CK signaling elements or CKs, in addition to plants with low CK concentrations, have a reduced capacity to adapt to drought conditions. SLs, in contrast, support enhanced resistance against water shortage by influencing the stomatal conductance and density (Nguyen *et al.*, 2016; Zhang *et al.*, 2018).

10.11.3 Auxin crosstalk with SLs

SLs and auxin (IAA) collectively regulate both root growth and shoot branching. In terms of root development, SLs regulate the growth of root hairs and lateral roots by regulating auxin transportation (Sun *et al.*, 2019; Ma *et al.*, 2020). IAA transportation is mainly dependent on IAA efflux proteins, also known as PINs, which generate basic maximum levels of IAA that are critical for starting and expanding roots (Zhang *et al.*, 2020). Basic helix–loop–helix transcription factors such as ROOT HAIR DEFECTIVE SIX-LIKE 4 (RSL4; auxin-associated element), positively activate genes required for the growth of root hairs, particularly under phosphorus deprivation (Zhu *et al.*, 2020). Despite this context, it is clear that RSL4 might serve as a primary coordinator, promoting interactions between IAA and SLs, regulating the elongation of root hairs (Marzol *et al.*, 2017). More research is required for a complete knowledge of how SLs and these kinds of hormones jointly influence the entire functioning and organization of different plant species. It will be important to precisely regulate natural plant hormone concentrations in order to improve the coordinated response to stressful conditions by these numerous hormones in various plant species.

10.11.4 Association of SLs with BRs

The application of pollen or spores from specific plants to different species' stigmas promotes the formation of parthenocarpic fruitlets. Growth-enhancing substances and pollen extracts have been shown to stimulate parthenocarpy. Ethelene synthesized from maize pollen greatly increased the length of internodes of entire bean plants when compared with treated plants or untreated plants with synthetic or natural auxin (Zullo and Bajguz,

2019). *B. napus* and *Alnus glutinosa* have the same effect on immature bean seeds.

BRs were first identified in the 1970s, when the pollen of *B. napus* was found to contain brassins (now known as brassinosteroids), which are plant growth regulators. When given in small amounts to another plant, brassins were considered to be translocatable organic compounds derived from plants that stimulated measurable inhibition of development (Kanwar *et al.*, 2017). Following the rapid publication of the first brassinolide (the most active form of brassins) and similar chemical syntheses, a micro-method for identifying them revealed compounds that resembled brassinolides in a number of species (Peres *et al.*, 2019). The discovery of additional compounds with structures similar to brassinolides, which are defined as '3-oxygenated 5-cholestane 22,23-diols or their derived substances obtained from plants, containing additional alkyl or oxy substituents,' in the following years led to identification of the family of BRs, are considered the sixth class of plant hormones (Brosa, 2020).

From the smallest category of algae to angiosperms, over 100 different plant species have been found to contain more than 60 phytohormone compounds. This suggests that these compounds are widely distributed throughout the plant kingdom. The BR family of plant growth regulators is involved in a number of processes, such as vascular differentiation, cell elongation and division, gene expression modification, and reproductive development (Manghwar *et al.*, 2022). BR-deficient and -insensitive mutants in *A. thaliana* exhibit dwarfism, delayed blooming, short petioles, and decreased reproduction (Matzke *et al.*, 2020). As well as monocots such as barley (*Hordeum vulgare*), rice (*O. sativa*), and maize (*Zea mays*), comparable mutants from several eudicot species have shown similar characteristics, including tomato, pea (*Pisum sativum*), and petunia (*Petunia hybrida*) (Peres *et al.*, 2019). Crosstalk between SLs and BRs has only recently been discovered, and more details about the concept are currently being discussed. Current knowledge describes the signaling pathways controlled by several SL and BR components. MAX2, an E3 ubiquitin ligase, is one SL signaling factor that stops shoot branching in *Arabidopsis*. Through the PEST domain, MAX2 continually interacts with BZR1/BES1 in *Arabidopsis* to prevent their degradation. Exogenous SL treatment inhibited shoot branching by degrading both BR transcription factors regulated by MAX2 (Q.-F. Li *et al.*, 2018; Hussain *et al.*, 2022). Thus, through MAX2, a relationship between SL and BR may affect the stability of BZR1 and BES1 over the course of development. Until recently, the SL signaling pathway has not had much attention. It is expected that future research on this unique class of phytohormone will shed more light on the hormonal interactions between SLs and BRs (Faizan *et al.*, 2020).

10.11.5 Association of SLs with SA

SA plays a role in the aging of leaves and supports the innate defenses of the entire plant. Resistance to *Botrytis cinerea* and *Alternaria* was investigated in tomato plants lacking SLs in order to investigate the role of SA in plant resistance and its association with SLs. It was discovered that the *Slccd8* mutant of tomato with SL deficiency had a lower SA concentration than the untreated group, suggesting a change in homeostasis of the hormone (Torres-Vera *et al.*, 2014; Khan *et al.*, 2015). SL-deficient tomato mutants with inherently inadequate amounts of exogenous SA have been reported to exhibit a resistance mechanism to disease. It was discovered that the SA-associated genes were triggered after infection with pathogenic bacteria in wild-type plants. Applications of GR24 significantly increased resistance against disease in both mutants and wild type, demonstrating that SLs have a triggering impact on SA-associated resistance of diseases (Kusajima *et al.*, 2022). The impact of SA and SLs on drought stress in wheat was studied, and it was observed that application of SA and *rac*-GR24 enhanced proline content, soluble sugar concentrations, and stomatal conductance. This boosts photosynthetic and transpiration rates, enabling the plant to tolerate drought. The combined applications of plant hormones enhanced osmotic balance, which may increase drought tolerance in wheat, revealing an understanding of the relationship between these two plant hormones. The precise process of drought resistance has not yet been reported. SA has the ability to boost wheat chlorophyll content

Table 10.1. Effect of exogenous application of strigolactones in enhancing tolerance against drought stress in different species of plants.

Plant	Effect of treatment with exogenous SLs	Reference(s)
Tall fescue	Treatment with 6-BA suppressed the drought-induced expression of SL signaling genes (*FaMAX2* and *FaD14*)	Rezaei Ghaleh *et al.* (2020)
Maize (seedlings)	Enhanced photosynthetic pigments, antioxidants, gas exchange, and water transport	Sattar *et al.* (2022)
Wheat cultivars	Increased the relative water content, decreased electrolyte leakage, limitation in leaf stomata, and maintained membrane stability index and activities of antioxidant enzymes	Sedaghat *et al.* (2017)
Grapevine	Decreased the MDA content, and improved the relative water content, photosynthesis rate, and chlorophyll content	Min *et al.* (2019)
Dracocephalum kotschyi	Lower electrolyte leakage, H_2O_2, MDA, total antioxidant activity, antioxidant power, and total phenol content; improved dry and fresh weights	Shirani Bidabadi and Sharifi (2021)
Elephant grass	Improved photosynthesis, growth, water-use efficiency, photosynthetic enzyme functions, and root development	Li *et al.* (2022)
Field mustard	Improved photosynthesis parameters, antioxidants activity, and growth; reduced MDA content	Ali *et al.* (2023)
Rice	Gene regulation, maintenance of photosynthesis, closure of stomata, and flowering regulation	Du *et al.* (2018); Haider *et al.* (2018)
Tomato	Increased stomatal conductance, higher membrane stability, higher enzymatic activity, higher soluble sugar and proline content, increased photosynthesis rate, and higher root growth	Visentin *et al.* (2016)

6-BA, 6-Benzylaminopurine; H_2O_2, hydrogen peroxide; MDA, malondialdehyde.

and photosynthetic rate (Sedaghat *et al.*, 2020; Khalvandi *et al.*, 2021). SA protects cell membranes and increases the action of ROS scavengers. SA and SLs exhibit a synergistic impact, and applying both plant hormones at the same time may help plants cope with drought stress more successfully. In the future, similar plant hormones are projected to be used increasingly on various cash crops, medicinal plants, and even horticultural plants (Table 10.1).

10.11.6 Association of SLs with karrikins

SLs and karrikins are closely related due to their chemical structures, which have similar modified enol-ether moieties and lactone butenolide that are essential to the growth process of plants. Instead of being biosynthesized in plants, karrikins are produce by the burning of various plant components such as carbohydrates and cellulose (Antala *et al.*, 2019; Guercio *et al.*, 2023). As a result, it was discovered that karrikins are found in particles of burned plant smoke and are required for the germination of latent seeds. Smoke improves the germination of latent seeds in approximately 80 genera and 1200 species, notably *Arabidopsis*, according to earlier studies (Kountche *et al.*, 2019; Uddin *et al.*, 2023). The primary part of smoke that promotes the germination process is the butanolide 3-methyl-2H-furo(2,3-c)pyran-2-one. Although the karrikin pyran moiety is necessary to promote seed germination, other alterations are acceptable (de Cuyper *et al.*, 2017;

Table 10.2. Various reported strigolactone (SL)-associated genes and their function under drought stress in different species of plants.

Plant	Function of SL-associated genes under drought stress	Reference
Rice (*Oryza sativa*)	Expression of various genes (*D27*, *D17*, *D14*, and *D10*) is reduced	Du *et al.* (2018)
Wheat (*Triticum aestivum*)	Combined application of salicylic acid and GR24 induced peroxidase activity	Sedaghat *et al.* (2017)
Tomato (*Solanum lycopersicum*)	Decreased/increased expression of *CCD7* genes in tomato	Mostofa *et al.* (2018)
Thale cress (*Arabidopsis thaliana*)	After the GR24 application, antioxidant activity and the formation of anthocyanin was enhanced	Wang *et al.* (2019)
Grapevine (*Vitis vinifera*)	Exogenous GR24 application increases the ABA, proline, peroxidase, and MDA content	Min *et al.* (2019)
Tall fescue (*Festuca arundinacea*)	Upregulate the expression of the *FaD17* gene	Zhuang *et al.* (2017)
Alfalfa (*Medicago sativa*)	Exogenous application of GR24 enhanced the proline, ABA, MDA, and peroxidase activity	Wang *et al.* (2022)
Tomato (*Solanum lycopersicum*)	Up- and downregulation of the *CCD8* and *CCD7* genes	Visentin *et al.* (2016)
Grapevine (*Vitis vinifera*)	Exogenous GR24 application increases the ABA, proline, peroxidase, and MDA content	Min *et al.* (2019)

ABA, abscisic acid; *CCD, carotenoid cleavage dioxygenase; D, DWARF*; MDA, malondialdehyde.

Banerjee *et al.*, 2019). To date, smoke from plants has yielded roughly six karrikins or karrikin molecules, named KAR1–KAR6 based on their distinct methyl substitutions. Regardless of their diverse origins, karrikins and SLs share the signaling factor MAX2 (Morffy *et al.*, 2016; Waters, 2017). The proteasome 26S more frequently targets D53/SMAX1-LIKE family members for breakdown by the KARRIKIN INSENSITIVE 2 (KAI2) hydrolases and DWARF14 (D14), which function as karrikin and SL receptors, respectively. SUPPRESSOR OF MAX2 1 (SMAX1)-LIKE/ D53 family members are specifically targeted via the 26S proteasome for cleavage by KAI2 hydrolases and D14, which function as KAR and SL receptors, respectively. Therefore, for karrikins to function in *Arabidopsis* and for SLs to function in rice, a chaperone-type protein is necessary. According to recent research, SLs and/or karrikins also control how plants react to a variety of abiotic stimuli, including nutrient depletion, cold, salt, and drought, as well as how well they can withstand them (Table 10.2) (Yang *et al.*, 2019; Liu *et al.*, 2023a; Wani *et al.*, 2023).

10.11.7 Association of SLs with GAs

Numerous publications have addressed the transcriptional changes occurring as a result of GR24 and GA interactions in overlapping gene sequences. For example, Lantzouni *et al.* (2017) suggested that GAs show an antagonistic effect on SL-activated modification of the ABC class of transporters, although subsequent *in silico* research revealed that SL biosynthesis is under the control of GAs, with experimental proof confirming this discovery via a GID1–DELLA signal mechanism (Ito *et al.*, 2015). GA-deficient rice mutants revealed enhanced expression of *ORYZA SATIVA HOMEOBOX 1* (*OSH1*), suggesting a similar signaling mechanism for SLs and GAs. The exact molecular links between GA and SL signaling pathways have proved difficult to establish. In *O. sativa*, SLR1 combines with D14, a type of DELLA protein that acts as a GA signaling repressor and downstream signal component (Omoarelojie *et al.*, 2019; Sun *et al.*, 2023).

The most recent findings on the molecular relationship between SL function and GA signaling have revealed that these growth-regulating molecules are functionally unconnected from each other in plants. A notable instance is that SL signaling and perception do not impact the stability and concentration of DELLA proteins (Lantzouni *et al.*, 2017; Selwal *et al.*, 2023).

The experimental research has still failed to indicate an influence of SLs on the strong signaling GA components such as GIBBERELLLIN INSENSITIVE DWARF1 and 2 (GID1 and 2), SNEEZY (SNE), GIBBERELLIC ACID INSENSITIVE (GAI), and SLEEPY1 (SLY1). A favorable divergence of SLs and GAs is supported by findings that SLs stimulate internode length without interacting with GAs, resulting in a higher cell number rather than a longer segment. Many studies have found that connections between GAs and SLs occur at a particular level and are enhanced by the interactions of signaling elements. The role of SL level modification has been shown by blocking GAs with SLs, although further study is required to fully understand the natural association between SLs and GAs (Bhoi *et al.*, 2021; Raza *et al.*, 2021; Rehman *et al.*, 2021).

10.11.8 SL crosstalk with ethylene signaling

Ethylene signaling regulates several developmental and growth processes, which are additionally controlled by SLs. These include germination of seeds, elongation of root hairs, hypocotyl growth, and senescence of leaves. In order to control hypocotyl growth, the COP1–HY5 complex integrates light and hormone signals during development. Ethylene increases COP1-mediated HY5 degradation to promote hypocotyl elongation, whereas SLs suppress hypocotyl extension by upregulating the HY5 level of expression in a MAX2-dependent way (Yu *et al.*, 2013; Jia *et al.*, 2014; Omoarelojie *et al.*, 2019). All of these findings point to the possibility that SLs and ethylene signaling interact in an antagonistic way to control the process of growth in the hypocotyl. How SLs counteract ethylene's promoting effects on hypocotyl growth in light is a crucial question because hypocotyl growth is quickly stopped in light. Ethylene signaling is also required for the SL regulation of elongation in the root hairs. This was demonstrated by the fact that the *Atein* and *Atetr* ethylene signaling mutants showed decreased sensitivity to GR24 treatment, and the effect of SL on root-hair elongation was eliminated when ethylene production was blocked. Additionally, GR24 treatment increased the transcription of the ethylene biosynthesis-related gene *AtACS2*. The same study's observations also revealed that SLs utilized ethylene, which may attract IAA signaling. This showed that root-hair development is controlled through the interaction of these three hormones. Moreover, the discovery that a transcriptional complex containing a beneficial modulator of root-hair cells and an ethylene-driven factor stimulates a hair-length-associated gene (*ROOT HAIR DEFECTIVE 6-LIKE*) and other related genes suggests that ethylene acts on its own and may not always be require IAA association to affect root hairs (Feng *et al.*, 2017; Omoarelojie *et al.*, 2019). An early research study reported that SLs promoted ethylene production in *Striga* seeds before germination, indicating an SL–ethylene interaction. Although ethylene-independent routes may be utilized, SLs also accelerated leaf aging by stimulating ethylene-mediated senescence signals. Together, these findings not only revealed that ethylene functions downstream of SLs but also demonstrate that modifying the amount of ethylene in tissues triggers SL-mediated reactions that depend on ethylene signaling. Ethylene's influence on the beginning of adventitious root formation sites in the hypocotyl of *Arabidopsis*, which is unaffected by SLs, gives support to the hypothesis that ethylene may work downstream of the various SL functions (Ueda and Kusaba, 2015; Rasmussen *et al.*, 2017; Omoarelojie *et al.*, 2019), although it is unclear which molecular channels SLs use to activate the transcriptional control of ethylene synthesis. It is known that certain SL responses are induced by phytochrome and cryptochrome. Light is an effective stimulator of ethylene production through a mechanism controlled by phytochrome and phytochrome-associated factors; it is important to explore the potential that SLs regulate the signaling of ethylene via a phytochrome-dependent pattern. As with GAs, evidence from experiments is insufficient for the formulation of a comprehensive and clear model of ethylene–SL association in modulating growing phases of plants. Understanding the function of SL–ethylene signaling and hormonal association in mechanisms including senescence, ripening of fruit, and root development involves assessing how SMXL/D53 activity and MAX2/D14 signaling impact crucial ethylene-responsive transcription factors and ethylene signaling components, as well as vice versa (Jia *et al.*, 2014; Zdarska *et al.*, 2015; Omoarelojie *et al.*, 2019).

10.11.9 SL crosstalk with JA signaling

JAs are a type of plant oxylipin that mediate a range of physiological activities such as secondary metabolism, wound healing, plant–pathogen and plant–insect interactions, and vegetative development. There is currently insufficient experimental data to adequately characterize the nature of SL–JA crosstalk. However, SLs and JA may play a role in disease tolerance, as evidenced by the finding that a tomato mutant (*Slccd8*) with SL deficiency had lower levels of JA and PINII, a JA-dependent gene that confers resistance to tomato against *B. cinerea*. It has been shown that tobacco NtPDR6 (PDR6), a transporter of the SL PhPDR1 transporter gene, is regulated by MeJA, a plant signaling defensive molecule, which lends credence to this connection (Torres-Vera *et al.*, 2014; Xie *et al.*, 2015; Omoarelojie *et al.*, 2019). NtPDR6 shows no association with SL transportation, although it is substantially stimulated by phosphate deficiency and engaged in shoot-branching control. A conflicting finding was recently identified, which revealed that GR24 application had no effect on the accumulation of JA in wild-type *A. thaliana* plants, *Atmax1*, and *Atmax2*, and nor did inoculation with *Mucor* sp. Although these findings are insufficient to establish solid conclusions about the function of JA–SL interactions, the observation that both hormones elicit reactions in analogous developmental pathways, which include senescence, plant–microbe interactions, and mesocotyl elongation, suggests that JA–SL crosstalk is probably going to participate in these processes and is impossible to rule out entirely (Xie *et al.*, 2015; Rozpądek *et al.*, 2018; Omoarelojie *et al.*, 2019).

10.12 Future Perspectives

Studies on the workings of SLs in plants have revealed that SLs are necessary for the developmental and growth phases of crops during abiotic stress. However, more investigation is required to determine the function of SLs in activating cell-signaling pathways in response to diverse abiotic stressors in different crops. Under severe conditions, chloroplasts and mitochondria produce the majority of ROS. In light of their function as a signaling molecule in ROS elimination, SLs have been revealed to have considerable potential as an interorganelle signaling or communications molecule in drought-stress environments. Transgenic crops with SL-related genes are resistant to different types of abiotic stressors. As a result, it is critical to understand how overexpression or ectopic expression of SL-associated genes modifies physiological mechanisms and exacerbates drought stress in various crops. More research is needed into the molecular mechanism of SL gene expression in diverse crops under stress. The functional characterization of SL-associated silence/knockout mutants has been confirmed in a variety of crops under different drought-stress conditions. More studies are required to fully comprehend the function of SLs in boosting crop quality and production under abiotic stress, particularly drought stress.

10.13 Conclusion

SLs are a class of carotenoid-derived phytohormones, and are plant secondary metabolite compounds. They regulate different plant growth and development mechanisms, such as leaf senescence, branching of shoots, root development, seed germination, and root development. SLs may also play an important role in plant resilience to various environmental stresses. In recent years, SLs have gained attention because of their crucial involvement in adaptation to different kinds of stresses, including nutritional shortage, heavy-metal stress, heat, salinity, drought, high temperature, and chilling stress, along with the regulation of several physiological and molecular factors. This chapter has summarized SLs types, their morphological, physiological, biochemical, and molecular regulation, and the association SLs with many other plant hormones, such as auxin and CKs, under drought conditions. The use of SLs improves antioxidant enzymatic activities, gas-exchange features, and chlorophyll pigments, therefore alleviating drought conditions. Therefore, under the scenario of climate change, SLs may be proposed as osmoprotectants to increase the development and growth of plants during drought conditions. This understanding will be important in future research into the pathways of SL regulation and the exogenous application of SLs to deal with detrimental abiotic stresses. As a result, the use of SLs may offer considerable potential for establishing new processes and procedures that are in accordance with the emerging concepts of sustainable agriculture.

References

Abdelaal, K., Alsubeie, M.S., Hafez, Y., Emeran, A., Moghanm, F. *et al.* (2022) Physiological and biochemical changes in vegetable and field crops under drought, salinity and weeds stresses: Control strategies and management. *Agriculture* 12(12), 2084. DOI: 10.3390/agriculture12122084.

Abe, S., Sado, A., Tanaka, K., Kisugi, T., Asami, K. *et al.* (2014) Carlactone is converted to carlactonoic acid by MAX1 in *Arabidopsis* and its methyl ester can directly interact with AtD14 *in vitro*. *Proceedings of the National Academy of Sciences USA* 111(50), 18084–18089. DOI: 10.1073/pnas.1410801111.

Abobatta, W.F. (2020) Plant responses and tolerance to combined salt and drought stress. In: Hasanuzzaman, M. and Tanveer, M. (eds) *Salt and Drought Stress Tolerance in Plants*. Springer Nature, Cham, Switzerland, pp. 17–52. DOI: 10.1007/978-3-030-40277-8_2.

Abuauf, H., Haider, I., Jia, K.-P., Ablazov, A., Mi, J. *et al.* (2018) The *Arabidopsis* DWARF27 gene encodes an all-trans-/9-cis-β-carotene isomerase and is induced by auxin, abscisic acid and phosphate deficiency. *Plant Science* 277, 33–42. DOI: 10.1016/j.plantsci.2018.06.024.

Abu-Shahba, M.S., Mansour, M.M., Mohamed, H.I. and Sofy, M.R. (2021) Comparative cultivation and biochemical analysis of iceberg lettuce grown in sand soil and hydroponics with or without microbubbles and macrobubbles. *Journal of Soil Science and Plant Nutrition* 21(1), 389–403. DOI: 10.1007/s42729-020-00368-x.

Alder, A., Jamil, M., Marzorati, M., Bruno, M., Vermathen, M. *et al.* (2012) The path from β-carotene to carlactone, a strigolactone-like plant hormone. *Science* 335(6074), 1348–1351. DOI: 10.1126/science.1218094.

Ali, A., Shah, T., Haider, G., Awan, M.I., Gohar, M. *et al.* (2023) Strigolactone-mediated oxidative stress alleviation in *Brassica rapa* through upregulating antioxidant system under water deficit conditions. *Journal of Plant Growth Regulation* 42(8), 4675–4687. DOI: 10.1007/s00344-023-10925-0.

An, J.-P., Zhao, L., Cao, Y.-P., Ai, D., Li, M.-Y. *et al.* (2024) The SMXL8–AGL9 module mediates crosstalk between strigolactone and gibberellin to regulate strigolactone-induced anthocyanin biosynthesis in apple. *Plant Cell* 36(10), 4404–4425. DOI: 10.1093/plcell/koae191.

Antala, M., Sytar, O., Rastogi, A. and Brestic, M. (2019) Potential of karrikins as novel plant growth regulators in agriculture. *Plants* 9(1), 43. DOI: 10.3390/plants9010043.

Badr, A. and Brüggemann, W. (2020) Comparative analysis of drought stress response of maize genotypes using chlorophyll fluorescence measurements and leaf relative water content. *Photosynthetica* 58(Special Issue), 638–645. DOI: 10.32615/ps.2020.014.

Banerjee, A. and Roychoudhury, A. (2018) Strigolactones: Multi-level regulation of biosynthesis and diverse responses in plant abiotic stresses. *Acta Physiologiae Plantarum* 40(5), 86. DOI: 10.1007/s11738-018-2660-5.

Banerjee, A., Tripathi, D.K. and Roychoudhury, A. (2019) The karrikin "calisthenics": Can compounds derived from smoke help in stress tolerance? *Physiologia Plantarum* 165(2), 290–302. DOI: 10.1111/ppl.12836.

Bartoli, C.G., Buet, A., Gergoff Grozeff, G., Galatro, A., Simontacchi, M. *et al.* (2017) Ascorbate–glutathione cycle and abiotic stress tolerance in plants. In: Hossain, M.A., Munné-Bosch, S., Burritt, D.J., Diaz-Vivancos, P. and Fujita, M. (eds) *Ascorbic Acid in Plant Growth, Development and Stress Tolerance*. Springer, Cham, Switzerland, pp. 177–200. DOI: 10.1007/978-3-319-74057-7_7.

Bennett, T., Liang, Y., Seale, M., Ward, S., Müller, D. *et al.* (2016) Strigolactone regulates shoot development through a core signalling pathway. *Biology Open* 5(12), 1806–1820. DOI: 10.1242/bio.021402.

Bhatt, D., Nath, M., Sharma, M., Bhatt, M.D., Bisht, D.S. *et al.* (2020) Role of growth regulators and phytohormones in overcoming environmental stress. In: Roychoudhury, A. and Tripathi, D.K. (eds) *Protective Chemical Agents in the Amelioration of Plant Abiotic Stress: Biochemical and Molecular Perspectives*. Wiley, Hoboken, New Jersey, pp. 254–279. DOI: 10.1002/9781119552154.

Bhatt, M.D. and Bhatt, D. (2020) Strigolactones in overcoming environmental stresses. In: Roychoudhury, A. and Tripathi, D.K. (eds) *Protective Chemical Agents in the Amelioration of Plant Abiotic Stress: Biochemical and Molecular Perspectives*. Wiley, Hoboken, New Jersey, pp. 327–341. DOI: 10.1002/9781119552154.

Bhoi, A., Yadu, B., Chandra, J. and Keshavkant, S. (2021) Contribution of strigolactone in plant physiology, hormonal interaction and abiotic stresses. *Planta* 254(2), 28. DOI: 10.1007/s00425-021-03678-1.

Boyer, F.-D., Germain, A., Pouvreau, J.-B., Clavé, G., Pillot, J.-P. *et al.* (2014) New strigolactone analogs as plant hormones with low activities in the rhizosphere. *Molecular Plant* 7(4), 675–690. DOI: 10.1093/mp/sst163.

Braun, N., Germain, A., Pillot, J.-P., Boutet-Mercey, S., Dalmais, M. *et al.* (2012) The pea TCP transcription factor PsBRC1 acts downstream of strigolactones to control shoot branching. *Plant Physiology* 158(1), 225–238. DOI: 10.1104/pp.111.182725.

Brewer, P.B., Koltai, H. and Beveridge, C.A. (2013) Diverse roles of strigolactones in plant development. *Molecular Plant* 6(1), 18–28. DOI: 10.1093/mp/sss130.

Brewer, P.B., Yoneyama, K., Filardo, F., Meyers, E., Scaffidi, A. *et al.* (2016) LATERAL BRANCHING OXIDOREDUCTASE acts in the final stages of strigolactone biosynthesis in *Arabidopsis*. *Proceedings of the National Academy of Sciences USA* 113(22), 6301–6306. DOI: 10.1073/pnas.1601729113.

Brosa, C. (2020) Biological effects of brassinosteroids. In: Parish, E.J. and Nes, W.D. (eds) *Biochemistry and Function of Sterols*. CRC Press, Boca Raton, Florida, pp. 201–220. DOI: 10.1201/9781003068754-15.

Bruno, M., Vermathen, M., Alder, A., Wüst, F., Schaub, P. *et al.* (2017) Insights into the formation of carlactone from in-depth analysis of the CCD8-catalyzed reactions. *FEBS Letters* 591(5), 792–800. DOI: 10.1002/1873-3468.12593.

Bürger, M. and Chory, J. (2020) The many models of strigolactone signaling. *Trends in Plant Science* 25(4), 395–405. DOI: 10.1016/j.tplants.2019.12.009.

Caparas, M., Zobel, Z., Castanho, A.D.A. and Schwalm, C.R. (2021) Increasing risks of crop failure and water scarcity in global breadbaskets by 2030. *Environmental Research Letters* 16(10), 104013. DOI: 10.1088/1748-9326/ac22c1.

Cardinale, F., Krukowski, P.K., Schubert, A. and Visentin, I. (2018) Strigolactones: Mediators of osmotic stress responses with a potential for agrochemical manipulation of crop resilience. *Journal of Experimental Botany* 69(9), 2291–2303. DOI: 10.1093/jxb/erx494.

Carlsson, G.H., Hasse, D., Cardinale, F., Prandi, C. and Andersson, I. (2018) The elusive ligand complexes of the DWARF14 strigolactone receptor. *Journal of Experimental Botany* 69(9), 2345–2354. DOI: 10.1093/jxb/ery036.

Ćavar, S., Zwanenburg, B. and Tarkowski, P. (2015) Strigolactones: Occurrence, structure, and biological activity in the rhizosphere. *Phytochemistry Reviews* 14(4), 691–711. DOI: 10.1007/s11101-014-9370-4.

Charnikhova, T.V., Gaus, K., Lumbroso, A., Sanders, M., Vincken, J.-P. *et al.* (2017) Zealactones. Novel natural strigolactones from maize. *Phytochemistry* 137, 123–131. DOI: 10.1016/j.phytochem.2017.02.010.

Chen, D., Mubeen, B., Hasnain, A., Rizwan, M., Adrees, M. *et al.* (2022) Role of promising secondary metabolites to confer resistance against environmental stresses in crop plants: Current scenario and future perspectives. *Frontiers in Plant Science* 13, 881032. DOI: 10.3389/fpls.2022.881032.

Chen, G.-T.E., Wang, J.Y., Jamil, M., Braguy, J. and Al-Babili, S. (2022) 9-*cis*-β-Apo-10′-carotenal is the precursor of strigolactones in planta. *Planta* 256(5), 88. DOI: 10.1007/s00425-022-03999-9.

Chesterfield, R.J., Vickers, C.E. and Beveridge, C.A. (2020) Translation of strigolactones from plant hormone to agriculture: Achievements, future perspectives, and challenges. *Trends in Plant Science* 25(11), 1087–1106. DOI: 10.1016/j.tplants.2020.06.005.

Choukri, H., Hejjaoui, K., El-Baouchi, A., El Haddad, N., Smouni, A. *et al.* (2020) Heat and drought stress impact on phenology, grain yield, and nutritional quality of lentil (*Lens culinaris* Medikus). *Frontiers in Nutrition* 7, 596307. DOI: 10.3389/fnut.2020.596307.

Cook, C.E., Whichard, L.P., Turner, B., Wall, M.E. and Egley, G.H. (1966) Germination of witchweed (*Striga lutea* Lour.): Isolation and properties of a potent stimulant. *Science* 154(3753), 1189–1190. DOI: 10.1126/science.154.3753.1189.

Crosino, A. and Genre, A. (2022) Peace talks: Symbiotic signaling molecules in arbuscular mycorrhizas and their potential application. *Journal of Plant Interactions* 17(1), 824–839. DOI: 10.1080/17429145.2022.2108150.

de Cuyper, C., Struk, S., Braem, L., Gevaert, K., de Jaeger, G. *et al.* (2017) Strigolactones, karrikins and beyond. *Plant, Cell & Environment* 40(9), 1691–1703. DOI: 10.1111/pce.12996.

Dell'Oste, V., Spyrakis, F. and Prandi, C. (2021) Strigolactones, from plants to human health: Achievements and challenges. *Molecules* 26(15), 4579. DOI: 10.3390/molecules26154579.

Du, H., Huang, F., Wu, N., Li, X., Hu, H. *et al.* (2018) Integrative regulation of drought escape through ABA-dependent and -independent pathways in rice. *Molecular Plant* 11(4), 584–597. DOI: 10.1016/j.molp.2018.01.004.

El-Beltagi, H.S., Mohamed, H.I., Abdelazeem, A.S., Youssef, R. and Safwat, G. (2019) GC-MS analysis, antioxidant, antimicrobial and anticancer activities of extracts from *Ficus sycomorus* fruits and leaves. *Notulae Botanicae Horti Agrobotanici Cluj-Napoca* 47(2), 493–505. DOI: 10.15835/nbha47211405.

El-Sheshtawy, H.S., Mahdy, H.M., Sofy, A.R. and Sofy, M.R. (2022) Production of biosurfactant by *Bacillus megaterium* and its correlation with lipid peroxidation of *Lactuca sativa*. *Egyptian Journal of Petroleum* 31(2), 1–6. DOI: 10.1016/j.ejpe.2022.03.001.

Fahad, S., Bajwa, A.A., Nazir, U., Anjum, S.A., Farooq, A. *et al.* (2017) Crop production under drought and heat stress: Plant responses and management options. *Frontiers in Plant Science* 8, 1147. DOI: 10.3389/fpls.2017.01147.

Faizan, M., Faraz, A., Sami, F., Siddiqui, H., Yusuf, M. *et al.* (2020) Role of strigolactones: Signalling and crosstalk with other phytohormones. *Open Life Sciences* 15(1), 217–228. DOI: 10.1515/biol-2020-0022.

Fan, X., Li, H., Guo, Y., Sun, H., Wang, S. *et al.* (2022) Integrated multi-omics analysis uncovers roles of mdm-miR164b-*MdORE1* in strigolactone-mediated inhibition of adventitious root formation in apple. *Plant, Cell & Environment* 45(12), 3582–3603. DOI: 10.1111/pce.14422.

Feng, Y., Xu, P., Li, B., Li, P., Wen, X. *et al.* (2017) Ethylene promotes root hair growth through coordinated EIN3/EIL1 and RHD6/RSL1 activity in *Arabidopsis*. *Proceedings of the National Academy of Sciences USA* 114(52), 13834–13839. DOI: 10.1073/pnas.1711723115.

Ferrero, M., Pagliarani, C., Novák, O., Ferrandino, A., Cardinale, F. *et al.* (2018) Exogenous strigolactone interacts with abscisic acid-mediated accumulation of anthocyanins in grapevine berries. *Journal of Experimental Botany* 69(9), 2391–2401. DOI: 10.1093/jxb/ery033.

Flematti, G.R., Scaffidi, A., Waters, M.T. and Smith, S.M. (2016) Stereospecificity in strigolactone biosynthesis and perception. *Planta* 243(6), 1361–1373. DOI: 10.1007/s00425-016-2523-5.

Foo, E., Ross, J.J., Jones, W.T. and Reid, J.B. (2013) Plant hormones in arbuscular mycorrhizal symbioses: An emerging role for gibberellins. *Annals of Botany* 111(5), 769–779. DOI: 10.1093/aob/mct041.

Fouda, H. and Sofy, M. (2022) Effect of biological synthesis of nanoparticles from *Penicillium chrysogenum* as well as traditional salt and chemical nanoparticles of zinc on canola plant oil productivity and metabolic activity. *Egyptian Journal of Chemistry* 65(3), 507–516. DOI: 10.21608/ejchem.2021.95120.4469.

Ghanbary, E., Fathizadeh, O., Pazhouhan, I., Zarafshar, M., Tabari, M. *et al.* (2021) Drought and pathogen effects on survival, leaf physiology, oxidative damage, and defense in two middle eastern oak species. *Forests* 12(2), 247. DOI: 10.3390/f12020247.

Ghosh, S., Kamble, N.U., Verma, P., Salvi, P., Petla, B.P. *et al.* (2020) *Arabidopsis* protein l-ISOASPARTYL METHYLTRANSFERASE repairs ISOASPARTYL damage to antioxidant enzymes and increases heat and oxidative stress tolerance. *Journal of Biological Chemistry* 295(3), 783–799. DOI: 10.1016/S0021-9258(17)49935-2.

Gong, J., Wang, R., Liu, B., Zhu, T., Li, H. *et al.* (2024) Regulatory mechanism of strigolactone in tall fescue to low-light stress. *Plant Physiology and Biochemistry* 215, 109054. DOI: 10.1016/j.plaphy.2024.109054.

Goossens, H., Heugebaert, T.S.A., Dereli, B., van Overtveldt, M., Karahan, O. *et al.* (2015) Elucidating the structural isomerism of fluorescent strigolactone analogue CISA-1. *European Journal of Organic Chemistry* 2015(6), 1211–1217. DOI: 10.1002/ejoc.201403457.

Guercio, A.M., Palayam, M. and Shabek, N. (2023) Strigolactones: Diversity, perception, and hydrolysis. *Phytochemistry Reviews* 22(2), 339–360. DOI: 10.1007/s11101-023-09853-4.

Guo, X., Yuan, L. and Chen, X. (2025) Spatial–temporal variations and influencing factors of vegetation net primary productivity: A case study of Yunnan province, China. *Polish Journal of Environmental Studies* 34(3), 2141–2156. DOI: 10.15244/pjoes/188039.

Ha, C.V., Leyva-González, M.A., Osakabe, Y., Tran, U.T., Nishiyama, R. *et al.* (2014) Positive regulatory role of strigolactone in plant responses to drought and salt stress. *Proceedings of the National Academy of Sciences USA* 111(2), 851–856. DOI: 10.1073/pnas.1322135111.

Haider, I., Andreo-Jimenez, B., Bruno, M., Bimbo, A., Floková, K. *et al.* (2018) The interaction of strigolactones with abscisic acid during the drought response in rice. *Journal of Experimental Botany* 69(9), 2403–2414. DOI: 10.1093/jxb/ery089.

Hoque, M.N., Tahjib-Ul-Arif, M., Hannan, A., Sultana, N., Akhter, S. *et al.* (2021) Melatonin modulates plant tolerance to heavy metal stress: Morphological responses to molecular mechanisms. *International Journal of Molecular Sciences* 22(21), 11445. DOI: 10.3390/ijms222111445.

Hu, W., Ren, T., Meng, F., Cong, R., Li, X. *et al.* (2019) Leaf photosynthetic capacity is regulated by the interaction of nitrogen and potassium through coordination of CO_2 diffusion and carboxylation. *Physiologia Plantarum* 167(3), 418–432. DOI: 10.1111/ppl.12919.

Hussain, A., Nazir, F. and Qazi, F. (2022) Brassinosteroids and strigolactone signaling in plants. In: Khan, M.T.A., Yusuf, M., Qazi, F. and Ahmad, A. (eds) *Brassinosteroids Signalling: Intervention with Phytohormones and Their Relationship in Plant Adaptation to Abiotic Stresses*. Springer Nature Sinapore, Singapore, pp. 81–99. DOI: 10.1007/978-981-16-5743-6_5.

Igboeli, E.E., Chukwuka, O., Ochege, F.U., Onyekwelu, C.A., Ling, Q. *et al.* (2025) Combined impacts of land change and climate variability on ecosystem net primary productivity in arid regions. *Global and Planetary Change* 245, 104682. DOI: 10.1016/j.gloplacha.2024.104682.

Iqbal, S., Wang, X., Mubeen, I., Kamran, M., Kanwal, I. *et al.* (2022) Phytohormones trigger drought tolerance in crop plants: Outlook and future perspectives. *Frontiers in Plant Science* 12, 799318. DOI: 10.3389/fpls.2021.799318.

Ito, S., Nozoye, T., Sasaki, E., Imai, M., Shiwa, Y. *et al.* (2015) Strigolactone regulates anthocyanin accumulation, acid phosphatases production and plant growth under low phosphate condition in *Arabidopsis*. *PLOS ONE* 10(3), e0119724. DOI: 10.1371/journal.pone.0119724.

Ito, S., Braguy, J., Wang, J.Y., Yoda, A., Fiorilli, V. *et al.* (2022) Canonical strigolactones are not the major determinant of tillering but important rhizospheric signals in rice. *Science Advances* 8(44), eadd1278. DOI: 10.1126/sciadv.add1278.

Jamil, M., Kountche, B.A., Wang, J.Y., Haider, I., Jia, K.-P. *et al.* (2020) A new series of carlactonoic acid based strigolactone analogs for fundamental and applied research. *Frontiers in Plant Science* 11, 434. DOI: 10.3389/fpls.2020.00434.

Jia, K.-P., Luo, Q., He, S.-B., Lu, X.-D. and Yang, H.-Q. (2014) Strigolactone-regulated hypocotyl elongation is dependent on cryptochrome and phytochrome signaling pathways in *Arabidopsis*. *Molecular Plant* 7(3), 528–540. DOI: 10.1093/mp/sst093.

Jia, K.-P., Kountche, B.A., Jamil, M., Guo, X., Ntui, V.O. *et al.* (2016) Nitro-phenlactone, a carlactone analog with pleiotropic strigolactone activities. *Molecular Plant* 9(9), 1341–1344. DOI: 10.1016/j.molp.2016.05.017.

Jia, K.-P., Baz, L. and Al-Babili, S. (2018) From carotenoids to strigolactones. *Journal of Experimental Botany* 69(9), 2189–2204. DOI: 10.1093/jxb/erx476.

Jiang, D., Fang, J., Lou, L., Zhao, J., Yuan, S. *et al.* (2015) Characterization of a null allelic mutant of the rice NAL1 gene reveals its role in regulating cell division. *PLOS ONE* 10(2), e0118169. DOI: 10.1371/journal.pone.0118169.

Jiang, L., Matthys, C., Marquez-Garcia, B., Cuyper, C., Smet, L. *et al.* (2016) Strigolactones spatially influence lateral root development through the cytokinin signaling network. *Journal of Experimental Botany* 67(1), 379–389. DOI: 10.1093/jxb/erv478.

Kanwar, M.K., Bajguz, A., Zhou, J. and Bhardwaj, R. (2017) Analysis of brassinosteroids in plants. *Journal of Plant Growth Regulation* 36(4), 1002–1030. DOI: 10.1007/s00344-017-9732-4.

Kapoor, D., Bhardwaj, S., Landi, M., Sharma, A., Ramakrishnan, M. *et al.* (2020) The impact of drought in plant metabolism: How to exploit tolerance mechanisms to increase crop production. *Applied Sciences* 10(16), 5692. DOI: 10.3390/app10165692.

Khalvandi, M., Siosemardeh, A., Roohi, E. and Keramati, S. (2021) Salicylic acid alleviated the effect of drought stress on photosynthetic characteristics and leaf protein pattern in winter wheat. *Heliyon* 7(1), e05908. DOI: 10.1016/j.heliyon.2021.e05908.

Khan, M.I.R., Fatma, M., Per, T.S., Anjum, N.A. and Khan, N.A. (2015) Salicylic acid-induced abiotic stress tolerance and underlying mechanisms in plants. *Frontiers in Plant Science* 6, 462. DOI: 10.3389/fpls.2015.00462.

Kim, H.I., Kisugi, T., Khetkam, P., Xie, X., Yoneyama, K. *et al.* (2014) Avenaol, a germination stimulant for root parasitic plants from *Avena strigosa*. *Phytochemistry* 103, 85–88. DOI: 10.1016/j.phytochem.2014.03.030.

Kohlen, W., Charnikhova, T., Liu, Q., Bours, R., Domagalska, M.A. *et al.* (2011) Strigolactones are transported through the xylem and play a key role in shoot architectural response to phosphate deficiency in nonarbuscular mycorrhizal host *Arabidopsis*. *Plant Physiology* 155(2), 974–987. DOI: 10.1104/pp.110.164640.

Korek, M. and Marzec, M. (2023) Strigolactones and abscisic acid interactions affect plant development and response to abiotic stresses. *BMC Plant Biology* 23(1), 314. DOI: 10.1186/s12870-023-04332-6.

Kountche, B.A., Jamil, M., Yonli, D., Nikiema, M.P., Blanco-Ania, D. *et al.* (2019) Suicidal germination as a control strategy for *Striga hermonthica* (Benth.) in smallholder farms of Sub-Saharan Africa. *Plants, People, Planet* 1(2), 107–118. DOI: 10.1002/ppp3.32.

Kumari, V.V., Banerjee, P., Verma, V.C., Sukumaran, S., Chandran, M.A.S. *et al.* (2022) Plant nutrition: An effective way to alleviate abiotic stress in agricultural crops. *International Journal of Molecular Sciences* 23(15), 8519. DOI: 10.3390/ijms23158519.

Kusajima, M., Fujita, M., Soudthedlath, K., Nakamura, H., Yoneyama, K. *et al.* (2022) Strigolactones modulate salicylic acid-mediated disease resistance in *Arabidopsis thaliana*. *International Journal of Molecular Sciences* 23(9), 5246. DOI: 10.3390/ijms23095246.

Lantzouni, O., Klermund, C. and Schwechheimer, C. (2017) Largely additive effects of gibberellin and strigolactone on gene expression in *Arabidopsis thaliana* seedlings. *Plant Journal* 92(5), 924–938. DOI: 10.1111/tpj.13729.

Li, Q.-F., Lu, J., Yu, J.-W., Zhang, C.-Q., He, J.-X. *et al.* (2018) The brassinosteroid-regulated transcription factors BZR1/BES1 function as a coordinator in multisignal-regulated plant growth. *Biochimica et Biophysica Acta—Gene Regulatory Mechanisms* 1861(6), 561–571. DOI: 10.1016/j.bbagrm.2018.04.003.

Li, W., Nguyen, K.H., Tran, C.D., Watanabe, Y., Tian, C. *et al.* (2020) Negative roles of strigolactone-related SMXL6, 7 and 8 proteins in drought resistance in *Arabidopsis*. *Biomolecules* 10(4), 607. DOI: 10.3390/biom10040607.

Li, Y., Li, S., Feng, Q., Zhang, J., Han, X. *et al.* (2022) Effects of exogenous strigolactone on the physiological and ecological characteristics of *Pennisetum purpureum* Schum. seedlings under drought stress. *BMC Plant Biology* 22(1), 578. DOI: 10.1186/s12870-022-03978-y.

Lian, Y., Lian, C., Wang, L., Li, Z., Yuan, G. *et al.* (2023) Suppressor of MAX2 like 6, 7, and 8 interact with DDB1 BINDING WD REPEAT DOMAIN HYPERSENSITIVE TO ABA DEFICIENT 1 to regulate the drought tolerance and target SUCROSE NONFERMENTING 1 RELATED PROTEIN KINASE 2.3 to abscisic acid response in *Arabidopsis*. *Biomolecules* 13(9), 1406. DOI: 10.3390/biom13091406.

Ligerot, Y., Germain, A., Waldie, T., Troadec, C., Citerne, S. *et al.* (2017) The pea branching RMS2 gene encodes the PsAFB4/5 auxin receptor and is involved in an auxin–strigolactone regulation loop. *PLOS Genetics* 13(12), e1007089. DOI: 10.1371/journal.pgen.1007089.

Lin, J., Wang, Y., Sun, S., Mu, C. and Yan, X. (2017) Effects of arbuscular mycorrhizal fungi on the growth, photosynthesis and photosynthetic pigments of *Leymus chinensis* seedlings under salt–alkali stress and nitrogen deposition. *Science of the Total Environment* 576, 234–241. DOI: 10.1016/j.scitotenv.2016.10.091.

Ling, F., Su, Q., Jiang, H., Cui, J., He, X. *et al.* (2020) Effects of strigolactone on photosynthetic and physiological characteristics in salt-stressed rice seedlings. *Scientific Reports* 10(1), 6183. DOI: 10.1038/s41598-020-63352-6.

Liu, J., He, H., Vitali, M., Visentin, I., Charnikhova, T. *et al.* (2015a) Osmotic stress represses strigolactone biosynthesis in *Lotus japonicus* roots: Exploring the interaction between strigolactones and ABA under abiotic stress. *Planta* 241(6), 1435–1451. DOI: 10.1007/s00425-015-2266-8.

Liu, Z., Li, Y., Wang, J., He, X. and Tian, C. (2015b) Different respiration metabolism between mycorrhizal and non-mycorrhizal rice under low-temperature stress: A cry for help from the host. *Journal of Agricultural Science* 153(4), 602–614. DOI: 10.1017/S0021859614000434.

Liu, H., Li, C., Yan, M., Zhao, Z., Huang, P. *et al.* (2022) Strigolactone is involved in nitric oxide-enhanced the salt resistance in tomato seedlings. *Journal of Plant Research* 135(2), 337–350. DOI: 10.1007/s10265-022-01371-2.

Liu, M., Shan, Q., Ding, E., Gu, T. and Gong, B. (2023a) Karrikin increases tomato cold tolerance via strigolactone and the abscisic acid signaling network. *Plant Science* 332, 111720. DOI: 10.1016/j.plantsci.2023.111720.

Liu, X., Xu, Y., Sun, W., Wang, J., Gao, Y. *et al.* (2023b) Strigolactones modulate stem length and diameter of cherry rootstocks through interaction with other hormone signaling pathways. *Frontiers in Plant Science* 14, 1092654. DOI: 10.3389/fpls.2023.1092654.

López, R., Ramírez-Valiente, J.A. and Pita, P. (2022) How plants cope with heatwaves in a drier environment. *Flora* 295, 152148. DOI: 10.1016/j.flora.2022.152148.

López-Ráez, J.A. (2016) How drought and salinity affect arbuscular mycorrhizal symbiosis and strigolactone biosynthesis. *Planta* 243(6), 1375–1385. DOI: 10.1007/s00425-015-2435-9.

Lu, T., Yu, H., Li, Q., Chai, L. and Jiang, W. (2019) Improving plant growth and alleviating photosynthetic inhibition and oxidative stress from low-light stress with exogenous GR24 in tomato (*Solanum lycopersicum* L.) seedlings. *Frontiers in Plant Science* 10, 490. DOI: 10.3389/fpls.2019.00490.

Lv, S., Zhang, Y., Li, C., Liu, Z., Yang, N. *et al.* (2018) Strigolactone-triggered stomatal closure requires hydrogen peroxide synthesis and nitric oxide production in an abscisic acid-independent manner. *New Phytologist* 217(1), 290–304. DOI: 10.1111/nph.14813.

Ma, N., Hu, C., Wan, L., Hu, Q., Xiong, J. *et al.* (2017) Strigolactones improve plant growth, photosynthesis, and alleviate oxidative stress under salinity in rapeseed (*Brassica napus* L.). *Frontiers in Plant Science* 8, 1671. DOI: 10.3389/fpls.2017.01671.

Ma, N., Wan, L., Zhao, W., Liu, H., Li, J. *et al.* (2020) Exogenous strigolactones promote lateral root growth by reducing the endogenous auxin level in rapeseed. *Journal of Integrative Agriculture* 19(2), 465–482. DOI: 10.1016/S2095-3119(19)62810-8.

Ma, C., Bian, C., Liu, W., Sun, Z., Xi, X. *et al.* (2022) Strigolactone alleviates the salinity–alkalinity stress of *Malus hupehensis* seedlings. *Frontiers in Plant Science* 13, 901782. DOI: 10.3389/fpls.2022.901782.

Manghwar, H., Hussain, A., Ali, Q. and Liu, F. (2022) Brassinosteroids (BRs) role in plant development and coping with different stresses. *International Journal of Molecular Sciences* 23(3), 1012. DOI: 10.3390/ijms23031012.

Manna, M., Achary, V.M.M. and Reddy, M.K. (2019) ROS signaling and its role in plants. In: Sopory, S. (ed.) *Sensory Biology of Plants*. Springer Nature, Singapore, pp. 361–388. DOI: 10.1007/978-981-13-8922-1_14.

Marquez-Garcia, B., Njo, M., Beeckman, T., Goormachtig, S. and Foyer, C.H. (2014) A new role for glutathione in the regulation of root architecture linked to strigolactones. *Plant, Cell & Environment* 37(2), 488–498. DOI: 10.1111/pce.12172.

Marzec, M. (2017) Strigolactones and gibberellins: A new couple in the phytohormone world. *Trends in Plant Science* 22(10), 813–815. DOI: 10.1016/j.tplants.2017.08.001.

Marzec, M. and Melzer, M. (2018) Regulation of root development and architecture by strigolactones under optimal and nutrient deficiency conditions. *International Journal of Molecular Sciences* 19(7), 1887. DOI: 10.3390/ijms19071887.

Marzec, M., Gruszka, D., Tylec, P. and Szarejko, I. (2016) Identification and functional analysis of the HvD14 gene involved in strigolactone signaling in *Hordeum vulgare*. *Physiologia Plantarum* 158(3), 341–355. DOI: 10.1111/ppl.12460.

Marzec, M., Situmorang, A., Brewer, P.B. and Brąszewska, A. (2020a) Diverse roles of MAX1 homologues in rice. *Genes* 11(11), 1348. DOI: 10.3390/genes11111348.

Marzec, M., Daszkowska-Golec, A., Collin, A., Melzer, M., Eggert, K. *et al.* (2020b) Barley strigolactone signalling mutant *hvd14.d* reveals the role of strigolactones in abscisic acid-dependent response to drought. *Plant, Cell & Environment* 43(9), 2239–2253. DOI: 10.1111/pce.13815.

Marzol, E., Borassi, C., Denita Juárez, S.P., Mangano, S. and Estevez, J.M. (2017) RSL4 takes control: Multiple signals, one transcription factor. *Trends in Plant Science* 22(7), 553–555. DOI: 10.1016/j.tplants.2017.04.007.

Mashiguchi, K., Seto, Y. and Yamaguchi, S. (2021) Strigolactone biosynthesis, transport and perception. *Plant Journal* 105(2), 335–350. DOI: 10.1111/tpj.15059.

Matzke, C.M., Shore, J.S., Neff, M.M. and McCubbin, A.G. (2020) The Turnera style S-locus gene TsBAHD possesses brassinosteroid-inactivating activity when expressed in *Arabidopsis thaliana*. *Plants* 9(11), 1566. DOI: 10.3390/plants9111566.

Mayzlish-Gati, E., LekKala, S.P., Resnick, N., Wininger, S., Bhattacharya, C. *et al.* (2010) Strigolactones are positive regulators of light-harvesting genes in tomato. *Journal of Experimental Botany* 61(11), 3129–3136. DOI: 10.1093/jxb/erq138.

Mehrabi, S.S., Sabokdast, M., Bihamta, M.R. and Dedičová, B. (2024) The coupling effects of PGPR inoculation and foliar spraying of strigolactone in mitigating the negative effect of salt stress in wheat plants: Insights from phytochemical, growth, and yield attributes. *Agriculture* 14(5), 732. DOI: 10.3390/agriculture14050732.

Miller, G., Suzuki, N., Ciftci-Yilmaz, S. and Mittler, R. (2010) Reactive oxygen species homeostasis and signalling during drought and salinity stresses. *Plant, Cell & Environment* 33(4), 453–467. DOI: 10.1111/j.1365-3040.2009.02041.x.

Min, Z., Li, R., Chen, L., Zhang, Y., Li, Z. *et al.* (2019) Alleviation of drought stress in grapevine by foliar-applied strigolactones. *Plant Physiology and Biochemistry* 135, 99–110. DOI: 10.1016/j.plaphy.2018.11.037.

Mishra, S., Upadhyay, S. and Shukla, R.K. (2017) The role of strigolactones and their potential cross-talk under hostile ecological conditions in plants. *Frontiers in Physiology* 7, 691. DOI: 10.3389/fphys.2016.00691.

Morffy, N., Faure, L. and Nelson, D.C. (2016) Smoke and hormone mirrors: Action and evolution of karrikin and strigolactone signaling. *Trends in Genetics* 32(3), 176–188. DOI: 10.1016/j.tig.2016.01.002.

Mostofa, M.G., Li, W., Nguyen, K.H., Fujita, M. and Tran, L.P. (2018) Strigolactones in plant adaptation to abiotic stresses: An emerging avenue of plant research. *Plant, Cell & Environment* 41(10), 2227–2243. DOI: 10.1111/pce.13364.

Mostofa, M.G., Rahman, M.M., Nguyen, K.H., Li, W., Watanabe, Y. *et al.* (2021) Strigolactones regulate arsenate uptake, vacuolar-sequestration and antioxidant defense responses to resist arsenic toxicity in rice roots. *Journal of Hazardous Materials* 415, 125589. DOI: 10.1016/j.jhazmat.2021.125589.

Moursi, Y.S., Thabet, S.G., Amro, A., Dawood, M.F.A., Baenziger, P.S. *et al.* (2020) Detailed genetic analysis for identifying QTLs associated with drought tolerance at seed germination and seedling stages in barley. *Plants* 9(11), 1425. DOI: 10.3390/plants9111425.

Mujahid, N., Shahbaz, M., Kiran, A. and Wahid, M.A. (2023) Modulations induced by seed priming of strigolactone (GR24) in morpho-physiological and biochemical attributes of ajwain (*Trachyspermum ammi* L.) under salt stress. *Journal of Plant Growth Regulation* 42(5), 2893–2906. DOI: 10.1007/s00344-022-10755-6.

Nguyen, K.H., Ha, C.V., Nishiyama, R., Watanabe, Y., Leyva-González, M.A. *et al.* (2016) Arabidopsis type B cytokinin response regulators ARR1, ARR10, and ARR12 negatively regulate plant responses to drought. *Proceedings of the National Academy of Sciences USA* 113(11), 3090–3095. DOI: 10.1073/pnas.1600399113.

Nisa, Z., Wang, Y., Ali, N., Chen, C., Zhang, X. *et al.* (2023) Strigolactone signaling gene from soybean GmMAX2a enhances the drought and salt-alkaline resistance in Arabidopsis via regulating transcriptional profiles of stress-related genes. *Functional & Integrative Genomics* 23(3), 216. DOI: 10.1007/s10142-023-01151-8.

Noctor, G., Mhamdi, A. and Foyer, C.H. (2014) The roles of reactive oxygen metabolism in drought: Not so cut and dried. *Plant Physiology* 164(4), 1636–1648. DOI: 10.1104/pp.113.233478.

Omoarelojie, L.O., Kulkarni, M.G., Finnie, J.F. and van Staden, J. (2019) Strigolactones and their crosstalk with other phytohormones. *Annals of Botany* 124(5), 749–767. DOI: 10.1093/aob/mcz100.

Omoarelojie, L.O., Kulkarni, M.G., Finnie, J.F. and van Staden, J. (2021) Strigolactone analog (rac-GR24) enhances chilling tolerance in mung bean seedlings. *South African Journal of Botany* 140, 173–181. DOI: 10.1016/j.sajb.2021.03.044.

Pandey, A., Sharma, M. and Pandey, G.K. (2016) Emerging roles of strigolactones in plant responses to stress and development. *Frontiers in Plant Science* 7, 434. DOI: 10.3389/fpls.2016.00434.

Pandya-Kumar, N., Shema, R., Kumar, M., Mayzlish-Gati, E., Levy, D. *et al.* (2014) Strigolactone analog GR24 triggers changes in PIN2 polarity, vesicle trafficking and actin filament architecture. *New Phytologist* 202(4), 1184–1196. DOI: 10.1111/nph.12744.

Peres, A.L.G.L., Soares, J.S., Tavares, R.G., Righetto, G., Zullo, M.A.T. *et al.* (2019) Brassinosteroids, the sixth class of phytohormones: A molecular view from the discovery to hormonal interactions in plant development and stress adaptation. *International Journal of Molecular Sciences* 20(2), 331. DOI: 10.3390/ijms20020331.

Phour, M. and Sindhu, S.S. (2023) Soil salinity and climate change: Microbiome-based strategies for mitigation of salt stress to sustainable agriculture. In: Parray, J.A. (ed.) *Climate Change and Microbiome Dynamics: Carbon Cycle Feedbacks*. Springer Nature, Cham, Switzerland, pp. 191–243. DOI: 10.1007/978-3-031-21079-2_13.

Qi, J., Mao, Y., Cui, J., Lu, X., Xu, J. *et al.* (2024) The role of strigolactones in resistance to environmental stress in plants. *Physiologia Plantarum* 176(4), e14419. DOI: 10.1111/ppl.14419.

Qiu, C.-W., Zhang, C., Wang, N.-H., Mao, W. and Wu, F. (2021) Strigolactone GR24 improves cadmium tolerance by regulating cadmium uptake, nitric oxide signaling and antioxidant metabolism in barley (*Hordeum vulgare* L.). *Environmental Pollution* 273, 116486. DOI: 10.1016/j.envpol.2021.116486.

Rasmussen, A., Mason, M.G., Cuyper, C., Brewer, P.B., Herold, S. *et al.* (2012) Strigolactones suppress adventitious rooting in *Arabidopsis* and pea. *Plant Physiology* 158(4), 1976–1987. DOI: 10.1104/pp.111.187104.

Rasmussen, A., Hu, Y., Depaepe, T., Vandenbussche, F., Boyer, F.-D. *et al.* (2017) Ethylene controls adventitious root initiation sites in *Arabidopsis* hypocotyls independently of strigolactones. *Journal of Plant Growth Regulation* 36(4), 897–911. DOI: 10.1007/s00344-017-9692-8.

Raza, A., Javed, R., Zahid, Z., Sharif, R., Hafeez, M.B. *et al.* (2021) Strigolactones for sustainable plant growth and production under adverse environmental conditions. In: Husen, A. (ed.) *Plant Performance Under Environmental Stress*. Springer Nature, Cham, Switzerland, pp. 129–166. DOI: 10.1007/978-3-030-78521-5_6.

Rehman, N.U., Li, X., Zeng, P., Guo, S., Jan, S. *et al.* (2021) Harmony but not uniformity: Role of strigolactone in plants. *Biomolecules* 11(11), 1616. DOI: 10.3390/biom11111616.

Ren, C.-G., Kong, C.-C. and Xie, Z.-H. (2018) Role of abscisic acid in strigolactone-induced salt stress tolerance in arbuscular mycorrhizal *Sesbania cannabina* seedlings. *BMC Plant Biology* 18(1), 74. DOI: 10.1186/s12870-018-1292-7.

Rezaei Ghaleh, Z., Sarmast, M.K. and Atashi, S. (2020) 6-benzylaminopurine (6-BA) ameliorates drought stress response in tall fescue via the influencing of biochemicals and strigolactone-signaling genes. *Plant Physiology and Biochemistry* 155, 877–887. DOI: 10.1016/j.plaphy.2020.08.009.

Rhaman, M.S., Imran, S., Karim, M.M., Chakrobortty, J., Mahamud, M.A. *et al.* (2021) 5-Aminolevulinic acid-mediated plant adaptive responses to abiotic stress. *Plant Cell Reports* 40(8), 1451–1469. DOI: 10.1007/s00299-021-02690-9.

Rhaman, M.S., Karim, M.M., Sagar, A., Asaduzzaman, M., Ye, W. *et al.* (2024) Phytohormone strigolactone: Involvement in guard cell signaling and abiotic stress tolerance in plants. *Journal of Plant Growth Regulation* 43(12), 4621–4634. DOI: 10.1007/s00344-024-11421-9.

Rodenburg, J., Demont, M., Zwart, S.J. and Bastiaans, L. (2016) Parasitic weed incidence and related economic losses in rice in Africa. *Agriculture, Ecosystems & Environment* 235, 306–317. DOI: 10.1016/j.agee.2016.10.020.

Rozpądek, P., Domka, A.M., Nosek, M., Ważny, R., Jędrzejczyk, R.J. *et al.* (2018) The role of strigolactone in the cross-talk between *Arabidopsis thaliana* and the endophytic Fungus mucor s. *Frontiers in Microbiology* 9, 441. DOI: 10.3389/fmicb.2018.00441.

Saeed, W., Naseem, S. and Ali, Z. (2017) Strigolactones biosynthesis and their role in abiotic stress resilience in plants: A critical review. *Frontiers in Plant Science* 8, 1487. DOI: 10.3389/fpls.2017.01487.

Saini, P., Gani, M., Kaur, J.J., Godara, L.C., Singh, C. *et al.* (2018) Reactive oxygen species (ROS): A way to stress survival in plants. In: Zargar, S.M. and Zargar, M.Y. (eds) *Abiotic Stress-Mediated Sensing and Signaling in Plants: An Omics Perspective*. Springer Singapore, Singapore, pp. 127–153. DOI: 10.1007/978-981-10-7479-0_4.

Salvi, P., Manna, M., Kaur, H., Thakur, T., Gandass, N. *et al.* (2021) Phytohormone signaling and crosstalk in regulating drought stress response in plants. *Plant Cell Reports* 40(8), 1305–1329. DOI: 10.1007/s00299-021-02683-8.

Samynathan, R., Venkidasamy, B., Shariati, M.A., Muthuramalingam, P. and Thiruvengadam, M. (2024) Biosynthesis, functional perspectives, and agricultural applications of strigolactones. *Brazilian Journal of Botany* 47(3), 751–770. DOI: 10.1007/s40415-023-00958-w.

Sarwar, Y. and Shahbaz, M. (2020) Modulation in growth, photosynthetic pigments, gas exchange attributes and inorganic ions in sunflower (*Helianthus annuus* L.) by strigolactones (GR24) achene priming under saline conditions. *Pakistan Journal of Botany* 52(1), 23–31. DOI: 10.30848/PJB2020-1(4).

Sasse, J., Simon, S., Gübeli, C., Liu, G.-W., Cheng, X. *et al.* (2015) Asymmetric localizations of the ABC transporter PaPDR1 trace paths of directional strigolactone transport. *Current Biology* 25(5), 647–655. DOI: 10.1016/j.cub.2015.01.015.

Sattar, A., Ul-Allah, S., Ijaz, M., Sher, A., Butt, M. *et al.* (2022) Exogenous application of strigolactone alleviates drought stress in maize seedlings by regulating the physiological and antioxidants defense mechanisms. *Cereal Research Communications* 50(2), 263–272. DOI: 10.1007/s42976-021-00171-z.

Saxena, S.C., Salvi, P., Kamble, N.U., Joshi, P.K., Majee, M. *et al.* (2020) Ectopic overexpression of cytosolic ascorbate peroxidase gene (Apx1) improves salinity stress tolerance in *Brassica juncea* by strengthening antioxidative defense mechanism. *Acta Physiologiae Plantarum* 42(4), 45. DOI: 10.1007/s11738-020-3032-5.

Sedaghat, M., Tahmasebi-Sarvestani, Z., Emam, Y. and Mokhtassi-Bidgoli, A. (2017) Physiological and antioxidant responses of winter wheat cultivars to strigolactone and salicylic acid in drought. *Plant Physiology and Biochemistry* 119, 59–69. DOI: 10.1016/j.plaphy.2017.08.015.

Sedaghat, M., Sarvestani, Z.T., Emam, Y., Bidgoli, A.M. and Sorooshzadeh, A. (2020) Foliar-applied GR24 and salicylic acid enhanced wheat drought tolerance. *Russian Journal of Plant Physiology* 67(4), 733–739. DOI: 10.1134/S1021443720040159.

Sedaghat, M., Emam, Y., Mokhtassi-Bidgoli, A., Hazrati, S., Lovisolo, C. *et al.* (2021) The potential of the synthetic strigolactone analogue GR24 for the maintenance of photosynthesis and yield in winter wheat under drought: Investigations on the mechanisms of action and delivery modes. *Plants* 10(6), 1223. DOI: 10.3390/plants10061223.

Selwal, N., Wani, A.K., Akhtar, N., Kaur, M. and Jassal, P.S. (2023) Molecular insights of strigolactone biosynthesis, signalling pathways, regulatory roles, and hormonal crosstalks in plant systems. *South African Journal of Botany* 160, 9–22. DOI: 10.1016/j.sajb.2023.06.046.

Seto, Y., Sado, A., Asami, K., Hanada, A., Umehara, M. *et al.* (2014) Carlactone is an endogenous biosynthetic precursor for strigolactones. *Proceedings of the National Academy of Sciences USA* 111(4), 1640–1645. DOI: 10.1073/pnas.1314805111.

Shakespear, S., Sivaji, M., Kumar, V., Arumugam Pillai, M., Wani, S.H. *et al.* (2025) Navigating through harsh conditions: Coordinated networks of plant adaptation to abiotic stress. *Journal of Plant Growth Regulation* 44(4), 1396–1414. DOI: 10.1007/s00344-023-11224-4.

Sharma, A., Kumar, V., Shahzad, B., Ramakrishnan, M., Singh Sidhu, G.P. *et al.* (2020) Photosynthetic response of plants under different abiotic stresses: A review. *Journal of Plant Growth Regulation* 39(2), 509–531. DOI: 10.1007/s00344-019-10018-x.

Shim, J.S., Oh, N., Chung, P.J., Kim, Y.S., Choi, Y.D. *et al.* (2018) Overexpression of OsNAC14 improves drought tolerance in rice. *Frontiers in Plant Science* 9, 310. DOI: 10.3389/fpls.2018.00310.

Shirani Bidabadi, S. and Sharifi, P. (2021) Strigolactone and methyl jasmonate-induced antioxidant defense and the composition alterations of different active compounds in dracocephalum kotschyi boiss under drought stress. *Journal of Plant Growth Regulation* 40(2), 878–889. DOI: 10.1007/s00344-020-10157-6.

Soares, C., Carvalho, M.E.A., Azevedo, R.A. and Fidalgo, F. (2019) Plants facing oxidative challenges—a little help from the antioxidant networks. *Environmental and Experimental Botany* 161, 4–25. DOI: 10.1016/j.envexpbot.2018.12.009.

Soliman, S., Wang, Y., Han, Z., Pervaiz, T. and El-Kereamy, A. (2022) Strigolactones in plants and their interaction with the ecological microbiome in response to abiotic stress. *Plants* 11(24), 3499. DOI: 10.3390/plants11243499.

Soundappan, I., Bennett, T., Morffy, N., Liang, Y., Stanga, J.P. *et al.* (2015) SMAX1-LIKE/D53 family members enable distinct MAX2-dependent responses to strigolactones and karrikins in *Arabidopsis*. *Plant Cell* 27(11), 3143–3159. DOI: 10.1105/tpc.15.00562.

Su, L., Fang, L., Zhu, Z., Zhang, L., Sun, X. *et al.* (2020) The transcription factor VaNAC17 from grapevine (*Vitis amurensis*) enhances drought tolerance by modulating jasmonic acid biosynthesis in transgenic *Arabidopsis*. *Plant Cell Reports* 39, 621–634. DOI: 10.1007/s00299-020-02519-x.

Sun, H., Xu, F., Guo, X., Wu, D., Zhang, X. *et al.* (2019) A strigolactone signal inhibits secondary lateral root development in rice. *Frontiers in Plant Science* 10, 1527. DOI: 10.3389/fpls.2019.01527.

Sun, H., Li, W., Burritt, D.J., Tian, H., Zhang, H. *et al.* (2022) Strigolactones interact with other phytohormones to modulate plant root growth and development. *Crop Journal* 10(6), 1517–1527. DOI: 10.1016/j.cj.2022.07.014.

Sun, H., Guo, X., Zhu, X., Gu, P., Zhang, W. *et al.* (2023) Strigolactone and gibberellin signaling coordinately regulate metabolic adaptations to changes in nitrogen availability in rice. *Molecular Plant* 16(3), 588–598. DOI: 10.1016/j.molp.2023.01.009.

Tai, Z., Yin, X., Fang, Z., Shi, G., Lou, L. *et al.* (2017) Exogenous GR24 alleviates cadmium toxicity by reducing cadmium uptake in switchgrass (*Panicum virgatum*) seedlings. *International Journal of Environmental Research and Public Health* 14(8), 852. DOI: 10.3390/ijerph14080852.

Tippani, R., Porika, M., Sura, S.P., Kagithoju, P. and Porika, S.S.S. (2025) Green nanoparticles used in agriculture to reduce ROS and increase plant resilience against abiotic stress. In: Pabbati, R.,

Chepuri, K., Reddy, K.V. and Maddela, N.R. (eds) *Plant-Based Nanoparticle Synthesis for Sustainable Agriculture*. CRC Press, Boca Raton, Florida. DOI: 10.1201/9781003477730-8.

Tiwari, S., Lata, C., Chauhan, P.S., Prasad, V. and Prasad, M. (2017) A functional genomic perspective on drought signalling and its crosstalk with phytohormone-mediated signalling pathways in plants. *Current Genomics* 18(6), 469–482. DOI: 10.2174/1389202918666170605083319.

Torres-Vera, R., García, J.M., Pozo, M.J. and López-Ráez, J.A. (2014) Do strigolactones contribute to plant defence? *Molecular Plant Pathology* 15(2), 211–216. DOI: 10.1111/mpp.12074.

Tsukano, C., Yasui, M. and Takemoto, Y. (2024) Total synthesis of avenaol. In: Nakada, M., Tanino, K., Nagasawa, K. and Yokoshima, S. (eds) *Modern Natural Product Synthesis: Overcoming Difficulties*. Springer Nature Singapore, Singapore, pp. 381–411. DOI: 10.1007/978-981-97-1619-7_18.

Turner, N.C. (2018) Turgor maintenance by osmotic adjustment: 40 years of progress. *Journal of Experimental Botany* 69(13), 3223–3233. DOI: 10.1093/jxb/ery181.

Uddin, M., Singh, S., Chishti, A.S., Singh, S., Bhat, U.H. *et al.* (2023) Rhizomes priming with smoke–water and karrikinolide enhance the production of essential oil and bioactive constituents of *Mentha arvensis* L. via improving the growth and physiological attributes of plant. *Journal of Plant Growth Regulation* 42(11), 7023–7037. DOI: 10.1007/s00344-023-10994-1.

Ueda, H. and Kusaba, M. (2015) Strigolactone regulates leaf senescence in concert with ethylene in *Arabidopsis*. *Plant Physiology* 169(1), 138–147. DOI: 10.1104/pp.15.00325.

Ueno, K., Furumoto, T., Umeda, S., Mizutani, M., Takikawa, H. *et al.* (2014) Heliolactone, a non-sesquiterpene lactone germination stimulant for root parasitic weeds from sunflower. *Phytochemistry* 108, 122–128. DOI: 10.1016/j.phytochem.2014.09.018.

Varadharajan, V., Rajendran, R., Muthuramalingam, P., Runthala, A., Madhesh, V. *et al.* (2025) Multi-omics approaches against abiotic and biotic stress—A review. *Plants* 14(6), 865. DOI: 10.3390/plants14060865.

Visentin, I., Vitali, M., Ferrero, M., Zhang, Y., Ruyter-Spira, C. *et al.* (2016) Low levels of strigolactones in roots as a component of the systemic signal of drought stress in tomato. *New Phytologist* 212(4), 954–963. DOI: 10.1111/nph.14190.

Visentin, I., Pagliarani, C., Deva, E., Caracci, A., Turečková, V. *et al.* (2020) A novel strigolactone–miR156 module controls stomatal behaviour during drought recovery. *Plant, Cell & Environment* 43(7), 1613–1624. DOI: 10.1111/pce.13758.

Waadt, R., Seller, C.A., Hsu, P.-K., Takahashi, Y., Munemasa, S. *et al.* (2022) Plant hormone regulation of abiotic stress responses. *Nature Reviews Molecular Cell Biology* 23(10), 680–694. DOI: 10.1038/s41580-022-00479-6.

Wakabayashi, T., Shida, K., Kitano, Y., Takikawa, H., Mizutani, M. *et al.* (2020) CYP722C from *Gossypium arboreum* catalyzes the conversion of carlactonoic acid to 5-deoxystrigol. *Planta* 251(5), 97. DOI: 10.1007/s00425-020-03390-6.

Wakabayashi, T., Ishiwa, S., Shida, K., Motonami, N., Suzuki, H. *et al.* (2021) Identification and characterization of sorgomol synthase in sorghum strigolactone biosynthesis. *Plant Physiology* 185(3), 902–913. DOI: 10.1093/plphys/kiaa113.

Waldie, T., McCulloch, H. and Leyser, O. (2014) Strigolactones and the control of plant development: Lessons from shoot branching. *Plant Journal* 79(4), 607–622. DOI: 10.1111/tpj.12488.

Wang, Q., Ni, J., Shah, F., Liu, W., Wang, D. *et al.* (2019) Overexpression of the stress-inducible SsMAX2 promotes drought and salt resistance via the regulation of redox homeostasis in *Arabidopsis*. *International Journal of Molecular Sciences* 20(4), 837. DOI: 10.3390/ijms20040837.

Wang, X., Kang, W., Wu, F., Miao, J. and Shi, S. (2022) Comparative transcriptome analysis reveals new insight of alfalfa (*Medicago sativa* L.) cultivars in response to abrupt freezing stress. *Frontiers in Plant Science* 13, 798118. DOI: 10.3389/fpls.2022.798118.

Wang, Z., Gerstein, M. and Snyder, M. (2009) RNA-seq: A revolutionary tool for transcriptomics. *Nature Reviews Genetics* 10(1), 57–63. DOI: 10.1038/nrg2484.

Wani, A.K., Akhtar, N., Prakash, A., Garg, S.S., Mir, T. *et al.* (2023) Karrikins: Molecular association with strigolactones: Role in plant growth regulation and quorum sensing modulation. In: Mukherjee, S. and Aftab, T. (eds) *Strigolactones, Alkamides and Karrikins in Plants*. CRC Press, Boca Raton, Florida. DOI: 10.1201/9781003224525-19.

Waters, M.T. (2017) From little things big things grow: Karrikins and new directions in plant development. *Functional Plant Biology* 44(4), 373–385. DOI: 10.1071/FP16405.

Waters, M.T., Gutjahr, C., Bennett, T. and Nelson, D.C. (2017) Strigolactone signaling and evolution. *Annual Review of Plant Biology* 68, 291–322. DOI: 10.1146/annurev-arplant-042916-040925.

Xie, X., Yoneyama, K. and Yoneyama, K. (2010) The strigolactone story. *Annual Review of Phytopathology* 48(1), 93–117. DOI: 10.1146/annurev-phyto-073009-114453.

Xie, X., Yoneyama, K., Kisugi, T., Uchida, K., Ito, S. *et al.* (2013) Confirming stereochemical structures of strigolactones produced by rice and tobacco. *Molecular Plant* 6(1), 153–163. DOI: 10.1093/mp/sss139.

Xie, X., Wang, G., Yang, L., Cheng, T., Gao, J. *et al.* (2015) Cloning and characterization of a novel *Nicotiana tabacum* ABC transporter involved in shoot branching. *Physiologia Plantarum* 153(2), 299–306. DOI: 10.1111/ppl.12267.

Xie, X., Kisugi, T., Yoneyama, K., Nomura, T., Akiyama, K. *et al.* (2017) Methyl zealactonoate, a novel germination stimulant for root parasitic weeds produced by maize. *Journal of Pesticide Science* 42(2), 58–61. DOI: 10.1584/jpestics.D16-103.

Xu, J., Zha, M., Li, Y., Ding, Y., Chen, L. *et al.* (2015) The interaction between nitrogen availability and auxin, cytokinin, and strigolactone in the control of shoot branching in rice (*Oryza sativa* L.). *Plant Cell Reports* 34(9), 1647–1662. DOI: 10.1007/s00299-015-1815-8.

Xu, J., Li, L., Liu, Y., Yu, Y., Li, H. *et al.* (2023) Molecular and physiological mechanisms of strigolactones-mediated drought stress in crab apple (*Malus hupehensis* Rehd.) seedlings. *Scientia Horticulturae* 311, 111800. DOI: 10.1016/j.scienta.2022.111800.

Yamada, Y. and Umehara, M. (2015) Possible roles of strigolactones during leaf senescence. *Plants* 4(3), 664–677. DOI: 10.3390/plants4030664.

Yamada, Y., Furusawa, S., Nagasaka, S., Shimomura, K., Yamaguchi, S. *et al.* (2014) Strigolactone signaling regulates rice leaf senescence in response to a phosphate deficiency. *Planta* 240(2), 399–408. DOI: 10.1007/s00425-014-2096-0.

Yang, T., Lian, Y. and Wang, C. (2019) Comparing and contrasting the multiple roles of butenolide plant growth regulators: Strigolactones and karrikins in plant development and adaptation to abiotic stresses. *International Journal of Molecular Sciences* 20(24), 6270. DOI: 10.3390/ijms20246270.

Yang, Y., Gu, M., Chen, J., Zhang, R., Liu, Z. *et al.* (2023) Comparative transcriptomes reveal the mitigation effect of GR24 in alfalfa under drought stress. *Journal of Plant Growth Regulation* 42(5), 3150–3161. DOI: 10.1007/s00344-022-10779-y.

Yoneyama, K. and Brewer, P.B. (2021) Strigolactones, how are they synthesized to regulate plant growth and development? *Current Opinion in Plant Biology* 63, 102072. DOI: 10.1016/j.pbi.2021.102072.

Yoneyama, K., Mori, N., Sato, T., Yoda, A., Xie, X. *et al.* (2018) Conversion of carlactone to carlactonoic acid is a conserved function of MAX1 homologs in strigolactone biosynthesis. *New Phytologist* 218(4), 1522–1533. DOI: 10.1111/nph.15055.

Yoneyama, K., Akiyama, K., Brewer, P.B., Mori, N., Kawano-Kawada, M. *et al.* (2020) Hydroxyl carlactone derivatives are predominant strigolactones in *Arabidopsis*. *Plant Direct* 4(5), e00219. DOI: 10.1002/pld3.219.

Yu, Y., Wang, J., Zhang, Z., Quan, R., Zhang, H. *et al.* (2013) Ethylene promotes hypocotyl growth and HY5 degradation by enhancing the movement of COP1 to the nucleus in the light. *PLOS Genetics* 9(12), e1004025. DOI: 10.1371/journal.pgen.1004025.

Zdarska, M., Dobisová, T., Gelová, Z., Pernisová, M., Dabravolski, S. *et al.* (2015) Illuminating light, cytokinin, and ethylene signalling crosstalk in plant development. *Journal of Experimental Botany* 66(16), 4913–4931. DOI: 10.1093/jxb/erv261.

Zhang, J., Hou, J., Zhang, H., Meng, C., Zhang, X. *et al.* (2019) Low soil temperature inhibits yield of rice under drip irrigation. *Journal of Soil Science and Plant Nutrition* 19(1), 228–236. DOI: 10.1007/s42729-019-0008-x.

Zhang, J., Mazur, E., Balla, J., Gallei, M., Kalousek, P. *et al.* (2020) Strigolactones inhibit auxin feedback on PIN-dependent auxin transport canalization. *Nature Communications* 11(1), 3508. DOI: 10.1038/s41467-020-17252-y.

Zhang, X.-H., Ma, C., Zhang, L., Su, M., Wang, J. *et al.* (2022) GR24-mediated enhancement of salt tolerance and roles of H2O2 and Ca2+ in regulating this enhancement in cucumber. *Journal of Plant Physiology* 270, 153640. DOI: 10.1016/j.jplph.2022.153640.

Zhang, Y., Lv, S. and Wang, G. (2018) Strigolactones are common regulators in induction of stomatal closure in planta. *Plant Signaling & Behavior* 13(3), e1444322. DOI: 10.1080/15592324.2018.1444322.

Zhang, Y., Hartinger, C., Wang, X. and Friml, J. (2020) Directional auxin fluxes in plants by intramolecular domain–domain coevolution of PIN auxin transporters. *New Phytologist* 227(5), 1406–1416. DOI: 10.1111/nph.16629.

Zhou, H., Shi, H., Yang, Y., Feng, X., Chen, X. *et al.* (2024) Insights into plant salt stress signaling and tolerance. *Journal of Genetics and Genomics* 51(1), 16–34. DOI: 10.1016/j.jgg.2023.08.007.

Zhu, S., Martínez Pacheco, J., Estevez, J.M. and Yu, F. (2020) Autocrine regulation of root hair size by the RALF–FERONIA–RSL4 signaling pathway. *New Phytologist* 227(1), 45–49. DOI: 10.1111/nph.16497.

Zhuang, L., Wang, J. and Huang, B. (2017) Drought inhibition of tillering in *Festuca arundinacea* associated with axillary bud development and strigolactone signaling. *Environmental and Experimental Botany* 142, 15–23. DOI: 10.1016/j.envexpbot.2017.07.017.

Zorrilla, J.G., Rial, C., Varela, R.M., Molinillo, J.M.G. and Macías, F.A. (2022) Strategies for the synthesis of canonical, non-canonical and analogues of strigolactones, and evaluation of their parasitic weed germination activity. *Phytochemistry Reviews* 21(5), 1627–1659. DOI: 10.1007/s11101-022-09801-8.

Zullo, M.A.T. and Bajguz, A. (2019) The brassinosteroids family—structural diversity of natural compounds and their precursors. In: Hayat, S., Yusuf, M., Bhardwaj, R. and Bajguz, A. (eds) *Brassinosteroids: Plant Growth and Development*. Springer Singapore, Singapore, pp. 1–44. DOI: 10.1007/978-981-13-6058-9_1.

Zwanenburg, B., Pospíšil, T. and Ćavar Zeljković, S. (2016) Strigolactones: New plant hormones in action. *Planta* 243(6), 1311–1326. DOI: 10.1007/s00425-015-2455-5.

11 Strigolactones: A Key Regulator of Cold Stress in Plants

Usman Tufail[1], Ansa Asghar[1*], Rizwana Kanwal[1], Ahmed Raza[1], Asima Asghar[2], Amina Ameer[3], Anum Javaid[1], Ummar Iqbal[4], Sana Maqsood[1] and Sana Basharat[5*]

[1]Department of Botany, University of Agriculture Faisalabad, Faisalabad, Pakistan; [2]Department of Zoology, Lahore College for Women University, Lahore, Pakistan; [3]College of Geographical Sciences, Faculty of Geographical Science and Engineering, Henan University, Zhengzhou, China; [4]College of Biological Sciences, China Agricultural University Beijing, China; [5]School of Breeding and Multiplication (Sanya Institute of Breeding and Multiplication), School of Tropical Agriculture and Forestry, Hainan University, Sanya, China

Abstract

Strigolactones (SLs) are plant-derived signaling molecules that are derivatives of carotenoids. The physiological and adaptative functions of many plants are regulated by SLs. They are responsible for several facets of stress reduction, particularly with regard to cold stress. Low-temperature-grown plants have increased SL production, which regulates plant development and encourages regulation of genes that react to stress, increasing plants' ability to withstand abiotic challenges. SLs take part in control mechanisms of plant stress adaptation, as evidenced by their interaction with other plant hormones as a result of environmental stresses, including abscisic acid (ABA), salicylic acid, and jasmonic acid. These phytohormone are produced in plant tissues in response to a number of environmental stresses, including freezing stress. SLs help plants overcome harsh environmental restrictions and lessen the negative effects on crop output by regulating a number of hormone-sensitive pathways. Because of its higher bioaccumulation, exogenous administration of the SL analog GR24 is a possible way to help plants establish various environmental responses. With a focus on their interactions with ABA, gibberellins, auxins, and microRNAs, this chapter examines the growing role of SLs as key regulators in cold-stress adaptation. It also highlights potential future applications of SL-mediated signaling in crop-improvement strategies for cold-prone environments.

Keywords: carotenoid, *CBF* genes, cold signaling, crop improvement, phytohormones, SLs

11.1 Introduction

One important environmental stress is chilling stress. It restricts vegetative growth and development and determines species' geographical spread (Bhattacharya, 2022). Plants have developed complex systems to detect and respond to freezing stress. The most thoroughly researched fundamental elements of cold acclimation in plants are transcription factors called C-REPEAT

*Corresponding authors: ansaasghar1122@gmail.com and sanabasharat143@gmail.com

DOI: 10.1079/9781836990598.0011

BINDING (CRT) FACTOR/DEHYDRATION RESPONSE ELEMENT (DRE) BINDING FACTOR 1 (CBF/DREB1). Chilling stress quickly activates the production of *CBF* genes, and the CBF proteins they encode attach to the conserved *CRT/DRE* motif CCGAC in the promoters of CBF-targeted *COLD-RESPONSIVE* (*COR*) genes (Song *et al.*, 2021b). Increased activity of downstream *COR* genes results in improved resistance to freezing stress in a variety of plant species, including *Arabidopsis thaliana*, *Brassica napus*, *Secale cereale*, *Triticum aestivum*, *Zea mays*, and *Solanum lycopersicum*. Therefore, the CBF–COR pathway is essential for plants to acclimate to cold (Białoskórska *et al.*, 2024).

Plant hormones also control how plants grow, develop, and react to stress (Waadt *et al.*, 2022). Research has shown that cold signaling and phytohormones interact. A crucial factor in the ethylene signaling cascade, ETHYLENE INSENSITIVE 3 (EIN3), suppresses the expression of *CBF* genes and adversely affects the ability of *Arabidopsis* to withstand cold temperatures (Shi *et al.*, 2012). The transcriptional activator of *CBF* genes, INDUCER OF CBF EXPRESSION 1 (ICE1), is directly impacted by and inhibited by JASMONATE ZIM-DOMAIN (JAZ) proteins, which are inhibitors of the jasmonic acid (JA) pathway. Plant freezing tolerance is increased when JA builds up during cold weather, which causes JAZ breakdown and increases CBF expression. It has also been demonstrated that a number of essential elements of brassinosteroid signaling alter plant cold responses. By directly binding to their promoters, transcription factors BRASSINAZOLE-RESISTANT1 (BZR1, also called as BRI1-EMS-SUPPRESSOR1 (BES1)) and CESTA, which are implicated in the brassinosteroid signaling cascade, facilitate the expression of *CBF* genes in freezing stress (Cao *et al.*, 2024). Furthermore, in order to modulate CBF regulation as a result of low-temperature stress, the kinase BRASSINOSTEROID-INSENSITIVE2 (BIN2) phosphorylates ICE1 and adversely regulates its stability. In plants under cold stress, auxin plays a crucial role in regulating DNA damage-triggered selective cell death. Because cold stress damages DNA in root stem cells and their progeny, auxin maxima are restored in the quiescent center to stop columella stem cells from dividing further. This "sacrifice-for-survival" mechanism aids plants in strengthening their roots' resistance to and capacity to adjust to cold conditions (Ding and Yang, 2022).

Strigolactones (SLs), which are phytohormones generated from carotenoid, were initially identified as strong germination promoters of the parasitic plant *Striga* (Jain *et al.*, 2025). SLs control root architecture and shoot branching. *MAX1*, *MAX2*, *MAX3*, and *MAX4* are members of the *MORE AXILLARY GROWTH* (*MAX*) family of genes involved in SL biosynthesis, perception, and signaling in *Arabidopsis* (Waters *et al.*, 2017). *Carotenoid cleavage dioxygenase 7* (*CCD7*; also known as *MAX3*) and *CCD8* (also known as *MAX4*) play crucial roles in the SL biosynthesis pathway. A cytochrome P450 mono-oxygenase that expressed downstream of *MAX3* and *MAX4* is encoded by *MAX1*. As a non-canonical phytohormone receptor, the α/β-hydrolase DWARF14 (D14) interacts with the F-box protein MAX2 in *Arabidopsis* (D3 in *Oryza sativa*) and produces and detects the active form of SL (Yao *et al.*, 2018). To initiate SL responses, MAX2 then recruits and targets the SUPPRESSOR OF MAX2-LIKE (SMXL; D53 in rice) for degradation (Lian *et al.*, 2023). The triple mutant *smxl6 smxl7 smxl8* inhibits the shoot branching of *max2* and *max3* in *Arabidopsis*, while GR24 causes the degradation of SMXL6, −7, and −8 in a D14- and MAX2-dependent manner. Plant reactions to abiotic stressors such drought, salt, cold, and heat are influenced by SLs. Nevertheless, the molecular mechanism by which SLs regulate plant cold tolerance is still unknown (Feng *et al.*, 2025).

MAX1–MAX4, which are fundamental elements of SL biosynthesis and transport, positively influence plants' ability to withstand freezing temperatures. It was found that the F-box protein MAX2 targets the transcription factor WRKY41 for degradation (Xu *et al.*, 2021). WRKY41 suppresses *CBF* expression under cold stress by binding to the W-box in *CBF* promoters. By facilitating the cold-induced degradation of WRKY41, MAX2 increases plant cold resistance by removing the restriction of CBFs. These findings thus demonstrate the molecular mechanism by which SLs control plant cold resistance by degradation of WRKY41, the negative modulator of CBF expression, in a MAX2-dependent manner (Wang *et al.*, 2023b).

11.1.1 Biosynthesis of SLs

In recent times, over 30 SLs and SL-like substances were found in the root secretions of different plants (Bhoi *et al.*, 2021). SLs are derived from carotenoids that have a preserved tricyclic lactone ring joined to an α,β-unsaturated furanone ring by an enol-ether bond. Cytosolic and plastidial enzymes work in tandem to produce SL carotenoids. In some instances, apocarotenoids such as SLs are formed by cleavage of C = C bonds to yield two carbonyl products; this is carried out by carotenoid-cleaving enzymes, which are typically members of the carotenoid cleavage oxygenase (CCO) family (Jia *et al.*, 2018). Several CCOs speed up the secondary cleavage of apocarotenoids. Although many of them have not yet been experimentally verified, CCOs are typically identified as carotenoid cleavage dioxygenases (CCDs) (Li *et al.*, 2023). CCDs are divided into five important subfamilies: CCD1, CCD4, CCD7, CCD8, and 9-*cis*-epoxycarotenoid dioxygenases (NCEDs). Because of its simple form and extensive prevalence among both monocotyledonous and dicotyledonous plants, 5-deoxystrigol is regarded as the ancestor of SLs. It was first identified from *Lotus japonicus* root exudates and has no substitutions on the A and B rings. Acetylation, ketolation, hydroxylation, decarboxylation, oxidation, dehydration, and epoxidation reactions of 5-deoxystrigol result in SLs. The primary location for biosynthesis of apocarotenoids is the plastid. All-*trans*-β-carotene is transformed into carlactone by action of the plastidial enzymes DWARF 27 (D27), CCD7, and CCD8. All-*trans*-β-carotene is first changed to 9-*cis*-β-carotene by carotenoid isomerase D27, and then CCD7 further converts it to 9-*cis*-β-apo-10′-carotene. Ultimately, this molecule is acted on by CCD8 to transform it to carlactone. Simultaneously, the furanone ring is formed. Through unknown rare stages that are performed by unique enzymes, carlactone is oxidized to a variety of SLs by MORE AXILLARY GROWTH 1 (MAX1), a cytochrome P450 monooxygenase (CYP711A1, in *Arabidopsis*), and its related derivatives. CCD7 and CCD8 (designated D17/HIGH TILLERING DWARF1 (HTD1) and D10 in rice, RAMOSUS 5 (RMS5) and RMS1 in peas, DECREASED APICAL DOMINANCE 3 (DAD3) and DAD1 in *Petunia*, MAX3 and MAX4 in *A. thaliana*, respectively) have been shown to operate in the SL biosynthesis pathway in a sequential manner (Wang *et al.*, 2023a). The generation of SLs is blocked or reduced when the CCD7 enzyme is knocked out or knocked down, which subsequently causes increased shoot branching in plants, indicating the role of SLs in control of branching (Chen *et al.*, 2023).

11.1.2 SL transporter

PLEIOTROPIC DRUG RESISTANCE1 (PDR1), the transporter of SLs, was initially identified in *Petunia hybrida*. This exporter belongs to the ABC exporter class and is identified as a functional carrier of SLs (Demurtas *et al.*, 2023). *Medicago truncatula* has been reported to use ABCG59, another ABC transporter, for the delivery of SLs. PDR1, which is connected to xylem filling in stems and is activate near the apical bud, is the mechanism by which SLas are secreted into the rhizosphere. It was observed that SLs are undetectable in the xylem fluids of *A. thaliana*, tomato and rice, indicating that instead of transport through the xylem, SLs are carried through the root to the stem both internally and externally (Wamhoff *et al.*, 2024).

11.2 Mechanism of Strigolactone Perception and Signal Transduction

The α/β-hydrolase (D14/AtD14/DAD2) and F-box (MAX2/D3/RMS4) proteins could be important in how SLs are detected in the plant (Fig. 11.1). A leucine (Leu)-rich-repeat F-box protein receives the signal from SLs through the α/β-fold hydrolase (Bhoi *et al.*, 2021). This protein functions as a recognition subunit in an SKP1–CUL1–F-box protein (SCF)-type ubiquitin ligase complex. 26S proteasome is further activated by SCF-type ubiquitin ligase to break down transcription repressors, including D53 in rice and SUPPRESSOR of MAX2 1-LIKE (SMXLs) in *Arabidopsis*. D14 comprises a highly conserved catalytic triad (Ser-His-Asp), which is important for hydrolysis and signal transduction of SLs. The fundamental elements of the SL signal-receiving complex are the F-box protein MAX2 and α/β-hydrolase (D14/DAD2) (Chevalier *et al.*,

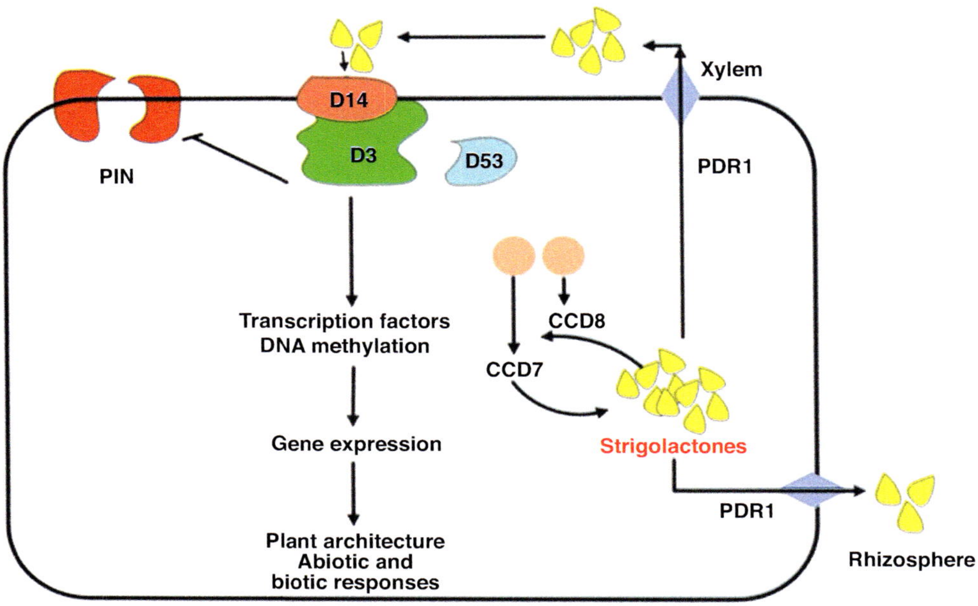

Fig. 11.1. Possible strigolactone signaling pathway.

2014). When SL binds with the "open state" pocket of the D14/AtD14 and α/β-hydrolase receptor, signal transduction starts. The important components of the SL-signaling route are inhibitors such as D53 (rice) and SMXL6, −7, and −8 (*A. thaliana*), and activators such as D14 and D3 (rice) and AtD14 and MAX2 (*A. thaliana*) (Alvi *et al.*, 2022). The conformational shift in D14/AtD14 ends in a "closed state," which initiates its interactivity with the D3/MAX2-based SCF complex, while deletion of the ABC ring by D14/AtD14 after binding with SL allows the furanone ring to stay firmly linked to the D14/AtD14 catalytic cavity. SLs are moved from the base of plants to the apex (acropetal movement) via transporter after they have been generated in the roots (Khuvung *et al.*, 2022).

11.3 Cold-stress Perception and Signal Transduction in Plants

Although calcium is the oldest recognized secondary carrier for stimuli interaction and response in a number of environmental signals, three primary hypotheses have been suggested explain how plants receive this cold signal (Xu *et al.*, 2022). As freezing-responsive *COR* gene activation is triggered by the artificial distortion of membrane fluidity, one primary idea behind cold sensing is membrane rigidification. The next concept focuses on the presence of a membrane-based receptor to detect a cold stimulus. Numerous investigations have been carried out to determine the cell-membrane receptors that sense cold, which are comparable to the ion-channel receptors found in mammals. However, the sole protein found in rice that can sense cold functions as the G-protein regulator, called COLD TOLERANCE DIVERGENCE1 (COLD1), and is found in the plasma membrane and endoplasmic reticulum. The G-protein α-subunit and COLD1 work together to activate nuclear *COR* genes, which modulate electrophysiological and cold signals (Wei *et al.*, 2021). The third theory suggests that calcium ions (Ca^{2+}), the most prevalent secondary messenger, are responsible for cold perception mediated by COLD1. Low-temperature stress causes calcium networks to open, increasing the concentration of Ca^{2+}in the cytosol and undoubtedly signaling the start of cold adaptation.

The cytosolic Ca^{2+} concentration rises quickly as a result of the release of Ca^{2+} from the intracellular compartments, and thus serves as a signaling reaction to cold stress (Gusain *et al.*, 2023). Calmodulin-binding proteins (CBLs) and calcium-dependent protein kinases (CDPKs) are subsequently activated by elevated Ca^{2+} concentrations. The nuclear-localized *CBF/DREB1* and *COR* genes are upstream of these kinases. A study demonstrated that CALMODULIN-BINDING TRANSCRIPTION ACTIVATOR (CAMTA) attaches itself to the *CBF2* gene promoter's regulatory region (Chao *et al.*, 2022). The translation of cytosolic calcium signals caused by low temperature into downstream regulation is directly mediated by the CAMTA proteins. MAP KINASE (MAPK3/6), MAP ERK KINASE (MEKK1-MEKK2-MAPK4), and CALMODULIN REGULATED RECEPTOR LIKE 1/2 (CRLK1/2) cascades are also activated by cold stress. By means of proteasome-mediated degradation, the phosphorylation of these MAPKs in conjunction with that of ICE1 controls the activity of CBF. Additionally, catechins in tea were discovered to play a role in cold reactions (Wang *et al.*, 2022), and it is proposed that ROS levels and Ca^{2+} influx are tightly related.

Table 11.1. List of transcription factors that are both autonomous and dependent on CBF.

Dependent on CBF	Independent of CBF
ICE1/2	SAG12
CAMTA1-5	WRKY33
CESTA	ERF5
B2R1/BES1	C2F1
CCA1/LHY	RAV1
SIZ1	CZF2
EBF1/2	STA1
BTFS	GCN5
PIF3/4/7	HD2C
14-3-3s	MYB73
CRK1	ZAT10
EIN3	HSFC1
PP-Rs	NPR1
MYB15	HSCFA1
MPK316	RCF1
PhyB	HOS15
HOS1	

11.4 Transcriptional Regulation of CBFs

The cold-signaling system in plants that is dependent on CBF and DREB1 has been thoroughly investigated. In barley, rice, tomato, wheat, and maize, many orthologs of CBF have been identified. Song *et al.* (2021a) observed that whereas CBF4 expression in *Arabidopsis* cannot be induced by cold, its overexpression may improve drought and cold resistance.

APETALA2/ETHYLENERESPONSIVE (AP2/ERF1)-type transcription factors and CBF1, −2, and −3, have been shown to stimulate activation of *COR* genes in cold environments by attaching to *CRT/DRE* motifs in the promoter region. Genes that are both independent of and reliant on CBF have been reported (Table 11.1). In addition, one study on tomatoes found that transduction of the ethylene signal in tomatoes was influenced by storage temperature (Yu *et al.*, 2023). Thus, cold storage occurred at the level of ethylene production, which may have improved ethylene signal transmission. The transcription factor APETALA2/ETHYLENE RESPONSE ELEMENT (AP2/ERF) is responsible for various stress reactions in rice. Several *AP2/ERF* genes in rice have now been fully described. It has been demonstrated that freezing tolerance in *Arabidopsis* can be increased by the heterologous production of CBFs from other plant species. Although several *CBF* genes are present in the cold-sensitive tomato species, Albornoz *et al.* (2024) verified that only *LeCBF1* is cold inducible and functional. Tomato plants that overexpressed *Arabidopsis CBF3* did not exhibit cold tolerance because of the existence of distinct *CBF* regulators in tomatoes, but overexpression of *LeCBF1* was demonstrated to provide freezing tolerance. The biological role of CBF1, −2, and −3 in regulating freezing tolerance is species specific and not highly conserved among plants. Because a mutation in the *ICE1* gene impairs CBF expression, freezing tolerance is reduced. A homolog of *ICE1*, *ICE2*, is likewise a powerful modulator of *CBF* expression (Hwarari *et al.*, 2022). The production of

Table 11.2. In plants, cold-tolerant signaling genes are linked to light, photoperiod, and the circadian rhythm.

Transcription factor gene	Mechanism and phenotype	Reference(s)
C-REPEAT/DRE-BINDING FACTOR (CBF)	CBF2 represses CBF1/3 and the *cbf2* mutant	Xin *et al.* (2025)
C-REPEAT/DRE-BINDING FACTOR 1 (CBF1)	CBF1/3; RNA interference lines are compromised by freezing resistance	Wang *et al.* (2022)
ELONGATED HYPOCOTYL 5 (HY5)	HY5 favorably controls the functioning of genes triggered by freezing stress	Jin *et al.* (2022)
PHYTOCHROME B (PHYB), PHYD	*phyB* and *phyD* mutants exhibit high cold resistance	Junior *et al.* (2021); Péter *et al.* (2024)
PSEUDO-RESPONSE REGULATOR 5/7/9 (PRR5/7/9)	Serve as CBF cascade negative regulators; the triple mutant *prr5/7/9* exhibits enhanced resistance to freezing	Jang *et al.* (2024)
CIRCADIAN CLOCK-ASSOCIATED 1 (CCA1)	CCA1 promotes freezing adaptation by binding to the *CBF* promoter	Yu *et al.* (2025)
LATE ELONGATED HYPOCOTYL (LHY)	Collaborates with CCA1 to regulate CBFs in a circadian rhythm	Liang *et al.* (2024)
GIGANTEA (GI)	Uses a route that is not dependent on CBF to effectively modulate cold resistance	Liu *et al.* (2024)
PHYTOCHROME-INTERACTING FACTOR 4 (PIF4), PIF7	PIF4 and PIF7 perform duplicate tasks in cold adaptation	Li *et al.* (2024)

COR genes, which are also said to be produced at low temperatures, depends on the activity of the Ca^{2+}-responsive protein calmodulins (CAMs). Because they include identical CAM-binding areas, CAMTAs have been demonstrated to trigger *CBF* transcription. *CBF1* and *CBF2* expression are efficiently regulated by CAMTA proteins, such as CAMTA1–5 (Dong *et al.*, 2021). For instance, CAMTA3 and CAMTA1 stimulate *CBF2* transcription, while CAMTA2 triggers *CBF1–3* transcription. Furthermore, in reaction to abrupt temperature drops, CAMTA3 and CAMTA5 control *CBF1* expression, although this is not always the case. A potential connection between calcium and cold signaling is provided by expression of the CRT/DRE motif by Ca^{2+}. CBF expression is negatively regulated by PHYTOCHROME-INTERACTING FACTOR3/4/7 (PIF3/4/7), a light-signaling inhibitory regulator that is blocked by the circadian clock (Manasa *et al.*, 2022) (Table 11.2).

CIRCADIAN CLOCK-ASSOCIATED 1 (CCA1) and LATE ELONGATED HYPOCOTYL (LHY) are both significant circadian promoters of *CBF* transcription and cold resilience (Table 11.2) (Liang *et al.*, 2024). The expression of *CBF* is also coordinated by many hormone-signaling components. JA signaling is suppressed by JASMONATE ZIM-DOMAINPROTEIN1/4 (JAZ1/4). By interacting with *ICE1/2* and controlling their transcription, they control the expression of *CBF*. According to reports, the ethylene signaling transcription factor ETHYLENE INSENSITIVE 3 (EIN3) adversely affects cold resistance, which lowers *CBF* transcription. By regulating the breakdown of EIN3 and PIF3, EIN3-BINDING F-BOX 1/2 (EBF1/2) further promote CBF transcription during cold stress. None of the protein levels were significantly decreased with chilling, despite the fact that the majority of ethylene signaling gene expression is decreased, suggesting that transcriptional and translational levels are differentially regulated, as reported by Zhai *et al.* (2022). Freezing tolerance is negatively regulated by brassinosteroids. The transcription factors BRI1-EMS-SUPPRESSOR1 (BES1) and BRASSINAZOLE-RESISTANT1 (BZR1) are upstream of BRASSINOSTEROID-INSENSITIVE2 (BIN2). These transcription factors have the ability to positively control freezing tolerance by attaching to the CBF1 and CBF2 regulators containing E-box binding spots. Therefore, it is important to remember that CBFs significantly regulate a number of interactions related to cold responses and plant development. In transgenic *Arabidopsis*, it has been discovered that the stress-responsive gene Dl-ICE1, which was identified from *Dimocarpus longan*, improves cold tolerance (Manasa *et al.*, 2022).

11.5 Role of Strigolactones During Chilling Stress

In many parts of the world, cold stress negatively affects agricultural productivity, resulting in losses (Manasa *et al.*, 2022). It has been shown that, in comparison with their wild-type equivalents, *A. thaliana max2-1* and *max4-1* and pea *rms4* mutants have rates of photosynthesis and leaf area that are significantly slowed down by cold temperatures. As SLs are associated with the *max* and *rms4* genes, it can be inferred that SLs effectively contribute to the ability of these plants to withstand low temperatures. Li *et al.* (2021) observed alterations in morphological characteristics after cold stress in turnip, despite the fact that GR24 administration greatly increased transpiration, cell survival, decreased protein and proline concentration and antioxidant enzymes, and reduced relative conductivity. Additionally, it was observed that GR24 effectively raised the activity of genes associated with freezing such as MAPK, NADPH oxidase, and antioxidant enzymes. According to Hayat *et al.* (2025), cooling treatment significantly reduces the quantum efficiency of photosystem II and promotes dehydration, which upsets the water status of *Vigna radiata*. In order to alleviate chilling stress, proline quantity, soluble sugar, and the activity of LOX and phenylpropanoid pathway enzymes were all increased by GR24 therapy, which also markedly reduced the rate of lipid peroxidation. Investigation of *O. sativa* gene expression revealed that the transcription of *D10* and *Os01g0701500* genes was downregulated at low temperatures, but the activity of *D27* and *D17* in the roots of non-mycorrhizal plants was similar at low temperatures. All of these genes, however, were shown to be significantly upregulated in mycorrhizal rice roots at minimum temperatures (Mitra *et al.*, 2023). It was recently demonstrated that the arbuscular mycorrhizal fungi (AMF) promote the buildup of SLs by increasing the activity of the genes responsible for their production in host plants. SLs may play a role in regulating plants' exposure to freezing stress but the role of SLs during cold stress is still poorly understood. The exact method that helps SLs reduce cold stress in plants will require further investigation.

11.6 Crosstalk Between Strigolactones and Other Phytohormones

11.6.1 Interactions between SLs and abscisic acid

Because of its ability to cooperate and regulate under a variety of abiotic stressors, abscisic acid (ABA) is considered a general stress plant hormone. In response to abiotic stresses such as drought, salt stress, pathogen assault, and damage, ABA controls stomatal closure and initiates long-distance communication (Zhang *et al.*, 2021). Changes in ABA levels not only cause stomatal closure to cope with stress but also stimulate the transcription of genes that encode proteins such as dehydrins, osmoprotectants, and genes linked to salinity and drought, which help plants tolerate stress. There have also been reports of ABA being synthesized from scratch in stressed leaves and roots. Both ABA and SLs are regarded as stress hormones and are derived from carotenoids (Fig. 11.2) (Jiang *et al.*, 2024). A relationship between ABA levels and SLs has previously been proposed, indicating that ABA plays a regulatory function in the production of SLs. The initial reaction in the ABA biosynthesis route is conversion of 9-*cis*-epoxycarotenoid to xanthoxin, catalyzed by NCED enzyme. When tomato was treated with certain NCED regulators that prevent ABA production, a lower concentration of SLs was observed. This decreased SL concentration in NCED variants seems to suggest that ABA, an outcome of NCED output, might be responsible for regulating SL production rather than NCED being directly connected to SL production (Korek and Marzec, 2023).

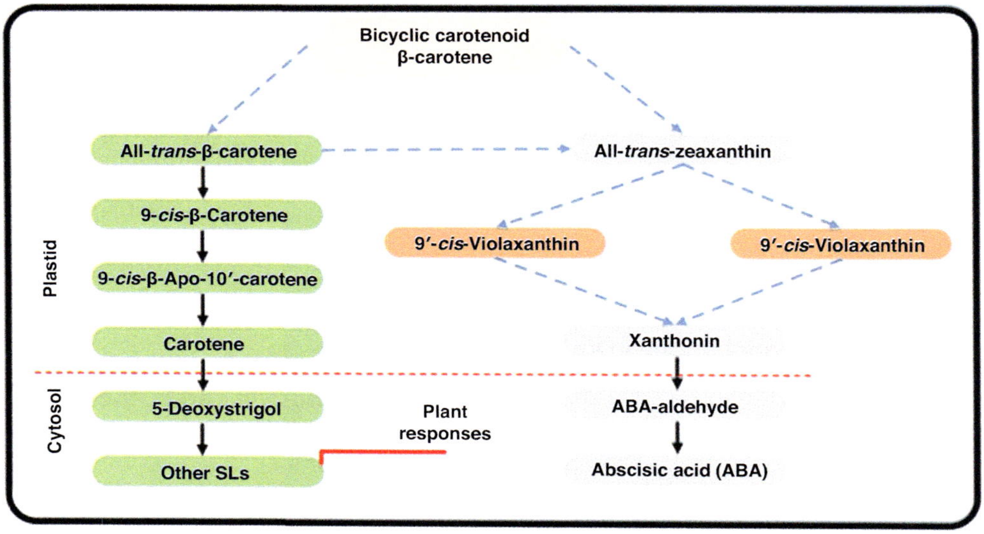

Fig. 11.2. Crosstalk between strigolactones and abscisic acid.

According to Samanta and Roychoudhury (2025), SLs favorably regulate the activity of stress and/or ABA-responsive genes associated with plant development and environmental stress reactions. In the *max2* variant of *A. thaliana*, it was shown that transcription of ABA importer genes (*ABCG22* and *ABCG40*) and a significant percentage of stress-inducible and/or ABA-inducible genes were negatively regulated by water stress. Numerous physiological processes have been widely documented to be regulated by ABA and SLs. Additionally, ABA is essential to AMF symbiosis. As well as its role in shielding mycorrhizal roots from stress, studies have demonstrated that ABA is essential for the establishment of symbiosis, the final stage of arbuscular development, and the promotion of long-term plant root colonization (Shi *et al.*, 2023). Therefore, increased ABA levels in mycorrhiza-associated stressed plants will promote stress tolerance while also strengthening and maintaining AMF symbiosis. Furthermore, control of the salt-stress response and development of symbiotic interactions between the host plant and AMF depend on SLs and ABA. Exogenous ABA administration may promote the accumulation of SLs under stressful circumstance (Feng *et al.*, 2024).

11.6.2 Interactions of SLs with JA signaling

A class of plant oxylipins known as JAs mediates a variety of physiological activities during wounding, secondary metabolism, relationships between plants and insects, and between plants and pathogens, and vegetative development (Browse, 2005). There is currently insufficient experimental evidence to adequately characterize the interaction between SLs and JAs. SLs and JAs may be related in disease tolerance, nevertheless, as evidenced by the finding that an SL-deficient *S. lycopersicum* variant (*Slccd8*) had lower levels of JA and PINII, a jasmonate-dependent gene that confers *S. lycopersicum* tolerance to *Botrytis cinerea*. The fact that tobacco PDR6 (NtPDR6), an ortholog of the SL transporter gene *PhPDR1*, is somewhat influenced by methyl jasmonate, further suggests this connection (Bienert *et al.*, 2014). NtPDR6 has not been implicated in SL transport, despite being substantially activated by phosphate deficiency and being responsible for the control of shoot branching. A somewhat contradictory finding was recently published, stating that neither GR24 application nor *Mucor* sp. inoculation had an impact on the buildup of JA in wild-type plants, *Atmax1* and *Atmax2* (Omoarelojie *et al.*, 2019). The finding that these two hormones cause reactions at comparable stages of development, such as mesocotyl expansion, leaf senescence, and pathogen interactions, provides evidence that SL–JA interactions are likely to play a vital part in these procedures and cannot be entirely ruled out, but these results alone are insufficient for drawing definitive conclusions about the purpose of SL–JA crosstalk.

11.7 Case Studies

Administration of an artificial SL analog boosts resistance against cold stress in *A. thaliana* and *Triticum aestivum*, and plants lacking SL production or genes for SL signaling exhibit reduced resistance in a number of plants (Saeed *et al.*, 2017). Many lines of evidence have shown that the reactions of tomato (*S. lycopersicum*) to heat and low-temperature stress are effectively regulated by SLs. Chi *et al.* (2021) observed that under cold stress, tomato roots accumulated more solanacol and the SL biosynthetic genes (*CCD7*, *CCD8*, *MAX1*) were activated; however, SL production was inhibited in the roots of non-mycorrhizal dicot plants, including *Lactuca sativa*, *S. lycopersicum*, and *L. japonicus*, by osmotic-related stressors such lack of water and excessive salts. This suggests that the regulation of SL biosynthesis may be influenced differently by various environmental stimuli. Pandey *et al.* (2016) suggested that although locally generated SLs may be partially responsible for the observed phenotypes, this issue cannot yet be resolved analytically as SL accumulation in plant shoots, such as those of *Arabidopsis* and tomatoes, is very low and often undetectable. Chi *et al.* (2021) found that relative electrolyte leakage and malondialdehyde concentrations in these species were higher compared with control species, indicating that CCD7 disruption or inhibition of CCD7, CCD8, MAX1, or MAX2 increased vulnerability to high and low-temperature stress. According to Chi *et al.* (2025), the damage from heat and cold stressors was reduced by applying the SL analog $GR24^{5DS}$ to the roots. These findings clearly show that SLs are responsible for tomato responses during stress conditions of low- and high-temperature stress.

It is well recognized that ABA has a role in how plants react to abiotic stressors such heat, salt, drought, and cold. When plants experience these stressors, they boost their antioxidant capacity and induce stomatal closure, which causes them to collect ABA. Gibberellins and SLs, two additional plant hormones involved in development, growth, and adaptation to stress, can interact with ABA (Singh and Roychoudhury, 2023). Recently, Rhaman *et al.* (2024) found that in comparison with wild-type plants, SL-deficient variants of *A. thaliana*, *L. japonicus*, and *S. lycopersicum* displayed either lower ABA accumulation or higher rates of water loss during dehydration, as well as diminished ABA responsiveness. It is interesting to note that *O. sativa d3*, *d10*, and *d17* variants have higher ABA levels and are more drought tolerant than the wild-type species, but the *d27* variants have lower ABA accumulation and less drought tolerance. The downregulation of several *Arabidopsis* genes that respond to stress and/or ABA is associated with reduced resistance in the *max2-3* genotype. Additionally, GR24 treatment can reduce stress-induced seed dormancy and trigger expression of ABA catabolism-related genes. Thus, ABA homeostasis and ABA sensitivity can be changed by SLs. Silencing the genes for SL production or *MAX2* led to greater stomatal conductance and more water loss in detached leaves, which is in line with earlier research in tomato and *Arabidopsis* (Feng *et al.*, 2025). pTRV is a tobacco rattle virus-based vector used for virus-induced gene silencing to downregulate target genes, and pTRV-CCD7, pTRV-CCD8, pTRV-MAX1, and pTRV-MAX2 are TRV constructs used to silence *CCD7*, *CCD8*, *MAX1*, and *MAX2* genes, respectively. Compared with pTRV-CCD7, pTRV-CCD8, and pTRV-MAX1 leaves, pTRV-MAX2 leaves exhibit higher stomatal conductance and sensitivity to dehydration. This is likely because MAX2 participates in the (KARRIKIN INSENSITIVE 2) KAI2-dependent signaling, as well as the SL pathway. While *ccd7* plants exhibited reduced transcript expression of the ABA biosynthesis gene and ABA-responsive genes *NCED6*, *Le4*, and *ABF4*, as well as reduced concentration of ABA in the leaves during low- and high-temperature stress, external $GR24^{5DS}$ application boosted the expression of these genes and prompted ABA buildup in both wild-type and *ccd7* plants (Malini *et al.*, 2023). Furthermore, it has been found that SLs are associated with high-temperature-induced HSP70 synthesis and freezing-induced *CBF1* gene expression, and that both of these responses were partly ABA dependent during either high- or low-temperature stress. However, the way SLs trigger ABA production and how ABA regulates *CBF1* and HSP70 protein expression during heat and cold stresses remain unclear (Ding *et al.*, 2025).

11.8 Future Perspectives on SLs in Agriculture

Particularly when it comes to controlling plant architecture, weak phenotypic variations may have better chances in breeding programs than robust mutants. Negative qualities are sometimes linked to agriculturally useful features. It may be possible to attain the desired feature while reducing undesirable qualities linked to the relevant signaling pathways by combining weaker phenotypes that operate through separate routes (Kim *et al.*, 2011). In this regard, it is expected that desirable breeding stock will be found in variations with altered SL profiles, such as *Lgs1* sorghum, altered SL signaling, such as overexpression of the SL exporter PDR1, or increased SL-responsive pathways, such as the gain-of-function *OsTB1* in *O. sativa*.

Although SLs have not been used commercially, they have been investigated as agrochemicals in field experiments. The lack of a reasonably priced supply of SLs and its alternatives is currently the main barrier to SL application (Aliche *et al.*, 2020). This is a major barrier to both the use of SLs as agrochemicals and the investigation of their possible uses. Because of its complex structure and stereochemistry, chemical manufacturing of organic SLs is impractical for commercial manufacturing; most SLs are currently unable to be produced at multi-gram levels and the procedure usually consists of over 20 stages. Biologically functional SLs analogs, which may be synthesized with higher production, are a significant source of SLs for research. Because of its potent ability to prevent branching and promote parasitic seed development, the commercially available SL analog GR24 is often utilized to investigate SL functions (Ruyter-Spira *et al.*, 2011). There have been reports of other physiologically functional SL alternatives with altered or simple structures that are easier to synthesize and have greater yields. For instance, it just takes two steps to manufacture the racemic mix of Nijmegen-1, which is employed in experiments for the suicidal germination of *Orobanche* and *Striga* spp. However, for large-scale applications, the cost is still prohibitive (Humphrey *et al.*, 2006).

11.9 Conclusion

SLs are secondary plant metabolite compounds generated from carotenoids. They influence several plant growth systems, such as root development, leaf senescence, shoot branching, and seed germination. Stress may affect SL synthesis, signaling, and responses to other phytohormones. Because of their important functions in controlling several physiological and molecular processes as well as helping plants adapt to different environmental stressors, including water, salt, nutritional, heat, cold, and heavy-metal stresses, SLs have gained much attention, and the results indicate how much of an impact SLs could have on modern agriculture. Thus, the application of SLs could offer significant opportunities for creating innovative approaches and technological breakthroughs that complement the evolving principles of sustainable agriculture.

References

Albornoz, K., Zhou, J., Zakharov, F., Grove, J., Wang, M. *et al.* (2024) Ectopic overexpression of *ShCBF1* and *SlCBF1* in tomato suggests an alternative view of fruit responses to chilling stress postharvest. *Frontiers in Plant Science* 15, 1429321. DOI: 10.3389/fpls.2024.1429321.

Aliche, E.B., Screpanti, C., de Mesmaeker, A., Munnik, T. and Bouwmeester, H.J. (2020) Science and application of strigolactones. *New Phytologist* 227(4), 1001–1011. DOI: 10.1111/nph.16489.

Alvi, A.F., Sehar, Z., Fatma, M., Masood, A. and Khan, N.A. (2022) Strigolactone: An emerging growth regulator for developing resilience in plants. *Plants* 11(19), 2604. DOI: 10.3390/plants11192604.

Bhattacharya, A. (2022) Effect of low-temperature stress on germination, growth, and phenology of plants: A review. In: *Physiological Processes in Plants under Low Temperature Stress*. Springer Nature Singapore, Singapore, pp. 1–106. DOI: 10.1007/978-981-16-9037-2_1.

Bhoi, A., Yadu, B., Chandra, J. and Keshavkant, S. (2021) Contribution of strigolactone in plant physiology, hormonal interaction and abiotic stresses. *Planta* 254(2), 28. DOI: 10.1007/s00425-021-03678-1.

Białoskórska, M., Rucińska, A. and Boczkowska, M. (2024) Molecular mechanisms underlying freezing tolerance in plants: Implications for cryopreservation. *International Journal of Molecular Sciences* 25(18), 10110. DOI: 10.3390/ijms251810110.

Bienert, M.D., Baijot, A. and Boutry, M. (2014) ABCG transporters and their role in the biotic stress response. In: Geisler, M. (ed.) *Plant ABC Transporters*. Springer, Cham, Switzerland, pp. 137–162.

Browse, J. (2005) Jasmonate: An oxylipin signal with many roles in plants. *Vitamins & Hormones* 72, 431–456. DOI: 10.1016/S0083-6729(05)72012-4.

Cao, X., Wei, Y., Shen, B., Liu, L. and Mao, J. (2024) Interaction of the transcription factors BES1/BZR1 in plant growth and stress response. *International Journal of Molecular Sciences* 25(13), 6836. DOI: 10.3390/ijms25136836.

Chao, L., Kim, Y., Gilmour, S.J. and Thomashow, M.F. (2022) Temperature modulation of CAMTA3 gene induction activity is mediated through the DNA binding domain. *Plant Journal* 112, 235–248. DOI: 10.1111/tpj.15944.

Chen, S., Song, X., Zheng, Q., Liu, Y., Yu, J. *et al.* (2023) Transcription factor SPL13 mediates strigolactone suppression of shoot branching by inhibiting cytokinin synthesis in *Solanum lycopersicum*. *Journal of Experimental Botany* 74(18), 5722–5735. DOI: 10.1093/jxb/erad303.

Chevalier, F., Nieminen, K., Sánchez-Ferrero, J.C., Rodríguez, M.L., Chagoyen, M. *et al.* (2014) Strigolactone promotes degradation of DWARF14, an α/β hydrolase essential for strigolactone signaling in *Arabidopsis*. *Plant Cell* 26(3), 1134–1150. DOI: 10.1105/tpc.114.122903.

Chi, C., Xu, X., Wang, M., Zhang, H., Fang, P. *et al.* (2021) Strigolactones positively regulate abscisic acid-dependent heat and cold tolerance in tomato. *Horticulture Research* 8(1), 237. DOI: 10.1038/s41438-021-00668-y.

Chi, C., Chen, X., Zhu, C., Cao, J., Li, H. *et al.* (2025) Strigolactones positively regulate HY5-dependent autophagy and the degradation of ubiquitinated proteins in response to cold stress in tomato. *New Phytologist* 245(3), 1106–1123. DOI: 10.1111/nph.20058.

Demurtas, O.C., Nicolia, A. and Diretto, G. (2023) Terpenoid transport in plants: How far from the final picture? *Plants* 12(3), 634. DOI: 10.3390/plants12030634.

Ding, Y. and Yang, S. (2022) Surviving and thriving: How plants perceive and respond to temperature stress. *Developmental Cell* 57(8), 947–958. DOI: 10.1016/j.devcel.2022.03.010.

Ding, F., Fan, X., Tian, R., Wang, M. and Sun, Z. (2025) Crosstalk of abscisic acid with other hormones and signaling molecules in tomato cold stress tolerance. *Horticulturae* 11(6), 647. DOI: 10.3390/horticulturae11060647.

Dong, F., Yang, F., Liu, Y., Jia, W., He, X. *et al.* (2021) Calmodulin-binding transcription activator (CAMTA)/ factors in plants. In: Upadhyay, S. (ed.) *Calcium Transport Elements in Plants*. Elsevier, London, pp. 249–266. DOI: 10.1016/B978-0-12-821792-4.00017-5.

Feng, D., Liu, W., Chen, K., Ning, S., Gao, Q. *et al.* (2024) Exogenous substances used to relieve plants from drought stress and their associated underlying mechanisms. *International Journal of Molecular Sciences* 25(17), 9249. DOI: 10.3390/ijms25179249.

Feng, Y., Li, Z., Kong, X., Khan, A., Ullah, N. *et al.* (2025) Plant coping with cold stress: Molecular and physiological adaptive mechanisms with future perspectives. *Cells* 14(2), 110. DOI: 10.3390/cells14020110.

Gusain, S., Joshi, S. and Joshi, R. (2023) Sensing, signalling, and regulatory mechanism of cold-stress tolerance in plants. *Plant Physiology and Biochemistry* 197, 107646. DOI: 10.1016/j.plaphy.2023.107646.

Hayat, U., Ul Din, K., Ahmad, M., Zulfiqar, U., Sajjad, M. *et al.* (2025) Salicylic acid confers cadmium tolerance in wheat by regulating photosynthesis, yield and ionic homeostasis. *Scientific Reports* 15(1), 3698. DOI: 10.1038/s41598-025-87236-9.

Humphrey, A.J., Galster, A.M. and Beale, M.H. (2006) Strigolactones in chemical ecology: Waste products or vital allelochemicals? *Natural Product Reports* 23(4), 592–614. DOI: 10.1039/b512776a.

Hwarari, D., Guan, Y., Ahmad, B., Movahedi, A., Min, T. *et al.* (2022) ICE–CBF–COR signaling cascade and its regulation in plants responding to cold stress. *International Journal of Molecular Sciences* 23(3), 1549. DOI: 10.3390/ijms23031549.

Jain, S., Kumar, R., Gunsola, D., Živković, S., Vasić, T. *et al.* (2025) Strigolactones and control of parasitic weeds. In: Aftab, T. and Wani, K.I. (eds) *Strigolactones: Emerging Plant Hormones*. Scrivener Publishing, Beverly, Massachusetts, pp. 261–279. DOI: 10.1002/9781394302826.

Jang, J., Lee, S., Kim, J.-I., Lee, S. and Kim, J. (2024) The roles of circadian clock genes in plant temperature stress responses. *International Journal of Molecular Sciences* 25(2), 918. DOI: 10.3390/ijms25020918.

Jia, K.-P., Baz, L. and Al-Babili, S. (2018) From carotenoids to strigolactones. *Journal of Experimental Botany* 69(9), 2189–2204. DOI: 10.1093/jxb/erx476.

Jiang, W., Fei Lu, C., Xu, X., Riaz, M.W., Lv, A. *et al.* (2024) Strigolactones: Biosynthetic regulation, hormonal interaction, and their involvement in abiotic stress adaption. *Scientia Horticulturae* 325, 112689. DOI: 10.1016/j.scienta.2023.112689.

Jin, J., Sun, W., Wu, J., Fang, Y., Li, X. *et al.* (2022) Hypocotyl elongation based on HY5 transcription factor in cold resistant winter rapeseed (*Brassica napus* L.). *Oil Crop Science* 7(1), 40–52. DOI: 10.1016/j.ocsci.2022.02.005.

Junior, C.A.S., d'Amico-Damião, V. and Carvalho, R.F. (2021) Phytochrome type B family: The abiotic stress responses signaller in plants. *Annals of Applied Biology* 178(2), 135–148. DOI: 10.1111/aab.12655.

Khuvung, K., Silva Gutierrez, F.A.O. and Reinhardt, D. (2022) How strigolactone shapes shoot architecture. *Frontiers in Plant Science* 13, 889045. DOI: 10.3389/fpls.2022.889045.

Kim, Y.-A., Wuchty, S. and Przytycka, T.M. (2011) Identifying causal genes and dysregulated pathways in complex diseases. *PLOS Computational Biology* 7(3), e1001095. DOI: 10.1371/journal.pcbi.1001095.

Korek, M. and Marzec, M. (2023) Strigolactones and abscisic acid interactions affect plant development and response to abiotic stresses. *BMC Plant Biology* 23(1), 314. DOI: 10.1186/s12870-023-04332-6.

Li, W., Che, J., Lian, Q., Wang, C., Dai, G. *et al.* (2023) Identification and expression pattern of the carotenoid cleavage oxygenase gene family in *Lycium* suggest *CCOs* respond to abiotic stress and promote carotenoids degradation. *Forests* 14(5), 983. DOI: 10.3390/f14050983.

Li, Y., Sun, Y., Ma, C., Kang, X., Wang, J. *et al.* (2021) 24-Epibrassinolide enhanced cold tolerance of winter turnip rape (*Brassica rapa* L.). *Biologia* 76(10), 2859–2877. DOI: 10.1007/s11756-021-00834-6.

Li, Z.-Y., Ma, N., Zhang, F.-J., Li, L.-Z., Li, H.-J. *et al.* (2024) Functions of phytochrome interacting factors (PIFs) in adapting plants to biotic and abiotic stresses. *International Journal of Molecular Sciences* 25(4), 2198. DOI: 10.3390/ijms25042198.

Lian, Y., Lian, C., Wang, L., Li, Z., Yuan, G. *et al.* (2023) SUPPRESSOR of MAX2 like 6, 7, and 8 interact with DDB1 binding WD repeat domain hypersensitive to ABA deficient 1 to regulate the drought tolerance and target *SUCROSE NONFERMENTING 1 RELATED PROTEIN KINASE 2.3* to abscisic acid response in *Arabidopsis*. *Biomolecules* 13(9), 1406. DOI: 10.3390/biom13091406.

Liang, T., Yu, S., Pan, Y., Wang, J. and Kay, S.A. (2024) The interplay between the circadian clock and abiotic stress responses mediated by ABF3 and CCA1/LHY. *Proceedings of the National Academy of Sciences* 121(7), e2316825121. DOI: 10.1073/pnas.2316825121.

Liu, L., Xie, Y., Yahaya, B.S. and Wu, F. (2024) *GIGANTEA* unveiled: Exploring its diverse roles and mechanisms. *Genes* 15(1), 94. DOI: 10.3390/genes15010094.

Malini, M.K., Karwa, S., Priyadarsini, P., Kumar, P., Nagar, S. *et al.* (2023) Abscisic-acid-modulated stomatal conductance governs high-temperature stress tolerance in rice accessions. *Agriculture* 13(3), 545. DOI: 10.3390/agriculture13030545.

Manasa, L.S., Panigrahy, M., Panigrahi, K.C.S. and Rout, G.R. (2022) Overview of cold stress regulation in plants. *Botanical Review* 88(3), 359–387. DOI: 10.1007/s12229-021-09267-x.

Mitra, D., Nayeri, F.D., Sansinenea, E., Ortiz, A., Bhatta, B.B. *et al.* (2023) Unraveling arbuscular mycorrhizal fungi interaction in rice for plant growth development and enhancing phosphorus use efficiency through recent development of regulatory genes. *Journal of Plant Nutrition* 46(13), 3184–3220. DOI: 10.1080/01904167.2023.2191638.

Omoarelojie, L., Kulkarni, M., Finnie, J. and van Staden, J. (2019) Strigolactones and their crosstalk with other phytohormones. *Annals of Botany* 124, 749–767. DOI: 10.1093/aob/mcz100.

Pandey, A., Sharma, M. and Pandey, G.K. (2016) Emerging roles of strigolactones in plant responses to stress and development. *Frontiers in Plant Science* 7, 434. DOI: 10.3389/fpls.2016.00434.

Péter, C., Ádám, É., Klose, C., Grézal, G., Hajdu, A. *et al.* (2024) Phytochrome C and low temperature promote the protein accumulation and red-light signaling of phytochrome D. *Plant and Cell Physiology* 65(10), 1717–1735. DOI: 10.1093/pcp/pcae089.

Rhaman, M.S., Karim, M.M., Sagar, A., Asaduzzaman, M., Ye, W. *et al.* (2024) Phytohormone strigolactone: Involvement in guard cell signaling and abiotic stress tolerance in plants. *Journal of Plant Growth Regulation* 43(12), 4621–4634. DOI: 10.1007/s00344-024-11421-9.

Ruyter-Spira, C., Kohlen, W., Charnikhova, T., van Zeijl, A., van Bezouwen, L. *et al.* (2011) Physiological effects of the synthetic strigolactone analog GR24 on root system architecture in *Arabidopsis*: Another belowground role for strigolactones? *Plant Physiology* 155(2), 721–734. DOI: 10.1104/pp.110.166645.

Saeed, W., Naseem, S. and Ali, Z. (2017) Strigolactones biosynthesis and their role in abiotic stress resilience in plants: A critical review. *Frontiers in Plant Science* 8, 1487. DOI: 10.3389/fpls.2017.01487.

Samanta, S. and Roychoudhury, A. (2025) Molecular crosstalk of jasmonate with major phytohormones and plant growth regulators during diverse stress responses. *Journal of Plant Growth Regulation* 44(1), 62–88. DOI: 10.1007/s00344-024-11412-w.

Shi, J., Wang, X. and Wang, E. (2023) Mycorrhizal symbiosis in plant growth and stress adaptation: From genes to ecosystems. *Annual Review of Plant Biology* 74(1), 569–607. DOI: 10.1146/annurev-arplant-061722-090342.

Shi, Y., Tian, S., Hou, L., Huang, X., Zhang, X. *et al.* (2012) Ethylene signaling negatively regulates freezing tolerance by repressing expression of *CBF* and type-A *ARR* genes in *Arabidopsis*. *Plant Cell* 24(6), 2578–2595. DOI: 10.1105/tpc.112.098640.

Singh, A. and Roychoudhury, A. (2023) Abscisic acid in plants under abiotic stress: Crosstalk with major phytohormones. *Plant Cell Reports* 42(6), 961–974. DOI: 10.1007/s00299-023-03013-w.

Song, Q., Wang, X., Li, J., Chen, T.H., Liu, Y. *et al.* (2021a) CBF1 and CBF4 in *Solanum tuberosum* L. differ in their effect on low-temperature tolerance and development. *Environmental and Experimental Botany* 185, 104416. DOI: 10.1016/j.envexpbot.2021.104416.

Song, Y., Zhang, X., Li, M., Yang, H., Fu, D. *et al.* (2021b) The direct targets of CBFs: In cold stress response and beyond. *Journal of Integrative Plant Biology* 63(11), 1874–1887. DOI: 10.1111/jipb.13161.

Waadt, R., Seller, C.A., Hsu, P.-K., Takahashi, Y., Munemasa, S. *et al.* (2022) Plant hormone regulation of abiotic stress responses. *Nature Reviews Molecular Cell Biology* 23(10), 680–694. DOI: 10.1038/s41580-022-00479-6.

Wamhoff, D., Gündel, A., Wagner, S., Ortleb, S., Borisjuk, L. *et al.* (2024) Anatomical limitations in adventitious root formation revealed by magnetic resonance imaging, infrared spectroscopy, and histology of rose genotypes with contrasting rooting phenotypes. *Journal of Experimental Botany* 75(16), 4784–4801. DOI: 10.1093/jxb/erae158.

Wang, H., Zhang, Y., Norris, A. and Jiang, C.-Z. (2022) S1-bZIP transcription factors play important roles in the regulation of fruit quality and stress response. *Frontiers in Plant Science* 12, 802802. DOI: 10.3389/fpls.2021.802802.

Wang, L., Di, T., Peng, J., Li, Y., Li, N. *et al.* (2022) Comparative metabolomic analysis reveals the involvement of catechins in adaptation mechanism to cold stress in tea plant (*Camellia sinensis* var. *sinensis*). *Environmental and Experimental Botany* 201, 104978. DOI: 10.1016/j.envexpbot.2022.104978.

Wang, L., Li, B., Dai, C., Ding, A., Wang, W. *et al.* (2023a) Genome-wide identification of MAXs genes for strigolactones synthesis/signaling in solanaceous plants and analysis of their potential functions in tobacco. *PeerJ* 11, e14669. DOI: 10.7717/peerj.14669.

Wang, X., Li, Z., Shi, Y., Liu, Z., Zhang, X. *et al.* (2023b) Strigolactones promote plant freezing tolerance by releasing the WRKY41-mediated inhibition of *CBF*/DREB1 expression. *EMBO Journal* 42(19), EMBJ2022112999. DOI: 10.15252/embj.2022112999.

Waters, M.T., Gutjahr, C., Bennett, T. and Nelson, D.C. (2017) Strigolactone signaling and evolution. *Annual Review of Plant Biology* 68, 291–322. DOI: 10.1146/annurev-arplant-042916-040925.

Wei, X., Liu, S., Sun, C., Xie, G. and Wang, L. (2021) Convergence and divergence: Signal perception and transduction mechanisms of cold stress in *Arabidopsis* and rice. *Plants* 10(9), 1864. DOI: 10.3390/plants10091864.

Xin, X., Wang, S., Pan, Y., Ye, L., Zhai, T. *et al.* (2025) MYB transcription factor CDC5 activates *CBF3* expression to positively regulate freezing tolerance via cooperating with ICE1 and histone modification in *Arabidopsis*. *Plant, Cell & Environment* 48(1), 97–108. DOI: 10.1111/pce.15144.

Xu, K., Wu, N., Yao, W., Li, X., Zhou, Y. *et al.* (2021) The biological function and roles in phytohormone signaling of the F-box protein in plants. *Agronomy* 11(11), 2360. DOI: 10.3390/agronomy11112360.

Xu, T., Niu, J. and Jiang, Z. (2022) Sensing mechanisms: Calcium signaling mediated abiotic stress in plants. *Frontiers in Plant Science* 13, 925863. DOI: 10.3389/fpls.2022.925863.

Yao, R., Wang, L., Li, Y., Chen, L., Li, S. *et al.* (2018) Rice DWARF14 acts as an unconventional hormone receptor for strigolactone. *Journal of Experimental Botany* 69(9), 2355–2365. DOI: 10.1093/jxb/ery014.

Yu, S., Li, P., Liu, H., Zhang, X., Gao, Y. *et al.* (2025) A CCA1-like MYB subfamily member CsMYB128 participates in chilling sensitivity and cold tolerance in tea plants (*Camellia sinensis*). *International Journal of Biological Macromolecules* 294, 139473. DOI: 10.1016/j.ijbiomac.2025.139473.

Yu, W., Ma, P., Sheng, J. and Shen, L. (2023) Postharvest fruit quality of tomatoes influenced by an ethylene signaling component during long-term cold storage. *Food Chemistry* 422, 136087. DOI: 10.1016/j.foodchem.2023.136087.

Zhai, Y., Fan, Z., Cui, Y., Gu, X., Chen, S. *et al.* (2022) APETALA2/ethylene responsive factor in fruit ripening: Roles, interactions and expression regulation. *Frontiers in Plant Science* 13, 979348. DOI: 10.3389/fpls.2022.979348.

Zhang, G., Kong, G. and Li, Y. (2021) Long-distance communication through systemic macromolecular signaling mediates stress defense responses in plants. *Physiologia Plantarum* 173(4), 1926–1934. DOI: 10.1111/ppl.13535.

12 Deciphering the Role of Strigolactones in the Alleviation of Waterlogging Stress

Ansa Asghar[1]*, Usman Tufail[1], Rizwana Kanwal[1], Asima Asghar[2], Ayesha Zafar[1], Farooq Ahmad[1], Mansoor Hameed[1], Muhammad Sajid Aqeel Ahmad[1], Sana Basharat[3]*, Wajid Saeed[3], Samavia Mobeen[3] and Waleed Hussain[4]

[1]Department of Botany, University of Agriculture Faisalabad, Faisalabad, Pakistan; [2]Department of Zoology, Lahore College for Women University, Lahore, Pakistan; [3]School of Breeding and Multiplication (Sanya Institute of Breeding and Multiplication), School of Tropical Agriculture and Forestry, Hainan University, Sanya, China; [4]Department of Botany, University of Narowal, Pakistan

Abstract

The development of vine growth and the fruit quality of grapes, which are economically significant fruit crops globally, are seriously endangered by the frequent occurrence of heavy rains and the escalation of waterlogging catastrophes brought on by global climate change. As a result, it is crucial to understand how plants react to waterlogging and to suggest suitable corrective actions. Although its role in plants' reactions to waterlogging stress is yet unknown, strigolactones (SLs), plant phytohormones, have been linked to abiotic stress tolerance. The morphology, physiology, and transcriptome alterations in the leaves and roots of plants following waterlogging and after GR24 (a synthetic analog of SL) treatments have been systematically analyzed. Transcriptome analysis has shown that while the main gene functions brought on by post-waterlogging stress varied over time, they were always linked to the metabolism of carbohydrates. To aid in the healing of the root system following waterlogging, it has been suggested that GR24 treatment improves the cytoskeleton and microtubules for cell formation. Such studies enable us to better understand how plants react to waterlogging and how GR24 helps them by providing insight into the transcriptome alterations that occur during these processes. Gaining insight into the function of SLs in waterlogging tolerance opens up new avenues for creating robust crop types and sustainable farming methods to address flooding stress brought on by climate change.

Keywords: carotenoids, GR24, hypoxic, lactone, strigolactones, waterlogging

12.1 Introduction

One important abiotic stressor that has an adverse effect on development of crops, growth, and yield is waterlogging (Zhang *et al.*, 2021). Waterlogging means the soil contains less available oxygen (O_2), which makes the root system's environment anoxic or anaerobic. As a result of this phenomena, the plant leaves gradually turn yellow, wilt, and shed, and the rhizomes gradually rot and

*Corresponding authors: ansaasghar1122@gmail.com and sanabasharat143@gmail.com

DOI: 10.1079/9781836990598.0012

become brown (Swetha and Sundararaj, 2022). Energy metabolism switches from aerobic to anaerobic respiration under hypoxic or anaerobic circumstances. The quantity of ATP generated per molecule of glucose through anaerobic respiration is considerably lower than that produced during aerobic respiration. This process lowers the amount of metabolizable sugar in crop roots, which impacts the likelihood that crops will survive waterlogging. Waterlogging also reduces the opening of leaf stomata, which lowers the rate at which carbon dioxide (CO_2) diffuses (Zhang *et al.*, 2025a). The activity of photosynthesis is adversely affected by the hypoxic environment. Enzymes drastically lower the amount of chlorophyll, which in turn reduces photosynthetic capacity. They also restrict the synthesis of chlorophyll and speed up its breakdown. Waterlogging causes reactive oxygen species (ROS), including hydrogen peroxide (H_2O_2) and O_2, to build up because it upsets the equilibrium of the plant photosystem and the Mehler reaction (Sachdev *et al.*, 2021). Waterlogging also causes an imbalance between anaerobic metabolic processes and aerobic respiration, which adds to the buildup of ROS, which can be harmful to plants. Under waterlogging stress, crops have developed specialized antioxidant defense mechanisms to lessen the negative effects of ROS. These methods include using active O_2-scavenging enzymes including catalase (CAT), peroxidase (POD), and superoxide dismutase (SOD), as well as non-enzymatic scavengers such as vitamin E and ascorbic acid, which lessen the damage that ROS cause to crops under waterlogging stress (Moursi *et al.*, 2021).

Due to their many physiological roles, strigolactones (SLs), a recently discovered family of plant hormones, have attracted much interest. They encourage branching of arbuscular mycorrhizal fungal hyphae and aid in the germination of parasitic plant seeds (Boyno *et al.*, 2023). Additionally, SLs prevent plant lateral buds from germinating both directly and indirectly. Notably, SLs show possible protective benefits in a range of stressful situations. For example, it has been demonstrated that GR24, a synthetic analog of SLs, lessens the negative effects of drought stress in wheat and *Arabidopsis*. For example, SLs reduce the physiological toxicity that pumpkin seedlings encounter under waterlogging stress although the precise mechanism is still unknown. The results of this study offer theoretical justification and useful recommendations for improving pumpkin seedlings' resistance to waterlogging in agricultural production (Liu *et al.*, 2024).

12.2 Strigolactone Biosynthesis

12.2.1 SLs are lactone-containing compounds

Over 30 SLs have been identified and structurally characterized using both qualitative and quantitative methods, such as tandem mass spectrometry using liquid chromatography. All SLs are made up of four rings (A–D), with an enol-ether bridge between the core tricyclic lactones (ABC-rings) and the D-ring, a chemical butenolide group (Fig. 12.1) (Xu *et al.*, 2021). SLs have one or more phenol or acetylenic members on the A- and B-rings, and one to two methyl chains on the A-ring. Because there are so many side groups on the C- and D-rings, they are the most preserved, while the A- and B-rings are the most varied. 5-Deoxystrigol, which was discovered in the root exudates of *Lotus japonicus* L., are the most basic SL, with no substitutions on the A- and B-rings (Zaidi *et al.*, 2024).

The presence or absence of the ABC-ring distinguishes two types of SLs. Canonical SLs such as strigol and orobanchol include the ABC-ring, while non-canonical SLs lack the ABC-ring and have only the D-ring. Carlactone, which has a β-ionone ring, is the precursor for SLs (Clark and Bennett, 2024).

5-Deoxystrigol consists of all four rings, A–D, containing hydrogen atoms in at the R^2 and R^3 positions on the A- and B-rings, respectively, and a methyl group in at the R^1 position on the A-ring. The D-ring has a methyl group at the C-4′ position, and the C- and D-rings are joined by an enol-ether bond. For SL action, the C- and D-rings are necessary according to studies on the structure–function relationship, and stereochemical structure at the D-ring's C-2′ position affects its bioactivity.

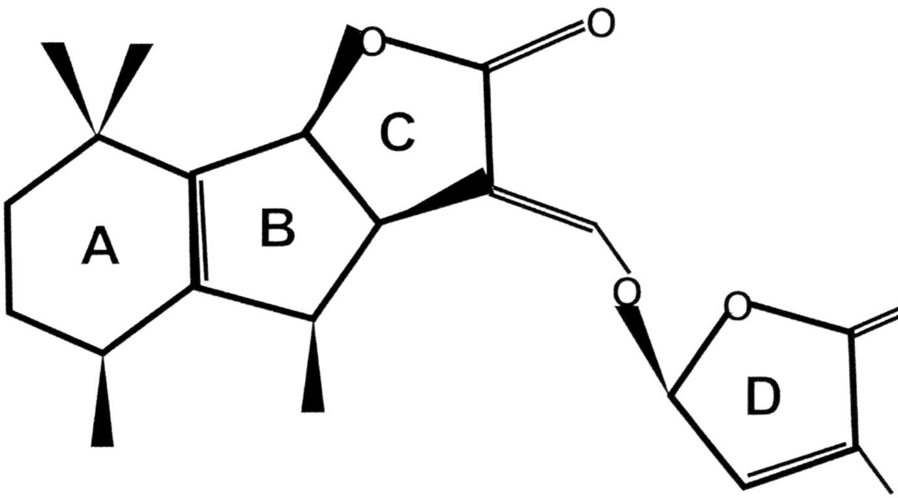

Fig. 12.1. Structure of strigol, a canonical strigolactone.

12.2.2 SLs are derived from carotenoids

All plant photosynthetic membranes contain carotenoids, which are isoprenoids made up of four terpene units. A variety of metabolites containing carbonyls, including retinoids, fungal pheromones, and abscisic acid (ABA) are produced when carotenoids break down into apocarotenoids, which can cleave at the C=C double bonds both enzymatically and non-enzymatically (Liang *et al.*, 2021). All SLs are sesquiterpene lactones chemically, and they are carotenoid derivatives based on their structural resemblance to terpenoids and isoprenoids. The primary processes in plants that generate terpenoids and isoprenoids involve mevalonic acid in the cytosol system and methylerythritol phosphate (MEP) in the plastidic route. Monoterpenes, diterpenes, carotenoids, gibberellins (GAs), ABA, and other substances are produced by the MEP route (Perreca *et al.*, 2022). To determine the components of the SL biosynthesis pathway, exudates were tested from the roots of wild-type plants treated with fluridone, an inhibitor of the carotenoid biosynthesis pathway, and maize carotenoid biosynthesis mutants, with the results providing credence to the theory that SLs are produced from carotenoids, most likely via the MEP route, and have germination-promoting properties.

By determining the enzymes needed for carotenoid degradation, such as carotenoid cleavage dioxygenases (CCDs), carotenoids were further proven to be the precursors for SLs. ABA is created when CCDs stimulate the oxidizing breakage of the double bond in 9-*cis*-epoxycarotenoids. By examining the maize mutant *viviparous14* (*vp14*), 9-*cis*-epoxycarotenoid dioxygenase (NCED) was the first CCD to be discovered (Qi *et al.*, 2023).

12.3 SL Signaling

A specific cellular mechanism that has previously been characterized is responsible for SL signaling. Fig. 12.2 shows a workable model of the SL signaling system based on these earlier explanations. DWARF14 (D14), an α/β-hydrolase detects SLs and allows a repeating leucine-rich F-box type protein to receive the signal. This protein can then bind to a complex of F-BOX–CULLIN–SKP (SCF). This interaction results in protein ubiquitination and initiates the degradation of transcription receptors, such as SUPPRESSOR OF MAX2 1-LIKE proteins (SMXLs) in *Arabidopsis* and DWARF53 (D53) in rice, by the 26S proteasome. When SLs are present, DWARF14 (D14) interacts with them

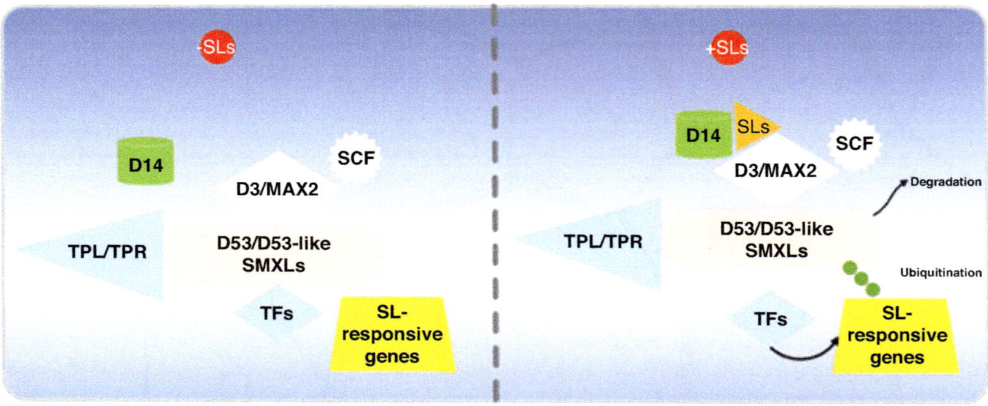

Fig. 12.2. Overview of the strigolactone (SL) signaling pathway. D14, DWARF14; D53, DWARF53; SCF, F-BOX–CULLIN–SKP; MAX2, MORE AXILLARY GROWTH 2; TPL/TPR, TOPLESS/TOPLESS-RELATED; TF, transcription factor.

to create a molecule from the D-ring (Sánchez Martín-Fontecha *et al.*, 2024). A nucleophilic assault then covalently seals the molecule in D14's catalytic active site. This link leads to a conformational shift of D14, which connects via the D53/D53-like SMXL enzymes and the D3/MAX2-derived SCF complex. As a result, important downstream genes such as D53 no longer undergo transcriptional suppression. When SLs are not present, D53 and D53-like SMXL proteins bind to TOPLESS/TOPLESS-RELATED (TPL/TPR) proteins, inhibiting downstream target genes and preventing the action of unknown transcription factors (Shen *et al.*, 2020).

12.4 Interactions Between Strigolactones and Other Phytohormones

For plants to grow normally, phytohormones must coordinate and communicate with one another. Certain essential elements of signaling pathways regulate hormone-to-hormone crosstalk (Kaleem *et al.*, 2025). Auxin, GAs, cytokinins (CKs), and ethylene are other plant hormones with which SLs communicate to carry out a number of crucial tasks (Nambara and van Wees, 2021). The following subsections outline studies that have demonstrated the connections between SLs and other plant hormones.

12.4.1 SLs and auxin

The essential signaling substance for growth and development of plants is auxin. There is a special feedback loop between auxins and SLs. Auxin partakes in SL-mediated events and controls SL production (Faizan *et al.*, 2020). In contrast, SLs increase auxin transport in plant stems, which increases the competition among axillary branches and axillary bud expansion. Auxin-induced secondary messengers are mediated by SLs, which travel upward to the buds and impede their expansion. Expression of the genes that produce SLs is regulated by auxin. However, auxin cannot suppress decapitation-induced shoot diverging without SLs. Consistent with the results of Agustí and Blázquez (2020), to increase vascular cambium activity, auxin requires SL biosynthesis and signaling. By increasing the rate at which the PIN-FORMED1 (PIN1) protein is removed, SLs restrict the formation of buds. Additionally, there is evidence that auxin and SLs interact to affect secondary growth, root formation, and tuberization. SLs affect auzin pathways either by promoting transcription of the auxin receptor TIR1 or by facilitating auxin transport (Kondhare *et al.*, 2021).

12.4.2 SLs and CKs

CKs are recognized to play a variety of roles in plant progress impacting several significant processes that are important agriculturally. Through its interactions with auxin, ABA, and other phytohormones including CKs and GAs, SLs regulate the growth and development of plants (Sosnowski *et al.*, 2023). While CKs promote bud expansion, SLs inhibit it. In *Pisum sativum*, a legume, the bud-specific gene *BRANCHED 1 (BRC1)* generates a transcription factor that prevents bud expansion and is impacted by both CKs and SLs. Auxin acts as a secondary messenger that counteracts CKs by raising SL levels. Exogenous SL application in *P. sativum* reduces axillary shoot length when decapitated or when CK is produced, which provides further evidence of interactions between SLs and CKs (Cao *et al.*, 2023).

12.4.3 SLs and ABA

SLs affect soil organisms, regulate plant morphology, and aid plants in adapting to unfavorable circumstances, possibly through ABA crosstalk (Soliman *et al.*, 2022). A possible biosynthetic relationship is suggested by the structural similarity between ABA and SLs. Both arise from a carotenoid biosynthesis pathway leading from all-*trans*-β-carotene to all-*trans*-violaxanthin. Thus, it may be concluded that the homeostasis of both hormones depends on one another. The DWARF27 (D27) protein's 9-*cis*/all-*trans*-β-carotene isomerization activity may indicate potential SL–ABA biosynthesis connections (Singh *et al.*, 2023). The considerably decreased germination of parasite seeds resulting from the extracts of roots from maize crops lacking ABA and harboring a deletion in the ABA-biosynthesis gene *ZmNCED1* is believed to be caused by low SL content, and the manufacture of SLs is suppressed in tomatoes when the concentration of ABA is decreased (either chemically or genetically) (Faizan *et al.*, 2020). Differences in the degree of sensitivity to SLs have an impact on ABA concentrations and ABA responses. Depending on the organism, SLs and ABA interact in different ways (Zhang *et al.*, 2023a). The decreased drought stress tolerance of SL biosynthesis mutants of *Lotus japonicus* and *Lycopersicon esculentum*, which is also connected to elevated ABA levels in the leaves, indicates a positive relationship between the two phytohormones. Osmotic stress rapidly and significantly reduced the amount of SLs in the tissues of wild-type *Lotus* roots, whereas drought stress raised the amount of ABA in *Lactuca sativa*, *Lotus japonicus*, and *Lycopersicon esculentum* while lowering the amount of SLs (Faizan *et al.*, 2020). Furthermore, it is known that stress alters SL metabolism and has an effect on ABA homeostasis in the roots and shoots (Abbas *et al.*, 2023).

12.5 Crosstalk Between Strigolactones and Abscisic Acid

Auxin instability in root tips may be correlated with SL signaling systems when lateral root initiation is disrupted. In conjunction with SL effects on meristem growth, Koren *et al.* (2013) claimed that SL signaling controls auxin flow into the root tip. Additionally, auxin and CK crosstalk regulates root growth by balancing cell partitioning and demarcation in the root tip meristem (Azarakhsh and Lebedeva, 2023). The auxin–CK balance in the root apical meristem is regulated by the gene *SHY2*. CKs stimulate *SHY2* transcription, whereas auxin suppresses *SHY2* activity. At the root tip, expression of *PIN1*, *PIN4*, *PIN3*, or *PIN2* transport is negatively impacted by SHY2's suppression of the *PIN* gene (Mandal *et al.*, 2024). Therefore, CKs increase the flow of auxin, which in turn influences cell division, via favorably regulating the activity of the *SHY2* gene. Furthermore, SHY2 signaling promotes the growth of pre-existing lateral roots but prevents their genesis by upsetting auxin homeostasis. SHY2 functions as a molecular switch in auxin transport, which mediates the effects of SLs on lateral root initiation (Jan *et al.*, 2024).

12.6 Strigolactones and Brassinosteroids

A thorough understanding of this inter-hormonal crosstalk is currently being developed, as the interaction between the SL and brassinosteroid (BR) signaling pathways has only recently been discovered. This signaling connection is known to be mediated by key components of the SL and BR signaling relays (Ravi *et al.*, 2023). The E3 ubiquitin ligase MAX2, which is known to limit shoot branching, is a crucial component of SL signaling. The transcription factors BZR1 and BES1, which are important modulators of BR-dependent gene expression, have direct interactions with MAX2 (Shi *et al.*, 2022). According to reports, BZR1 and BES1 are continuously engaging with the MAX2 E3 ubiquitin ligase. BZR1 and BES1 are immediately bound by MAX2, which degrades them. SL signaling prevents shoot branching by encouraging the MAX2-mediated degradation of the BZR1/BES1 transcription factors. Therefore, it has been suggested that by modifying BZR1/BES1 stability, BRs and SLs control the same developmental process (Sharma *et al.*, 2022).

12.7 Waterlogging Stress in Plants

Following waterlogging, there can be an abrupt increase in O_2 and light within the plant system, as plants experience a rapid environmental shift from water to air, which can cause serious harm to the plant (Zhang *et al.*, 2025a). This reoxygenation, often known as post-waterlogging injury, is the primary source of this harm. The overproduction of ROS following abrupt re-exposure to ambient O_2 is mostly caused by reoxygenation of plants. Photoinhibition is the term used to describe the harm that intense light after waterlogging stress does to the photosynthetic apparatus. The strong, brilliant light that generates ROS in photosystem I (PSI) and II (PSII), which are located in the thylakoids of the plastid, is primarily responsible for the inactivation of the reaction centers of the photosystems. Furthermore, photosystem II, which increases the production of H_2O_2 can be directly destroyed by high light. Additionally, ROS limit the synthesis of new proteins that repair photodamaged PSII (Foyer and Hanke, 2022).

The O_2-containing molecules known as ROS are more chemically reactive than O_2. The endoplasmic reticulum, mitochondria, peroxisome, and chloroplast are the four main plant organelles that produce ROS. ROS-producing enzymes are also found in the apoplast, cell wall, and cell membrane. The electron transport chains involved in photosynthesis in chloroplasts and respiration in mitochondria are the primary producers of ROS in aerobic metabolism in plants. Singlet O_2 ($^1O_2$), superoxide anion (O_2^-), H_2O_2, hydroxyl radical (OH^-), hydroperoxyl radical (HO_2), and ozone (O_3) are some of the radical and non-radical ROS found in plants. ROS can damage cellular nucleic acids, proteins, lipids, and subcellular structures (Edge and Truscott, 2021).

After waterlogging stress, plant shoot tissues experience drought-like stress due to either damaged or decreased root hydraulic conductivity. Plants under post-waterlogging stress exhibit comparable phenological and physiological symptoms to those under drought stress, including leaf rolling, withering, stomatal closure, and a drop in relative water content as a result of changes in membrane permeability. However, when osmotic stress and water deprivation are present, the expression of pheophorbide a oxygenase (*PAO*), stay-green (*SGR*) and pheophytinase (*PPH*) genes is markedly increased, which ultimately damages the amount of chlorophyll (Wahab *et al.*, 2022). As a result, photosynthesis is lost under waterlogging stress due to the breakdown of photosynthetic components, in addition to the decrease in CO_2 absorption caused by stomatal closure.

12.7.1 Hormonal modulation in the recovery from waterlogging stress

The main phytohormone involved in overcoming waterlogging and waterlogging stress is ethylene. Both submerged root and shoot tissues produce ethylene from its precursor, 1-amino cyclopropane-1-carboxylic acid (ACC) (Chen *et al.*, 2022). Thus, pre-existing ethylene and the creation of new ACC in the shoot tissues upon reoxygenation result in ethylene buildup during waterlogging stress. ERF IX (ERF1 and ERF2) is very active when plants are reoxygenated, whereas ERF VII (ethylene response factor VII) primarily controls plants' early response to waterlogging stress. Ethylene, which primarily inhibits ABA signaling, activates several dehydration-responsive element-binding (DREB) transcription factors, including C-repeat binding factor 1 (CBF1), TINY, and HARDY. One of the most significant transcription factor families in plants, AP2/ERFs (APELATA2/ethylene response factors), regulates a variety of biotic and abiotic stress-sensitive processes (Tang *et al.*, 2024). AP2/ERFs are further classified into AP2, RELATED TO ABSCISIC ACID INSENSITIVE3/ VIVIPAROUS1 (RAV), DREB (subgroups A1–A6), ERF (subgroups B1–B6), and others based on their DNA-binding domain. Expression of the genes responsible for tobacco dehydration stress is indicated by the presence of DREBs (Adel and Carels, 2023).

During reoxygenation, ethylene and jasmonic acid (JA) have antagonistic interactions. During reoxygenation, they keep cellular homeostasis and phytohormone responses in equilibrium, but this does not imply that they prevent one another from functioning. When antioxidants have eliminated as much ROS as possible, ethylene controls JA-mediated antioxidant production. JA causes the expression of a transcription factor known as MYELOCYTOMATOSIS 2 (MYC2), which is an essential regulator in the JA-mediated reoxygenation response (Song *et al.*, 2022). Another pathway involves MYC2 expression through the ABA-dependent differentiation pathway. The gene families for glutathione synthase (GSH) and VITAMIN C DEFECTIVE (VTC), which are in charge of the ascorbate and glutathione production pathways, are activated by MYC2. The ascorbate–glutathione cycle in plant cells uses two primary non-enzymatic antioxidants, ascorbate and glutathione, to detoxify ROS. This suggests that plants under reoxygenation stress are rescued by JA. During the osmotic stress of post-waterlogging stress, plants additionally release the plant hormone salicylic acid, which enhances the control of H_2O_2-metabolizing enzymes, such as ascorbate peroxidase, POD, and CAT, as well as SOD during dehydration stress.

Low O_2 levels resulting from waterlogging stress limit the generation of ATP by mitochondrial respiration, which results in a lack of energy in plants (Pan *et al.*, 2021). Therefore, in addition to storing energy, plants must look for other energy production pathways or non-TCA (tricarboxylic acid) cycle enzymes that will speed up the energy production under O_2-stressed conditions. Ethylene signaling provides these non-TCA cycle enzymes, which increase the TCA cycle rate during post-waterlogging reoxygenation stress. In the phase of reoxygenation following waterlogging, ethylene signaling controls the activity of glutamate dehydrogenase (GDH) and pyruvate phosphate dikinase (PPDK) (Yang *et al.*, 2023). To continue the TCA cycle, PPDK transforms phosphoenol pyruvate into pyruvate. Additionally, GDH is responsible for the conversion of glutamate to 2-oxoglutarate during reoxygenation stress, which indirectly provides pyruvates to speed up the TCA cycle as shown in Fig. 12.3.

The reaction of plants to dehydration pressure is significantly influenced by CKs. Under dryness stress, equal up- and downregulation of CKs has been demonstrated. In contrast to ABA, CKs delay the senescence of leaves as well as stomatal closure. The dehydration response regulates gene expression through both receptor-independent and -dependent CK receptors. Driven by a promoter unique to senescence, SENESCENCE-ASSOCIATED GENE 12 (SAG12), the CK biosynthesis gene *IPT* (isopentenyltransferase) generates CK in *Arabidopsis* (Glanz-Idan *et al.*, 2022). During post-submergence recovery, transgenic wheat plants displayed delayed senescence and elevated CK content. Thus, the ability of *Arabidopsis* to withstand dehydration stress is enhanced when CK levels are decreased.

The control of dehydration stress in plants is also influenced by other hormones, such as auxins, GAs, SLs, and BRs. Through the production of a gene linked to auxin, transgenic poplar plants demonstrated improved resistance to oxidative stress. By overexpressing the Bs biosynthesis gene *DWARF4* (*DWF4*), transgenic *Brassica napus* demonstrated superior resistance to heat stress and dehydration in comparison with wild type. SLs and H_2O_2/nitric oxide work together to activate *SLOW ANION CHANNEL-ASSOCIATED*

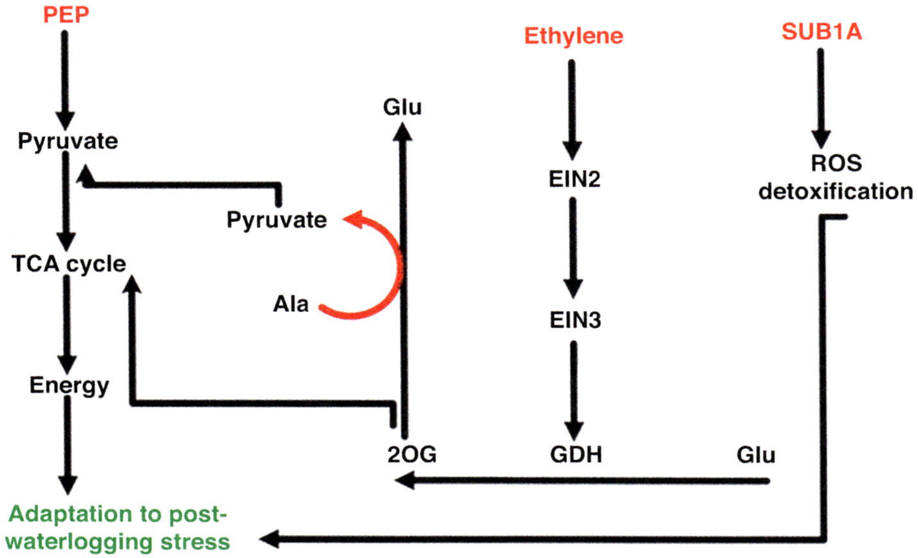

Fig. 12.3. Plant adaptability to post-waterlogging stress is influenced by the ethylene signaling system. 2OG, 2-oxoglutarate; EIN2/3, ETHYLENE INSENSITIVE 2/3; GDH, glutamate dehydrogenase; PEP, phosphoenol pyruvate; ROS, reactive oxygen species; SUB1A, SUBMERGENCE1A; TCA, tricarboxylic acid.

1 (*SLAC1*) without ABA, while stomatal closure is controlled by SLAC1 (Ma *et al.*, 2024).

Little is known about the interactions between the various plant hormones that regulate reoxygenation or dehydration stress in plants during post-waterlogging stress. The key takeaway, however, is that all of these hormones either open or close stomata, or detoxify ROS via various physiological and signaling routes to help plants that are experiencing reoxygenation or dehydration stress after waterlogging.

12.8 Beneficial Effects of Plant Growth-promoting Rhizobacteria-mediated Phytohormones

In addition to degrading natural resources, abiotic stressors such high soil salinity, drought, floods, oxidative stress, heat, cold, and heavy metals can hinder crop development and yield by lowering nutrient absorption (Kumari *et al.*, 2022). Plant nutrients have a significant impact on the physical and biological developments of plants, and a lack of these nutrients can hinder a plant's growth and development or even cause it to die. The best methods for creating stress-resistant plants are genetic engineering and plant breeding, yet despite tremendous efforts, little progress has been made in this area. One strategy to mitigate nutrient insufficiency in abiotic stress situations is to increase nutrient availability and absorption for plants. One promising option for protecting plants is the use of plant growth-promoting rhizobacteria (PGPR), which can be used to populating the rhizosphere and generate antagonistic antimicrobial compounds to protect the plant from pathogenic bacteria and fungi, thus supporting plant health and immunity (Hamid *et al.*, 2021).

PGPRs connected to plant roots enhance plant efficiency and protection in a number of ways, including the overproduction of useful phytohormones, which lead to increased root surface area and number of root tips, thus enhancing nutrient absorption from the soil and reduces stress-related damage (Khan *et al.*, 2021). In addition to producing growth-promoting hormones and converting atmospheric nitrogen into physiologically accessible nitrogen compounds, PGPR also helps stop plant pathogen infestation and boost plant resistance to infections. PGPR promote plant development by facilitating the synthesis and breakdown of the main groups of plant hormones. Plant hormones or their precursors are among the several possible substrates for rhizobacterial development found in plant root exudates. The local impacts of rhizobacterial mediation of plant hormones include root elongation and architecture, and improved water uptake (Aloo *et al.*, 2023).

Environmentalists across the world are pushing to lower chemical fertilizer levels below those advised for maximum yields, due to their environmental consequences and other environmental pressures in crop production. However, this would result nutritional stress in the plants. Therefore, using PGPR is a good way to increase crop output while using less fertilizer. According to Artyszak and Gozdowski (2021), the yield of wheat (*Triticum aestivum* L.) produced using a PGPR strain with 75% of the required dosages of NPK (nitrogen, phosphorus, potassium) fertilizers was equal to that achieved with the maximum dose of NPK fertilizer without PGPR.

There are two types of PGPR: intracellular and extracellular. Both encourage plant growth and development in unfavorable environments. According to Grover *et al.* (2021), plant growth and development can be directly enhanced by the profuse root formation that occurs in response to PGPR stimulation. The production of phytohormones such as indole-3-acetic acid (IAA), CKs, and GAs during drought, nutrient restriction, and salt and metal toxicity stressors affects root structure. A higher root surface area allows plants to absorb more nutrients from the soil, which aids plant development. Additionally, PGPR improve the efficiency of plant absorption of nutrients (nitrogen, phosphorus, and potassium). PGPB strains also enhance nitrogen fixation and help plants thrive in stressful soil circumstances (Hasan *et al.*, 2024).

In order to promote plant growth and development, PGPR neutralize or decrease stress-related hormones (ethylene via bacterial ACC deaminase activity) that are produced in plants under stressful conditions. They also decrease the injury caused by various pathogens, thus acting as environmental protectors, root colonizers, and biocontrol agents (Chaudhary *et al.*, 2022). Phytohormones are one of the primary signaling molecules generated by PGPR used to increase agricultural output. Plant immunity, biocontrol, and crop yield are stabilized by IAA and other hormones (natural, semi-synthetic, and synthetic) such as ABA, CK, ethylene, and others. Phytohormones are used both endogenously and exogenously to maintain a number of physiological traits in plants, including seed dormancy, seedling emergence, initiation, enhancement, elongation, and somatic embryogenesis, which are important to maintain sustainable agriculture and manage food production. Some PGPR have been found to release new signaling chemicals that aid in the development of plants.

12.9 Metabolic Responses to Waterlogging in Crop Species

Rapid rhizosphere O_2 depletion upsets soil chemistry and interferes with hormone and energy metabolism, which sets off the subsequent physiological and biochemical processes outlined in the preceding section (Chen *et al.*, 2024). Cotton is highly susceptible to waterlogging so enhancing the cell-level resistance to waterlogging is a logical target for adaptive responses. Four general categories may be used to classify the known metabolic reactions to O_2 deficiency:

1. Anaerobic polypeptides are induced, which facilitates the mobilization of carbohydrates and fermentation.
2. Acidosis controls the activity of several important enzymes (Zhao *et al.*, 2024), including pyruvate decarboxylase (PDC), nitrate reductase, and nicotinamide adenine dinucleotide (NADH)-dependent glutamate synthase, and it establishes a new "set point" for low-O_2 metabolism.
3. According to a recent study, the N-end rule pathway mediates O_2 sensing in barley. The activation of proteins that regulate ROS signaling and related components of the N-end rule pathway found in barley are potential targets for improving cotton's resilience to waterlogging (Yang *et al.*, 2023).

4. *In vivo* sensing is achieved using an amino-terminal cysteine residue, which improved resistance to waterlogging and the generation of genes associated with hypoxia.

12.10 Signaling Pathways and Gene Regulation

Although proteomic and genomic techniques have increased our understanding of responses to O_2 deficiencies, the entire range of responses that can confer waterlogging tolerance is still unknown (Table 12.1). Ethylene synthesis, nitrogen metabolism, and cell-wall disintegration are additional examples of often-seen gene-expression responses under hypoxia (Table 12.1) (Hartman *et al.*, 2021). In a wide variety of plant species that represent varying degrees of flood tolerance, it has been demonstrated that similar genes are upregulated in response to O_2 shortages. These continuous alterations show that evolutionary "solutions" to surviving this most demanding of environmental stressors originate from ancient progenitors, who are frequently prokaryotic.

Although many different plant species exhibit similar transcriptome responses to anoxia and hypoxia, individual plant species are also represented in the transcriptomic profiles. Hypoxic stress mostly affects the root tissues (Basit *et al.*, 2024). The reactions of the shoot (induction of hypoxia-responsive genes) may also be regulated by the migration of metabolites such as alanine and γ-aminobutyrate toward the shoot. These separate transcriptome and proteome investigations yield data sets that can be used to find hypoxia tolerance markers. Between 4% and 10% of known genes showed differential expression in response to hypoxia, according to early transcriptome differences in roots from cotton. As a result, a thorough examination of distinct tissues from a variety of organs (e.g. roots) throughout a wide variety of taxa is still required. Yang *et al.* (2023) used microarray analysis of waterlogged cotton roots and leaves to determine upregulation of

Table 12.1. Gene-expression investigations in a variety of higher plant species subjected to low-oxygen circumstances.

Species	Treatment	Genes upregulated	Genes downregulated	Reference
Cotton (*Gossypium hirsutum*)	Soil waterlogging	Glycolysis, ethylene synthesis, alanine synthesis	Cell-wall synthesis, synthesis of amino acids	Owusu *et al.* (2023)
Arabidopsis	Hypoxia (3% oxygen)	Glycolysis, ethylene synthesis, protein kinase activity	Cell-wall synthesis, nucleosome structures	Wang *et al.* (2025)
Poplar	Soil waterlogging	Glycolysis, fermentation, trehalose synthesis	Phenylalanine synthesis	Guo *et al.* (2024)
Rice	Anoxia	Glycolysis, ethylene response factors, heat-shock proteins, starch synthesis	Phosphoenol pyruvate carboxylase, sugar transporters	Gautam *et al.* (2022)
Sugarbeet	Waterlogging	Glycolysis phosphate cycle, seed-specific proteins	Cytoplasmic ribosomal proteins, seed storage proteins	Long *et al.* (2022)
Maize	Submergence	Glycolysis, ethanolic fermentation	Starch synthase, aminotransferase	Kim *et al.* (2024)
Soybean	Submergence	Photosynthesis, glycolysis, Ser-Gly-Cys group amino acid synthesis	Synthesis of phosphofructokinase glucosyl and glucuronyl transferase	Chatterjee *et al.* (2025)

the genes controlling metabolic activities such as fermentation, glycolysis, and mitochondrial electron transport pathways. This highlights how important residual respiratory activity and ethanolic fermentation are to plant survival in hypoxic environments. Genes involved in producing amino acids, flavonoids, and cell walls are examples of downregulated genes that may provide similarly useful insight into flood-tolerance processes. The best data sets for inferring such differences are those from cereals whose responses to flooding have been comparatively thoroughly examined. Development of breeding programs in cotton would be greatly aided by systematic analysis, especially when comparing domestic and wild cultivars.

12.11 Metabolic Adaptation

Through complex structural changes at the cellular and organ levels, waterlogging-tolerant plants may be able to escape O_2 deficiencies (Zhang *et al.*, 2025a). These changes are triggered by phytohormones, with ethylene, GAs, and ABA all known to play roles in cell-level reactions to decreased O_2. Quiescence has gained more prominence in recent years as a survival tactic during submersion, preserving energy during periods of limited O_2 supply. This is the most likely way to increase field-grown cotton's resistance to waterlogging.

According to Zenda (2025), the fermentation of alcohol is the most significant fermentative process in plants. Pyruvate decarboxylase first transforms pyruvate into acetaldehyde, and alcohol dehydrogenase subsequently transforms it into ethanol. Jayawardhane *et al.* (2021) confirmed that in hypoxic cotton seeds, the primary metabolic route for energy production is ethanolic fermentation. However, when compared with tolerant plants, radicle development was dramatically decreased at these very low fermentation rates (Table 12.2), suggesting that

Table 12.2. Changes in the production of ethanol in various plant organs when oxygen is scarce.

Species	Plant organ	Oxygen concentration	Ethanol synthesis rate	Reference
Cotton	Seeds	9 ± 4 mmol O_2 mol^{-1}	0.44 μmol h^{-1} seed^{-1}	Arroyo *et al.* (2021)
	Roots	Hypoxia (5% O_2)	0.05 μmol g^{-1} FW h^{-1}	Feng *et al.* (2022)
	Transgenic roots		0.06–0.1 μmol g^{-1} FW h^{-1}	Wang *et al.* (2024)
Rice	Whole plant (14 days)	Anoxia (N_2) 20 hour	28 μmol g^{-1} FW h^{-1}	Yemelyanov *et al.* (2022)
	Shoots (14 days)	Anoxia (N_2) 4 hour	50 μmol g^{-1} FW h^{-1}	Yemelyanov *et al.* (2022)
	Roots (14 days)	Hypoxia (3% O_2)	2.5 μmol g^{-1} FW h^{-1}	Liu *et al.* (2023)
	Coleoptiles	Anoxia (N_2)	5.2–8.3 μmol g^{-1} FW h^{-1}	Zuo *et al.* (2022)
			6.8–9.7 μmol g^{-1} FW h^{-1}	
Maize	Root tips (3 days pre-hypoxic)	Anoxia (N_2) 8 hour	15.7 μmol g^{-1} FW h^{-1}	Shao *et al.* (2024)
Lettuce	Roots (5 days)	Anoxia (N_2) 6 hour	1.8 μmol g^{-1} FW h^{-1}	Baiyin *et al.* (2023)
Wheat	Shoot (9 days)	Anoxia (N_2) 4 hour	1.1 μmol g^{-1} FW h^{-1}	Yang and Adkins (2021)
	Roots (9 days)		1.3 μmol g^{-1} FW h^{-1}	Saderi *et al.* (2024)
Arabidopsis	Shoots	Hypoxia (5% O_2)	0.23 μmol g^{-1} FW h^{-1}	Iqbal *et al.* (2021)
	Roots (4 weeks)		0.04 μmol g^{-1} FW h^{-1}	Xu *et al.* (2023)

FW, fresh weight.

either acetaldehyde toxicity inhibits growth or cotton seeds do not produce enough energy from fermentation under waterlogging. This does not mean that during waterlogging events, a higher degree of fermentation cannot be achieved in root apices or other tissues. Transgenic cotton lines that overexpress the genes for alcohol dehydrogenase and pyruvate decarboxylase were employed to improve ethanolic fermentation and thus resistance to O_2 deprivation. The transgenic line produced up to 80% more ethanol, but neither growth nor plant life showed any discernible increase in hypoxia or anoxia tolerance. This suggests that increased ethanol synthesis (and hence ATP production) did not confer tolerance. When compared with their respective controls, the same transgenic lines showed no increase in cotton output throughout the field tests, whether under waterlogged conditions or not. Thus, more physiological and biochemical research is required to ascertain the connection between anaerobic fermentation and the ability of cotton roots to withstand field waterlogging. The few genes controlling fermentation rate are probably not the only factors influencing plant anoxia resistance. The survival of cotton tissues under anoxia may depend on a variety of proteins (e.g. pumps and enzymes in primary metabolism); some proposed targets include ion transporters, ROS scavengers, and enzymes that mobilize carbohydrates. Attention should also be paid to tolerance to harmful substances such as metal ions and acetaldehyde (Mansoor *et al.*, 2023).

12.12 Strategies to Overcome Waterlogging Stress

Insufficient water drainage during flood or furrow irrigation is a common problem for cotton grown on fine-textured, clay-rich soils (Admas *et al.*, 2025). The issue is made worse in improperly leveled fields and after rainstorms, which cause soil waterlogging and O_2 shortages in a matter of hours to days during warm growing conditions. By employing appropriate field practices such foliar nutrients, irrigation timing, land leveling, elevating slope, and optimal layout design, recent advancements in production methods have significantly increased cotton crop output (Chang *et al.*, 2024). These management strategies are necessary for commercial cotton crops to increase yield in waterlogging-prone environments, but taking advantage of genotype–management–environment interactions might result in notable increases in waterlogging tolerance. Crop-management techniques should ideally guide breeding for enhanced stress tolerance by utilizing growing genomic sequence availability and increased understanding of processes (Derbyshire *et al.*, 2022).

12.12.1 Fertilizer application

Cotton growth and yield loss due to hypoxia may be caused by one or more of the following factors: (i) a lack of nutrients; (ii) increased ethylene buildup; or (iii) inadequate net carbon fixation and photosynthesis of leaf area per unit (Zhang *et al.*, 2023b). Depending on the severity and length of waterlogging, a sequence of chemical reactions occurs once the O_2 level in the soil drops, changing the pH as well as the availability and status of nutrients. Exogenous fertilizer application could theoretically aid in the recovery of plants. Nutrient ions may be forced into a damaged root system because waterlogging reduces the quantity of nutrients available to plants (Feng *et al.*, 2024). Therefore, attention should be paid to nutrient species, administration technique, rate, and timing in order to prevent the detrimental effects of nutrient imbalance on soil ecology. Waterlogging tolerance may result from post-waterlogging nitrogen fertilization ($240\,kg\ ha^{-1}$), as indicated by improvements in stomatal resistance, photosynthetic rate, and growth in waterlogged cotton plants, as well as improvements in root growth, vigor, and photosynthesis. It has been proposed that fertilizer application after waterlogging can lessen the negative impacts of hypoxia on growth and productivity. Because of the damaged roots' ineffective nitrogen absorption ability, fertilizer application during or shortly after waterlogging was less effective. Furthermore, because of the significant leaching risks in the wet soil, the applied nitrogen may no longer be accessible for plant absorption. Similarly, cotton may have excessive vegetative growth and harvesting issues if more nitrogen is administered during the late growth period (Zhang *et al.*, 2023b). To put it simply, the reaction must be in line with the yield and growth that are anticipated for the remainder of the season. Application of fertilizer 8 days after the end of waterlogging, when compared with immediate post-waterlogging application, resulted in increased recovery of cotton. Similar to this, applying an extra 20–30% fertilizer 5 days after waterlogging (beyond the usual rate) greatly boosted the waterlogged cotton's growth and yield in comparison with unfertilized control plants. Waterlogged plants should be given foliar fertilizer treatment because damage to the roots caused by hypoxia limits the uptake of nutrients from the soil, even in cases when there is an abundance of nutrients available (Pais *et al.*, 2022). Hodgson and MacLeod (1988) demonstrated the effectiveness of foliar nitrogen when they discovered that applying it before waterlogging considerably reduced the undesirable effects of waterlogging on cotton lint production. The detrimental effects of iron chlorosis were lessened by a foliar spray of iron sulfate ($FeSO_4$) before waterlogging, which restored the natural color of cotton leaves (Bhagawati and Ahngar, 2023).

12.12.2 Role of anti-ethylene agents

Young fruit abscission is caused by waterlogging-induced ethylene buildup in cotton, which is linked to a variety of wounds and stressors. It has been demonstrated that substances that prevent the production or perception of ethylene, such as cobalt and silver ions, aminoethoxyvinylglycine (AVG), 1-methylcyclopropene (1-MCP), and aminoethoxy acetic acid (AOA), can regulate the accumulation of ethylene by obstructing the biosynthetic route or ethylene action (Liu *et al.*, 2022). It has been proposed that applying AVG and 1-MCP can reduce the harm that ethylene causes to a variety of crops. Applying AVG is thought to have the potential to increase output by reducing fruit abscission, because there is a correlation between higher ethylene accumulation and early fruit shedding in stressed cotton. Zhang *et al.* (2025b) showed that applying different doses of AVG (62.5, 125, and 250g of active component ha^{-1}) right before waterlogging enhanced the boll number and seed cotton yield of waterlogged cotton. Similarly, 1-MCP has been shown to have a beneficial effect on water-stressed cotton plants, improving physiological functions include water potential, antioxidant enzyme activity, and stomatal resistance while inhibiting ethylene action. AVG and 1-MCP had a favorable impact on cotton seed and lint output in 2-year field research. It was proposed that the ideal way to increase cotton production under both stressed and unstressed situations is to apply AVG during the early reproductive phase. In a recent study, Zhang *et al.* (2025b) found that cotton yield and ethylene production were negatively correlated to waterlogging, indicating that controlling ethylene manufacture through the use of AVG can enhance waterlogged cotton's photosynthesis and fruit retention. It was also found that removing ethylene sensitivity (using an ethylene-insensitive mutant) can greatly enhance cotton behavior in both waterlogged and non-waterlogged conditions. As ethylene controls the generation of lint in cotton, creating plants that are ethylene insensitive may lead to a reduction in lint output. With transgenic (ethylene-insensitive) plants improving the ability of other plants to withstand abiotic stress, this strategy may have a wider range of uses (Nazir *et al.*, 2024).

12.13 Conclusion and Future Prospects

In order to understand the mechanisms that impact cotton development and production, this chapter makes use of our knowledge of how plants behave physically and biochemically to O_2 limitation. Waterlogging affects water relations in plants and hinders increases in biomass by lowering cellular respiration, nutrient availability, and O_2 diffusion. The developmental impacts of waterlogging, such as ethylene-induced floral abscission, significantly worsen yield losses. As studies on waterlogged cotton plants do not offer a single explanation for the decline in growth and yield, a more comprehensive analysis of gene-expression patterns and hormone physiology is required. Specially, the limited tolerance range observed across farmed cotton genotypes makes it more challenging to properly grasp the underlying genetic mechanisms of waterlogged soil hypoxia adaptation. Gene-expression patterns under brief hypoxia may offer insight into important energy-transducing mechanisms that confer flood resistance. Enhancing waterlogging tolerance throughout a cotton crop's life cycle will require determining the relationship between environmental cues such as soil O_2, temperature, and light levels, and gene expression (e.g. through promoter analysis), which will reveal particular physiological and biochemical processes that facilitate survival. Increasing waterlogging resistance throughout vegetative and reproductive development would be made easier by traditional and transgenic breeding methods if the hypoxic response and post-stress recovery time of cotton plants were better understood.

The manipulation of hormone physiology received less attention in earlier research, which concentrated on using fertilizer application to induce waterlogging tolerance in cotton. However, waterlogged cotton yield losses and fruit abscission are caused by increased ethylene production; thus, it is necessary to investigate the potential of anti-ethylene compounds to improve cotton's resistance to waterlogging. By altering the genes that control ACC synthesis, bioengineering may be able to reduce the buildup of ethylene. When combined with sensible crop-management techniques, these strategies may prove to be successful in combating the problems of waterlogging in plants.

References

Abbas, K., Li, J., Gong, B., Lu, Y., Wu, X. *et al.* (2023) Drought stress tolerance in vegetables: The functional role of structural features, key gene pathways, and exogenous hormones. *International Journal of Molecular Sciences* 24(18), 13876. DOI: 10.3390/ijms241813876.

Adel, S. and Carels, N. (2023) Plant tolerance to drought stress with emphasis on wheat. *Plants* 12(11), 2170. DOI: 10.3390/plants12112170.

Admas, H., Abbi, S. and Kassahun, T. (2025) Assessment of salt-affected soil extent and spatial variability using GIS and remote sensing in Asaita district, Northeastern Ethiopia. *Scientific Reports* 15(1), 28038. DOI: 10.1038/s41598-025-10969-0.

Agustí, J. and Blázquez, M.A. (2020) Plant vascular development: Mechanisms and environmental regulation. *Cellular and Molecular Life Sciences* 77(19), 3711–3728. DOI: 10.1007/s00018-020-03496-w.

Aloo, B., Dessureault-Rompré, J., Tripathi, V., Nyongesa, B.O. and Were, B. (2023) Signaling and crosstalk of rhizobacterial and plant hormones that mediate abiotic stress tolerance in plants. *Frontiers in Microbiology* 14, 1171104. DOI: 10.3389/fmicb.2023.1171104.

Arroyo, E., de Navascues, P., Gómez-Ramírez, A., Molina, R., Perea, Á. *et al.* (2021) Factors triggering germination in plasma-activated cotton seeds: Water imbibition vs. reactive species' formation. *Journal of Physics D: Applied Physics* 54(32), 325205. DOI: 10.1088/1361-6463/abfefc.

Artyszak, A. and Gozdowski, D. (2021) Is it possible to maintain the quantity and quality of winter wheat grain by replacing part of the mineral nitrogen dose by growth activators and plant growth-promoting rhizobacteria (PGPR)? *Sustainability* 13(11), 5834. DOI: 10.3390/su13115834.

Azarakhsh, M. and Lebedeva, M.A. (2023) Lateral root versus nodule: The auxin–cytokinin interplay. *Journal of Plant Growth Regulation* 42(11), 6903–6919. DOI: 10.1007/s00344-023-10983-4.

Baiyin, B., Xiang, Y., Hu, J., Tagawa, K., Son, J.E. *et al.* (2023) Nutrient solution flowing environment affects metabolite synthesis inducing root thigmomorphogenesis of lettuce (*Lactuca sativa* L.) in hydroponics. *International Journal of Molecular Sciences* 24, 16616. DOI: 10.3390/ijms242316616.

Basit, F., Khalid, M., El-Keblawy, A., Sheteiwy, M.S., Sulieman, S. *et al.* (2024) Hypoxia stress: Plant's sensing, responses, and tolerance mechanisms. *Environmental Science and Pollution Research* 31, 63458–63472. DOI: 10.1007/s11356-024-35439-4.

Bhagawati, S. and Ahngar, T.A. (2023) Production technology of groundnut (*Arachis hypogaea* L.). In: Dangi, R.S., Bajaj, S., Singh, V. and Ahirwal, A. (eds) *From Seed to Harvest a Comprehensive Kharif Crop Production*. Elite Publishing House, New Delhi, pp. 160–172.

Boyno, G., Rezaee Danesh, Y., Demir, S., Teniz, N., Mulet, J.M. *et al.* (2023) The complex interplay between arbuscular mycorrhizal fungi and strigolactone: Mechanisms, sinergies, applications and future directions. *International Journal of Molecular Sciences* 24(23), 16774. DOI: 10.3390/ijms242316774.

Cao, D., Chabikwa, T., Barbier, F., Dun, E.A., Fichtner, F. *et al.* (2023) Auxin-independent effects of apical dominance induce changes in phytohormones correlated with bud outgrowth. *Plant Physiology* 192(2), 1420–1434. DOI: 10.1093/plphys/kiad034.

Chang, Y., Li, D. and He, S. (2024) Effects of drip irrigation flow rate and layout designs on soil salt leaching and cotton growth under limited irrigation. *Agronomy* 14(7), 1499. DOI: 10.3390/agronomy14071499.

Chatterjee, Y., Tomar, S., Mishra, M., Pareek, A. and Singla-Pareek, S.L. (2025) OsLdh7 overexpression in rice confers submergence tolerance by regulating key metabolic pathways: Anaerobic glycolysis, ethanolic fermentation and amino acid metabolism. *Plant, Cell & Environment* 48(4), 2804–2820. DOI: 10.1111/pce.15358.

Chaudhary, P., Agri, U., Chaudhary, A., Kumar, A. and Kumar, G. (2022) Endophytes and their potential in biotic stress management and crop production. *Frontiers in Microbiology* 13, 933017. DOI: 10.3389/fmicb.2022.933017.

Chen, H., Wu, Q., Ni, M., Chen, C., Han, C. *et al.* (2022) Transcriptome analysis of endogenous hormone response mechanism in roots of *Styrax tonkinensis* under waterlogging. *Frontiers in Plant Science* 13, 896850. DOI: 10.3389/fpls.2022.896850.

Chen, Q., Song, Y., An, Y., Lu, Y. and Zhong, G. (2024) Mechanisms and impact of rhizosphere microbial metabolites on crop health, traits, functional components: A comprehensive review. *Molecules* 29(24), 5922. DOI: 10.3390/molecules29245922.

Clark, J. and Bennett, T. (2024) Cracking the enigma: Understanding strigolactone signalling in the rhizosphere. *Journal of Experimental Botany* 75(4), 1159–1173. DOI: 10.1093/jxb/erad335.

Derbyshire, M.C., Batley, J. and Edwards, D. (2022) Use of multiple 'omics techniques to accelerate the breeding of abiotic stress tolerant crops. *Current Plant Biology* 32, 100262. DOI: 10.1016/j.cpb.2022.100262.

Edge, R. and Truscott, T.G. (2021) The reactive oxygen species singlet oxygen, hydroxy radicals, and the superoxide radical anion—examples of their roles in biology and medicine. *Oxygen* 1(2), 77–95. DOI: 10.3390/oxygen1020009.

Faizan, M., Faraz, A., Sami, F., Siddiqui, H., Yusuf, M. *et al.* (2020) Role of strigolactones: Signalling and crosstalk with other phytohormones. *Open Life Sciences* 15, 217–228. DOI: 10.1515/biol-2020-0022.

Feng, D., Wang, L., Ning, S., Peng, D., Xu, H. *et al.* (2024) Exogenous chemicals used to alleviate or salvage plants under flooding/waterlogging stress: Their biochemical effects and perspectives. *Journal of Agricultural and Food Chemistry* 72(1), 65–79. DOI: 10.1021/acs.jafc.3c06897.

Feng, K., Wang, X., Zhou, Q., Dai, T., Cao, W. *et al.* (2022) Waterlogging priming enhances hypoxia stress tolerance of wheat offspring plants by regulating root phenotypic and physiological adaption. *Plants* 11(15), 1969. DOI: 10.3390/plants11151969.

Foyer, C.H. and Hanke, G. (2022) ROS production and signalling in chloroplasts: Cornerstones and evolving concepts. *Plant Journal* 111(3), 642–661. DOI: 10.1111/tpj.15856.

Gautam, H., Fatma, M., Sehar, Z., Iqbal, N., Albaqami, M. *et al.* (2022) Exogenously-sourced ethylene positively modulates photosynthesis, carbohydrate metabolism, and antioxidant defense to enhance heat tolerance in rice. *International Journal of Molecular Sciences* 23(3), 1031. DOI: 10.3390/ijms23031031.

Glanz-Idan, N., Lach, M., Tarkowski, P., Vrobel, O. and Wolf, S. (2022) Delayed leaf senescence by upregulation of cytokinin biosynthesis specifically in tomato roots. *Frontiers in Plant Science* 13, 922106. DOI: 10.3389/fpls.2022.922106.

Grover, M., Bodhankar, S., Sharma, A., Sharma, P., Singh, J. *et al.* (2021) PGPR mediated alterations in root traits: Way toward sustainable crop production. *Frontiers in Sustainable Food Systems* 4, 618230. DOI: 10.3389/fsufs.2020.618230.

Guo, Y., Wang, Q., Xie, H., Zhang, X., Huang, T. *et al.* (2024) Analysis of root response mechanism of quinoa seedlings to waterlogging stress based on metabolome. *Journal of Plant Growth Regulation* 43(7), 2251–2264. DOI: 10.1007/s00344-024-11260-8.

Hamid, B., Zaman, M., Farooq, S., Fatima, S., Sayyed, R.Z. *et al.* (2021) Bacterial plant biostimulants: A sustainable way towards improving growth, productivity, and health of crops. *Sustainability* 13(5), 2856. DOI: 10.3390/su13052856.

Hartman, S., Sasidharan, R. and Voesenek, L.A.C.J. (2021) The role of ethylene in metabolic acclimations to low oxygen. *New Phytologist* 229(1), 64–70. DOI: 10.1111/nph.16378.

Hasan, A., Tabassum, B., Hashim, M. and Khan, N. (2024) Role of plant growth promoting rhizobacteria (PGPR) as a plant growth enhancer for sustainable agriculture: A review. *Bacteria* 3(2), 59–75. DOI: 10.3390/bacteria3020005.

Hodgson, A.S. and MacLeod, D.A. (1988) Seasonal and soil fertility effects on the response of waterlogged cotton to foliar- applied nitrogen fertilizer. *Agronomy Journal* 80, 259–265. DOI: 10.2134/agronj1988.00021962008000020021x.

Iqbal, P., Ghani, M.A., Ali, B., Shahid, M., Iqbal, Q. *et al.* (2021) Exogenous application of glutamic acid promotes cucumber (*Cucumis sativus* L.) growth under salt stress conditions. *Emirates Journal of Food and Agriculture* 33, 407. DOI: 10.9755/ejfa.2021.v33.i5.2699.

Jan, M., Muhammad, S., Jin, W., Zhong, W., Zhang, S. *et al.* (2024) Modulating root system architecture: Cross-talk between auxin and phytohormones. *Frontiers in Plant Science* 15, 1343928. DOI: 10.3389/fpls.2024.1343928.

Jayawardhane, J., Wijesinghe, M.K.P.S., Bykova, N.V. and Igamberdiev, A.U. (2021) Metabolic changes in seed embryos of hypoxia-tolerant rice and hypoxia-sensitive barley at the onset of germination. *Plants* 10(11), 2456. DOI: 10.3390/plants10112456.

Kaleem, M., Jabeen, A., Anjum, L., Minhas, R.W., Mumtaz, A.S. *et al.* (2025) Signal perception and transduction in plant stress responses. In: Waseem, M., Liu, P. and Yang, G. (eds) *Plant Stress* Responses. Ethics International Press, Bradford, UK, pp. 1–46.

Khan, N., Ali, S., Shahid, M.A., Mustafa, A., Sayyed, R.Z. *et al.* (2021) Insights into the interactions among roots, rhizosphere, and rhizobacteria for improving plant growth and tolerance to abiotic stresses: A review. *Cells* 10(6), 1551. DOI: 10.3390/cells10061551.

Kim, J.W., Hong, S., Go, J., Park, J.S. and Yi, G. (2024) Transcriptomic changes reveal hypoxic stress response in submerged seeds of maize (*Zea mays* L.). *Applied Biological Chemistry* 67(1), 68. DOI: 10.1186/s13765-024-00922-6.

Kondhare, K.R., Patil, A.B. and Giri, A.P. (2021) Auxin: An emerging regulator of tuber and storage root development. *Plant Science* 306, 110854. DOI: 10.1016/j.plantsci.2021.110854.

Koren, D., Resnick, N., Gati, E.M., Belausov, E., Weininger, S. *et al.* (2013) Strigolactone signaling in the endodermis is sufficient to restore root responses and involves SHORT HYPOCOTYL 2 (SHY2) activity. *New Phytologist* 198(3), 866–874. DOI: 10.1111/nph.12189.

Kumari, V.V., Banerjee, P., Verma, V.C., Sukumaran, S., Chandran, M.A.S. *et al.* (2022) Plant nutrition: An effective way to alleviate abiotic stress in agricultural crops. *International Journal of Molecular Sciences* 23(15), 8519. DOI: 10.3390/ijms23158519.

Liang, M.-H., He, Y.-J., Liu, D.-M. and Jiang, J.-G. (2021) Regulation of carotenoid degradation and production of apocarotenoids in natural and engineered organisms. *Critical Reviews in Biotechnology* 41(4), 513–534. DOI: 10.1080/07388551.2021.1873242.

Liu, J., Islam, M.T. and Sherif, S.M. (2022) Effects of aminoethoxyvinylglycine (AVG) and 1-methylcyclopropene (1-MCP) on the pre-harvest drop rate, fruit quality, and stem-end splitting in 'gala' apples. *Horticulturae* 8(12), 1100. DOI: 10.3390/horticulturae8121100.

Liu, J., Wang, J., Zhang, T., Li, M., Yan, H. *et al.* (2023) Exogenous melatonin positively regulates rice root growth through promoting the antioxidant system and mediating the auxin signaling under root-zone hypoxia stress. *Agronomy* 13(2), 386. DOI: 10.3390/agronomy13020386.

Liu, Z., Sun, L., Liu, Z. and Li, X. (2024) Effect of exogenous melatonin on growth and antioxidant system of pumpkin seedlings under waterlogging stress. *PeerJ* 12, e17927. DOI: 10.7717/peerj.17927.

Long, J., Xing, W., Wang, Y., Wu, Z., Li, W. *et al.* (2022) Comparative proteomic analysis on chloroplast proteins provides new insights into the effects of low temperature in sugar beet. *Botanical Studies* 63(1), 18. DOI: 10.1186/s40529-022-00349-6.

Ma, Y., Huang, L., Zhang, Z., Liang, S. and Wang, L. (2024) Hydrogen sulfide induced by hydrogen peroxide mediates strigolactones-triggered stomatal closure in *Arabidopsis thaliana*. *Pakistan Journal of Botany* 56(2), 427–436. DOI: 10.30848/PJB2024-2(33).

Mandal, D., Datta, S., Mitra, S. and Nag Chaudhuri, R. (2024) ABSCISIC ACID INSENSITIVE 3 promotes auxin signalling by regulating *SHY2* expression to control primary root growth in response to dehydration stress. *Journal of Experimental Botany* 75(16), 5111–5129. DOI: 10.1093/jxb/erae237.

Mansoor, S., Ali, A., Kour, N., Bornhorst, J., AlHarbi, K. *et al.* (2023) Heavy metal induced oxidative stress mitigation and ROS scavenging in plants. *Plants* 12(16), 3003. DOI: 10.3390/plants12163003.

Moursi, Y.S., Dawood, M.F., Sallam, A., Thabet, S.G. and Alqudah, A.M. (2021) Antioxidant enzymes and their genetic mechanism in alleviating drought stress in plants. In: *Organic Solutes, Oxidative Stress, and Antioxidant Enzymes under Abiotic Stressors*. CRC Press, Boca Raton, Florida.

Nambara, E. and van Wees, S.C.M. (2021) Plant hormone functions and interactions in biological systems. *Plant Journal* 105(2), 287–289. DOI: 10.1111/tpj.15151.

Nazir, F., Peter, P., Gupta, R., Kumari, S., Nawaz, K. *et al*. (2024) Plant hormone ethylene: A leading edge in conferring drought stress tolerance. *Physiologia Plantarum* 176(1), E14151. DOI: 10.1111/ppl.14151.

Owusu, A.G., Lv, Y.-P., Liu, M., Wu, Y., Li, C.-L. *et al*. (2023) Transcriptomic and metabolomic analyses reveal the potential mechanism of waterlogging resistance in cotton (*Gossypium hirsutum* L.). *Frontiers in Plant Science* 14, 1088537. DOI: 10.3389/fpls.2023.1088537.

Pais, I.P., Moreira, R., Semedo, J.N., Ramalho, J.C., Lidon, F.C. *et al*. (2022) Wheat crop under waterlogging: Potential soil and plant effects. *Plants* 12(1), 149. DOI: 10.3390/plants12010149.

Pan, J., Sharif, R., Xu, X. and Chen, X. (2021) Mechanisms of waterlogging tolerance in plants: Research progress and prospects. *Frontiers in Plant Science* 11, 627331. DOI: 10.3389/fpls.2020.627331.

Perreca, E., Eberl, F., Santoro, M.V., Wright, L.P., Schmidt, A. *et al*. (2022) Effect of drought and methyl jasmonate treatment on primary and secondary isoprenoid metabolites derived from the MEP pathway in the white spruce *Picea glauca*. *International Journal of Molecular Sciences* 23(7), 3838. DOI: 10.3390/ijms23073838.

Qi, Z., Tong, X., Zhang, Y., Jia, S., Fang, X. *et al*. (2023) Carotenoid cleavage dioxygenase 1 and its application for the production of C13-apocarotenoids in microbial cell factories: A review. *Journal of Agricultural and Food Chemistry* 71(49), 19240–19254. DOI: 10.1021/acs.jafc.3c06459.

Ravi, B., Foyer, C.H. and Pandey, G.K. (2023) The integration of reactive oxygen species (ROS) and calcium signalling in abiotic stress responses. *Plant, Cell & Environment* 46(7), 1985–2006. DOI: 10.1111/pce.14596.

Sachdev, S., Ansari, S.A., Ansari, M.I., Fujita, M. and Hasanuzzaman, M. (2021) Abiotic stress and reactive oxygen species: Generation, signaling, and defense mechanisms. *Antioxidants* 10(2), 277. DOI: 10.3390/antiox10020277.

Saderi, S.Z., Abrishamchi, P., Ganjeali, A. and Radjabian, T. (2024) *Glomus mosseae* symbiosis improves disease resistance of *Melissa officinalis* to *Fusarium culmorum* through inducing antioxidant responses and plant hormones levels. *Scientia Horticulturae* 338, 113607. DOI: 10.1016/j.scienta.2024.113607.

Sánchez Martín-Fontecha, E., Cardinale, F., Bürger, M., Prandi, C. and Cubas, P. (2024) Novel mechanisms of strigolactone-induced DWARF14 degradation in *Arabidopsis thaliana*. *Journal of Experimental Botany* 75(22), 7145–7159. DOI: 10.1093/jxb/erae365.

Shao, J., Wang, Q., Liu, P., Zhao, B., Han, W. *et al*. (2024) The complex stress of waterlogging and high temperature accelerated maize leaf senescence and decreased photosynthetic performance at different growth stages. *Journal of Agronomy and Crop Science* 210(2), E12689. DOI: 10.1111/jac.12689.

Sharma, N., Kour, S., Kumar, D., Kaur, R., Khajuria, A. *et al*. (2022) Role of brassinosteroids (brs) in modulating antioxidative defense mechanism in plants growing under abiotic and biotic stress conditions. In: Aftab, T. and Hakeem, K.R. (eds) *Antioxidant Defense in Plants: Molecular Basis of Regulation*. Springer Nature Singapore, Singapore, pp. 325–367. DOI: 10.1007/978-981-16-7981-0_15.

Shen, R., Ma, X. and Wang, H. (2020) SMXL6/7/8: Dual-function transcriptional repressors of strigolactone signaling. *Molecular Plant* 13(9), 1244–1246. DOI: 10.1016/j.molp.2020.08.002.

Shi, H., Li, X., Lv, M. and Li, J. (2022) BES1/BZR1 family transcription factors regulate plant development via brassinosteroid-dependent and independent pathways. *International Journal of Molecular Sciences* 23(17), 10149. DOI: 10.3390/ijms231710149.

Singh, N., Chattopadhyay, D. and Gupta, S.K. (2023) Updating the impact of drought on root exudation: A strigolactones perspective. *Journal of Plant Growth Regulation* 42(8), 5131–5151. DOI: 10.1007/s00344-023-11061-5.

Soliman, S., Wang, Y., Han, Z., Pervaiz, T. and El-Kereamy, A. (2022) Strigolactones in plants and their interaction with the ecological microbiome in response to abiotic stress. *Plants* 11(24), 3499. DOI: 10.3390/plants11243499.

Song, C., Cao, Y., Dai, J., Li, G., Manzoor, M.A. *et al*. (2022) The multifaceted roles of MYC2 in plants: Toward transcriptional reprogramming and stress tolerance by jasmonate signaling. *Frontiers in Plant Science* 13, 868874. DOI: 10.3389/fpls.2022.868874.

Sosnowski, J., Truba, M. and Vasileva, V. (2023) The impact of auxin and cytokinin on the growth and development of selected crops. *Agriculture* 13(3), 724. DOI: 10.3390/agriculture13030724.

Swetha, P. and Sundararaj, R. (2022) Diseases of medicinal plants cultivated in Karnataka and their management. In: Kumar, S. (ed.) *Medicinal Plants*. Intech Open, London. DOI: 10.5772/intechopen.104632.

Tang, Q., Wei, S., Zheng, X., Tu, P. and Tao, F. (2024) APETALA2/ethylene-responsive factors in higher plant and their roles in regulation of plant stress response. *Critical Reviews in Biotechnology* 44(8), 1533–1551. DOI: 10.1080/07388551.2023.2299769.

Wahab, A., Abdi, G., Saleem, M.H., Ali, B., Ullah, S. *et al.* (2022) Plants' physio-biochemical and phyto-hormonal responses to alleviate the adverse effects of drought stress: A comprehensive review. *Plants* 11(13), 1620. DOI: 10.3390/plants11131620.

Wang, G., Yang, X. and Xue, X. (2024) Transgenic tobacco with the *BADH* gene shows enhanced photo-synthesis resistance to drought stress induced by PEG-6000. *Agronomy* 14(4), 690. DOI: 10.3390/agronomy14040690.

Wang, L.N., Wang, W.C., Liao, K., Xu, L.J., Xie, D.X. *et al.* (2025) Survival mechanisms of plants under hypoxic stress: Physiological acclimation and molecular regulation. *Journal of Integrative Plant Biology* 67(3), 440–454. DOI: 10.1111/jipb.13880.

Xu, E., Chai, L., Zhang, S., Yu, R., Zhang, X. *et al.* (2021) Catabolism of strigolactones by a carboxylester-ase. *Nature Plants* 7(11), 1495–1504. DOI: 10.1038/s41477-021-01011-y.

Xu, Z., Pan, J., Ullah, N., Duan, Y., Hao, R. *et al.* (2023) 5-Aminolevulinic acid mitigates the chromium-induced changes in *Helianthus annuus* L. as revealed by plant defense system enhancement. *Plant Physiology and Biochemistry* 198, 107701. DOI: 10.1016/j.plaphy.2023.107701.

Yang, L. and Adkins, S. (2021) Changes of oxidative metabolism in the roots of wheat (*Triticum aestivum* L.) seedlings in response to elevated ammonium concentrations. *Journal of Integrative Agriculture* 20, 1216–1228. DOI: 10.1016/S2095-3119(20)63216-6.

Yang, L., Li, N., Liu, Y., Miao, P., Liu, J. *et al.* (2023) Updates and prospects: Morphological, physiological, and molecular regulation in crop response to waterlogging stress. *Agronomy* 13(10), 2599. DOI: 10.3390/agronomy13102599.

Yemelyanov, V.V., Lastochkin, V.V., Prikaziuk, E.G. and Chirkova, T.V. (2022) Activities of catalase and peroxidase in wheat and rice plants under conditions of anoxia and post-anoxic aeration. *Russian Journal of Plant Physiology* 69, 117. DOI: 10.1134/S1021443722060036.

Zaidi, S., Arif, Y., Imtiaz, H., Shiraz, M. and Hayat, S. (2024) Structural chemistry, biosynthesis, and signal-ing of multifaceted plant growth regulator: Strigolactone. *Journal of Plant Growth Regulation* 43(8), 2489–2502. DOI: 10.1007/s00344-024-11285-z.

Zenda, T. (2025) Ethanol (C_2H_5OH) involvement in plant abiotic stress response. *Discover Plants* 2(1), 268. DOI: 10.1007/s44372-025-00346-6.

Zhang, R., Dong, Y., Li, Y., Ren, G., Chen, C. *et al.* (2023a) SLs signal transduction gene *CSMAX2* of cucumber positively regulated to salt, drought and ABA stress in *Arabidopsis thaliana* L. *Gene* 864, 147282. DOI: 10.1016/j.gene.2023.147282.

Zhang, Y., Liu, G., Dong, H. and Li, C. (2021) Waterlogging stress in cotton: Damage, adaptability, allevia-tion strategies, and mechanisms. *Crop Journal* 9(2), 257–270. DOI: 10.1016/j.cj.2020.08.005.

Zhang, Y., Chen, X., Geng, S. and Zhang, X. (2025a) A review of soil waterlogging impacts, mechanisms, and adaptive strategies. *Frontiers in Plant Science* 16, 1545912. DOI: 10.3389/fpls.2025.1545912.

Zhang, Y., Qiu, S., Liang, T., Xu, S., Li, Z. *et al.* (2025b) Mitigating waterlogging-induced yield loss in cotton through removal of early fruits: Agronomic and physiological mechanisms. *Field Crops Research* 331, 109996.

Zhang, Z., Ahmed, S., Chattha, M.S., Liu, A., Liu, J. *et al.* (2023b) Managing plant density and nitrogen fer-tilizer to reduce nitrogen input without yield reduction of late-sown cotton after wheat by improving light interception and sink nitrogen partitioning in a double cropping system. *Field Crops Research* 295, 108875. DOI: 10.1016/j.fcr.2023.108875.

Zhao, C., Yang, L., Li, H. and Deng, Z. (2024) Nitrogen metabolism during anaerobic fermentation of actual food waste under different pH conditions. *Fermentation* 10(3), 129. DOI: 10.3390/fermentation10030129.

Zuo, S., Li, J., Gu, W. and Wei, S. (2022) Exogenous proline alleviated low temperature stress in maize embryos by optimizing seed germination, inner proline metabolism, respiratory metabolism and a hormone regulation mechanism. *Agriculture* 12(4), 548. DOI: 10.3390/agriculture12040548.

13 Strigolactones in Nutrient-use Efficiency Regulation in Crop Plants

Emmaculate Akinyi Ogutu[1], Sabavat Raju Naik[2,3], Dasari Venu Gopal[4], Vasavi Nagula[4] and Ashwini Talakayala[2*]

[1]Department of Agricultural Production, College of Agriculture and Environmental Sciences, Makerere University, Kampala, Uganda; [2]International Crops Research Institute for Semi-Arid Tropics, Patancheru, Hyderabad, Telangana, India; [3]Department of Genetics and Plant Breeding, Vasantrao Naik Marathwada Krishi Vidyapeeth, Parbhani, Maharashtra, India; [4]Loyola Academy, Degree & P.G. College, Hyderabad, Telangana, India

Abstract

Strigolactones (SLs) are a fascinating class of plant hormones with a pivotal role in regulating nutrient-use efficiency (NUE) in crop plants, primarily through their profound effects on root architecture and symbiotic interactions. SLs influence root development, nutrient uptake, and overall plant growth, potentially leading to sustainable agricultural practices by reducing fertilizer dependency. The SL-induced establishment of arbuscular mycorrhizal fungal symbiosis significantly expands the plant's root–soil interface, improving the uptake of poorly mobile nutrients such as phosphate. The integration of SL signaling with other hormonal and transcriptional networks enables plants to finely tune growth in response to nutrient availability. Harnessing the regulatory functions of SLs offers promising strategies to improve NUE, reduce fertilizer dependency, and enhance crop resilience under low-input agricultural systems. This chapter explores the molecular mechanisms by which SLs influence nutrient uptake, their interactions with other signaling molecules, and their potential applications in sustainable crop improvement.

Keywords: crop resilience, nutrient uptake, nutrient-use efficiency (NUE), root architecture, strigolactones

13.1 Introduction

13.1.1 Importance of nutrient-use efficiency in crop production

Modern agriculture faces the vital challenge of sustaining an increasing population while mitigating environmental impacts (Fróna *et al.*, 2019; Yang *et al.*, 2024). Nutrient-use efficiency (NUE) is the amount of crop yield produced per unit of nutrient input, and is important for sustainable crop production (Congreves *et al.*, 2021; Govindasamy *et al.*, 2023). Currently, the overall NUE of agriculture is low because plants only take up about 10–55% of the nutrients that are applied, the rest being lost. Global estimates show that only

*Corresponding author: Ashwini.Talakayala@icrisat.org

DOI: 10.1079/9781836990598.0013

about 46% of nitrogen fertilizer that is put on crops is actually taken up by the plants (Lassaletta *et al.*, 2014; Sharma and Bali, 2017; Zhang *et al.*, 2021). These kinds of inefficiencies not only waste farmers' financial resources but also cause substantial environmental problems. For instance, excessive nitrogen and other nutrients can leach into waterways, causing eutrophication, and can also break down into gases such as nitrous oxide (N_2O), a potent greenhouse gas. Poor NUE leads to pollution of water, soil, and the climate (Lal, 2009, 2012; Govindasamy *et al.*, 2023). Thus, enhancing NUE is crucial for augmenting crop yield and minimizing the reliance on excessive fertilizer applications, consequently reducing production expenses and environmental impact (Langholtz *et al.*, 2021; Govindasamy *et al.*, 2023). Better management practices and breeding or engineering crops that can take in more nutrients and recycle them inside the plant can help with this. In this context, plant hormonal control of root growth and nutrient signaling is a promising way to improve NUE.

13.1.2 Overview of strigolactones as phytohormones

Strigolactones (SLs) are an emerging class of phytohormones derived from carotenoids that have multiple functions in rhizosphere signaling and plant development (Alvi *et al.*, 2022). SLs were initially discovered as root exudates that promote arbuscular mycorrhizal fungal (AMF) hyphal branching and induce germination of parasitic weeds, particularly of the genera *Striga*, and *Orobanche* (Vurro *et al.*, 2019; Chesterfield *et al.*, 2020; Soto-Cruz *et al.*, 2021). Structurally, they share a conserved ABC-ring system that is connected to a butenolide D-ring by an enol-ether bridge (Brun *et al.*, 2018; Nelson, 2021; Kee *et al.*, 2023). Endogenously, they regulate plant architecture by suppressing axillary bud outgrowth, by maintaining apical dominance, and influencing tiller number (Teichmann and Muhr, 2015; Yan *et al.*, 2020; Rehman *et al.*, 2021; Khuvung *et al.*, 2022; Kelly *et al.*, 2023), as demonstrated in *Arabidopsis* (Stirnberg *et al.*, 2007), pea (Foo *et al.*, 2013), rice (Arite *et al.*, 2009), and *Petunia* mutants (Jiang *et al.*, 2013) defective in SL biosynthesis or signaling. In response to nutrient status, SLs in roots modify the architecture of the root system, promote the elongation of root hairs, and regulate the formation of lateral roots (Marzec and Melzer, 2018). Under phosphate starvation, their production increases dramatically, leading to increased SL exudation, which improves nutrient acquisition and AMF colonization.

SLs also have an impact on abiotic stress responses, leaf senescence, and the balance of shoot–root growth (Saeed *et al.*, 2017; Banerjee and Roychoudhury, 2018). By integrating endogenous developmental cues and external environmental signals, SLs coordinate growth regulation with ecological interactions and link the plant's nutritional state to its capacity for resource optimization, symbiotic relationships, and competitive interactions in the rhizosphere.

13.1.3 Role of SLs in sustainable agriculture

SLs present a promising opportunities to enhance NUE and crop resilience for sustainable agriculture (Saeed *et al.*, 2017). SLs enhance phosphorus uptake and decrease dependence on chemical fertilizers by optimizing AMF symbiosis. Plants that have been treated with synthetic SL analogs or genetically modified to produce more SLs show more mycorrhizal colonization, better nutrient uptake, and healthier soil (Borghi *et al.*, 2021; Boyno *et al.*, 2023). By altering root development and stomatal function, SLs promote stress tolerance and aid in preserving yield in the face of salinity or drought. In order to optimize shoot architecture, which boosts root development, and increase NUE, researchers are investigating SL analogs and pathway engineering. Crop-breeding programs can increase sustainability, yield, and stress resilience by implementing SL-based techniques.

13.2 Biosynthesis and Signaling Pathways of Strigolactones

13.2.1 Biosynthetic pathway

SLs are form of terpenoid lactones derived from carotenoids and function as rhizosphere signals and internal developmental regulators (Al-Babili and Bouwmeester, 2015; Jia *et al.*, 2018). Biosynthesis of SLs starts in plastids, where the iron-binding isomerase DWARF27 (D27) transforms all-*trans*-β-carotene into 9-*cis*-β-carotene (Abuauf *et al.*, 2018; Jia *et al.*, 2018; Yang *et al.*, 2023). The activity of D27 depends on Mg^{2+}, and when there is no phosphate, transcript levels can increase by as much as five times. Carotenoid cleavage dioxygenase 7 (CCD7; also known as MAX3, RMS5, and D17 in different species) cleaves 9-*cis*-β-carotene to make β-ionone (C_{13}) and 9-*cis*-β-apo-10′-carotenal (C_{27}) (Fig. 13.1) (Bruno *et al.*, 2014). Seto *et al.* (2014) showed that CCD8 (also known as MAX4, RMS1, and D10) catalyzes a series of oxidation, cyclization, and rearrangement reactions that produce carlactone, which is a central, low-bioactivity intermediate that has the ABC tricyclic lactone scaffold.

Cytochrome P450 enzymes from the CYP711A/MORE AXILLARY GROWTH 1 (MAX1) clade support downstream diversification (Yoneyama and Brewer, 2021; Alvi *et al.*, 2022). In *Arabidopsis*, a single MAX1 oxidizes carlactone to carlactonoic acid, which is then methylated to methyl carlactonoate and modified by LATERAL BRANCHING OXIDOREDUCTASE (LBO) into a bioactive, although structurally unresolved, SL (Moreno *et al.*, 2021; Mashiguchi *et al.*, 2022). In monocotyledonous plants such rice, duplicated *MAX1* genes divide up their tasks: *Os900* converts carlactone to 4-deoxyorobanchol, and *Os1400* converts 4-deoxyorobanchol to orobanchol (Fig. 13.1) (Wang *et al.*, 2023a, b; Li *et al.*, 2024). This branching builds different canonical (orobanchol-type and strigol-type) and non-canonical SLs, and the stereochemistry changes how well they bind to receptors (Fig. 13.1). The source of carotenoids connects to the abscisic acid (ABA) pathway through shared precursors that come before 9-*cis*-β-carotene (Fig. 13.1). Transcription factors that respond to nutrient and developmental signals control regulation. For example, when roots are starved of nutrients, the levels of *MAX3* and *MAX4* transcripts can rise three- to sixfold, which increases SL exudation and helps mycorrhizal colonization. Mutant analyses (*max*, *rms*, *dad*) combined with liquid chromatography/mass spectrometry-based metabolite profiling have elucidated enzymatic order, substrate preferences, and tissue-specific fluxes, establishing the pathway as a strategic target for breeding crop architecture and stress adaptation (Pandey *et al.*, 2016; Haq *et al.*, 2017; Kumar *et al.*, 2019).

13.2.2 Perception and signal transduction

Perception of SLs is thought to occur through the α/β-hydrolase receptor DWARF14 (D14), the F-box adapter MAX2/D3 and transcriptional repressors of the SUPPRESSOR OF MAX2-LIKE (SMXL)/D53 family (Soundappan *et al.*, 2015; Waters and Nelson, 2023). D14 sequesters SLs in the cytoplasm and cleaves the D-ring to generate a covalent intermediate with its catalytic serine. This triggers a conformational change that allows D14 to dock on to MAX2, a component of the SCF^{MAX2} E3 ubiquitin ligase complex (Skp–Cullin–F-box (SCF)). MAX2 targets repressors such as D53 in rice or SMXL6/7/8 in *Arabidopsis* to SCF for ubiquitination and 26S proteasome degradation (Fig. 13.2) (Temmerman *et al.*, 2022).

SL-responsive transcriptional processes, such as *BRANCHED1* (*BRC1*) activation in buds and genes related to root architecture (e.g. *WOX5*, *PLT1*), are depressed by repressor turnover (Wang *et al.*, 2019; Guan *et al.*, 2025). By stabilizing the receptor–F-box–substrate complex until degradation takes place, the SL molecule functions as a molecular bridge. The catalytic triad is positioned for D-ring hydrolysis and structural investigations have revealed that SL binding causes the hydrophobic pocket of D14 to close (Fig. 13.2) (Kagiyama *et al.*, 2013; Zhao *et al.*, 2015).

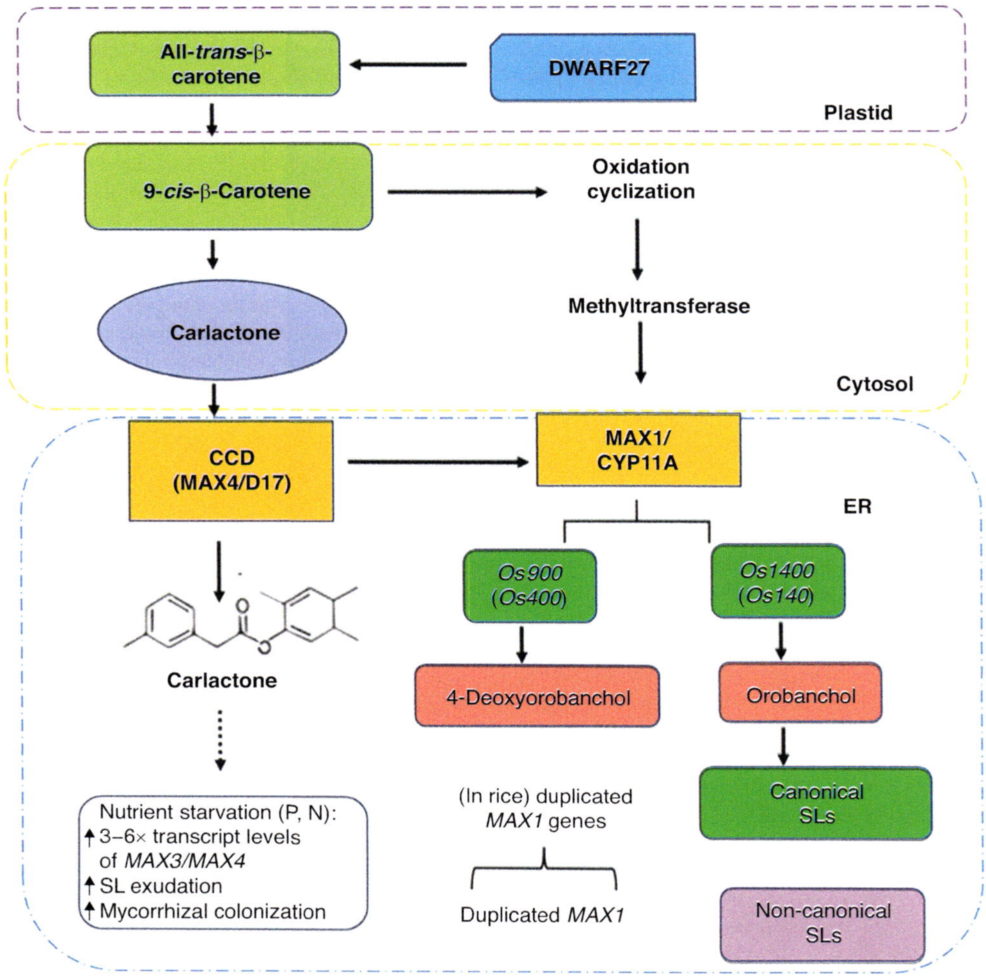

Fig. 13.1. The synthesis pathway of strigolactones (SLs). Carotenoids are converted into SLs through a series of enzymatic processes in the plastid, cytosol, and endoplasmic reticulum (ER). DWARF27 isomerizes all-*trans*-β-carotene in plastids to 9-*cis*-β-carotene, which is cleaved by carotenoid cleavage dioxygenase 7 (CCD7) to produce 9-*cis*-β-apo-10′-carotenal. This intermediate is transformed by CCD8 into carlactone, the central hub, which is then transformed by MAX1 homologs into canonical (strigol-type, orobanchol-type) and non-canonical SLs. The production of 4-deoxyorobanchol and orobanchol is driven by duplicated MAX1 genes (*Os900*, *Os1400*) in rice, whereas carlactone is oxidized to carlactonoic acid and then further changed by LATERAL BRANCHING OXIDOREDUCTASE (LBO) in *Arabidopsis*. Under nutrient deprivation, the process is activated and shares precursors with abscisic acid, which promotes mycorrhizal colonization and SL exudation.

Unlike most plant hormone receptors, D14 is a single-use catalytic receptor that degrades alongside its target, resulting in transitory signal output. MAX2 is multi-functional, mediating karrikin signaling through the KAI2 receptor, indicating some overlap in downstream proteolytic machinery (Fig. 13.2) (Wang *et al.*, 2020b; Temmerman *et al.*, 2022). Natural allelic variations of *D14* in rice exhibit enhanced SL sensitivity, suggesting that perception efficiency can evolve quickly under selection for branching or stress-tolerance traits (Yao *et al.*, 2018).

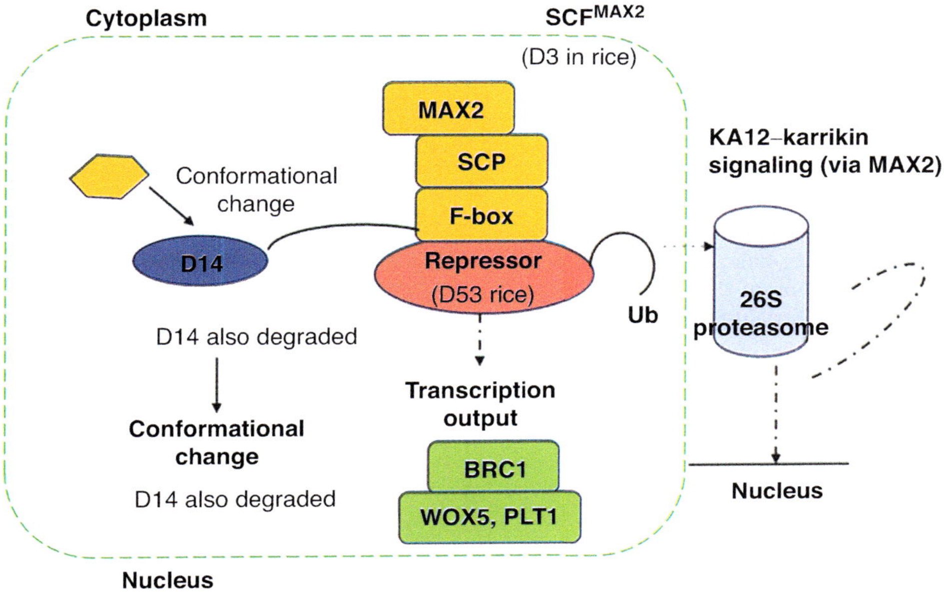

Fig. 13.2. Perception and signaling of strigolactones (SL). Docking to MAX2/D3 within the SCF^{MAX2} ubiquitin ligase complex is made possible by SL binding to the receptor D14, which causes hydrolysis and conformational change. The 26S proteasome ubiquitinates and breaks down repressors such as D53 (rice) and SMXL6/7/8 (*Arabidopsis*), releasing transcriptional regulation of genes including *BRC1* and *WOX5/PLT1*. While MAX2 is also involved in karrikin signaling, D14 is a single-use receptor that is broken down along with its substrate. Ub, ubiquitin.

13.2.3 Hormonal crosstalk and network integration

SL–auxin coordination

SL–auxin interaction forms a bidirectional regulatory loop that connects apical dominance to nutrient sensing (Andreo-Jimenez *et al.*, 2015). In plants, shoot-derived auxin activates CCD7 and CCD8 in the stem vasculature, increasing SL production under low-phosphate conditions (Pandey *et al.*, 2016; Mazzoni-Putman *et al.*, 2021). SLs modulate polar auxin transport by changing the quantity and polarity of PINFORMED 1 (PIN1)/PIN3, which is commonly mediated by PID/WAG kinase phosphorylation (Barbosa *et al.*, 2018). In *Arabidopsis max4* mutants, PIN1 retention in axillary buds is associated with increased bud auxin efflux and branching; application of the synthetic SL analog GR24 reduces PIN polarization after 12 hour (Shinohara *et al.*, 2013; Zhang *et al.*, 2020).

Monocots, such as rice, exhibit increased auxin-induced SL production in roots, consistent with their increased dependence on SL for mycorrhizal recruitment (Kobae *et al.*, 2018; Yao *et al.*, 2018). Transcriptomic investigations of *max4 pin* double mutants show that this loop also regulates root meristem identity genes (*PLT*, *WOX*), extending crosstalk beyond branching to the root-system architecture (Josse, 2022).

SL–cytokinin antagonism

SLs and cytokinins (CKs) interact antagonistically during BRC1-mediated bud dormancy (Wang *et al.*, 2019; Dong *et al.*, 2023). SL signaling activates cytokinin oxidases (CKX1, CKX3) and inhibits ISOPENTENYLTRANSFERASE (IPT)-dependent CK production in nodes, reducing CK availability (Duan *et al.*, 2019; Zha *et al.*,

2022). Hormone profiling has revealed that *max4* mutants accumulate more than twice as much of the CK *trans*-zeatin in nodes, which is eliminated by GR24 treatment (Hu *et al.*, 2014). When phosphate is available, CK treatment in rice can partially override SL suppression, emphasizing nutrient-dependent changes in hormonal dominance (Wu *et al.*, 2022). Under phosphate limitation, SLs promote primary root elongation and root-hair formation, whereas CKs reallocate carbon to the shoots, striking a balance between nutrient foraging and canopy expansion.

SL–ABA interplay

SLs and ABA interact differently depending on the circumstances, although they both share carotenoid precursors (Negin and Moshelion, 2016; Korek and Marzec, 2023). ABF2 and AREB1 are ABA-responsive basic leucine zippers that bind *MAX3/MAX4* promoters during drought, increasing the biosynthesis of SLs (Shen *et al.*, 2012; Yoshida *et al.*, 2015). Drought causes ABA accumulation in tomatoes, which is followed by a two- to threefold increase in SL exudation; SL-deficient mutants exhibit increased wilting, delayed stomatal closure, and approximately 30% lower ABA levels (Korek and Marzec, 2023). SLs encourage CYP707A1-mediated ABA catabolism during germination, which relieves dormancy by lowering ABA in radicles by around 40% (Tuan *et al.*, 2018; Ali *et al.*, 2022). When ABA signaling is compromised, SLs can also trigger SLOW ANION CHANNEL 1 (SLAC1)-dependent stomatal closure without the help of ABA (Merilo *et al.*, 2015), supplying a backup guard-cell control mechanism. They integrate nutrient status and environmental cues to shape root architecture. Under low phosphorus and nitrogen conditions, SL production increases, promoting deeper soil exploration through primary root elongation, lateral root formation, and longer root hairs (Santoro *et al.*, 2020; Naz *et al.*, 2025). SLs also enhance symbiosis with AMF and rhizobia, improving nutrient uptake (Naseer *et al.*, 2024). These responses involve crosstalk with auxin, CKs, and ethylene to regulate root growth and branching, while under nutrient sufficiency SLs may suppress adventitious roots, showing context-dependent roles (Sun *et al.*, 2022; Wu *et al.*, 2022).

13.3 Strigolactone-mediated Regulation of Root Architecture

13.3.1 Root-system modifications under nutrient deficiency and the impact on lateral root formation and root-hair development

Root-system modifications help plants to cope with limited nutrients, by reallocating growth to enhance soil exploration and uptake. Under phosphorus deficiency, typical shortening of primary root occurs with increasing lateral root density and root-hair length/density, increasing the absorptive surface near the top soil where phosphorus is most available (i.e. TOR signaling with reprogramming of root-system architecture under limited phosphorus) (Choi *et al.*, 2025). Nitrogen deficiency promotes deeper rooting through primary root elongation as well as reduced proliferation of laterals unless local nitrate patches are detected with a localized nitrogen supply, which stimulates lateral root initiation/elongation for foraging plasticity. Potassium deficiency reduces primary root growth and thus enhances lateral root formation and root-hair growth; KUP9, a potassium transporter, and auxin signaling help to adapt responses, with natural variation within *Arabidopsis* spp. (Justamante *et al.*, 2025). Under phosphorus and nitrogen deficiency, the production and secretion of SLs increase and influence primary root development, lateral root density, and the elongation of root hairs through interactions with auxin and ethylene signaling pathways via exogenous SL application (GR24), which alleviates the effect of low phosphorus on the root-system architecture (Mitra *et al.*, 2024; Guan *et al.*, 2025). Various impacts of nutrient deficiency strongly shape the development of lateral roots and root hairs, thus enhancing soil exploration and nutrient uptake. Under nutrient stress, lateral root development is modulated by nitrogen and phosphorus availability: nitrogen limitation suppresses lateral roots except in localized nitrate-rich zones, while phosphorus deficiency increases root density in the upper soil layers, and SLs, together with auxin and CKs, fine-tune these branching responses (Koevoets *et al.*, 2016; Santoro *et al.*, 2020). Root hairs are cost-efficient structures for nutrient capture, whereby under phosphate and potassium deficiency, plants elongate root hairs and increase their density to enlarge the absorptive surface area (Takatsuka *et al.*, 2024). SL signaling is central to these processes: exogenous GR24 enhances root-hair elongation under low phosphate, and SL-deficient mutants show impaired responses (Sun *et al.*, 2019; Santoro *et al.*, 2020). Therefore, SLs regulate root architecture by controlling primary root growth, lateral root initiation, and root-hair development to enhance nutrient foraging (Mayzlish-Gati *et al.*, 2012; Santoro *et al.*, 2020). Their biosynthesis increases under phosphate and nitrogen deficiency, linking root plasticity with nutrient uptake (Giehl and von Wirén, 2014; Yuan *et al.*, 2023). Through crosstalk with auxin and CKs, SLs fine-tune root branching and hair elongation for efficient resource use (Jiang *et al.*, 2021).

13.3.2 Enhancing nutrient foraging through root plasticity

SLs promote root plasticity by modifying root-system characteristics and facilitating symbiotic interactions. On the developmental side, SLs control lateral root growth and root-hair elongation, increasing the surface area accessible for nutrient absorption. On the symbiotic side, SLs function as chemical signals that activate AMF, causing spore germination and hyphal branching. These early signals boost the frequency of AMF colonization and broaden the root–fungus contact. Once plants have been colonized, SLs facilitate the coordinated functioning of phosphate transporters, such as those in the PHOSPHATE TRANSPORTER 1 (PHT1) family, which control phosphate transfer from fungal hyphae into plant tissues (Foo *et al.*, 2013; Kodama *et al.*, 2022). Through this dual action of remodeling root architecture and encouraging AMF symbiosis, SLs enable plants to maximize nutrient intake under changing soil conditions. This adaptability is an important element of nutrient foraging, allowing plants to maintain nutrient availability and efficiency, even when external inputs are limited.

13.4 Strigolactones in Nitrogen-use Efficiency

13.4.1 Response of SLs to nitrogen deficiency

Nitrogen is vital for plant growth, and its deficiency triggers adaptive responses mediated by SLs. Under nitrogen starvation, SL biosynthesis genes (*D27*, *CCD7*, *CCD8*, *MAX1*) are upregulated, raising SL levels in roots; a nitrate resupply quickly reduces them (Luo, 2022). In maize, sorghum, and wheat, nitrogen limitation also enhances SL exudation, such as 5-deoxystrigol, confirming their conserved role as nutrient scarcity signals (Yoneyama *et al.*, 2012). The biological significance of these responses is reinforced by studies with SL-deficient mutants. Maize lines lacking *CCD8* displayed heightened sensitivity to nitrogen deficiency, exhibiting reduced chlorophyll content, decreased protein accumulation, and impaired shoot development (Quaggiotti *et al.*, 2024). These findings highlight the importance of SLs in maintaining nitrogen homeostasis, as they sustain photosynthesis, regulate carbon–nitrogen balance, and stabilize amino acid metabolism (Luo, 2022). In their absence, plants struggle to mobilize reserves or adjust metabolic pathways, resulting in severe growth penalties. Thus, SLs are more than nutrient-deficiency signals; they function as master regulators that synchronize developmental adjustments with metabolic reprogramming. Beyond developmental effects, SLs also impact symbiotic interactions under nitrogen stress. For example, increased SL production promotes the recruitment of AMF, which improve nitrogen acquisition through their extensive hyphal networks (Yoneyama, 2019). In legumes, SLs may also influence nodulation efficiency and symbiotic nitrogen fixation, linking external nitrogen availability with internal resource allocation. Collectively, these findings highlight that SLs play pivotal roles at the molecular, physiological, and ecological levels during nitrogen stress.

13.4.2 Interaction of SLs with nitrate transporters and assimilation pathways

A crucial mechanism through which SLs enhance nitrogen-use efficiency lies in their regulation of transport and assimilation. SLs stimulate the expression of nitrate transporter genes such as *NRT1.1*, *NRT1.2*, and *NRT2.1*, which are vital for nitrate uptake under limited conditions (Quaggiotti *et al.*, 2024). Without SLs, as observed in mutant studies, these transporters are not sufficiently activated, leading to inefficient nitrogen acquisition. Alongside transporter regulation, SLs support the enzymatic machinery for assimilation, notably nitrate reductase and glutamine synthetase, ensuring that absorbed nitrate is effectively converted into amino acids and proteins.

The impact of SLs enhances nitrogen use through crosstalk with gibberellins, which stabilize GROWTH-REGULATING FACTOR 4 (GRF4) to activate nitrogen assimilation and photosynthesis genes in rice (Wang *et al.*, 2020a). Synthetic SL analogs such as GR24 restore growth in SL-deficient plants, underscoring their role in nitrogen pathways and the potential for improving nitrogen-use efficiency in crops. Another emerging aspect is how SLs interact with carbon metabolism during nitrogen stress. Because carbon and nitrogen metabolisms are tightly coupled, SLs appear to ensure that energy and carbon skeletons are available for amino acid biosynthesis when nitrogen is scarce (Yuan *et al.*, 2023). By aligning nitrogen uptake with carbon supply, SLs optimize growth under resource limitations.

13.4.3 Integration of SLs with nitrogen signaling networks

The role of SLs extends beyond single pathways, as they operate at the crossroads of multiple signaling networks. Under nitrogen limitation, SLs redirect auxin transport to promote deeper root elongation and suppress shoot branching (Luo, 2022; Yuan *et al.*, 2023). This architectural adjustment enhances belowground nutrient acquisition while minimizing aboveground resource expenditure. At the same time, SLs act antagonistically to CKs, which otherwise stimulate shoot proliferation, thereby maintaining an energy-conserving balance during stress. SLs interact with ABA and ethylene to enhance drought and salinity tolerance, indirectly supporting nitrogen uptake (Wu *et al.*, 2022). They also regulate sulfur and iron transport in maize (Quaggiotti *et al.*, 2024), and coordinate nitrogen acquisition with nodulation in legumes (Yoneyama, 2019). As integrators of nutrient and hormonal signals, SLs optimize growth and stress resilience, making them key targets for improving nitrogen-use efficiency and yield stability under low-input conditions.

13.5 Strigolactones in Phosphorus-use Efficiency

13.5.1 Role of SLs in phosphate starvation responses

Phosphorus is a vital macronutrient for plants, serving as a key component in several essential processes for growth and development. It is prominent in the plant's energy and genetic machinery, such as energy transfer through ATP, nucleic acid synthesis, membrane structure formation, and intracellular signaling (Sharma *et al.*, 2025). Plants often cannot access enough phosphorus because it binds with minerals in the soil, such as iron, aluminum and calcium, a process known as phosphorus fixation. This creates insoluble compounds that roots cannot absorb, leading to deficiency. Plants have developed complex adaptive mechanisms to sense phosphate limitation and respond efficiently, and SLs have emerged as central component of these responses (Czarnecki *et al.*, 2013). SLs help coordinate a wide range of physiological, molecular, and developmental changes that improve phosphate acquisition and use efficiency (Mayzlish-Gati *et al.*, 2012). SLs are synthesized primarily in the roots through the action of CCD7, CCD8, MAX1, and associated regulators (López-Ráez *et al.*, 2008; Kohlen *et al.*, 2012). When plants experience phosphate deficiency, the expression of these enzymes increases, leading to enhanced production and exudation of SLs (Kodama *et al.*, 2022). SLs are subsequently transported throughout the plant and released into the rhizosphere, where they alter the root-system architecture by encouraging primary root growth, promoting the formation of lateral roots, and boosting the density of root hairs (Mayzlish-Gati *et al.*, 2012; Santoro *et al.*, 2021). These changes enhance soil exploration and facilitate better phosphate uptake. At the molecular level, SLs are essential for combining developmental and stress-response pathways with phosphate signaling. They alter the expression of genes involved in phosphate uptake, transport, and remobilization by interacting with transcription factors that respond to phosphate starvation, such as *PHOSPHATE STARVATION RESPONSE 1* (*PHR1*) (Khan *et al.*, 2023). Furthermore, SLs participate in wide range of hormonal interactions; for example, they work in tandem with auxins and CKs to optimize root elongation and branching in response to nutrient stress (Marro *et al.*, 2022). SLs connect phosphate acquisition with the plant's overall physiological state and environmental challenges by interacting with ABA and reactive oxygen species-mediated pathways in addition to nutrient signaling (Czarnecki *et al.*, 2013). Under phosphate limitation, SLs mobilize phosphate from older to actively growing tissues and coordinate carbon allocation by limiting shoot branching and promoting root growth, ensuring efficient nutrient use and continued development (Kohlen *et al.*, 2011; Kodama *et al.*, 2022). Some of the emerging evidence suggests that SLs extend their regulatory influence beyond root architecture and nutrient allocation to encompass epigenetic and metabolic responses under phosphate starvation. SLs appear to modulate chromatin accessibility and the expression of key phosphate-responsive genes, potentially enabling plants to retain a 'memory' of prior phosphate stress and respond more effectively to subsequent deficiencies (Khan *et al.*, 2023). In parallel, SLs contribute to metabolic reprogramming, including the enhanced production of organic acids and secondary metabolites, which can facilitate phosphate solubilization and availability in the rhizosphere (Santoro *et al.*, 2021). Collectively, these multi-faceted roles illustrate how SLs integrate environmental sensing, root-system modulation, gene regulation, and nutrient allocation, allowing plants to optimize growth and maintain physiological homeostasis under phosphate-limited conditions (Marro *et al.*, 2022).

13.5.2 SLs in mycorrhizal symbiosis for phosphorus acquisition

AMF enhance phosphate acquisition by extending beyond the root zone to access soil phosphate (Balzergue *et al.*, 2011). Under low-phosphate conditions, plants exude SLs, which stimulate AMF spore germination, hyphal growth, and branching, promoting root colonization. Within the roots, arbuscules form as sites of efficient phosphate transfer from fungus to plant (Foo *et al.*, 2013; Kodama *et al.*, 2022). SLs not only promote AMF colonization but also enhance symbiotic efficiency by regulating phosphate transporter expression at the peri-arbuscular membrane, optimizing phosphate transfer. In return, plants supply carbon to the fungi, creating a mutually beneficial system that boosts growth and reproduction in phosphate-poor soils (Balzergue *et al.*, 2011; Foo *et al.*, 2013; Kodama *et al.*, 2022). SLs enhance AMF symbiosis by regulating fungal colonization and phosphate transfer, interacting with signaling molecules such as JAs and nitric oxide, and influencing soil microbial communities to improve nutrient cycling and stress tolerance. Their production is phosphate dependent: low phosphate triggers SL synthesis to recruit fungi, while sufficient phosphate reduces it, ensuring efficient symbiotic investment (Kodama *et al.*, 2022). Coordination with root architecture and mycorrhizal interactions is by promoting lateral root and root-hair growth, increasing the interface for fungal colonization and enhancing phosphate uptake. They integrate fungal-derived phosphate with systemic nutrient allocation, ensuring a supply to actively growing tissues and organs, thus linking root development, symbiosis, and nutrient distribution under phosphate-limited conditions (Mayzlish-Gati *et al.*, 2012; Santoro *et al.*, 2021; Marro *et al.*, 2022). SLs enhance plant resilience beyond phosphate uptake by supporting AMF symbiosis, which improves tolerance to stresses such as drought and salinity, enabling plants to maintain growth under challenging conditions (Kodama *et al.*, 2022).

13.5.3 SL-dependent modulation of phosphate transporter genes

SLs enhance phosphate-use efficiency by upregulating PHT1 transporters in root cells, boosting uptake under low-phosphate conditions (Czarnecki *et al.*, 2013; Santoro *et al.*, 2021). They also increase transporter activity at the peri-arbuscular membrane in mycorrhizal roots, coordinating nutrient exchange with fungi and integrating with auxin, CKs, and ABA to optimize root growth and phosphorus balance (Foo *et al.*, 2013; Kodama *et al.*, 2022; Marro *et al.*, 2022). Beyond uptake, SLs mobilize phosphate from older to growing tissues, regulate storage and redistribution genes, and adjust root architecture, ensuring efficient phosphate acquisition and homeostasis (Kohlen *et al.*, 2011). By reducing fertilizer needs and improving yields in poor soils, SLs thus represent key targets for sustainable crop management.

13.6 Role of Strigolactones in Uptake of Other Nutrients: Potassium, Sulfur, and Micronutrients

13.6.1 Potential roles in potassium, sulfur, and micronutrient acquisition

Potassium, sulfur, and micronutrients (zinc, iron, manganese, copper) are crucial for plant growth, metabolism, and stress tolerance, but deficiencies limit yield (Singh *et al.*, 2025). Soil microbes enhance their availability: potassium-solubilizing bacteria (*Bacillus*, *Paenibacillus*, *Frateuria*) release acids to mobilize potassium, sulfur-oxidizing bacteria (*Thiobacillus*, *Acidithiobacillus*) convert sulfur to sulfate, and AMF extend hyphae to improve uptake, soil structure, and water retention (Ahmed *et al.*, 2025). These microbes also produce phytohormones (indole-3-acetic acid, gibberellins, CKs) and induce nutrient transporter genes, boosting root growth and nutrient acquisition (Zeng *et al.*, 2025). Additionally, potassium-solubilizing bacteria, soil-beneficial microorganisms, and AMF stimulate stress-related hormones (ABA, JA, SA) and antioxidant activity, strengthening plant resilience to drought, salinity, and pathogens (Kumar *et al.*, 2025).

Some of the research studies indicate that microbial consortia can synergistically enhance nutrient acquisition. Soil-beneficial microorganisms modify soil pH and redox to solubilize potassium and micronutrients, while potassium-solubilizing bacteria shape the rhizosphere community, promoting beneficial microbes that improve nutrient cycling, soil fertility, and crop productivity (Giri *et al.*, 2025). Enhancing nutrient availability and decomposing organic matter from microorganisms, thus releasing potassium, sulfur, and micronutrients, enzymes, and metabolites, combined with organic amendments and sustainable practices, means that microbial inoculants can be used to reduce fertilizer dependence while maintaining soil fertility and microbial diversity (Alzate Zuluaga *et al.*, 2024). Microbes enhance nutrient uptake while mitigating toxicity by sequestering harmful ions and mobilizing essential nutrients, ensuring safe and efficient plant nutrition, highlighting their potential as natural biofertilizers for sustainable agriculture (Nunes *et al.*, 2024).

13.6.2 Emerging evidence and knowledge gaps

Despite their benefits, microbial efficacy in nutrient acquisition varies with soil properties, environmental conditions, and crop genotype. Factors such as soil chemistry, native microbial populations, and root-exudate profiles can affect colonization and symbiosis, highlighting the need for tailored microbial inoculants for consistent agricultural performance (Sahoo *et al.*, 2025). Microbial consortia enhance nutrient acquisition more effectively than single strains. Potassium-solubilizing bacteria, soil-beneficial microorganisms, AMF, and other microbes interact to solubilize multiple nutrients, alter root exudates, activate nutrient transporter genes, and boost soil enzyme activity, improving plant growth and stress resilience (Upadhayay *et al.*, 2025). These findings highlight their potential as integrated biofertilizers. However, the long-term ecological impacts and practical challenges of repeated microbial inoculation on soil diversity, nutrient cycling, and crop performance remain unclear, highlighting the need for comprehensive studies (Pandey and Saharan, 2025). Microbial activity fluctuates with season, soil conditions, and plant growth stages, affecting nutrient solubilization. Timing the inoculant application to correlate with peak microbial activity can optimize nutrient availability during critical growth phases (Singh *et al.*, 2025). Climate change affects microbial diversity, activity, and symbiosis, impacting nutrient availability. Developing climate-resilient microbial formulations is key to ensuring stable nutrient acquisition and crop performance (Giri *et al.*, 2025). Integrating microbial inoculants with precision agriculture using soil mapping, nutrient sensors, and targeted applications can optimize nutrient delivery, reduce fertilizer use, and enhance crop productivity and soil health sustainably. Microbial exudates modulate plant gene expression, activating nutrient transport, stress tolerance, and growth pathways. Understanding these plant–microbe signaling networks can enable the development of next-generation biofertilizers that boost productivity and resilience in an environmentally friendly way (Ahmed *et al.*, 2025).

13.7 Molecular and Genetic Regulation of Strigolactone-mediated Nutrient-use Efficiency

SLs modulate NUE via complex molecular and genetic networks that connect hormonal, transcriptional, and nutritional signaling pathways. At the molecular level, SL signaling influences root-system design by interacting with important transcription factors and nutrient-responsive regulators, such as NRT1.1-mediated nitrate sensing and transport systems. This control fine-tunes lateral root initiation and elongation, improving nutrient foraging capability. In parallel, SLs regulate the coordination of carbon–nitrogen allocation, ensuring that metabolic resources are balanced between growth and stress tolerance. Genetic studies show that SL-responsive genes are part of a larger signaling hub that overlaps with the auxin, CK, and ABA pathways, highlighting the complexities of hormonal interaction in nutrient regulation. Collectively, these interactions provide a strong genetic underpinning for SL-mediated NUE augmentation, indicating possible options for breeding crop varieties that maintain high productivity with lower fertilizer inputs (Jia *et al.*, 2022).

13.7.1 Key genes and transcription factors

The biosynthesis of SLs begins with carotenoids which are transformed by the enzymes CCD7, CCD8, and MAX1, a cytochrome P450 mono-oxygenase (López-Ráez *et al.*, 2008). CCD7 cleaves 9-*cis*-β-carotene to generate 9-*cis*-β-apo-10′-carotenal. CCD8 subsequently catalyzes a series of reactions to produce carlactone, which serves as a key intermediate in the SL biosynthetic pathway (Alder *et al.*, 2012). MAX1 further oxidizes carlactone into bioactive SLs, such as 5-deoxystrigol, which are then diversified into species-specific SL derivatives (Zhang *et al.*, 2014). The expression of *CCD7*, *CCD8*, and *MAX1* is tightly regulated by nutrient status. Under nitrogen deficiency, these genes are significantly upregulated, leading to higher SL accumulation. Elevated SL levels act as a molecular signal that links nutrient scarcity to adaptive developmental responses, including suppression of shoot branching, stimulation of primary root elongation, and enhanced lateral root formation in nutrient-rich patches (Ueda and Kusaba, 2015). Thus, SL biosynthetic genes function as integrators of nutrient sensing and developmental plasticity, contributing to improved nitrogen-use efficiency and adaptive nutrient foraging strategies in plants (Giehl and von Wirén, 2014).

SLs are detected and transmitted through unique signaling mechanisms that connect environmental nutrient signals to the developmental reactions of plants. The D14 receptor identifies SL molecules and, upon binding with the ligand, experiences a conformational shift that allows it to interact with the F-box protein MAX2 (known as D3 in rice). MAX2, which is part of the SCF ubiquitin ligase

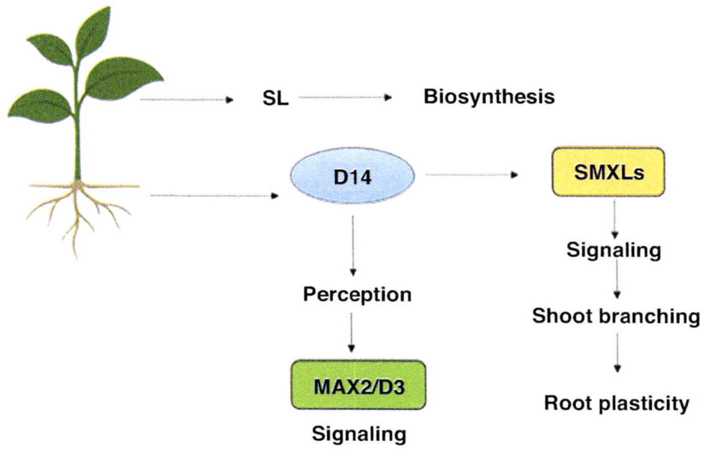

Fig. 13.3. Schematic representation of the strigolactone (SL) signaling pathway in plants. SLs are synthesized in the roots and perceived by the receptor DWARF14 (D14), which triggers downstream signaling through MAX2/D3. This perception and signaling cascade modulates SMXL proteins, leading to regulation of shoot branching and root plasticity. Arrows indicate the direction of signal flow from SL biosynthesis to the physiological responses in both shoots and roots.

complex, directs SMXL proteins acting as inhibitors of SL responses toward proteasomal degradation. This D14–MAX2–SMXL pathway constitutes the fundamental framework of SL perception and signaling, enabling plants to modulate shoot branching, root formation, and nutrient acquisition in reaction to environmental factors such as nitrogen and phosphorus availability (Fig. 13.3) (Hamiaux *et al.*, 2012; Stanga *et al.*, 2013; Yao *et al.*, 2016).

13.7.2 Transcriptional regulation under nutrient deficiency

SLs play a crucial role in coordinating plant development and responses to environmental cues, particularly nutrient availability. Among the various nutrients, phosphate and nitrogen deficiencies are the most prominent triggers for SL biosynthesis, largely through transcriptional reprogramming of key SL biosynthetic genes. Phosphate deficiency triggers strong upregulation of the SL biosynthesis genes *D27*, *CCD7*, and *CCD8* in the roots. The transcription factor PHR1 drives this response by activating phosphate-responsive genes, while SPX proteins modulate PHR1 activity, relieving repression under phosphate deficiency and indirectly enhancing SL production (He *et al.*, 2023). Nitrogen deficiency enhances SL production by transcriptionally upregulating *CCD7* and *CCD8*, although the response is generally weaker than that for phosphate deficiency. Nitrogen-responsive transcription factors such as NIN-like proteins (NLPs) may integrate nitrate signaling with SL biosynthesis (Wang *et al.*, 2023a, b). Nutrient deficiencies induce root-specific transcriptional activation of SL biosynthesis genes, as roots function as primary sensors of soil nutrient status and regulate SL production under nutrient stress. The root-derived SLs then act systemically, traveling to the shoots to suppress lateral bud outgrowth, optimizing resource allocation during nutrient stress (Mayzlish-Gati *et al.*, 2012). The downstream effects of transcriptional changes are upregulated SL biosynthesis under nutrient deficiency, promoting root-hair elongation, enhancing mycorrhizal associations, and suppressing shoot branching to optimize nutrient uptake and resource allocation (Yoneyama, 2019).

13.7.3 Genetic studies and mutant analysis

SLs regulate various aspects of plant growth and development, and genetic studies and mutant analyses have been instrumental in elucidating their biosynthesis and signaling pathways, providing insights into their roles in plant physiology. Kumar *et al.* (2019) reported that *Arabidopsis* mutants *max1*, *max3*, and *max4* showed reduced branching, while *max2* exhibited increased branching, reflecting disruptions in SL biosynthesis and signaling pathways. Karniel *et al.* (2024) in tomato SB mutants (*sb1* and *sb2* (biosynthesis) and *sb3* (perception)) showed overbranching and reduced yield, underscoring the role of SLs in plant architecture. Duan *et al.* (2019) studied the rice *osckx9* mutant, which affects cytokinin oxidase/dehydrogenase 9, in combination with SL-pathway mutants to understand the interplay between CK and SL signaling in regulating tillering and plant height. These studies have provided valuable insights into the biosynthesis and signaling pathways of SLs and their interactions with other hormonal pathways, and continued research in this area holds promise for developing crops with improved growth and yield characteristics through the manipulation of SL signaling pathways (Table 13.1).

13.7.4 Crosstalk with other hormonal pathways in nutrient regulation

SLs regulate nutrient responses not in isolation but through integration with other hormones. Under phosphate and nitrogen scarcity, SL biosynthesis and signaling are upregulated, and their effects are mediated through crosstalk with auxin, CKs, ABA, and gibberellins. aligning developmental outcomes with nutrient availability (Umehara *et al.*, 2008; Yoneyama *et al.*, 2012). Auxin enhances SL biosynthesis by upregulating CCD7 and CCD8, while SLs restrict auxin transport to suppress shoot branching and promote root foraging, forming a feedback loop that optimizes nutrient stress responses (Bennett *et al.*, 2006; Shinohara *et al.*, 2013). CKs promote branching, whereas SLs suppress it under nutrient deficiency; rice *osckx9* and SL-deficient mutants highlight their antagonistic balance in regulating tillering and NUE (Duan

Table 13.1. Mutant-based insights into strigolactone (SL) regulation across different plant species. Key *Arabidopsis*, tomato, and rice mutants have been instrumental in dissecting the roles of SL biosynthesis and signaling.

Plant	Mutants	Affected process	Phenotype/key insight	Reference
Arabidopsis	*max1*, *max2*, *max3*, *max4*	SL biosynthesis (*max1*, *max3*, *max4*), SL signaling (*max2*)	*max1*, *max3*, *max4*: reduced branching; *max2*: increased branching Helped define SL biosynthesis and perception pathways	Kumar *et al.* (2019)
Tomato	*sb1*, *sb2* (biosynthesis), *sb3* (perception)	SL biosynthesis and signaling	Overbranching and reduced yield Confirmed role of SLs in regulating plant architecture	Karniel *et al.* (2024)
Rice	*osckx9* (cytokinin oxidase/ dehydrogenase 9) with SL mutants	Cytokinin–SL signaling crosstalk	Altered tillering and plant height Revealed interplay between cytokinin and SL in growth regulation	Duan *et al.* (2019)

et al., 2019; Wang et al., 2020a). SLs interact with ABA under nutrient and drought stress, enhancing ABA sensitivity to regulate stomata and promote mycorrhizal symbiosis (Ha et al., 2014; Faizan et al., 2020). SLs interact with gibberellins to regulate internode elongation and branching, coordinating plant architecture and resource allocation under nutrient-limiting conditions (de Saint Germain et al., 2013; Marzec, 2017). Overall, SLs act as central integrators of nutrient signals, coordinating with auxin, CKs, ABA, and gibberellins to balance growth, optimize nutrient use, and maintain developmental plasticity, providing a basis for improving crop resilience and efficiency. Understanding this crosstalk not only clarifies the ecological roles of SLs but also provides a molecular basis for designing crop varieties with improved resilience and NUE.

13.8 Strigolactones in Stress-induced Nutrient Imbalance

13.8.1 Role of SLs in drought, salinity, and nutrient stress crosstalk

Nutrient uptake rarely occurs in isolation, as environmental stresses such as drought and salinity often exacerbate nutrient deficiencies. SLs emerge as crucial modulators of these interactions, balancing morphological adjustments, physiological processes, and symbiotic relationships. Under abiotic stress, SL biosynthesis and root exudation increase significantly, which aids in sustaining root growth and adjusting shoot architecture (Smith and Read, 2010; Marzec et al., 2013). This involves suppressing excessive tillering while stimulating root elongation, an adaptive strategy to improve water and nutrient capture from depleted soils. One major contribution of SLs during stress is their regulation of plant–microbe interactions. By enhancing AMF symbiosis, SLs facilitate the uptake of both phosphorus and nitrogen under restricted soil availability (Yoneyama, 2019; Mitra et al., 2024). This function is particularly vital during drought, when nutrient diffusion in soil is limited. SL-deficient plants not only display impaired nutrient uptake but also show accelerated leaf senescence, reduced chlorophyll content, and compromised photosynthetic capacity (Quaggiotti et al., 2024). In contrast, plants with intact SL pathways can maintain pigment stability, delay senescence, and extend photosynthetic activity under stress (Luqman et al., 2023). This highlights the role of SLs in maintaining metabolic efficiency and yield potential during combined stress scenarios. SLs also contribute to balancing micronutrient acquisition. Altered expression of sulfate and iron transporter genes in SL-deficient maize mutants suggested that these hormones influence a wide array of nutrient pathways beyond nitrogen and phosphate (Quaggiotti et al., 2024). This ability to coordinate macro- and micronutrient uptake underscores their role as central regulators of nutritional homeostasis during stress.

13.8.2 Maintaining nitrogen-use efficiency under abiotic stress conditions

SLs maintain nitrogen-use efficiency under combined stresses by regulating root growth, nitrate transporter expression, and nutrient uptake. Synthetic SL analogs such as GR24 mitigate nitrogen-deficiency effects (Sedaghat et al., 2021), while in sorghum and wheat, SLs enhance AMF symbiosis under nitrogen and phosphate scarcity, improving multi-nutrient acquisition (Yoneyama et al., 2012; Mitra et al., 2024). SLs also interact with ABA and ethylene to strengthen antioxidant defenses and osmotic regulation under drought and salinity (Marzec et al., 2013; Hajam et al., 2024). Genetic manipulation of SL biosynthesis via CRISPR-Cas9 (clustered regularly interspaced short palindromic repeats/CRISPR-associated protein 9) can optimize plant architecture and resilience (Butt et al., 2018; Shi et al., 2025), making SLs key integrators of nutrient and stress signals for sustainable crop improvement.

13.9 Applications in Crop Improvement

SL are phytohormones that regulate symbiotic relationships, root development, and plant architecture, especially in nutrient-limited environments (Marzec *et al.*, 2013). Wang *et al.* (2020a) in rice identified SL biosynthesis genes as key factors in increased crop production, which increased molecular understanding of the green revolution in rice. The number of tillers in rice plants is directly increased by a partial loss-of-function allele of the *HIGH-TILLERING DWARF 1* (*HTD1*) gene (*D17*/*CCD7*). Decreasing the function of *HTD1* results in a favorable increase in branching, which increases larger grain yield because of SLs, which suppress branching. Breeders co-selected a particular *HTD1* allele with the *SEMIDWARF 1* (*SD1*). The allele produce a robust yield where SD1 enhancing lodging resistance through semi-dwarf phenotype *HDT1* allele, with increased tillering. Yuan *et al.* (2023) elucidated the connecting role in control of SL production and low phosphorus availability in rice, whereby nuclear proteins (NSP1 and NSP2) form a heterodimer when there is a phosphorus shortage. As a transcriptional regulator, the NSP1–NSP2 complex selectively targets and activates genes related to the SL production pathway with modification of root and shoot architecture with the rise of SLs within the plant. Thus, rice plants can more efficiently forage with limited phosphorus by changed root branching and shoot tillering; this mechanism thus underscores the intricate link between nutrient sensing and phytohormone signaling for optimizing their growth for NUE. Shi *et al.* (2025) reported elucidation of the resistance mechanism in sorghum at the molecular level whereby two ABCG family transporters, SbSLT1 and SbSLT2, are responsible for export of SLs from the roots; expression of these transporter is elevated in phosphorus-deficient soils where growth of the weed *Striga* is a challenge; an increase in SL exudation in turn triggers *Striga* germination. SL exudation was significantly reduced in mutant lines with suppression of *Striga* parasitism and germination. Under field conditions, decreases in the growth of this parasite results in significant increase in grain yield or output. Mitra *et al.* (2024) showed that SLs enhance nutrient uptake under phosphorus limitation, and that combined application of AMF and GR24 alleviated phosphorus deficiency in aerobic rice. SLs trigger hyphal branching of AMF in the rhizosphere and promote colonization, and thus the hyphal network increases the plant surface for nutrient absorption which helps the plant absorbs nutrients such as phosphorus; this adaptive SL-mediated response enables rice to maximize its phosphorus usage. Guo *et al.* (2023) showed that SLs have enhanced the activity of antioxidant enzymes in plants. For instance, a study on wheat under drought stress found that the activities of key antioxidant enzymes, especially superoxide dismutase and ascorbate peroxidase, increased notably when soil water content decreased to 50% of field capacity. This effect was even more pronounced in plants treated with GR24. This indicates that an enhanced antioxidant defense system plays crucial role in enabling exogenous GR24 to reduce oxidative stress and enhance stress tolerance. Luqman *et al.* (2023) recently investigated the impact of SLs, especially GR24, which is emerging as a growth regulator that can mitigate stress effects, by exogenous GR24 foliar application on drought-stressed maize hybrids (HY-1898 and FH-1046). The plants were exposed to drought conditions at 60% field capacity, and after 40 days, foliar sprays of GR24 were applied at concentrations of 0, 0.001, 0.01, and 0.1 mg l^{-1}. The results are described in Table 13.2), together with the results of other studies discussed in this section.

13.10 Challenges and Future Perspectives

13.10.1 Current limitations in SLs

SLs helps in regulating plant architecture to facilitate interactions with soil microbes and parasitic plants. While research in this field has made significant strides, several key limitations and challenges remain, as SLs are produced in extremely in low quantities and exhibit high structural diversity, with over 25 natural variants differing in structure and stereochemistry. This poses significant challenges for both research and application: their minute concentrations make detection and quantification in biological samples technically demanding, while the complex synthesis of natural and synthetic analogs is time-consuming and limits their availability for large-scale studies and agricultural use (Dell'Oste *et al.*, 2021). The chemical instability of SLs under light, heat,

Table 13.2. Key studies on strigolactones (SLs) in nutrient-use efficiency and stress tolerance.

Crop	Gene/treatment	Mechanism/finding	Outcome	Reference
Rice	*HTD1/D17* (*CCD7*) allele + *SD1* (semi-dwarf gene)	Partial loss of function in *HTD1* increases tillering; combined with *SD1* provides semi-dwarf stature	Increased branching, higher grain yield, robust plant architecture (green revolution trait)	Wang *et al.* (2020a)
Rice	Nuclear proteins NSP1 + NSP2	Under phosphate deficiency, NSP1–NSP2 heterodimer activates SL biosynthesis genes	Elevated SLs → altered root branching and shoot tillering → improved P foraging and nutrient-use efficiency	Yuan *et al.* (2023)
Sorghum	Transporters SbSLT1, SbSLT2 (ABCG family)	Export of SLs from roots; expression rises in P-deficient soils; mutant lines show reduced SL exudation	Mutants suppress *Striga* germination → reduced parasitism → increased grain yield	Shi *et al.* (2025)
Rice (aerobic)	GR24	GR24 triggers AMF hyphal branching and root colonization → improved symbiosis	Enhanced P uptake under deficit → maximized phosphorus-use efficiency	Mitra *et al.* (2024)
Wheat (dryland)	Foliar GR24 under drought	Increased antioxidant enzyme activity (SOD, APX) at 50% field capacity	Improved antioxidant defense, reduced oxidative stress, enhanced drought tolerance and yield	Guo *et al.* (2023)
Maize (HY-1898, FH-1046 hybrids)	Foliar GR24 (0–0.1 mg l^{-1}) under drought (60% field capacity)	GR24 improved chlorophyll, photosynthesis, antioxidant activity, osmolyte accumulation (proline, glycine betaine)	Increased cob weight, cob diameter, grain number; optimal at 0.01 mg l^{-1}	Luqman *et al.* (2023)

AMF, arbuscular mycorrhizal fungi; APX, ascorbate peroxidase; GR24, synthetic SL analog; P, phosphate; SOD, superoxide dismutase.

and moisture further restricts their effectiveness and creates challenges for the development of stable, economical formulations for agricultural uses such as parasitic weed control and crop improvement (Jamil *et al.*, 2022). Although important genes such as *CCD7*, *CCD8*, and *MAX2* have been identified, the full network of SL biosynthesis and signaling pathways is still unknown (Hajam *et al.*, 2024). Crosstalk with other hormones (auxins, CKs, and ABA) further complicates their roles, and the regulatory mechanisms governing SL production under adverse conditions such as drought and nutrient deficiency remain incompletely understood (Wu *et al.*, 2022). Application of SLs in agriculture shows great promise for sustainable agriculture, but is currently limited by high synthesis costs, species-specific responses due to structural diversity, and challenges in implementing the suicidal germination strategy against parasitic weeds, largely due to instability and the need for precise field application protocols (Jamil *et al.*, 2022). Although SLs are linked to improved tolerance under drought and salinity stress, the precise mechanisms remain unclear and need further investigation across diverse crop species (Luqman *et al.*, 2023).

13.10.2 Opportunities for translational research in crops

There are promising opportunities to transform SLs into agricultural applications using breeding and marker-assisted selection to identify natural variations in SL-related genes (e.g. *HTD/D17*, *D14*, *MAX2*, *SbSLT1/2*), providing targets for developing nutrient-efficient and stress-resilient cultivars (Gouda *et al.*, 2020; Wang *et al.*, 2020a; Shi *et al.*, 2025). Using new breeding tools such as CRISPR-Cas9, precise modification of SL biosynthetic or signaling genes can optimize tillering, root architecture, and nutrient uptake without yield penalties (Butt *et al.*, 2018). SLs could be used as biostimulants, where synthetic analogs or mimicking compounds could be developed as foliar sprays or as soil amendments to improve nutrient foraging and stress tolerance (Borghi *et al.*, 2021; Sedaghat *et al.*, 2021; Popa *et al.*, 2025). SLs could be integrated with AMF, where co-application of AMF inoculants could synergistically improve phosphorus acquisition in low-input systems (Besserer *et al.*, 2006; Mitra *et al.*, 2024; Naseer *et al.*, 2024). Lastly, application in systems biology and omics approaches using transcriptomics, proteomics, and metabolomics could help unravel SL-mediated nutrient signaling networks and identify biomarkers for NUE (Zhou *et al.*, 2020).

13.10.3 Future directions for enhancing NUE through SLs

To use SLs for improving NUE, future research should prioritize and should focus on three key areas: optimizing the plant–microbe symbiosis, engineering crops for improved NUE, and developing next-generation agricultural products (Borghi *et al.*, 2021; Sun *et al.*, 2023; Naseer *et al.*, 2024). Future research should focus first on engineering crop alleles for beneficial use of SLs to promote nutrient signaling and AMF symbiosis, and on their harmful function of triggering the germination of parasite weeds such as *Striga*, with the aim of developing plants that retain the positive traits while reducing the exudation of specific SLs that acts as *Striga* germination stimulants (Huang and Qi, 2025). Secondly, field-scale validation of SL analogs will involve developing stable, cost-effective, and crop-specific formulations of SL analogs in the agricultural world, and these products will need to be tested under various environmental conditions to prove their efficacy and economic viability for farmers (Jamil *et al.*, 2025). Exploiting natural allelic diversity for screening global germplasm of SL-related genes linked to NUE and stress adaptation will enable breeding of more resilient and high-yielding crop varieties (Wang *et al.*, 2020a). SLs interact with nutrient signaling pathways, particularly nitrogen, phosphorus, and potassium. Thus, studies of SL–nutrient crosstalk will help to optimize with nutrient uptake and its overall use efficiency. Bouwmeester *et al.* (2007) evaluated the long-term ecological and regulatory impacts of SL analogs, including their need for soil microbiota, non-target plants, and ecosystem stability, to ensure safe and sustainable use in agriculture.

13.11 Conclusion

SLs have become important modulators of NUE of crop plants, combining developmental processes, hormonal signals, and environmental cues to maximize plant growth and nutrient uptake. SLs facilitate the effective uptake and distribution of nitrogen, phosphorus, potassium, sulfur, and micronutrients by altering the architecture of the root system, increasing transporter function, and interacting with hormones including auxin, CKs, and ABA. In order to further enhance nutrient availability and plant resilience, they also interact with beneficial soil microbes such as nutrient-solubilizing bacteria and AMF. By promoting antioxidant defenses, osmotic control, and stress-responsive signaling pathways, SLs also help to maintain NUE in the face of abiotic stressors including salinity and drought. Developments in genome editing, molecular genetics, and the use of synthetic SL analogs provide useful methods to improve the use of SLs.

References

Abuauf, H., Haider, I., Jia, K.-P., Ablazov, A., Mi, J. *et al.* (2018) The *Arabidopsis* DWARF27 gene encodes an all-trans-/9-cis-β-carotene isomerase and is induced by auxin, abscisic acid and phosphate deficiency. *Plant Science* 277, 33–42. DOI: 10.1016/j.plantsci.2018.06.024.

Ahmed, N., Li, J., Li, Y., Deng, L., Deng, L. *et al.* (2025) Symbiotic synergy: How arbuscular mycorrhizal fungi enhance nutrient uptake, stress tolerance, and soil health through molecular mechanisms and hormonal regulation. *IMA Fungus* 16, e144989. DOI: 10.3897/imafungus.16.144989.

Al-Babili, S. and Bouwmeester, H.J. (2015) Strigolactones, a novel carotenoid-derived plant hormone. *Annual Review of Plant Biology* 66, 161–186. DOI: 10.1146/annurev-arplant-043014-114759.

Alder, A., Jamil, M., Marzorati, M., Bruno, M., Vermathen, M. *et al.* (2012) The path from β-carotene to carlactone, a strigolactone-like plant hormone. *Science* 335(6074), 1348–1351. DOI: 10.1126/science.1218094.

Ali, F., Qanmber, G., Li, F. and Wang, Z. (2022) Updated role of ABA in seed maturation, dormancy, and germination. *Journal of Advanced Research* 35, 199–214. DOI: 10.1016/j.jare.2021.03.011.

Alvi, A.F., Sehar, Z., Fatma, M., Masood, A. and Khan, N.A. (2022) Strigolactone: An emerging growth regulator for developing resilience in plants. *Plants* 11(19), 2604. DOI: 10.3390/plants11192604.

Alzate Zuluaga, M.Y., Fattorini, R., Cesco, S. and Pii, Y. (2024) Plant–microbe interactions in the rhizosphere for smarter and more sustainable crop fertilization: The case of PGPR-based biofertilizers. *Frontiers in Microbiology* 15, 1440978. DOI: 10.3389/fmicb.2024.1440978.

Andreo-Jimenez, B., Ruyter-Spira, C., Bouwmeester, H.J. and Lopez-Raez, J.A. (2015) Ecological relevance of strigolactones in nutrient uptake and other abiotic stresses, and in plant–microbe interactions below-ground. *Plant and Soil* 394(1–2), 1–19. DOI: 10.1007/s11104-015-2544-z.

Arite, T., Umehara, M., Ishikawa, S., Hanada, A., Maekawa, M. *et al.* (2009) D14, a strigolactone-insensitive mutant of rice, shows an accelerated outgrowth of tillers. *Plant & Cell Physiology* 50(8), 1416–1424. DOI: 10.1093/pcp/pcp091.

Balzergue, C., Puech-Pagès, V., Bécard, G. and Rochange, S.F. (2011) The regulation of arbuscular mycorrhizal symbiosis by phosphate in pea involves early and systemic signalling events. *Journal of Experimental Botany* 62(3), 1049–1060. DOI: 10.1093/jxb/erq335.

Banerjee, A. and Roychoudhury, A. (2018) Strigolactones: Multi-level regulation of biosynthesis and diverse responses in plant abiotic stresses. *Acta Physiologiae Plantarum* 40(5), 86. DOI: 10.1007/s11738-018-2660-5.

Barbosa, I.C.R., Hammes, U.Z. and Schwechheimer, C. (2018) Activation and polarity control of PIN-FORMED auxin transporters by phosphorylation. *Trends in Plant Science* 23(6), 523–538. DOI: 10.1016/j.tplants.2018.03.009.

Bennett, T., Sieberer, T., Willett, B., Booker, J., Luschnig, C. *et al.* (2006) The *arabidopsis MAX* pathway controls shoot branching by regulating auxin transport. *Current Biology* 16(6), 553–563. DOI: 10.1016/j.cub.2006.01.058.

Besserer, A., Puech-Pagès, V., Kiefer, P., Gomez-Roldan, V., Jauneau, A. *et al.* (2006) Strigolactones stimulate arbuscular mycorrhizal fungi by activating mitochondria. *PLOS Biology* 4(7), e226. DOI: 10.1371/journal.pbio.0040226.

Borghi, L., Screpanti, C., Lumbroso, A., Lachia, M., Gübeli, C. *et al.* (2021) Efficiency and bioavailability of new synthetic strigolactone mimics with potential for sustainable agronomical applications. *Plant and Soil* 465(1–2), 109–123. DOI: 10.1007/s11104-021-04943-8.

Bouwmeester, H.J., Roux, C., Lopez-Raez, J.A. and Bécard, G. (2007) Rhizosphere communication of plants, parasitic plants and AM fungi. *Trends in Plant Science* 12(5), 224–230. DOI: 10.1016/j.tplants.2007.03.009.

Boyno, G., Rezaee Danesh, Y., Demir, S., Teniz, N., Mulet, J.M. *et al.* (2023) The complex interplay between arbuscular mycorrhizal fungi and strigolactone: Mechanisms, sinergies, applications and future directions. *International Journal of Molecular Sciences* 24(23), 16774. DOI: 10.3390/ijms242316774.

Brun, G., Braem, L., Thoiron, S., Gevaert, K., Goormachtig, S. *et al.* (2018) Seed germination in parasitic plants: What insights can we expect from strigolactone research? *Journal of Experimental Botany* 69(9), 2265–2280. DOI: 10.1093/jxb/erx472.

Bruno, M., Hofmann, M., Vermathen, M., Alder, A., Beyer, P. *et al.* (2014) On the substrate- and stereospecificity of the plant carotenoid cleavage dioxygenase 7. *FEBS Letters* 588(9), 1802–1807. DOI: 10.1016/j.febslet.2014.03.041.

Butt, H., Jamil, M., Wang, J.Y., Al-Babili, S. and Mahfouz, M. (2018) Engineering plant architecture via CRISPR/Cas9-mediated alteration of strigolactone biosynthesis. *BMC Plant Biology* 18(1), 174. DOI: 10.1186/s12870-018-1387-1.

Chesterfield, R.J., Vickers, C.E. and Beveridge, C.A. (2020) Translation of strigolactones from plant hormone to agriculture: Achievements, future perspectives, and challenges. *Trends in Plant Science* 25(11), 1087–1106. DOI: 10.1016/j.tplants.2020.06.005.

Choi, I., Cho, H., Brandizzi, F. and Rouached, H. (2025) Phosphorus deficiency and root growth: The role of TOR signaling in adaptive responses. *Journal of Experimental Botany* (in press), eraf204. DOI: 10.1093/jxb/eraf204.

Congreves, K.A., Otchere, O., Ferland, D., Farzadfar, S., Williams, S. *et al.* (2021) Nitrogen use efficiency definitions of today and tomorrow. *Frontiers in Plant Science* 12, 637108. DOI: 10.3389/fpls.2021.637108.

Czarnecki, O., Yang, J., Weston, D.J., Tuskan, G.A. and Chen, J.G. (2013) A dual role of strigolactones in phosphate acquisition and utilization in plants. *International Journal of Molecular Sciences* 14(4), 7681–7701. DOI: 10.3390/ijms14047681.

de Saint Germain, A., Ligerot, Y., Dun, E.A., Pillot, J.-P., Ross, J.J. *et al.* (2013) Strigolactones stimulate internode elongation independently of gibberellins. *Plant Physiology* 163(2), 1012–1025. DOI: 10.1104/pp.113.220541.

Dell'Oste, V., Spyrakis, F. and Prandi, C. (2021) Strigolactones, from plants to human health: Achievements and challenges. *Molecules* 26(15), 4579. DOI: 10.3390/molecules26154579.

Dong, H., Wang, J., Song, X., Hu, C., Zhu, C. *et al.* (2023) HY5 functions as a systemic signal by integrating BRC1-dependent hormone signaling in tomato bud outgrowth. *Proceedings of the National Academy of Sciences USA* 120(16), e2301879120. DOI: 10.1073/pnas.2301879120.

Duan, J., Yu, H., Yuan, K., Liao, Z., Meng, X. *et al.* (2019) Strigolactone promotes CYTOKININ degradation through transcriptional activation of *CYTOKININ OXIDASE/DEHYDROGENASE 9* in rice. *Proceedings of the National Academy of Sciences USA* 116(28), 14319–14324. DOI: 10.1073/pnas.1810980116.

Faizan, M., Faraz, A., Sami, F., Siddiqui, H., Yusuf, M. *et al.* (2020) Role of strigolactones: Signalling and crosstalk with other phytohormones. *Open Life Sciences* 15(1), 217–228. DOI: 10.1515/biol-2020-0022.

Foo, E., Yoneyama, K., Hugill, C.J., Quittenden, L.J. and Reid, J.B. (2013) Strigolactones and the regulation of pea symbioses in response to nitrate and phosphate deficiency. *Molecular Plant* 6(1), 76–87. DOI: 10.1093/mp/sss115.

Fróna, D., Szenderák, J. and Harangi-Rákos, M. (2019) The challenge of feeding the world. *Sustainability* 11(20), 5816. DOI: 10.3390/su11205816.

Giehl, R.F.H. and von Wirén, N. (2014) Root nutrient foraging. *Plant Physiology* 166(2), 509–517. DOI: 10.1104/pp.114.245225.

Giri, S., Panwar, H., Singh, K. and Kumar, M. (2025) Microbial contributions to crop adaptation: Innovation for climate-resilient agriculture. In: Mann, A., Kumar, N., Kumar, A., Chandra, P., Sanwal, S.K. *et al.* (eds) *Cutting Edge Technologies for Developing Future Crop Plants*. Springer Nature Singapore, Singapore, pp. 349–364. DOI: 10.1007/978-981-96-2508-6_17.

Gouda, G., Gupta, M.K., Donde, R., Mohapatra, T., Vadde, R. *et al.* (2020) Marker-assisted selection for grain number and yield-related traits of rice (*Oryza sativa* L.). *Physiology and Molecular Biology of Plants* 26(5), 885–898. DOI: 10.1007/s12298-020-00773-7.

Govindasamy, P., Muthusamy, S.K., Bagavathiannan, M., Mowrer, J., Jagannadham, P.T.K. *et al.* (2023) Nitrogen use efficiency—a key to enhance crop productivity under a changing climate. *Frontiers in Plant Science* 14, 1121073. DOI: 10.3389/fpls.2023.1121073.

Guan, Y., Li, L., Wang, D., Zhou, J., Qi, W. *et al.* (2025) Multifaceted functions of strigolactones in annual and perennial plants: Developmental regulation, phytohormone crosstalk and abiotic stresses. *Plant Stress* 16, 100895. DOI: 10.1016/j.stress.2025.100895.

Guo, S., Wei, X., Ma, B., Ma, Y., Yu, Z. *et al.* (2023) Foliar application of strigolactones improves the desiccation tolerance, grain yield and water use efficiency in dryland wheat through modulation of non-hydraulic root signals and antioxidant defense. *Stress Biology* 3(1), 54. DOI: 10.1007/s44154-023-00127-9.

Ha, C.V., Leyva-González, M.A., Osakabe, Y., Tran, U.T., Nishiyama, R. *et al.* (2014) Positive regulatory role of strigolactone in plant responses to drought and salt stress. *Proceedings of the National Academy of Sciences USA* 111(2), 851–856. DOI: 10.1073/pnas.1322135111.

Hajam, A.H., Ali, M.S., Singh, S.K., Bashri, G. and Dubey, G. (2024) Implications of strigolactones in plant biology: Achievements, future perspectives, and challenges. Strigolactones 221–236. DOI: 10.1016/B978-0-443-13521-7.00012-9.

Hamiaux, C., Drummond, R.S.M., Janssen, B.J., Ledger, S.E., Cooney, J.M. *et al.* (2012) DAD2 is an α/β hydrolase likely to be involved in the perception of the plant branching hormone, strigolactone. *Current Biology* 22(21), 2032–2036. DOI: 10.1016/j.cub.2012.08.007.

Haq, B.U., Ahmad, M.Z., Ur Rehman, N., Wang, J., Li, P. *et al.* (2017) Functional characterization of soybean strigolactone biosynthesis and signaling genes in *Arabidopsis* MAX mutants and GmMAX3 in soybean nodulation. *BMC Plant Biology* 17(1), 259. DOI: 10.1186/s12870-017-1182-4.

He, K., Du, J., Han, X., Li, H., Kui, M. *et al.* (2023) PHOSPHATE STARVATION RESPONSE1 (PHR1) interacts with JASMONATE ZIM-DOMAIN (JAZ) and MYC2 to modulate PHOSPHATE deficiency-induced JASMONATE signaling in *Arabidopsis*. *Plant Cell* 35(6), 2132–2156. DOI: 10.1093/plcell/koad057.

Hu, Z., Yamauchi, T., Yang, J., Jikumaru, Y., Tsuchida-Mayama, T. *et al.* (2014) Strigolactone and cytokinin act antagonistically in regulating rice mesocotyl elongation in darkness. *Plant and Cell Physiology* 55(1), 30–41. DOI: 10.1093/pcp/pct150.

Huang, Z. and Qi, F. (2025) Engineering strigolactone signaling: Toward crops that resist parasites without sacrificing symbiosis. *Crop Health* 3(1), 13. DOI: 10.1007/s44297-025-00053-4.

Jamil, M., Wang, J.Y., Yonli, D., Patil, R.H., Riyazaddin, M. *et al.* (2022) A new formulation for strigolactone suicidal germination agents, towards successful *Striga* management. *Plants* 11(6), 808. DOI: 10.3390/plants11060808.

Jamil, M., Mutinda, S., Wang, J.Y., Barminga, D., Mwihaki, A. *et al.* (2025) Evaluation of formulated strigolactone analogs for *Striga* management in Kenyan agriculture. *Journal of Agriculture and Food Research* 21, 101921. DOI: 10.1016/j.jafr.2025.101921.

Jia, K.P., Baz, L. and Al-Babili, S. (2018) From carotenoids to strigolactones. *Journal of Experimental Botany* 69(9), 2189–2204. DOI: 10.1093/jxb/erx476.

Jiang, L., Liu, X., Xiong, G., Liu, H., Chen, F. *et al.* (2013) DWARF 53 acts as a repressor of strigolactone signalling in rice. *Nature* 504(7480), 401–405. DOI: 10.1038/nature12870.

Jiang, X., Jia, X., Gao, S., Jiang, Y., Wei, N. *et al.* (2021) Plant nutrient contents rather than physical traits are coordinated between leaves and roots in a desert shrubland. *Frontiers in Plant Science* 12, 734775. DOI: 10.3389/fpls.2021.734775.

Jia, Z., Giehl, R.F.H. and von Wirén, N. (2022) Nutrient-hormone relations: Driving root plasticity in plants. *Molecular Plant* 15(1), 86–103. DOI: 10.1016/j.molp.2021.12.004.

Josse, M.R. (2022) The role of strigolactone and KAI2 signalling in seedling development. PhD thesis, University of Leeds, Leeds, UK.

Justamante, M.S., Larriba, E., Luque, A., Nicolás-Albujer, M. and Pérez-Pérez, J.M. (2025) A systematic review to identify target genes that modulate root system architecture in response to abiotic stress. *Scientific Reports* 15(1), 13219. DOI: 10.1038/s41598-025-97266-y.

Kagiyama, M., Hirano, Y., Mori, T., Kim, S.Y., Kyozuka, J. *et al.* (2013) Structures of D14 and D14L in the strigolactone and karrikin signaling pathways. *Genes to Cells* 18(2), 147–160. DOI: 10.1111/gtc.12025.

Karniel, U., Koch, A., Bar Nun, N., Zamir, D. and Hirschberg, J. (2024) Tomato mutants reveal root and shoot strigolactone involvement in branching and broomrape resistance. *Plants* 13(11), 1554. DOI: 10.3390/plants13111554.

Kee, Y.J., Ogawa, S., Ichihashi, Y., Shirasu, K. and Yoshida, S. (2023) Strigolactones in rhizosphere communication: Multiple molecules with diverse functions. *Plant & Cell Physiology* 64(9), 955–966. DOI: 10.1093/pcp/pcad055.

Kelly, J.H., Tucker, M.R. and Brewer, P.B. (2023) The strigolactone pathway is a target for modifying crop shoot architecture and yield. *Biology* 12(1), 95. DOI: 10.3390/biology12010095.

Khan, F., Siddique, A.B., Shabala, S., Zhou, M. and Zhao, C. (2023) Phosphorus plays key roles in regulating plants' physiological responses to abiotic stresses. *Plants* 12(15), 2861. DOI: 10.3390/plants12152861.

Khuvung, K., Silva Gutierrez, F.A.O. and Reinhardt, D. (2022) How strigolactone shapes shoot architecture. *Frontiers in Plant Science* 13, 889045. DOI: 10.3389/fpls.2022.889045.

Kobae, Y., Kameoka, H., Sugimura, Y., Saito, K., Ohtomo, R. *et al.* (2018) Strigolactone biosynthesis genes of rice are required for the punctual entry of arbuscular mycorrhizal fungi into the roots. *Plant & Cell Physiology* 59(3), 544–553. DOI: 10.1093/pcp/pcy001.

Kodama, K., Rich, M.K., Yoda, A., Shimazaki, S., Xie, X. *et al.* (2022) An ancestral function of strigolactones as symbiotic rhizosphere signals. *Nature Communications* 13(1), 3974. DOI: 10.1038/s41467-022-31708-3.

Koevoets, I.T., Venema, J.H., Elzenga, J.T. and Testerink, C. (2016) Roots withstanding their environment: Exploiting root system architecture responses to abiotic stress to improve crop tolerance. *Frontiers in Plant Science* 7, 215317. DOI: 10.3389/fpls.2016.01335.

Kohlen, W., Charnikhova, T., Liu, Q., Bours, R., Domagalska, M.A. *et al.* (2011) Strigolactones are transported through the xylem and play a key role in shoot architectural response to phosphate deficiency in nonarbuscular mycorrhizal host *Arabidopsis*. *Plant Physiology* 155(2), 974–987. DOI: 10.1104/pp.110.164640.

Kohlen, W., Charnikhova, T., Lammers, M., Pollina, T., Tóth, P. *et al.* (2012) The tomato *CAROTENOID CLEAVAGE DIOXYGENASE8 (SlCCD8)* regulates rhizosphere signaling, plant architecture and affects reproductive development through strigolactone biosynthesis. *New Phytologist* 196(2), 535–547. DOI: 10.1111/j.1469-8137.2012.04265.x.

Korek, M. and Marzec, M. (2023) Strigolactones and abscisic acid interactions affect plant development and response to abiotic stresses. *BMC Plant Biology* 23(1), 314. DOI: 10.1186/s12870-023-04332-6.

Kumar, M., Kim, I., Kim, Y.K., Heo, J.B., Suh, M.C. *et al.* (2019) Strigolactone signaling genes showing differential expression patterns in *Arabidopsis max* mutants. *Plants* 8(9), 352. DOI: 10.3390/plants8090352.

Kumar, R., Farda, B., Mignini, A., Djebaili, R., Koolman, L. *et al.* (2025) Microbial solutions in agriculture: Enhancing soil health and resilience through bio-inoculants and bioremediation. *Bacteria* 4(3), 28. DOI: 10.3390/bacteria4030028.

Lal, R. (2009) Soil degradation as a reason for inadequate human nutrition. *Food Security* 1(1), 45–57. DOI: 10.1007/s12571-009-0009-z.

Lal, R. (2012) Climate change and soil degradation mitigation by sustainable management of soils and other natural resources. *Agricultural Research* 1(3), 199–212. DOI: 10.1007/s40003-012-0031-9.

Langholtz, M., Davison, B.H., Jager, H.I., Eaton, L., Baskaran, L.M. *et al.* (2021) Increased nitrogen use efficiency in crop production can provide economic and environmental benefits. *Science of the Total Environment* 758, 143602. DOI: 10.1016/j.scitotenv.2020.143602.

Lassaletta, L., Billen, G., Grizzetti, B., Anglade, J. and Garnier, J. (2014) 50 year trends in nitrogen use efficiency of world cropping systems: The relationship between yield and nitrogen input to cropland. *Environmental Research Letters* 9(10), 105011. DOI: 10.1088/1748-9326/9/10/105011.

Li, C., Haider, I., Wang, J.Y., Quinodoz, P. and Suarez Duran, H.G. (2024) OsCYP706C2 diverts rice strigolactone biosynthesis to a noncanonical pathway branch. *Science Advances* 10(35), eadq3942. DOI: 10.1126/sciadv.adq3942.

López-Ráez, J.A., Charnikhova, T., Gómez-Roldán, V., Matusova, R., Kohlen, W. *et al.* (2008) Tomato strigolactones are derived from carotenoids and their biosynthesis is promoted by phosphate starvation. *New Phytologist* 178(4), 863–874. DOI: 10.1111/j.1469-8137.2008.02406.x.

Luo, L. (2022) Is strigolactone signaling a key player in regulating tiller formation in response to nitrogen? *Frontiers in Plant Science* 13, 1081740. DOI: 10.3389/fpls.2022.1081740.

Luqman, M., Shahbaz, M., Maqsood, M.F., Farhat, F., Zulfiqar, U. *et al.* (2023) Effect of strigolactone on growth, photosynthetic efficiency, antioxidant activity, and osmolytes accumulation in different maize (*Zea mays* L.) hybrids grown under drought stress. *Plant Signaling & Behavior* 18(1), 2262795. DOI: 10.1080/15592324.2023.2262795.

Marzec, M. (2017) Strigolactones and gibberellins: A new couple in the phytohormone world? *Trends in Plant Science* 22(10), 813–815. DOI: 10.1016/j.tplants.2017.08.001.

Marzec, M. and Melzer, M. (2018) Regulation of root development and architecture by strigolactones under optimal and nutrient deficiency conditions. *International Journal of Molecular Sciences* 19(7), 1887. DOI: 10.3390/ijms19071887.

Marro, N., Lidoy, J., Chico, M.Á., Rial, C., García, J. *et al.* (2022) Strigolactones: New players in the nitrogen–phosphorus signalling interplay. *Plant, Cell & Environment* 45(2), 512–527. DOI: 10.1111/pce.14212.

Marzec, M., Muszynska, A. and Gruszka, D. (2013) The role of strigolactones in nutrient-stress responses in plants. *International Journal of Molecular Sciences* 14(5), 9286–9304. DOI: 10.3390/ijms14059286.

Mashiguchi, K., Seto, Y., Onozuka, Y., Suzuki, S., Takemoto, K. *et al.* (2022) A carlactonoic acid methyltransferase that contributes to the inhibition of shoot branching in *Arabidopsis*. *Proceedings of the National Academy of Sciences USA* 119(14), e2111565119. DOI: 10.1073/pnas.2111565119.

Mayzlish-Gati, E., De-Cuyper, C., Goormachtig, S., Beeckman, T., Vuylsteke, M. *et al.* (2012) Strigolactones are involved in root response to low phosphate conditions in *Arabidopsis*. *Plant Physiology* 160(3), 1329–1341. DOI: 10.1104/pp.112.202358.

Mazzoni-Putman, S.M., Brumos, J., Zhao, C., Alonso, J.M. and Stepanova, A.N. (2021) Auxin interactions with other hormones in plant development. *Cold Spring Harbor Perspectives in Biology* 13(10), a039990. DOI: 10.1101/cshperspect.a039990.

Merilo, E., Jalakas, P., Laanemets, K., Mohammadi, O., Hõrak, H. *et al.* (2015) Abscisic acid transport and homeostasis in the context of stomatal regulation. *Molecular Plant* 8(9), 1321–1333. DOI: 10.1016/j.molp.2015.06.006.

Mitra, D., Panneerselvam, P., Chidambaranathan, P., Nayak, A.K., Priyadarshini, A. *et al.* (2024) Strigolactone GR24-mediated mitigation of phosphorus deficiency through mycorrhization in aerobic rice. *Current Research in Microbial Sciences* 6, 100229. DOI: 10.1016/j.crmicr.2024.100229.

Moreno, J.C., Mi, J., Alagoz, Y. and Al-Babili, S. (2021) Plant apocarotenoids: From retrograde signaling to interspecific communication. *Plant Journal* 105(2), 351–375. DOI: 10.1111/tpj.15102.

Naseer, M.A., Zhang, Z.Q., Mukhtar, A., Asad, M.S., Wu, H.Y. *et al.* (2024) Strigolactones: A promising tool for nutrient acquisition through arbuscular mycorrhizal fungi symbiosis and abiotic stress tolerance. *Plant Physiology and Biochemistry* 215, 109057. DOI: 10.1016/j.plaphy.2024.109057.

Naz, M., Afzal, M.R., Raza, M.A., Tariq, M., Yan, M. *et al.* (2025) The significant effects of strigolactones on plant growth and microbe interactions: A review. *Plant Growth Regulation* 105, 937–957. DOI: 10.1007/s10725-025-01325-3.

Negin, B. and Moshelion, M. (2016) The evolution of the role of ABA in the regulation of water-use efficiency: From biochemical mechanisms to stomatal conductance. *Plant Science* 251, 82–89. DOI: 10.1016/j.plantsci.2016.05.007.

Nelson, D.C. (2021) The mechanism of host-induced germination in root parasitic plants. *Plant Physiology* 185(4), 1353–1373. DOI: 10.1093/plphys/kiab043.

Pandey, K. and Saharan, B.S. (2025) Soil microbiomes: A promising strategy for boosting crop yield and advancing sustainable agriculture. *Discover Agriculture* 3(1), 54. DOI: 10.1007/s44279-025-00208-5.

Nunes, P.S.O., Lacerda-Junior, G.V., Mascarin, G.M., Guimarães, R.A., Medeiros, F.H.V. *et al.* (2024) Microbial consortia of biological products: Do they have a future? *Biological Control* 188, 105439. DOI: 10.1016/j.biocontrol.2024.105439.

Pandey, A., Sharma, M. and Pandey, G.K. (2016) Emerging roles of strigolactones in plant responses to stress and development. *Frontiers in Plant Science* 7, 434. DOI: 10.3389/fpls.2016.00434.

Popa, D.G., Tritean, N., Georgescu, F., Lupu, C., Shaposhnikov, S. *et al.* (2025) SL-6 mimic is a biostimulant for *Chlorella sorokiniana* and enhances the plant biostimulant effect of microalgal extract. *Plants* 14(7), 1010. DOI: 10.3390/plants14071010.

Quaggiotti, S., Buzzicotti, L., Koch, K.E., Guan, J.C., Trevisan, S. *et al.* (2024) Strigolactone roles in maize tolerance to low nitrogen involve shifts in acquisition and partitioning of protein, sulfur, and iron. *Plant and Soil* 505(1–2), 41–64. DOI: 10.1007/s11104-024-06561-6.

Rehman, N.U., Li, X., Zeng, P., Guo, S., Jan, S. *et al.* (2021) Harmony but not uniformity: Role of strigolactone in plants. *Biomolecules* 11(11), 1616. DOI: 10.3390/biom11111616.

Saeed, W., Naseem, S. and Ali, Z. (2017) Strigolactones biosynthesis and their role in abiotic stress resilience in plants: A critical review. *Frontiers in Plant Science* 8, 1487. DOI: 10.3389/fpls.2017.01487.

Sahoo, A., Yadav, G., Mehta, T., Meena, M. and Swapnil, P. (2025) Omics-driven insights into plant growth-promoting microorganisms for sustainable agriculture. *Discover Sustainability* 6(1), 1–23. DOI: 10.1007/s43621-025-01582-2.

Santoro, V., Schiavon, M., Gresta, F., Ertani, A., Cardinale, F. *et al.* (2020) Strigolactones control root system architecture and tip anatomy in *Solanum lycopersicum* L. plants under P starvation. *Plants* 9(5), 612. DOI: 10.3390/plants9050612.

Santoro, V., Schiavon, M., Visentin, I., Constán-Aguilar, C., Cardinale, F. *et al.* (2021) Strigolactones affect phosphorus acquisition strategies in tomato plants. *Plant, Cell & Environment* 44(11), 3628–3642. DOI: 10.1111/pce.14169.

Sedaghat, M., Emam, Y., Mokhtassi-Bidgoli, A., Hazrati, S., Lovisolo, C. *et al.* (2021) The potential of the synthetic strigolactone analogue GR24 for the maintenance of photosynthesis and yield in winter wheat under drought: Investigations on the mechanisms of action and delivery modes. *Plants* 10(6), 1223. DOI: 10.3390/plants10061223.

Seto, Y., Sado, A., Asami, K., Hanada, A., Umehara, M. *et al.* (2014) Carlactone is an endogenous biosynthetic precursor for strigolactones. *Proceedings of the National Academy of Sciences USA* 111(4), 1640–1645. DOI: 10.1073/pnas.1314805111.

Sharma, L.K. and Bali, S.K. (2017) A review of methods to improve nitrogen use efficiency in agriculture. *Sustainability* 10(1), 51. DOI: 10.3390/su10010051.

Sharma, N., Chaudhary, A., Sharma, N., Shukla, S., Joshi, N.C. *et al.* (2025) Plant nutrient requirements and nutrient homeostasis in plants. In: Joshi, N.C., Leustek, T. and Dingh, P.K. (eds) *Soil Health and Nutrition Management*. CAB International, Wallingford, UK, pp. 1–20. DOI: 10.1079/9781800624597.0001.

Shen, H., Zhu, L., Bu, Q.-Y. and Huq, E. (2012) MAX2 affects multiple hormones to promote photomorphogenesis. *Molecular Plant* 5(3), 750–762. DOI: 10.1093/mp/sss029.

Shi, J., Mei, C., Ge, F., Hu, Q., Ban, X. *et al.* (2025) Resistance to *Striga* parasitism through reduction of strigolactone exudation. *Cell* 188(7), 1955–1966. DOI: 10.1016/j.cell.2025.01.022.

Shinohara, N., Taylor, C. and Leyser, O. (2013) Strigolactone can promote or inhibit shoot branching by triggering rapid depletion of the auxin efflux protein PIN1 from the plasma membrane. *PLOS Biology* 11(1), e1001474. DOI: 10.1371/journal.pbio.1001474.

Singh, B.K., Hu, H.W., Macdonald, C.A. and Xiong, C. (2025) Microbiome-facilitated plant nutrient acquisition. *Cell Host & Microbe* 33(6), 869–881. DOI: 10.1016/j.chom.2025.05.005.

Smith, S.E. and Read, D.J. (2010) *Mycorrhizal Symbiosis*, 3rd edn. Academic Press, Netherlands.

Soto-Cruz, F.J., Zorrilla, J.G., Rial, C., Varela, R.M., Molinillo, J.M.G. *et al.* (2021) Allelopathic activity of strigolactones on the germination of parasitic plants and arbuscular mycorrhizal fungi growth. *Agronomy* 11(11), 2174. DOI: 10.3390/agronomy11112174.

Soundappan, I., Bennett, T., Morffy, N., Liang, Y., Stanga, J.P. *et al.* (2015) SMAX1-LIKE/D53 family members enable distinct MAX2-dependent responses to strigolactones and karrikins in *Arabidopsis*. *Plant Cell* 27(11), 3143–3159. DOI: 10.1105/tpc.15.00562.

Stanga, J.P., Smith, S.M., Briggs, W.R. and Nelson, D.C. (2013) *SUPPRESSOR OF MORE AXILLARY GROWTH2 1* controls seed germination and seedling development in *Arabidopsis*. *Plant Physiology* 163(1), 318–330. DOI: 10.1104/pp.113.221259.

Stirnberg, P., Furner, I.J. and Leyser, H.M.O. (2007) MAX2 participates in an SCF complex which acts locally at the node to suppress shoot branching. *Plant Journal* 50(1), 80–94. DOI: 10.1111/j.1365-313X.2007.03032.x.

Sun, H., Xu, F., Guo, X., Wu, D., Zhang, X. *et al.* (2019) A strigolactone signal inhibits secondary lateral root development in rice. *Frontiers in Plant Science* 10, 1527. DOI: 10.3389/fpls.2019.01527.

Sun, H., Li, W., Burritt, D.J., Tian, H., Zhang, H. *et al.* (2022) Strigolactones interact with other phytohormones to modulate plant root growth and development. *Crop Journal* 10(6), 1517–1527. DOI: 10.1016/j.cj.2022.07.014.

Sun, H., Guo, X., Zhu, X., Gu, P., Zhang, W. *et al.* (2023) Strigolactone and gibberellin signaling coordinately regulate metabolic adaptations to changes in nitrogen availability in rice. *Molecular Plant* 16(3), 588–598. DOI: 10.1016/j.molp.2023.01.009.

Takatsuka, H., Amari, T. and Umeda, M. (2024) Cytokinin signaling is involved in root hair elongation in response to phosphate starvation. *Plant Signaling & Behavior* 19(1), 2305030. DOI: 10.1080/15592324.2024.2305030.

Teichmann, T. and Muhr, M. (2015) Shaping plant architecture. *Frontiers in Plant Science* 6, 233. DOI: 10.3389/fpls.2015.00233.

Temmerman, A., Guillory, A., Bonhomme, S., Goormachtig, S. and Struk, S. (2022) Masks start to drop: Suppressor of MAX2 1-like proteins reveal their many faces. *Frontiers in Plant Science* 13, 887232. DOI: 10.3389/fpls.2022.887232.

Tuan, P.A., Kumar, R., Rehal, P.K., Toora, P.K. and Ayele, B.T. (2018) Molecular mechanisms underlying abscisic acid/gibberellin balance in the control of seed dormancy and germination in cereals. *Frontiers in Plant Science* 9, 668. DOI: 10.3389/fpls.2018.00668.

Ueda, H. and Kusaba, M. (2015) Strigolactone regulates leaf senescence in concert with ethylene in *Arabidopsis*. *Plant Physiology* 169(1), 138–147. DOI: 10.1104/pp.15.00325.

Umehara, M., Hanada, A., Yoshida, S., Akiyama, K., Arite, T. *et al.* (2008) Inhibition of shoot branching by new terpenoid plant hormones. *Nature* 455(7210), 195–200. DOI: 10.1038/nature07272.

Upadhayay, V.K., Gangola, S., Taj, G., Gaurav, K., Rani, A. *et al.* (2025) Zinc-solubilizing bacterial consortia: A promising approach for zinc biofortification of crops. *Frontiers in Microbiology* 16, 1575514. DOI: 10.3389/fmicb.2025.1575514.

Vurro, M., Boari, A., Thiombiano, B. and Bouwmeester, H. Koltai, H. and Prandi, C. (eds). (2019) *Strigolactones and parasitic plants*. Springer Nature, Cham, Switzerland, pp. 89–120. DOI: 10.1007/978-3-030-12153-2_3.

Wang, F., Wang, Y., Ying, L., Lu, H., Liu, Y. *et al.* (2023a) Integrated transcriptomic analysis identifies coordinated responses to nitrogen and phosphate deficiency in rice. *Frontiers in Plant Science* 14, 1164441. DOI: 10.3389/fpls.2023.1164441.

Wang, J.Y., Chen, G.T.E., Braguy, J., Jamil, M., Berqdar, L. *et al.* (2023b) Disruption of the cytochrome *CYP711A5* gene reveals *MAX1* redundancy in rice strigolactone biosynthesis. *Journal of Plant Physiology* 287, 154057. DOI: 10.1016/j.jplph.2023.154057.

Wang, L., Wang, B., Yu, H., Guo, H., Lin, T. *et al.* (2020a) Transcriptional regulation of strigolactone signalling in *Arabidopsis*. *Nature* 583(7815), 277–281. DOI: 10.1038/s41586-020-2382-x.

Wang, L., Xu, Q., Yu, H., Ma, H., Li, X. *et al.* (2020b) Strigolactone and karrikin signaling pathways elicit ubiquitination and proteolysis of SMXL2 to regulate hypocotyl elongation in *Arabidopsis*. *Plant Cell* 32(7), 2251–2270. DOI: 10.1105/tpc.20.00140.

Wang, M., Le Moigne, M.A., Bertheloot, J., Crespel, L., Perez-Garcia, M.D. *et al.* (2019) BRANCHED1: A key hub of shoot branching. *Frontiers in Plant Science* 10, 76. DOI: 10.3389/fpls.2019.00076.

Waters, M.T. and Nelson, D.C. (2023) Karrikin perception and signalling. *New Phytologist* 237(5), 1525–1541. DOI: 10.1111/nph.18598.

Wu, F., Gao, Y., Yang, W., Sui, N. and Zhu, J. (2022) Biological functions of strigolactones and their crosstalk with other phytohormones. *Frontiers in Plant Science* 13, 821563. DOI: 10.3389/fpls.2022.821563.

Yan, Y., Zhao, N., Tang, H., Gong, B. and Shi, Q. (2020) Shoot branching regulation and signaling. *Plant Growth Regulation* 92(2), 131–140. DOI: 10.1007/s10725-020-00640-1.

Yang, Y., Abuauf, H., Song, S., Wang, J.Y., Alagoz, Y. *et al.* (2023) The arabidopsis D27-LIKE1 is a *cis/cis/trans*-β-carotene isomerase that contributes to strigolactone biosynthesis and negatively impacts ABA level. *Plant Journal* 113(5), 986–1003. DOI: 10.1111/tpj.16095.

Yang, Y., Tilman, D., Jin, Z., Smith, P., Barrett, C.B. *et al.* (2024) Climate change exacerbates the environmental impacts of agriculture. *Science* 385(6713), eadn3747. DOI: 10.1126/science.adn3747.

Yao, R., Ming, Z., Yan, L., Li, S., Wang, F. *et al.* (2016) DWARF14 is a non-canonical hormone receptor for strigolactone. *Nature* 536(7617), 469–473. DOI: 10.1038/nature19073.

Yao, R., Wang, L., Li, Y., Chen, L., Li, S. *et al.* (2018) Rice DWARF14 acts as an unconventional hormone receptor for strigolactone. *Journal of Experimental Botany* 69(9), 2355–2365. DOI: 10.1093/jxb/ery014.

Yoneyama, K. (2019) How do strigolactones ameliorate nutrient deficiencies in plants? *Cold Spring Harbor Perspectives in Biology* 11(8), a034686. DOI: 10.1101/cshperspect.a034686.

Yoneyama, K. and Brewer, P.B. (2021) Strigolactones, how are they synthesized to regulate plant growth and development? *Current Opinion in Plant Biology* 63, 102072. DOI: 10.1016/j.pbi.2021.102072.

Yoneyama, K., Xie, X., Kim, H.I., Kisugi, T., Nomura, T. *et al.* (2012) How do nitrogen and phosphorus deficiencies affect strigolactone production and exudation? *Planta* 235(6), 1197–1207. DOI: 10.1007/s00425-011-1568-8.

Yoshida, T., Fujita, Y., Maruyama, K., Mogami, J., Todaka, D. *et al.* (2015) Four *Arabidopsis* AREB/ABF transcription factors function predominantly in gene expression downstream of snRK2 kinases in abscisic acid signalling in response to osmotic stress. *Plant, Cell & Environment* 38(1), 35–49. DOI: 10.1111/pce.12351.

Yuan, K., Zhang, H., Yu, C., Luo, N., Yan, J. *et al.* (2023) Low phosphorus promotes NSP1–NSP2 heterodimerization to enhance strigolactone biosynthesis and regulate shoot and root architecture in rice. *Molecular Plant* 16(11), 1811–1831. DOI: 10.1016/j.molp.2023.09.022.

Zeng, W., Xiang, D., Li, X., Gao, Q., Chen, Y. *et al.* (2025) Effects of combined inoculation of arbuscular mycorrhizal fungi and plant growth-promoting rhizosphere bacteria on seedling growth and rhizosphere microecology. *Frontiers in Microbiology* 15, 1475485. DOI: 10.3389/fmicb.2024.1475485.

Zha, M., Zhao, Y., Wang, Y., Chen, B. and Tan, Z. (2022) Strigolactones and cytokinin interaction in buds in the control of rice tillering. *Frontiers in Plant Science* 13, 837136. DOI: 10.3389/fpls.2022.837136.

Zhang, J., Mazur, E., Balla, J., Gallei, M., Kalousek, P. *et al.* (2020) Strigolactones inhibit auxin feedback on PIN-dependent auxin transport canalization. *Nature Communications* 11(1), 3508. DOI: 10.1038/s41467-020-17252-y.

Zhang, X., Zou, T., Lassaletta, L., Mueller, N.D., Tubiello, F.N. *et al.* (2021) Quantification of global and national nitrogen budgets for crop production. *Nature Food* 2(7), 529–540. DOI: 10.1038/s43016-021-00318-5.

Zhang, Y., van Dijk, A.D.J., Scaffidi, A., Flematti, G.R., Hofmann, M. *et al.* (2014) Rice cytochrome P450 MAX1 homologs catalyze distinct steps in strigolactone biosynthesis. *Nature Chemical Biology* 10(12), 1028–1033. DOI: 10.1038/nchembio.1660.

Zhao, L.-H., Zhou, X.E., Yi, W., Wu, Z., Liu, Y. *et al.* (2015) Destabilization of strigolactone receptor DWARF14 by binding of ligand and E3-ligase signaling effector DWARF3. *Cell Research* 25(11), 1219–1236. DOI: 10.1038/cr.2015.122.

Zhou, X., Liu, L., Li, Y., Li, K., Liu, X. *et al.* (2020) Integrative metabolomic and transcriptomic analyses reveal metabolic changes and its molecular basis in rice mutants of the strigolactone pathway. *Metabolites* 10(11), 425. DOI: 10.3390/metabo10110425.

14 Reprogramming of Salt Stress under the Influence of Strigolactones

Rizwana Kanwal[1], Ansa Asghar[1*], Usman Tufail[1], Asima Asghar[2], Farooq Ahmad[1], Sana Basharat[3*], Amina Ameer[4], Jazab Shafqat[1], Fouzia Shabbir[1], Muhammad Sufyan[5], Nimra Shahid[6] and Waleed Hussain[6]

[1]Department of Botany, University of Agriculture Faisalabad, Faisalabad, Pakistan; [2]Department of Zoology, Lahore College for Women University, Lahore, Pakistan; [3]School of Breeding and Multiplication (Sanya Institute of Breeding and Multiplication), School of Tropical Agriculture and Forestry, Hainan University, Sanya 572025, China; [4]College of Geographical Sciences, Faculty of Geographical Science and Engineering, Henan University, Zhengzhou, China; [5]College of Biological Sciences, China Agricultural University, Beijing, China; [6]Department of Botany, University of Narowal, Pakistan

Abstract

Phytohormones are essential for enhancing the resistance of plants to a range of abiotic stressors. The main phytohormones are brassinosteroids, ethylene, cytokinins, gibberellins, jasmonic acid, salicylic acid, auxins, and abscisic acid. In addition to these, strigolactones (SLs) are recently discovered phytohormones that regulate a number of developmental processes in plants that help them adapt to several abiotic stressors, especially salinity, heat, drought, cold, and waterlogging, offering new opportunities for agriculture. Phytohormones regulate plants, enabling them to grow and adapt to ecological constraints. The adaptation of integrated stress responses in plants has been linked to phytohormone crosstalk. This chapter discusses the production, signaling, and crosstalk of phytohormones and the roles of SLs under abiotic stress conditions. It is evident that comprehensive analysis of the roles of SLs will be crucial in overcoming the challenges of crop damage brought on by harsh environments.

Keywords: brassinosteroids, pesticide, phytohormones, photosynthetic pigment, photosystem, strigolactones

14.1 Introduction

The main environmental stressor limiting agricultural sustainability and productivity in arid and semi-arid regions is salinity (Naorem *et al.*, 2023). Currently, excessive fertilizer and pesticide use, illogical irrigation, and the consequences of climate change are making soil salinization worse globally. Because of this, the yields of the majority of salt-sensitive crops such as maize, cucumbers, and peppers have been decreasing annually. Plant metabolic equilibrium can be upset by salt stress, which can also prevent photosynthesis and

*Corresponding authors: ansaasghar1122@gmail.com and sanabasharat143@gmail.com

DOI: 10.1079/9781836990598.0014

respiration. Prior research has shown that salt stress can have either direct (limiting stomata diffusion movement and altering the metabolism of photosynthetic energy) or secondary (oxidative stress brought on by many additive pressures) impacts on photosynthesis (Tarafdar *et al.*, 2022). A reduction in photochemical efficiency was noted as a result of the breakdown of the photosynthetic machinery and the deterioration of photosynthetic pigment, ultimately causing the growth and development of plants to be delayed (Simkin *et al.*, 2022).

In rice (*Oryza sativa*), salt stress harms the electron transport chain of photosynthesis, limits photochemical efficiency and photosynthetic rate, and dramatically decreases stomatal conductance, photochemical quenching coefficient, and the transpiration rate. Complexes of protein that include photosynthesis-related proteins and photosystem I (PSI) and PSII response centers, under negative conditions, had an impact on electron transport, lowering the maximum photochemical reactions of PSII in watermelon seedlings exposed to short-term salt stress (Muhammad *et al.*, 2021). Additionally, sodium (Na^+) and chloride (Cl^-) ions surged into the cells of plants under salt stress, accumulating Na^+ and influencing the plant's ability to absorb potassium ions (K^+), resulting in ionic toxicity. As a result, during prolonged salt stress, reactive oxygen species (ROS) buildup, chloroplast membrane structural degradation, and photosynthetic suppression were seen. Plants' oxidative defense mechanism can be destroyed by salt stress in addition to preventing photosynthesis (Mushtaq *et al.*, 2020).

As a result of oxygen consumption, ROS can operate as an additional messenger and are vital for the transmission of signals controlling plant development and growth. However, an overabundance of ROS will upset cell homeostasis, impede cell development and metabolism, and eventually kill mitochondria in plants (Nadarajah, 2020). Plants have a number of efficient ROS-scavenging systems to preserve the equilibrium of intracellular ROS. Root respiratory burst oxidase homolog-dependent hydrogen peroxide production was used to modulate Na^+ restriction and closure of stomatal cells, increasing the ability of grafted cucumber to tolerate salt. Furthermore, by controlling oxidative homeostasis and boosting oxidative defense, mitogen-activated protein kinase cascade pathways also play a role in the way plants responded to stress (He *et al.*, 2020). Initially, strigolactones (SLs) were thought to be host-derived germination stimulants for parasitic weeds belonging to the genera *Orobanche* and *Striga*. SLs were later discovered to be hyphal-branching and host-detection signals for arbuscular mycorrhizal fungi (AMF). As well as their initial dual function as rhizosphere signaling molecules, SLs were shown to represent a novel family of branch-inhibiting phytohormones that control the general structure of land plants (Hossain *et al.*, 2021).

As more research organizations show interest in the physiological and biological role of SLs, new functions will probably be discovered. Over the past 10 years, research on SLs has grown exponentially, resulting in an abundance of knowledge about the molecular and biological aspects of SL activity (Ahmad *et al.*, 2025).

14.2 Strigolactones: Biosynthesis and Perception

14.2.1 Biosynthesis

Exudates from the roots of a wide variety of plants have been found to contain over 20 SL and SL-like chemicals. They all share the same retained tricyclic lactone structure, which is made up of three rings known as ABC-rings joined by an enol-ether bridge to an invariant α,β-unsaturated furanone moiety known as the D-ring (Daignan-Fornier *et al.*, 2024). The bioactiphore is located in the area that joins the D-ring to the core; the size of the A-ring, the stereochemistry of the BC-ring junction, and the patterns of A- and B-ring substitution provide the chemical diversity (Meanwell, 2023).

Many of the SL biosynthesis (and signaling) mechanisms currently understood have been determined by analysis of a large collection of branching mutants found in *O.*

sativa (*dwarf* (*d*) or *high-tillering dwarf* (*htd*) mutants), among others (Cheng *et al.*, 2022).

14.2.2 Perception and signal transduction

SL signaling processes are supported by proteosomal degradation, just like the majority of controllers of plant growth (e.g. jasmonic acid (JA), gibberellins (GAs), and auxin). The SL signaling machinery consists of an F-box leucine-rich protein (Xu *et al.*, 2021), MORE AXILLARY GROWTH 2 (MAX2; also known as RMS4/D3), the α/β-fold hydrolase known as DWARF14 (D14; also known as DAD2/RMS3), and the repressor protein DWARF53 (D53), which is a member of a small protein family and shares certain characteristics with class I Clp ATPase enzymes. It is interesting to note that parts of the perception and signal transduction machinery for JA, auxin, GAs, and salicylic acid (SA) are identical to either D14, MAX2, or both. As molecular glues, phytohormones facilitate formation of the active receptor/signaling complex and cause the proteasome to degrade the negative moderators of signaling in each of the aforementioned routes. D14 functions in a signaling cascade that resembles that of GAs and involves the F-box protein MAX2, according to the current molecular theory of SL perception (Xu *et al.*, 2021).

Consequently, upon contact, the hydrolyzed ligand (as its bioactiphore molecule) and the target molecule are joined together, making them inaccessible for additional sensory events (Sescil *et al.*, 2025). D14 differs from other phytohormones in that it functions as both an enzyme and a receptor. The catalytic triad Ser/His/Asp is necessary for the binding and hydrolysis of SLs in order for D14 to perceive them. It has been suggested that MAX2/D3 functions as a recognition subunit for downstream repressors and SL-loaded D14 in the SKP1–CULLIN–F-BOX PROTEIN (SCF) complex, as shown in Fig. 14.1.

D14 then interacts with and changes SLs mostly through its water-phobic ligand-interacting pocket and, more precisely, the triad's nucleophiles, which target the D-ring at the butenolide carbonyl and detach it from the ABC-ring (Daignan-Fornier *et al.*, 2024). It is believed that conformational changes in the D14–SL complex are primarily determine SL signal transduction. It has been demonstrated that the synthetic SL analog GR24 thermally destabilizes D14 *in vitro*; therefore, the catalytic triad is required. Following its recruitment to SCF^{MAX2}, D14 loaded with SL triggers further proteasome degradation of target proteins, such as D53 (Sánchez Martín-Fontecha *et al.*, 2024).

Late proteasomal degradation of D14 itself is believed to strengthen tissue desensitization to SLs. Several orthologs of D53 have been found in *Arabidopsis*, although it is still unclear which molecular partners are responsible for suppressing shoot branching and other SL-regulated activities (Hu *et al.*, 2020). Due to its EAR motifs, D53 is thought to interact with TOPLESS-related (TPR) proteins, which are transcriptional suppressors. Although it has not yet been proven, the D53–TPR complex may suppress the transcription of SL targets. Various SUPPRESSOR OF MAX2 1-LIKE (SMXL) family members may be in charge of regulating various facets of SL action, although D53 seems to be the main regulator of the shoot branching impact of SLs (Kerr *et al.*, 2021).

14.2.3 Regulation of SL biosynthesis

A thorough understanding of the regulatory processes in SL biosynthesis will provide the means to modify genes involved in the response pathway and SL production (Mashiguchi *et al.*, 2021). Prior research has demonstrated that the several external and endogenous variables discussed in this section affect SL production at both the post-transcriptional and post-translational levels (Wang *et al.*, 2020).

14.3 Plant Hormone-mediated Regulation

In rice, *Arabidopsis*, pea, and chrysanthemum (*Dendranthema grandiflorum*), auxin stimulates the expression of carotenoid cleavage dioxygenase 7 (CCD7) (Sun *et al.*, 2021). RMS1 (CCD8) and RMS5 (CCD7) transcript levels in peas are

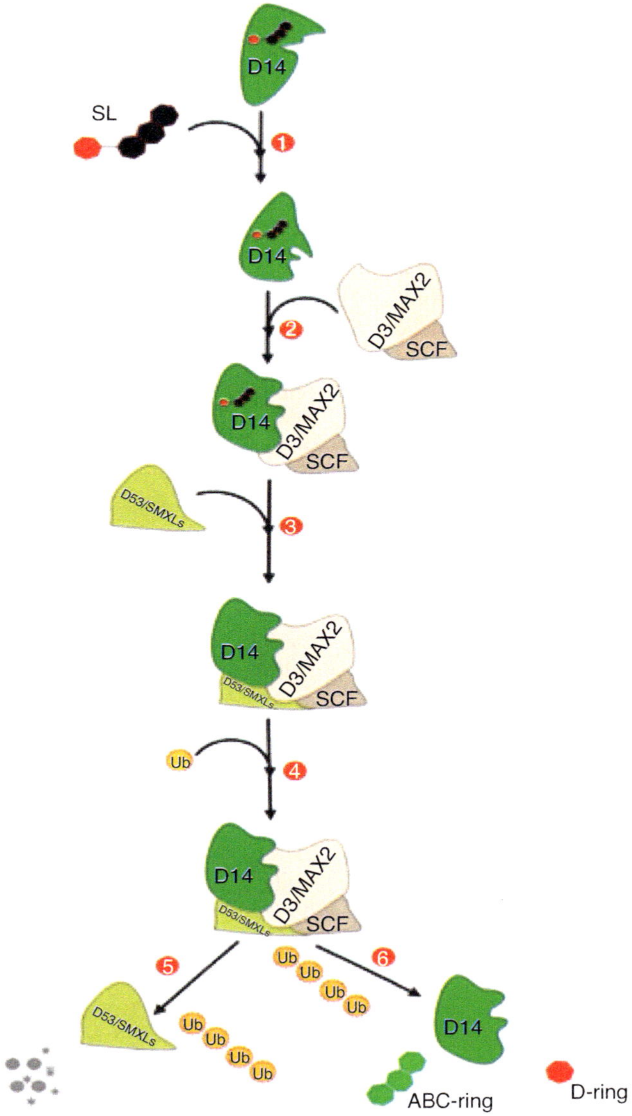

Fig. 14.1. Pathway showing the interaction of SLs, D14 and the SCF complex. D14, DWARF 14; SCF, SKP1–CULLIN–F-BOX PROTEIN complex; Ub, ubiquitin.

favorably regulated by auxin, as evidenced by tests including exogenous auxin delivery, apical shoot tip decapitation, and the use of auxin transport inhibitors. Comparable outcomes were observed in *Arabidopsis*, where auxins preserve SL levels and the transcript levels of *AtCCD7/MAX3* and *AtCCD8/MAX4*. *axr1* and *bdl* mutants had lower levels of *AtCCD7* and *AtCCD8* transcripts, where *BODENLOS* (*BDL*) encodes the transcription repressor *INDOLEACETIC ACID RESPONSE12* (*IAA12*), and the protein encoded by *AUXIN RESISTANT 1* (*AXR1*) is strikingly similar to the ubiquitin-activated enzyme E1. The ubiquitin ligase complex stability of SKP1–CUL1 E3, also known as $SCF^{TIR1/AFB}$, which is made up of the F-box protein AUXIN-RELATED F-BOX (AFB), which depends on the AXR1 protein (Jing *et al.*, 2023). AXR1 mutations alter this complex and the auxin-regulated downstream processes, which may affect SL biosynthesis.

The SCF$^{TIR1/AFB}$ complex largely facilitates the breakdown of BDL proteins. These findings have been validated in part by phenotypic investigation of several combinations of auxin mutations and SL signaling pathways. Exogenous SL treatment inhibits the branching phenotype in quadruple mutant (*tir1 afb1 afb2 afb3*), *Arabidopsis axr1*, and *bdl* mutant plants. Additionally, auxin-responsive regions were found in both the *CCD8* and *CCD7* genes by promoter analysis, suggesting that auxin controls the production of SLs (Hu *et al.*, 2023).

Studies on *axr1* mutant plants revealed that auxin depletion, acquired by decapitation and treatment with naphthylphthalamic acid, reduces expression levels of *D27*, *MAX3* and *MAX4*, and blocks auxin transport (Abas *et al.*, 2021). Furthermore, *MAX4* and *MAX3* transcript levels are elevated in *d27* mutants, suggesting that SLs are in charge of feedback regulation. While decapitation reduces the production of *CCD7* and *CCD8*, exogenous naphthaleneacetic acid injection increases *CCD8* expression in the inner cortex of the root apex propagation zone and the (pro)vascular tissue of the primary roots. While cytokine (CK) production is inhibited, auxin signaling through the AXR1 pathway promotes the gene expression involved in SL biosynthesis (Yan *et al.*, 2020).

The decreased rates of *Striga hermonthica* and *Phelipanche ramosa* seed germination in the presence of abscisic acid (ABA)-deficient tomato (*notabilis*) and maize (*viviparous 14*) provide more evidence that ABA regulates SL biosynthesis. A liquid chromatography-tandem mass spectrometry (LC-MS/MS) investigation of tomato root exudates indicated a lower SL level in these mutants, although it was unclear whether this was due to 9-*cis*-epoxycarotenoid dioxygenase or ABA. These findings lend support to the concept that ABA has a favorable effect on SL levels (Yan *et al.*, 2020).

14.3.1 AMF and nodulation-mediated regulation

When *SlCCD7* is activated in tomato, SL levels rise in reaction to AMF. The symbiosis between legumes and AMF, which most likely involves SLs, also induces nodulation (Li *et al.*, 2021). The first three members of the GRAS-type transcription factors REPRESSOR OF GAI (RGA), GIBBERELLIC-ACID INSENSITIVE (GAI), and SCARECROW (SCR) regulate nodule morphogenesis, phytochrome A signal transduction, GA signaling, and root and shoot development. Additionally, in rice and *Medicago truncatula*, NSP1 and NSP2 are necessary for SL production under non-symbiotic conditions (Nasir *et al.*, 2021). SLs are present in negligible amounts in double-knockout mutants of *O. sativa* (*Osnsp1Osnsp2*) and *M. truncatula* (*nsp1nsp2*). Additionally, *MtD27* and *OsD27* transcription is noticeably decreased in these lines. It is interesting to note that the calcium calmodulin kinase pathway, which is often necessary for nodulation in legumes, is not used by NSP1 or NSP2. Reduced AMF symbiosis is another characteristic of these plants, suggesting that AMF activates NSP1 and NSP2 to regulate SL biosynthesis (Li *et al.*, 2022a, b).

14.3.2 MicroRNA-mediated regulation

MicroRNAs are complementary sequences of 21–24 nucleotides that, when bound to the target mRNA, degrade and inhibit translation. All of the SL pathway genes in rice (*D3*, *D27*, *D10*) and Arabidopsis (*MAX1*, *MAX3*, *MAX4*) have miR156a-g target sequences (Moturu and Koča, 2020). Plant morphology and growth, which are impacted by the environment, are impacted by miR156. In *Arabidopsis*, rice, and maize, overexpression of miR156 results in increased shoot branching, a trait shared by SL biosynthesis and response mutants (Jeyakumar *et al.*, 2020). *D27* transcript levels are lower in rice plants overexpressing osa-miR156, according to reverse transcription-polymerase chain reaction studies, indicating that osa-miR156 adversely controls *D27*. Interestingly, these plants had higher amounts of *D3* and *D14* transcripts, indicating that the SL pathway genes are under feedback regulation. It is known that SQUAMOSA PROMOTER BINDING PROTEIN LIKE (SPL) proteins in rice and maize regulate morphological development (Sun *et al.*, 2021).

14.3.3 Regulation under nutrient deficiency

Non-organic nutrients, such as phosphate (PO_4^{3-}) and nitrate (NO_3^-), are vital providers of nitrogen and phosphorus and are known to influence the architecture and growth of plants, which are dependent on shoot and root development (Mishra *et al.*, 2024). The function of plant hormones in nutrient transmission pathways, such as auxin and CKs that alter plant shape, is strongly supported by the data. Researchers have linked SLs to nutritional signaling pathways over the past 10 years. Plants grown under standard laboratory settings have extremely low levels of SLs, whereas suboptimal conditions, such as a lack of nutrients, increase SL levels, most likely endowing plants with plasticity to adapt to such environments (Raza *et al.*, 2021). An LC-MS/MS study revealed an increased amount of SLs in roots and increased exudates from roots when there was a nitrate and phosphate deficiency. Leguminous plants, for example, respond to both phosphate and nitrogen deficiency by increasing their SL levels (Leoni *et al.*, 2021). Collectively, these results suggest that during nutritional limitation, SLs might function as secondary messenger molecules.

Auxins also control the production of SLs. RNA sequencing investigations of the maize root transition zone, where ADAPTOR PROTEIN 2 (AP2) and SL biosynthesis and transport genes are downregulated under nitrogen-deficient circumstances, further imply that auxin and SLs are cross-regulated. It has been demonstrated that SLs control auxin transport through the PIN-FORMED 1 (PIN1) and AP2 proteins. These findings suggest that the transition zone's early reaction to nutritional deficiency may be the downregulation of SLs.

Wild-type *Arabidopsis* plants exhibit increased amounts of orobanchol in their root and xylem sap, respectively, and suppression of lateral bud expansion when exposed to phosphate deficiency (Wang *et al.*, 2024). As *Arabidopsis* is not an AMF host, SLs most likely mainly improve phosphate utilization efficiency by altering plant architecture. Similar outcomes were shown in rice grown in a phosphate-deficient environment, where tiller bud development was impeded and 2′-epi-5-deoxystrigol levels were elevated. As *d3* and *d10* rice mutants do not exhibit this effect, it is possible that SL biosynthesis rather than the perception pathways is necessary in phosphate-deficient environments. With the exception of SL signaling mutants, GR24 restores phosphate deficiencies in all genotypes, suggesting that SL synthesis is crucial in these circumstances. Auxin is required for SL-mediated root architecture regulation because it influences the location of PIN proteins (Rehman *et al.*, 2021).

Phosphate-deficiency hallmark genes that show altered expression in *max2* and *max4* mutants include PHOSPHATE TRANSPORTER 1;5 (PHT1;5), TYPE 5 ACID PHOSPHATASE (ACP5), and PHT1;4. The reaction of SLs to low phosphate was demonstrated to involve the auxin transporter gene TRANSPORT INHIBITOR RESPONSE 1 (TIR1), indicating that auxins and SLs organize the reaction to low phosphate. Under phosphate deficiency, *Os900* expression in rice is increased but *Os1400* expression is unaffected. Recent research on *O. sativa* has demonstrated that when nitrogen and phosphate are insufficient, SL production is increased regulation. Low nitrogen and phosphate in wild-type plants result in longer roots and lower lateral root density (Mehmood *et al.*, 2025).

An expression study showed that in nitrogen- and phosphate-deficient circumstances, *D10*, *D27*, and *D3* were upregulated. Furthermore, all mutants except *d3* showed the insensitive phenotype recovered by exogenous administration of GR24, suggesting that phosphate and nitrogen shortage increase the production of SLs (Sun *et al.*, 2023).

14.4 How Strigolactones Affect Plant Responses and Salinity Adaptation

It has been shown that SLs are crucial in MAX2-dependent salinity tolerance. When exposed to elevated salt levels during the stages of germination and vegetative growth, *Arabidopsis* SL-deficient *max3* and *max4* mutants and SL-signaling *max2* mutants displayed salt-sensitive phenotypes (Mukherjee and Aftab, 2023). However, the roots of the *max* mutant plants did not differ in their level of damage compared with wild type. In a similar vein, *max2* mutant plants under salt stress showed a reduction in the rate of germinative greening that was dependent on both time and concentration when compared with wild type (Tarchoun *et al.*, 2022).

These findings suggest that lowering endogenous SL concentration or impairing SL signaling reduces the ability of plants to survive salt stress during different phases of development. In order to provide a proper mechanical understanding of the relationship between SL signaling and plant salt-stress resistance, functional assessments of D14 and KARRIKIN INSENSITIVE2 (KAI2) in plant responses to salt stress must be carried out concurrently, as has been done in research on drought tolerance (Li *et al.*, 2020).

In the context of AMF symbiosis, the function of SLs in salt tolerance has been demonstrated. For instance, it was discovered that, in the presence of AMF, SL production in lettuce roots was favorably connected with enhanced growth and PSII efficiency under a salt-stress gradient. Additionally, the researchers found that AMF-colonized plants had better stomatal conductance, which was associated with efficient carbon dioxide fixation and ideal lettuce plant growth (Jajoo and Mathur, 2021).

It has bene found that in AMF-colonized *Sesbania cannabina* seedlings, SL accumulation was necessary for hydrogen peroxide-induced salt tolerance. Even in the absence of stress, it has been noted that GR24 treatment in *Arabidopsis* led to a significant rise in decreased glutathione levels in the roots of wild-type, *max3*, and *max4* plants but not in *max2* mutant plants (Qi *et al.*, 2024). These results show that SLs have a *MAX2*-dependent role in redox homeostasis. Although the exact mechanism of how SLs and ROS interact is not yet known, it is speculated that the balance in plant design regulated by SLs, along with ROS metabolism, may help plants adapt to changing conditions (Hasanuzzaman *et al.*, 2020).

14.5 Interactions Between Strigolactones and Other Phytohormones

14.5.1 Crosstalk between SLs and auxin signaling

Endogenous auxins affect biological processes in plants through either antagonistic or synergistic interactions with other hormones (Fig. 14.2). Auxin and SL signaling have been linked in a number of studies. The primary way that auxin affects SL signaling is by controlling SL synthesis, as SLs can serve as a second messenger to trigger certain auxin-dependent reactions (Faizan *et al.*, 2020). The expression of *CCD7* and *CCD8* is upregulated by auxin, allowing it to govern shoot branching. Naz *et al.* (2025) discovered that these are AXR1-dependent processes. There is a lack of experimental data on the roles of auxin in SL flux and transport.

By interfering with polar auxin transport (PAT)/distribution and thus auxin canalization, SLs control shoot branching. Bud activation and growth are significantly influenced by auxin flux/canalization. The polar concentration of auxin on the plasma membrane also reduces the auxin sink strength of the PAT stream, which in turn suppresses bud development (Zou *et al.*, 2024). However, in pea plants with impaired auxin transport, SLs have been shown to limit bud development, and decreased auxin transport in pea had minimal suppressive effects on bud development in mutants lacking SL. Omoarelojie *et al.* (2019) suggested that, although SL-mediated branching modulation may employ PAT/canalization, SLs may also function without this mechanism.

Inconsistencies between the two theories of SL activity in bud activation have been tested experimentally. These inconsistencies include the following: the unresponsiveness of *FINE CULM 1* (*FC1*) expression to GR24, buds lacking *BRANCHED 1* (*BRC1*) transcription remain repressed and sensitive to SL inhibition; the uncoupling of the *TEOSINTE BRANCHED 1* (*TB1*) subnetwork from SL signaling in maize; the activity of buds with high *BRC1* transcripts; and the previously mentioned contradictory reports of the impact of SLs on auxin transport and canalization.

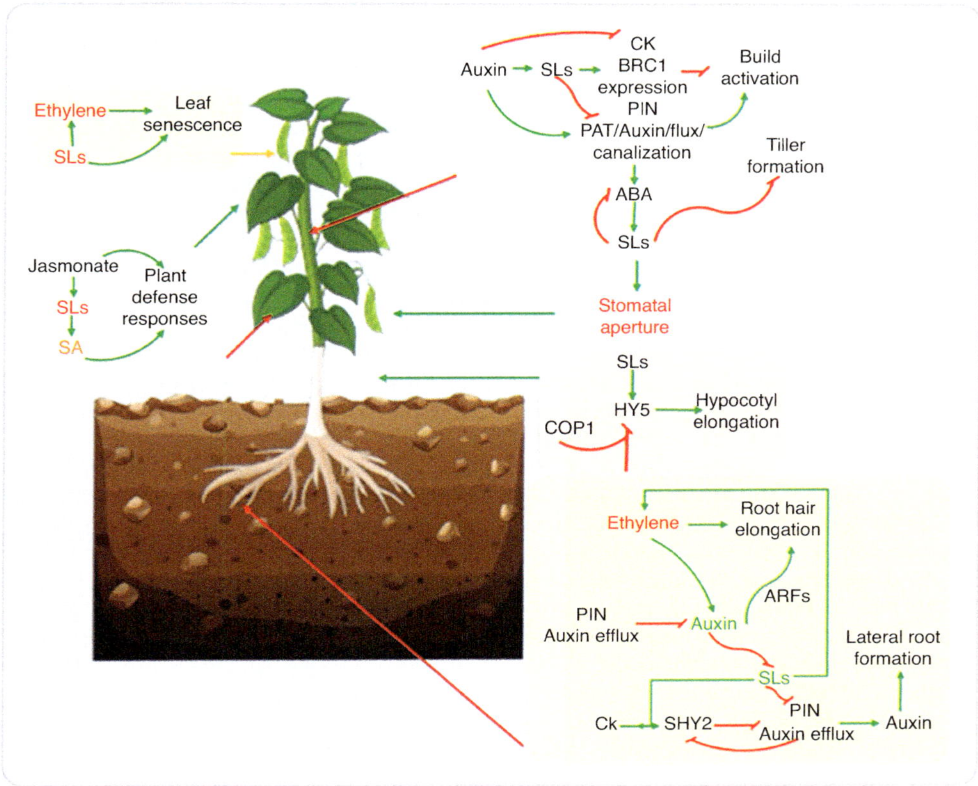

Fig. 14.2. Interactions between strigolactones (SLs) and other phytohormones.

All of this necessitates a rethinking of these models in order to create one that accounts for these limits. The occurrences that occur before, during, or after *BRC1/FC1/TB1* activities may indicate the way forward. Bud activation and development may involve auxin canalization and inhibition of branching factors, and whether an activated bud develops into a branch may depend on other processes regulated by these two hormones.

By affecting auxin sensitivity, PAT from root to shoot, and auxin flow within tissues of the root, SLs work with auxin to regulate the growth of roots, just as they do with shoot branching (Jan *et al.*, 2024). In this situation, auxin signal modules take action downstream if SLs. The initial phases of lateral root development are driven by specific auxin gradients and response maxima, which involve preparing the basal meristem's pericycle cells for lateral root initiation by priming them, moving from founder cells to the creation of lateral root primordium and initiation, and then maturing the primordium until lateral root emergence (Du and Scheres, 2018).

GR24 has been shown to affect the lateral root-forming ability, as well as the polarization and translocation of PIN proteins. As local auxin flow and homeostasis are necessary to create auxin response maxima, SLs most likely control these processes.

14.5.2 SLs and CK signaling interaction

SLs and CKs seem to interact differently depending on the physiological process. They work together to control lateral root growth, independently in adventitious roots, and antagonistically in bud stimulation and shoot branching

(Raspor *et al.*, 2021). The genes responsible for their biosynthesis and expression are regulated by both hormones. Although CK production and export from the root have been shown to be mediated by SL-associated genes, nothing is known about how CKs affect SL transport. SLs and CKs interact directly in buds to modulate bud activation and shoot branching, and their actions converge on the transcriptional regulation of *BRC1* in *Arabidopsis* and *FC1* in rice. Thus, bud growth will be impacted by any stimulation that modifies CK and SL homeostasis within the bud (Yuan *et al.*, 2023).

Nevertheless, components of CK signaling are necessary for the suppression of lateral root development caused by GR24. Through *SHORT HYPOCOTYL 2* (*SHY2*), ARABIDOPSIS HISTIDINE KINASE 3 (AHK3), ARABIDOPSIS RESPONSE REGULATOR 1 (ARR1), and ARR12 limit lateral root production by GR24 under sufficient phosphate by downregulating *PIN* gene expression. SHY2 inhibition of PIN activity inhibits lateral root development because auxin transport is necessary to generate an auxin gradient that is essential for initiating lateral roots from pericycle founder cells (Konstantinova *et al.*, 2021).

In terms of lateral root development, results have shown that the insensitivity of a loss-of-function mutant in *SHY2* causes unresponsiveness to GR24 (Faizan *et al.*, 2020). With auxin boosting its breakdown and CK promoting its expression, SHY2 is also essential for maintaining the balance between auxin and CK signaling (Jan *et al.*, 2024). Although it is unknown whether SLs employ crosstalk and the same channel to support lateral root development under phosphate-limiting circumstances, it is evident that SHY2 serves as a hub for the integration of CK, auxin, and SL signaling to control lateral root development in phosphate-sufficient situations (Crombez *et al.*, 2019).

14.5.3 SLs and ABA signaling interaction

In light of the growing experimental evidence that SLs actively mediate the responses of plants and their resistance to biotic and abiotic stresses, it is anticipated that SLs will work in tandem with ABA, either directly or indirectly, to control the physiological stress responses of plants (Manzoor *et al.*, 2023). Based on the information now available, three important points about ABA–SL interactions can be deduced. First, in order to elicit ABA-mediated reactions, SL production and signaling must be controlled and ABA acts upstream of SLs (Asad *et al.*, 2019). The findings that ABA-deficient plants exhibit markedly reduced SL content and SL biosynthesis gene expression lend credence to this theory. To mediate the resistance of *S. cannabina* to salt stress, ABA increased the activity of SL production and signaling genes (Ren *et al.*, 2018). However, this only resulted in temporary and partial improvements in plants' ability to withstand salt treated with an inhibitor of SL synthesis. SL-deficient and signaling mutants have decreased stomatal sensitivity to ABA, decreased sensitivity to exogenous ABA, and increased vulnerability to drought stress. In the SL-deficient lines, infection-induced increase in ABA levels also led to an increase in stomatal openings and loss of water (Ahmed, 2018).

Second, there is an antagonistic relationship between the regulation of ABA and SLs in terms of specific aspects such as fruit ripening and shoot architecture. To promote tiller formation, ABA inhibits the expression of signaling genes and SL biosynthesis (Bai *et al.*, 2024). However, exogenous GR24 administration significantly reduced ABA-induced sugar and anthocyanin accumulation in grape berries instead of increasing ABA activity.

Lastly, SLs may act as a feedback channel in regulating ABA signaling, in addition to being necessary for ABA-mediated responses (Mukherjee *et al.*, 2023).

14.5.4 SL crosstalk with GA signaling

Numerous experimental findings imply interactions of SLs and gibberellin, although these are still to be determined. For example, Lantzouni *et al.* (2017) demonstrated that a set of overlapping genes experienced a additive transcriptional change as a result of GA and GR24. According to Borghi *et al.* (2019), GAs might also work against the upregulation of several ABC-type transporters induced by SLs.

Additionally, GAs were shown to regulate the production of SLs in an *in silico* study; this turned out to be regulated by a GID1–DELLA signal pathway (Dahal *et al.*, 2025. Although this has not yet been shown experimentally, the finding that *ORYZA SATIVA HOMEOBOX 1 (OSH1)* transcription is increased in highly branching GA-deprived mutants and SL-signaling mutants implies that GAs and SLs share a comparable molecular mechanism (Hernández-García *et al.*, 2021). Experiments have tried to find a clear molecular connection between SLs and downstream GA signaling components. For example, D14 interacts in an SL-dependent way with SLENDER RICE1 (SLR1), a downstream signal component and DELLA protein that controls GA signaling (Sun *et al.*, 2023). However, there is currently no experimental proof of SL-induced SLR1 degradation, and little is understood about the relationship's functional importance or long-term effects.

The results of recent efforts to identify a molecular connection between GA signaling and SLs support the idea that the two hormones have independent functions. For example, the stability of DELLA proteins was unaffected by their concentration or by SL signaling (Nohales and Kay, 2019). The finding that SLs promote internode elongation independently of GAs by increasing cell quantity rather than length lends more credence to a functional diversification for both hormones.

In light of the aforementioned, it is probable that the crosstalk between GAs and SLs is minimal or may not be caused by direct interactions between the signaling elements, and that regulation by GAs of SL synthesis solely modifies SL content. In any event, more research is necessary to determine the actual nature of SL–gibberellin crosstalk (Castro-Camba *et al.*, 2022).

14.5.5 Interaction between ethylene signaling and SLs

Ethylene signals affect certain developmental and growth processes that are also controlled by SLs (Khan *et al.*, 2024). These include germination of seeds, extension of root hairs, hypocotyl development, and senescence of leaves. In order to control hypocotyl growth, the CONSTITUTIVELY PHOTOMORPHOGENIC 1 (COP1)–ELONGATED HYPOCOTYL 5 (HY5) complex integrates light and hormone signals during development. By upregulating HY5 expression in a MAX2-dependent manner, SLs prevent hypocotyl elongation in light, whereas ethylene encourages COP1-mediated HY5 breakdown to encourage hypocotyl elongation (Park *et al.*, 2024).

These results all suggest that SLs and ethylene signaling interact antagonistically to control hypocotyl development. How SLs counteract ethylene's promoting effects on hypocotyl growth in light is a crucial question because hypocotyl growth is quickly arrested in light. Ethylene signaling is also necessary for the SL regulation of root-hair elongation (Li *et al.*, 2022a). This was demonstrated by the fact that *At-ein* and *At-etr*, ethylene signaling mutants, showed decreased sensitivity to GR24 treatment, and that the effects of SLs on root-hair elongation were abolished when ethylene synthesis was blocked. Additionally, transcription of *At-ACS2*, a gene implicated in ethylene production, was enhanced by GR24 therapy (Nasir *et al.*, 2019). These observations also revealed that SLs act through ethylene, which may further interact with auxin signaling. Together, these results suggest that root-hair formation is regulated by a three-way hormonal interaction among SLs, ethylene, and auxin.

The finding that a gene that regulates length of root hairs and other genes linked to root hairs are co-activated by a regulatory complex made up of an ethylene-dependent factor and a positive regulator of root-hair cells implies that ethylene can affect the formation of root hairs independently and is not necessarily dependent on auxin signaling (Khan *et al.*, 2024).

14.5.6 Crosstalk between SLs and JA signaling

A class of plant oxylipins known as JAs mediates a variety of physiological activities during wounding, secondary metabolism, plant–insect and plant–pathogen interactions, and vegetative growth. The function of SL–JA crosstalk cannot yet be sufficiently described due to limited experimental evidence (Samanta and Roychoudhury, 2025). However, a connection between SLs and JAs in tolerance to disease is suggested by the discovery that an SL-deficient tomato mutant (*Slccd8*) showed reduced levels of the JA-dependent gene *PINII*, which gives tomatoes immunity against *Botrytis cinerea*. An ortholog of the SL transporter gene *PhPDR1*, *Nicotiana tabacum PDR6* (*NtPDR6*), is impacted by methyl jasmonate, a plant defense signaling molecule, which further supports this connection. *NtPDR6*, despite being substantially stimulated by phosphate shortage and engaged in shoot branching regulation, has not been linked to SL transport (Mehmood *et al.*, 2025). Recently, a contradictory finding showed that neither the GR24 supply nor *Mucor* sp. inoculation affected the accumulation of JA in wild-type plants, *Atmax1*, and *Atmax2* (Vogel et al., 2023). The fact that both hormones produce reactions in similar developmental stages, such as plant–microbe interactions, senescence, and mesocotyl elongation, suggests that SL–JA crosstalk cannot be completely ruled out and is most likely involved in these processes, even though these findings are insufficient to draw any meaningful conclusions about the nature of this crosstalk Omoarelojie *et al.* (2019).

14.5.7 SA signaling and SL crosstalk

SA is crucial for plant sensing, pathogen defense, and abiotic stress tolerance. Its control in this regard stems mostly from its effect on plant ROS levels. SA regulates drought tolerance, senescence, and stomatal closure/conductance, all of which fall under SL control. Only plant endophytic fungal contacts have demonstrated SA–SL interactions thus far, with GR24 increasing SA accumulation, while *max2*, an SL signaling mutant, showing reduced SA concentrations (López-Ruiz *et al.*, 2020).

This implies that SLs may influence how plants fight infections by inducing SA signals. Whether SLs generate SA accumulation by directly controlling the transcription of genes linked to SA biosynthesis or by inducing oxidative stressors or ROS, which are known to cause SA production, still needs to be determined (Roychowdhury *et al.*, 2024). Given the recent findings of SL activities in ABA-induced senescence and stomatal closure during drought conditions, examining SL–SA crosstalk in these areas may provide invaluable resources for breeding programs aimed at producing crops that can resist drought and have longer fruit shelf life (Manzoor *et al.*, 2023).

14.6 Strigolactone Signaling Regulatory Mechanisms

14.6.1 Trafficking

According to earlier studies, auxin flow is affected by SL signaling at the root tip, which in turn affects primary root meristem size, root-hair elongation, and lateral root growth. Auxin flux is then regulated by the location of the PIN protein, an auxin transporter, in the plasma membrane (Xie *et al.*, 2021b). Filamentous actin (F-actin) is essential for PIN1 localization and trafficking, and PIN1 trafficking is inhibited when F-actin is stabilized. The way in which SLs work was revealed by root elongation studies conducted in *Arabidopsis* mutant plants *max2*, *eir1* (PIN2), *tir3* and *der1* (actin), which showed an increase in PIN2 endocytosis, PIN2 transcription, and actin cytoskeleton remodeling. Following GR24 treatment, *max2* plants exhibit none of these characteristics (Pandya-Kumar *et al.*, 2014). When it comes to root-hair elongation, *der1* and *tir3* plants are less sensitive to GR24 (Elhiti *et al.*, 2021). Similar results were achieved by measuring PIN1 levels at the plasma membrane in *Arabidopsis* shoots. The shoot branching phenotype is brought on by accelerated clearance by SLs of PIN1 from the plasma membrane. The reduction of PIN1 levels in the plasma membrane of xylem cells of parenchyma in the stem is also mediated by clathrin. In summary, by rearranging the actin cytoskeleton and modifying auxin distribution, SLs function as positive regulators of PIN protein localization, translation, transcription, and trafficking. SL biosynthesis is therefore favorably regulated by auxin (Zhang *et al.*, 2020).

14.6.2 Transcription

A transcriptome comparison between *max2* and wild-type *Arabidopsis* plants grown in well-watered and dehydrated conditions revealed an interaction between SLs and plant hormones such as CKs and ABA (Yang *et al.*, 2019). AtNAC2, a NAC transcription factor, is downregulated in *max2* plants, which is induced by auxin signaling, ABA, ethylene, and CBL-INTERACTING PROTEIN KINASE 1 (CIPK1). Additionally, lateral root formation, which is also an SL-regulated process, is known to be influenced by AtNAC2. The ABA import genes *ABCG22* and *ABCG40* are expressed less in *max2* plants. ABA's beneficial regulatory function in drought signaling reactions was previously supported by the demonstration that *abcg22* and *abcg40* plants were prone to drought as a result of decreased stomata closure and increased transpiration (Feng *et al.*, 2023). Given that CKs are known to improve drought tolerance, it has been shown that SLs regulate the expression of the CYTOKININ OXIDASE/DEHYDROGENASE (*CRX*) gene, which is necessary for CK catabolism. *max2* plants exhibit downregulated *CRX* genes. DELLA (a GA-activated transcription factor) and BRI1-EMS-SUPPRESSOR 1 (BES1), two types of transcription factor, have been proposed by biochemical methods as downstream targets of MAX2, where the former interacts with MAX2. Moreover, the *max2-1* mutant branching phenotype is inhibited by BES1 knockdown, whereas BES1 degradation is promoted by AtD14, a putative SL receptor (Hu *et al.*, 2020). SLs control BRC1, another transcription factor of peas, which is a homolog of the *Arabidopsis* AtBRC1 and maize TB1. BRC1 acts downstream of SLs in *psbrc1* mutants due to an increase in the GR24-resistant branching phenotype. D53, a class I Clp ATPase, has been discovered in rice recently. Through structural analysis of D53, three amphiphilic repression (EAR) motifs linked with an ethylene-responsive element binding factor have been found. It has been demonstrated that EAR motifs control the expression of plant genes through transcriptional repression, and it has been suggested that D53 controls SL-responsive genes in a similar way. However, because D53 only binds to TPR proteins weakly, there are not enough data to support this mode of D53-mediated control. Additionally, EAR motifs have been linked to interactions with proteins that include CTLH domains. CTLH-domain proteins play a role in endocytosis and cytoskeleton reorganization in *Arabidopsis*. Through cytoskeleton rearrangements and trafficking, it has been demonstrated that SLs control auxin transport and distribution (Hossain *et al.*, 2021).

14.6.3 Proteolysis

RING-finger E3 ligases, such as the anaphase promoting complex (APC) and the SKP1-CULIN-F-BOX PROTEIN (SCF) complex, are used by a post-translational regulation mechanism known as proteolysis to preserve cellular homeostasis (Bialek *et al.*, 2023). The α/β-fold hydrolase D14/D88/HTD2 of rice and DAD2 of *Petunia*, which may be the likely SL receptor, and the leucine-rich repeat F-box protein, ORE9/MAX2/RMS4/D3, which is a substrate-attracting subunit of SCF-type ubiquitin E3 ligase, are required for the regulation mechanism of SL signaling.

Their roles in the SL perception pathway are supported by the highly branching phenotype and SL insensitivity displayed by both *max2* and *d14* mutants. Similar to GA signaling, these two proteins promote the development of the GA–GID1–DELLA complex, in which the DELLA protein functions as a transcription factor by binding to the GA receptor (Sarwar *et al.*, 2023). The SCFSLY 1/GID2 complex, which includes the F-box protein SLEEPY1 (SLY1), recognizes the GA–GID1–DELLA complex and recruits DELLA proteins for proteasomal destruction.

Research in *Arabidopsis* has demonstrated that ORE9/MAX2 possesses a functioning F-box domain, which further supports MAX2's position in the SCF complex. CaMV35S::MAX2 (deleted in the F-box domain) cannot complement the phenotype of the *max2* mutant. The core SCF subunits of *Arabidopsis* CULIN and ARABIDOPSIS SKP1-LIKE (ASK1) also interact with Myc-tagged MAX2. Similar to GID1 (GA-INSENSITIVE DWARF1) function, GR24 binds in *Petunia* and is hydrolyzed by DAD2 (D14) into ABC-OH and D-OH rings. Furthermore, the DAD2–GR24 complex's protein–protein interaction with PhMAX2 (the *Petunia* homolog of MAX2) requires the hydrolytic activity of DAD2. Mutant studies have verified the presence of three stable amino acid residues necessary for DAD2/D14 action, and the crystal structure of DAD2 revealed the existence of a catalytic cavity that SLs can occupy (Hamiaux *et al.*, 2019).

When SLs are hydrolyzed by D14, downstream targets such as the rice DELLA protein SLR1 are activated. It has been demonstrated that DELLA proteins interact with GID1 to negatively inhibit GA signaling (Bouré and Arnaud, 2023). Rice D53 is a member of the protein family Clp ATPase, which is well known for its role in the disintegration and breakdown of proteins. D53 is a homolog of the protein belonging to the *Arabidopsis* SMXL family, which is degraded by MAX2–D14–D53. D53 acts as negative regulator of SL signaling, as evidenced by the suppression of the *d3* (rice homolog of MAX2) and *d14* mutant phenotypes upon knockdown, while a specific deletion of a 5 amino acid motif in D53 produces a dominant SL-resistant enhanced branching phenotype. Yeast two-hybrid assays have verified a connection between D53, D14, and D3 (MAX2) (Zhu *et al.*, 2023). Furthermore, D53 degrades quickly when GR24 is present (Liu *et al.*, 2023).

14.6.4 Transport and localization

Early grafting studies revealed that SLs are moved from the roots to the shoots in the tomato and *Arabidopsis* xylem, which shed light on the control of SL signaling through localization and transport. The main location for SL production is in the roots, from which SLs are either transferred to other plant parts via the xylem or secreted into the rhizosphere (Kee *et al.*, 2023).

Curiously, in *pdr1* mutant *Petunia hybrida* plants, SL levels in root extract are comparable to those of the wild type but they are significantly lower in root exudates (Liu *et al.*, 2018). This suggests that PhPDR1 is involved in SL exudation and that feedback control is likely in place to maintain SL levels. PDR1 also facilitates SL migration across the plant, as evidenced by the phenotypically increased lateral bud growth and decreased contact with AMF seen in *pdr1* plants (Shiratake *et al.*, 2019).

The location of PaPDR1 in the plasma membrane of the subepidermal cells in the lateral roots of *Arabidopsis* provided more evidence for its function in the rhizosphere's SL efflux. It has been demonstrated that PaPDR1 localizes to the apical membrane of root hypodermal cells when it is co-expressed with DAD1 (CCD8), perhaps promoting acropetal transfer (Foster *et al.*, 2018). PaPDR1 is found in the lateral membranes of the hypodermal passage cells, which serve as a portal for AMF and are likely responsible for SL transfer to the soil.

SL transport to the rhizosphere and shoot tip is compromised in a *papdr1* mutant. When compared with wild type, GFP–PDR1-overexpression functioning showed that SL signaling pathways are responsible for the prevention of lateral branch development and for darker green plants as a result of photosynthetic pathway stimulation (Zhao *et al.*, 2021). Following GR24 treatment, PIN proteins in roots and stems decrease. Conversely, PIN2 apical and vacuolar localization is induced by GR24.

According to these findings, PhPIN1 is downregulated and PhPIN2 is upregulated in PDR1-overexpressing root tips, which is in line with the PDR1-overexpression shoot phenotype. Thus, by differentially regulating PIN1 and PIN2, SLs influence the pattern of its transport through auxin signaling in addition to directly inducing *PaPDR1* expression. The discovery of the ABCG transporter NtPDR6 in tobacco recently raises the possibility that SL transport and signaling are regulated similarly in all plant species (Xie *et al.*, 2021a).

14.6.5 Regulation of epigenetics

SL signaling has been linked to DNA methylation, particularly in the germination phase of *Orobanche ramosa* (hemp broomrape). The seeds need to be conditioned for 4 days in order to germinate (Brun *et al.*, 2025). During the germination phase, GR24 therapy activates the ABA catabolic gene *PrCYP707A1*, which lowers the amount of ABA, a hormone that causes seeds to become dormant. DNA methylation is involved in this activation step. The conditioning period is shortened by 5-azacytidine treatment (a hypomethylation reagent) (Li *et al.*, 2021). In contrast, *PrCYP707A1* expression and, consequently, seed germination are inhibited by treatment with hydroxyurea (a hypermethylation reagent). Most likely, GR24 hypomethylates *PrCYP707A1*, which lowers ABA levels and promotes seed germination.

14.7 Agricultural and Biotechnological Applications of Strigolactones Under Salt Stress

SLs are becoming more widely acknowledged as promising instruments for enhancing crop resistance to salt stress using biotechnology and agronomic approaches (Chesterfield *et al.*, 2020). Exogenous administration of synthetic SL analogs such as GR24 has been demonstrated to improve salt tolerance in agriculture by enhancing the architecture of the root system, preserving ion homeostasis, and triggering antioxidant defenses. In saline-prone areas, these treatments could be incorporated into crop-management plans to reduce output losses. Additionally, pre-treatment with SLs improves stress memory and primes seedlings for improved establishment in saline soils, suggesting that SLs have potential in seed-priming techniques. From a biotechnological standpoint, breeding crops that can withstand salt can be produced by genetically modifying SL production and signaling pathways. Increased stress-responsive gene expression, decreased Na^+ buildup, and increased water-use efficiency can result from overexpression of SL biosynthesis genes or their downstream transcription factors (Singhal *et al.*, 2023). Similarly, targeted changes to fine-tune SL signaling for stress adaptation may be possible by clustered regularly interspaced short palindromic repeats/CRISPR-associated protein (CRISPR/Cas)-based genome editing of SL-related genes. Breeding initiatives to create salt-resilient cultivars may also be accelerated via SL-related marker-assisted selection. SLs offer indirect uses in sustainable agriculture in addition to direct stress tolerance. They can increase the efficiency of nutrient uptake, especially for nitrogen and phosphorus, which is important in soils affected by salt when nutrient availability is restricted. Additionally, SLs control advantageous symbiotic relationships with AMF, improving resilience to stress and nutrient uptake. These diverse functions imply that SL-based methods may be incorporated into integrated management plans for crop production under saline conditions (Huang *et al.*, 2023).

14.8 Future Perspectives and Conclusion

The responses of plants to a range of abiotic factors, such as cold, heat, salt, and drought, are significantly regulated by SLs. SLs increase plant resistance by controlling the expression of genes that respond to stress at the molecular level (Lu *et al.*, 2023). Moreover, genes involved in SL production and signaling pathways are essential for controlling abiotic stress responses.

By maintaining photosynthetic efficiency, SLs improve tolerance both physiologically and biochemically, regulating stomatal activity to managing ROS metabolism to lessen oxidative damage, and encouraging osmotic adjustment (Khan *et al.*, 2019). SLs optimize plant structure for stress tolerance at the morphological level by influencing root architecture, shoot development, and leaf morphology. The interactions between SLs and other phytohormones have been studied in detail. A number of biotechnological techniques have been used to quicken SL-based agricultural development. The importance of CCD7, CCD8, MAX1, and MAX2 in abiotic stress adaptation has been highlighted by the demonstration that their genetic disruption or silencing by CRISPR/Cas9 or virus-induced gene silencing reduces tolerance of tomatoes to heat and cold (Zulfiqar *et al.*, 2023). Together, these results show how SL pathways can be altered through molecular breeding and genetic engineering to increase stress tolerance. Integrative multi-omics approaches are crucial to fully understand the complexity of SL-mediated stress regulation.

An understanding of how SLs orchestrate networks of physiological and environmental responses will be provided by the confluence of transcriptomics, proteomics, phenomics, epigenomics, and metabolomics (Tolani *et al.*, 2021). Advanced methods such as single-cell RNA sequencing and spatial transcriptomics can also be used to reveal the dynamics of SL synthesis and signaling during abiotic stress. Stress tolerance, crop productivity and quality, and overall agricultural sustainability are all impacted by SLs and their synthetic analogs (Bhattacharjee *et al.*, 2023). Analogs including methyl phenlactonoates and Nijmegen-1, in addition to the widely used GR24, have demonstrated strong biological activity, especially in promoting self-harming germination of parasitic weeds, supporting ecologically friendly weed-management methods.

Despite their promise, a number of obstacles prevent these analogs from being widely used in agriculture including chemical instability, regulatory limits, and the need for cost-effective synthesis procedures (Chojnacka, 2024). Their future agronomic use will depend on how well these constraints are addressed through structural optimization and regulatory support.

In conclusion, SLs show promise as a goal to enhance agricultural ability to withstand abiotic stress and reduce the negative consequences of climate change on livestock output. Gaining a better knowledge of SL-centered transmission networks will help create sustainable and creative ways to improve crop performance, which will support global food security in the face of increasingly difficult environmental circumstances.

References

Abas, L., Kolb, M., Stadlmann, J., Janacek, D.P., Lukic, K. *et al.* (2021) Naphthylphthalamic acid associates with and inhibits pin auxin transporters. *Proceedings of the National Academy of Sciences USA* 118(1), e2020857118. DOI: 10.1073/pnas.2020857118.

Ahmad, M., Tahir, M., Hong, Z., Zia, M.A., Rafeeq, H. *et al.* (2025) Plant and marine-derived natural products: Sustainable pathways for future drug discovery and therapeutic development. *Frontiers in Pharmacology* 15, 1497668. DOI: 10.3389/fphar.2024.1497668.

Ahmed, N.M. (2018) The role of strigolactones in resistance, tolerance and control of striga infection in sorghum. PhD thesis, Wageningen University, Wageningen, Netherlands.

Asad, M.A.U., Zakari, S.A., Zhao, Q., Zhou, L., Ye, Y. *et al.* (2019) Abiotic stresses intervene with ABA signaling to induce destructive metabolic pathways leading to death: Premature leaf senescence in plants. *International Journal of Molecular Sciences* 20(2), 256. DOI: 10.3390/ijms20020256.

Bai, J., Guo, H., Xiong, H., Xie, Y., Gu, J. *et al.* (2024) Strigolactone and abscisic acid synthesis and signaling pathways are enhanced in the wheat oligo-tillering mutant *ot1*. *Molecular Breeding* 44(2), 12. DOI: 10.1007/s11032-024-01450-3.

Bhattacharjee, A., Dubey, S. and Sharma, S. (2023) "Next-generation bioformulations" for plant growth promotion and stress mitigation: A promising approach for sustainable agriculture. *Journal of Plant Growth Regulation* 42(10), 6741–6759. DOI: 10.1007/s00344-023-10996-z.

Bialek, W., Collawn, J.F. and Bartoszewski, R. (2023) Ubiquitin-dependent and independent proteasomal degradation in host-pathogen interactions. *Molecules* 28(18), 6740. DOI: 10.3390/molecules28186740.

Borghi, L., Kang, J. and de Brito Francisco, R. (2019) Filling the gap: Functional clustering of ABC proteins for the investigation of hormonal transport in *planta*. *Frontiers in Plant Science* 10, 422. DOI: 10.3389/fpls.2019.00422.

Bouré, N. and Arnaud, N. (2023) Molecular GA pathways as conserved integrators for adaptive responses. *Plant Biology* 25(5), 649–660. DOI: 10.1111/plb.13549.

Brun, G., Pöhl, J. and Wicke, S. (2025) Aging seeds of weedy broomrapes and witchweeds lose sensitivity to strigolactones as DNA demethylates. *Plants, People, Planet* 7(2), 476–485. DOI: 10.1002/ppp3.10465.

Castro-Camba, R., Sánchez, C., Vidal, N. and Vielba, J. (2022) Interactions of gibberellins with phytohormones and their role in stress responses. *Horticulturae* 8(3), 241. DOI: 10.3390/horticulturae8030241.

Cheng, X., Huang, Y., Tan, Y., Tan, L., Yin, J. *et al.* (2022) Potentially useful dwarfing or semi-dwarfing genes in rice breeding in addition to the *sd1* gene. *Rice* 15(1), 66. DOI: 10.1186/s12284-022-00615-y.

Chesterfield, R.J., Vickers, C.E. and Beveridge, C.A. (2020) Translation of strigolactones from plant hormone to agriculture: Achievements, future perspectives, and challenges. *Trends in Plant Science* 25(11), 1087–1106. DOI: 10.1016/j.tplants.2020.06.005.

Chojnacka, K. (2024) Sustainable chemistry in adaptive agriculture: A review. *Current Opinion in Green and Sustainable Chemistry* 46, 100898. DOI: 10.1016/j.cogsc.2024.100898.

Crombez, H., Motte, H. and Beeckman, T. (2019) Tackling plant phosphate starvation by the roots. *Developmental Cell* 48(5), 599–615. DOI: 10.1016/j.devcel.2019.01.002.

Dahal, P., Wang, Y., Hu, J., Park, J., Forker, K. *et al.* (2025) Structural insights into proteolysis-dependent and -independent suppression of the master regulator DELLA by the gibberellin receptor. *Proceedings of the National Academy of Sciences USA* 122(32), e2511012122. DOI: 10.1073/pnas.2511012122.

Daignan-Fornier, S., Keita, A. and Boyer, F.-D. (2024) Chemistry of strigolactones, key players in plant communication. *Chembiochem* 25(12), e202400133. DOI: 10.1002/cbic.202400133.

Du, Y. and Scheres, B. (2018) Lateral root formation and the multiple roles of auxin. *Journal of Experimental Botany* 69(2), 155–167. DOI: 10.1093/jxb/erx223.

Elhiti, M., Mira, M.M., So, K.K.Y., Stasolla, C. and Hebelstrup, K.H. (2021) Synthetic strigolactone GR24 improves *Arabidopsis* somatic embryogenesis through changes in auxin responses. *Plants* 10(12), 2720. DOI: 10.3390/plants10122720.

Faizan, M., Faraz, A., Sami, F., Siddiqui, H., Yusuf, M. *et al.* (2020) Role of strigolactones: Signalling and crosstalk with other phytohormones. *Open Life Sciences* 15, 217–228. DOI: 10.1515/biol-2020-0022.

Feng, Z., Liang, X., Tian, H., Watanabe, Y., Nguyen, K.H. *et al.* (2023) Suppressor of MAX2 1 (SMAX1) and SMAX1-Like2 (SMXL2) negatively regulate drought resistance in *Arabidopsis thaliana*. *Plant and Cell Physiology* 63(12), 1900–1913. DOI: 10.1093/pcp/pcac080.

Foster, T.M., Ledger, S.E., Janssen, B.J., Luo, Z., Drummond, R.S.M. *et al.* (2018) Expression of *MdCCD7* in the scion determines the extent of sylleptic branching and the primary shoot growth rate of apple trees. *Journal of Experimental Botany* 69(9), 2379–2390. DOI: 10.1093/jxb/erx404.

Hamiaux, C., Larsen, L., Lee, H.W., Luo, Z., Sharma, P. *et al.* (2019) Chemical synthesis and characterization of a new quinazolinedione competitive antagonist for strigolactone receptors with an unexpected binding mode. *Biochemical Journal* 476(12), 1843–1856. DOI: 10.1042/BCJ20190288.

Hasanuzzaman, M., Bhuyan, M.H.M.B., Parvin, K., Bhuiyan, T.F., Anee, T.I. *et al.* (2020) Regulation of ROS metabolism in plants under environmental stress: A review of recent experimental evidence. *International Journal of Molecular Sciences* 21(22), 8695. DOI: 10.3390/ijms21228695.

He, X., Wang, C., Wang, H., Li, L. and Wang, C. (2020) The function of MAPK cascades in response to various stresses in horticultural plants. *Frontiers in Plant Science* 11, 952. DOI: 10.3389/fpls.2020.00952.

Hernández-García, J., Briones-Moreno, A. and Blázquez, M.A. (2021) Origin and evolution of gibberellin signaling and metabolism in plants. *Seminars in Cell & Developmental Biology* 109, 46–54. DOI: 10.1016/j.semcdb.2020.04.009.

Hu, J., Sun, S. and Wang, X. (2020) Regulation of shoot branching by strigolactones and brassinosteroids: Conserved and specific functions of *Arabidopsis* BeS1 and rice BzR1. *Molecular Plant* 13(6), 808–810. DOI: 10.1016/j.molp.2020.03.008.

Hu, J., Li, X. and Sun, T.-P. (2023) Four class A auxin response factors promote tomato fruit growth despite suppressing fruit set. *Nature Plants* 9(5), 706–719. DOI: 10.1038/s41477-023-01396-y.

Hossain, A., Raza, A., Maitra, S., Asaduzzaman, M., Islam, M.R. *et al.* (2021) Strigolactones: A novel carotenoid-derived phytohormone–biosynthesis, transporters, signalling, and mechanisms in abiotic stress. In: Aftab, T. and Hakeem, K.R. (eds) *Plant Growth Regulators: Signalling under Stress Conditions*. Springer, Cham, Switzerland, pp. 275–303. DOI: 10.1007/978-3-030-61153-8_13.

Huang, H., Xiao, Y., Ding, G., Liao, L., Yan, C. *et al.* (2023) Comprehensive evaluation of island habitat quality based on the invest model and terrain diversity: A case study of Haitan Island, China. *Sustainability* 15(14), 11293. DOI: 10.3390/su151411293.

Jajoo, A. and Mathur, S. (2021) Role of arbuscular mycorrhizal fungi as an underground saviour for protecting plants from abiotic stresses. *Physiology and Molecular Biology of Plants* 27(11), 2589–2603. DOI: 10.1007/s12298-021-01091-2.

Jan, M., Muhammad, S., Jin, W., Zhong, W., Zhang, S. *et al.* (2024) Modulating root system architecture: Cross-talk between auxin and phytohormones. *Frontiers in Plant Science* 15, 1343928. DOI: 10.3389/fpls.2024.1343928.

Jeyakumar, J.M.J., Ali, A., Wang, W.-M. and Thiruvengadam, M. (2020) Characterizing the role of the miR156-SPL network in plant development and stress response. *Plants* 9(9), 1206. DOI: 10.3390/plants9091206.

Jing, H., Wilkinson, E.G., Sageman-Furnas, K. and Strader, L.C. (2023) Auxin and abiotic stress responses. *Journal of Experimental Botany* 74(22), 7000–7014. DOI: 10.1093/jxb/erad325.

Kee, Y.J., Ogawa, S., Ichihashi, Y., Shirasu, K. and Yoshida, S. (2023) Strigolactones in rhizosphere communication: Multiple molecules with diverse functions. *Plant & Cell Physiology* 64(9), 955–966. DOI: 10.1093/pcp/pcad055.

Kerr, S.C., Patil, S.B., de Saint Germain, A., Pillot, J.-P., Saffar, J. *et al.* (2021) Integration of the SMXL/D53 strigolactone signalling repressors in the model of shoot branching regulation in *Pisum sativum*. *Plant Journal* 107(6), 1756–1770. DOI: 10.1111/tpj.15415.

Khan, M.N., Zhang, J., Luo, T., Liu, J., Rizwan, M. *et al.* (2019) Seed priming with melatonin coping drought stress in rapeseed by regulating reactive oxygen species detoxification: Antioxidant defense system, osmotic adjustment, stomatal traits and chloroplast ultrastructure perseveration. *Industrial Crops and Products* 140, 111597. DOI: 10.1016/j.indcrop.2019.111597.

Khan, S., Alvi, A.F. and Khan, N.A. (2024) Role of ethylene in the regulation of plant developmental processes. *Stresses* 4(1), 28–53. DOI: 10.3390/stresses4010003.

Konstantinova, N., Korbei, B. and Luschnig, C. (2021) Auxin and root gravitropism: Addressing basic cellular processes by exploiting a defined growth response. *International Journal of Molecular Sciences* 22(5), 2749. DOI: 10.3390/ijms22052749.

Lantzouni, O., Klermund, C. and Schwechheimer, C. (2017) Largely additive effects of gibberellin and strigolactone on gene expression in *Arabidopsis thaliana* seedlings. *Plant Journal* 92(5), 924–938. DOI: 10.1111/tpj.13729.

Leoni, F., Hazrati, H., Fomsgaard, I.S., Moonen, A.-C. and Kudsk, P. (2021) Determination of the effect of co-cultivation on the production and root exudation of flavonoids in four legume species using LC-MS/MS analysis. *Journal of Agricultural and Food Chemistry* 69(32), 9208–9219. DOI: 10.1021/acs.jafc.1c02821.

Li, W., Gupta, A., Tian, H., Nguyen, K.H., Tran, C.D. *et al.* (2020) Different strategies of strigolactone and karrikin signals in regulating the resistance of *Arabidopsis thaliana* to water-deficit stress. *Plant Signaling & Behavior* 15(9), 1789321. DOI: 10.1080/15592324.2020.1789321.

Li, Z., Hu, Y., Chang, M., Kashif, M.H., Tang, M. *et al.* (2021) 5-Azacytidine pre-treatment alters DNA methylation levels and induces genes responsive to salt stress in kenaf (*Hibiscus cannabinus* L.). *Chemosphere* 271, 129562. DOI: 10.1016/j.chemosphere.2021.129562.

Li, M., Zhu, Y., Li, S., Zhang, W., Yin, C. *et al.* (2022a) Regulation of phytohormones on the growth and development of plant root hair. *Frontiers in Plant Science* 13, 865302. DOI: 10.3389/fpls.2022.865302.

Li, X., Gao, M., Guo, Y., Zhang, Z., Zhang, Z. *et al.* (2022b) 6-Benzyladenine alleviates NaCl stress in watermelon (*Citrullus lanatus*) seedlings by improving photosynthesis and upregulating antioxidant defences. *Functional Plant Biology* 50, 230–241. DOI: 10.1071/FP22047.

Liu, G., Pfeifer, J., de Brito Francisco, R., Emonet, A., Stirnemann, M. *et al.* (2018) Changes in the allocation of endogenous strigolactone improve plant biomass production on phosphate-poor soils. *New Phytologist* 217(2), 784–798. DOI: 10.1111/nph.14847.

Liu, S., Wang, J., Song, B., Gong, X., Liu, H. *et al.* (2023) Conformational dynamics of the D53–D3–D14 complex in strigolactone signaling. *Plant & Cell Physiology* 64(9), 1046–1056. DOI: 10.1093/pcp/pcad067.

López-Ruiz, B.A., Zluhan-Martínez, E., Sánchez, M. de la P., Álvarez-Buylla, E.R. and Garay-Arroyo, A. (2020) Interplay between hormones and several abiotic stress conditions on *Arabidopsis thaliana* primary root development. *Cells* 9(12), 2576. DOI: 10.3390/cells9122576.

Lu, X., Liu, X., Xu, J., Liu, Y., Chi, Y. *et al.* (2023) Strigolactone-mediated trehalose enhances salt resistance in tomato seedlings. *Horticulturae* 9(7), 770. DOI: 10.3390/horticulturae9070770.

Manzoor, M.A., Xu, Y., lv, Z., Xu, J., Wang, Y. *et al.* (2023) Fruit crop abiotic stress management: A comprehensive review of plant hormones mediated responses. *Fruit Research* 3, 30. DOI: 10.48130/FruRes-2023-0030.

Mashiguchi, K., Seto, Y. and Yamaguchi, S. (2021) Strigolactone biosynthesis, transport and perception. *Plant Journal* 105(2), 335–350. DOI: 10.1111/tpj.15059.

Meanwell, N.A. (2023) Applications of bioisosteres in the design of biologically active compounds. *Journal of Agricultural and Food Chemistry* 71(47), 18087–18122. DOI: 10.1021/acs.jafc.3c00765.

Mehmood, I., Wani, K.I. and Aftab, T. (2025) Strigolactone interplay with other phytohormones under stressed and normal conditions. In: Aftab, T. and Wani, K.I. (eds) *Strigolactones: Emerging Plant Hormones*. Scrivener Publishing, Beverly, Massachusetts, pp. 161–191. DOI: 10.1002/9781394302826.ch8.

Mishra, S., Levengood, H., Fan, J. and Zhang, C. (2024) Plants under stress: Exploring physiological and molecular responses to nitrogen and phosphorus deficiency. *Plants* 13(22), 3144. DOI: 10.3390/plants13223144.

Moturu, T. and Koča, J. (2020) Deciphering the role of strigolactone hormone in plant developmental processes. PhD thesis, Masaryk University, Brno, Czechia.

Muhammad, I., Shalmani, A., Ali, M., Yang, Q.-H., Ahmad, H. *et al.* (2021) Mechanisms regulating the dynamics of photosynthesis under abiotic stresses. *Frontiers in Plant Science* 11, 615942. DOI: 10.3389/fpls.2020.615942.

Mukherjee, A., Dwivedi, S., Bhagavatula, L. and Datta, S. (2023) Integration of light and aba signaling pathways to combat drought stress in plants. *Plant Cell Reports* 42(5), 829–841. DOI: 10.1007/s00299-023-02999-7.

Mukherjee, S. and Aftab, T. (eds) (2023) *Strigolactones, Alkamides and Karrikins in Plants: Recent Updates and Future Prospects*. CRC Press, Boca Raton, Florida.

Mushtaq, Z., Faizan, S. and Gulzar, B. (2020) Salt stress, its impacts on plants and the strategies plants are employing against it: A review. *Journal of Applied Biology & Biotechnology* 8(3), 81–91. DOI: 10.7324/JABB.2020.80315.

Nadarajah, K.K. (2020) ROS homeostasis in abiotic stress tolerance in plants. *International Journal of Molecular Sciences* 21(15), 5208. DOI: 10.3390/ijms21155208.

Naorem, A., Jayaraman, S., Dang, Y.P., Dalal, R.C., Sinha, N.K. *et al.* (2023) Soil constraints in an arid environment—challenges, prospects, and implications. *Agronomy* 13(1), 220. DOI: 10.3390/agronomy13010220.

Nasir, F., Tian, L., Shi, S., Chang, C., Ma, L. *et al.* (2019) Strigolactones positively regulate defense against *Magnaporthe oryzae* in rice (*Oryza sativa*). *Plant Physiology and Biochemistry* 142, 106–116. DOI: 10.1016/j.plaphy.2019.06.028.

Nasir, F., Bahadur, A., Lin, X., Gao, Y. and Tian, C. (2021) Novel insights into host receptors and receptor-mediated signaling that regulate arbuscular mycorrhizal symbiosis. *Journal of Experimental Botany* 72(5), 1546–1557. DOI: 10.1093/jxb/eraa538.

Naz, M., Afzal, M.R., Raza, M.A., Tariq, M., Yan, M. *et al.* (2025) The significant effects of strigolactones on plant growth and microbe interactions: A review. *Plant Growth Regulation* 105(4), 937–957. DOI: 10.1007/s10725-025-01325-3.

Nohales, M.A. and Kay, S.A. (2019) GIGANTEA gates gibberellin signaling through stabilization of the DELLA proteins in *Arabidopsis*. *Proceedings of the National Academy of Sciences USA* 116(43), 21893–21899. DOI: 10.1073/pnas.1913532116.

Omoarelojie, L.O., Kulkarni, M.G., Finnie, J.F. and Van Staden, J. (2019) Strigolactones and their crosstalk with other phytohormones. *Annals of Botany* 124(5), 749–767. DOI: 10.1093/aob/mcz100.

Pandya-Kumar, N., Shema, R., Kumar, M., Mayzlish-Gati, E., Levy, D. *et al.* (2014) Strigolactone analog GR24 triggers changes in PIN2 polarity, vesicle trafficking and actin filament architecture. *New Phytologist* 202(4), 1184–1196. DOI: 10.1111/nph.12744.

Park, Y.J., Nam, B.E. and Park, C.M. (2024) Environmentally adaptive reshaping of plant photomorphogenesis by karrikin and strigolactone signaling. *Journal of Integrative Plant Biology* 66(5), 865–882. DOI: 10.1111/jipb.13602.

Qi, J., Mao, Y., Cui, J., Lu, X., Xu, J. *et al.* (2024) The role of strigolactones in resistance to environmental stress in plants. *Physiologia Plantarum* 176(4), e14419. DOI: 10.1111/ppl.14419.

Raspor, M., Motyka, V., Kaleri, A.R., Ninković, S., Tubić, L. *et al.* (2021) Integrating the roles for cytokinin and auxin in *de novo* shoot organogenesis: From hormone uptake to signaling outputs. *International Journal of Molecular Sciences* 22(16), 8554. DOI: 10.3390/ijms22168554.

Raza, A., Javed, R., Zahid, Z., Sharif, R., Hafeez, M.B. *et al.* (2021) Strigolactones for sustainable plant growth and production under adverse environmental conditions. In: Husen, A. (ed.) *Plant Performance Under Environmental Stress: Hormones, Biostimulants and Sustainable Plant Growth Management*. Springer Nature, Cham, Switzerland, pp. 129–166. DOI: 10.1007/978-3-030-78521-5_6.

Rehman, N.U., Li, X., Zeng, P., Guo, S., Jan, S. *et al.* (2021) Harmony but not uniformity: Role of strigolactone in plants. *Biomolecules* 11(11), 1616. DOI: 10.3390/biom11111616.

Ren, C.-G., Kong, C.-C. and Xie, Z.-H. (2018) Role of abscisic acid in strigolactone-induced salt stress tolerance in arbuscular mycorrhizal *Sesbania cannabina* seedlings. *BMC Plant Biology* 18(1), 74. DOI: 10.1186/s12870-018-1292-7.

Roychowdhury, R., Mishra, S., Anand, G., Dalal, D., Gupta, R. *et al.* (2024) Decoding the molecular mechanism underlying salicylic acid (SA)-mediated plant immunity: An integrated overview from its biosynthesis to the mode of action. *Physiologia Plantarum* 176(3), e14399. DOI: 10.1111/ppl.14399.

Samanta, S. and Roychoudhury, A. (2025) Molecular crosstalk of jasmonate with major phytohormones and plant growth regulators during diverse stress responses. *Journal of Plant Growth Regulation* 44(1), 62–88. DOI: 10.1007/s00344-024-11412-w.

Sánchez Martín-Fontecha, E., Cardinale, F., Bürger, M., Prandi, C. and Cubas, P. (2024) Novel mechanisms of strigolactone-induced DWARF14 degradation in *Arabidopsis thaliana*. *Journal of Experimental Botany* 75(22), 7145–7159. DOI: 10.1093/jxb/erae365.

Sarwar, R., Zhu, K.M., Jiang, T., Ding, P., Gao, Y. *et al.* (2023) DELLAs directed gibberellins responses orchestrate crop development: A brief review. *Crop Science* 63(1), 1–28. DOI: 10.1002/csc2.20825.

Sescil, J., Havens, S.M. and Wang, W. (2025) Principles and design of molecular tools for sensing and perturbing cell surface receptor activity. *Chemical Reviews* 125(5), 2665–2702. DOI: 10.1021/acs.chemrev.4c00582.

Shiratake, K., Notaguchi, M., Makino, H., Sawai, Y. and Borghi, L. (2019) Petunia PLEIOTROPIC DRUG RESISTANCE 1 is a strigolactone short-distance transporter with long-distance outcomes. *Plant & Cell Physiology* 60(8), 1722–1733. DOI: 10.1093/pcp/pcz081.

Simkin, A.J., Kapoor, L., Doss, C.G.P., Hofmann, T.A., Lawson, T. *et al.* (2022) The role of photosynthesis related pigments in light harvesting, photoprotection and enhancement of photosynthetic yield in *planta*. *Photosynthesis Research* 152(1), 23–42. DOI: 10.1007/s11120-021-00892-6.

Singhal, R.K., Fahad, S., Kumar, P., Choyal, P., Javed, T. *et al.* (2023) Beneficial elements: New players in improving nutrient use efficiency and abiotic stress tolerance. *Plant Growth Regulation* 100(2), 237–265. DOI: 10.1007/s10725-022-00843-8.

Sun, D., Zhang, L., Yu, Q., Zhang, J., Li, P. *et al.* (2021) Integrated signals of jasmonates, sugars, cytokinins and auxin influence the initial growth of the second buds of chrysanthemum after decapitation. *Biology* 10(5), 440. DOI: 10.3390/biology10050440.

Sun, H., Guo, X., Qi, X., Feng, F., Xie, X. *et al.* (2021) SPL14/17 act downstream of strigolactone signalling to modulate rice root elongation in response to nitrate supply. *Plant Journal* 106(3), 649–660. DOI: 10.1111/tpj.15188.

Sun, H., Guo, X., Zhu, X., Gu, P., Zhang, W. *et al.* (2023) Strigolactone and gibberellin signaling coordinately regulate metabolic adaptations to changes in nitrogen availability in rice. *Molecular Plant* 16(3), 588–598. DOI: 10.1016/j.molp.2023.01.009.

Tarafdar, M., Bahadur, V., Rana, S. and Singh, R.K. (2022) A review: Abiotic stress on transpiration, stomatal diffusive resistance and photosynthetic rate. *Pharma Innovation Journal* 11, 1632–1635.

Tarchoun, N., Saadaoui, W., Mezghani, N., Pavli, O.I., Falleh, H. *et al.* (2022) The effects of salt stress on germination, seedling growth and biochemical responses of Tunisian squash (*Cucurbita maxima* Duchesne) germplasm. *Plants* 11(6), 800. DOI: 10.3390/plants11060800.

Tolani, P., Gupta, S., Yadav, K., Aggarwal, S. and Yadav, A.K. (2021) Big data, integrative omics and network biology. *Advances in Protein Chemistry and Structural Biology* 127, 127–160. DOI: 10.1016/bs.apcsb.2021.03.006.

Vogel, D., Hills, P. and Moore, J.P. (2023) Strigolactones GR-24 and Nijmegen applications result in reduced susceptibility of tobacco and grapevine plantlets to *Botrytis cinerea* infection. *Plants* 12(18), 3202. DOI: 10.3390/plants12183202.

Wang, J.Y., Chen, G.-T.E., Braguy, J. and Al-Babili, S. (2024) Distinguishing the functions of canonical strigolactones as rhizospheric signals. *Trends in Plant Science* 29(8), 925–936. DOI: 10.1016/j.tplants.2024.02.013.

Wang, M., Zang, L., Jiao, F., Perez-Garcia, M.-D., Ogé, L. *et al.* (2020) Sugar signaling and post-transcriptional regulation in plants: An overlooked or an emerging topic? *Frontiers in Plant Science* 11, 578096. DOI: 10.3389/fpls.2020.578096.

Xie, X., Cao, P., Wang, Z., Gao, J., Wu, M. *et al.* (2021a) Genome-wide characterization and expression profiling of the *PDR* gene family in tobacco (*Nicotiana tabacum*). *Gene* 788, 145637. DOI: 10.1016/j.gene.2021.145637.

Xie, X., Wang, Y., Datla, R. and Ren, M. (2021b) Auxin and target of rapamycin spatiotemporally regulate root organogenesis. *International Journal of Molecular Sciences* 22(21), 11357. DOI: 10.3390/ijms222111357.

Xu, K., Wu, N., Yao, W., Li, X., Zhou, Y. *et al.* (2021) The biological function and roles in phytohormone signaling of the F-box protein in plants. *Agronomy* 11(11), 2360. DOI: 10.3390/agronomy11112360.

Yan, Y., Zhao, N., Tang, H., Gong, B. and Shi, Q. (2020) Shoot branching regulation and signaling. *Plant Growth Regulation* 92(2), 131–140. DOI: 10.1007/s10725-020-00640-1.

Yuan, Y., Khourchi, S., Li, S., Du, Y. and Delaplace, P. (2023) Unlocking the multifaceted mechanisms of bud outgrowth: Advances in understanding shoot branching. *Plants* 12(20), 3628. DOI: 10.3390/plants12203628.

Yang, T., Lian, Y. and Wang, C. (2019) Comparing and contrasting the multiple roles of butenolide plant growth regulators: Strigolactones and karrikins in plant development and adaptation to abiotic stresses. *International Journal of Molecular Sciences* 20(24), 6270. DOI: 10.3390/ijms20246270.

Zhang, J., Mazur, E., Balla, J., Gallei, M., Kalousek, P. *et al.* (2020) Strigolactones inhibit auxin feedback on pin-dependent auxin transport canalization. *Nature Communications* 11(1), 3508. DOI: 10.1038/s41467-020-17252-y.

Zhao, B., Liu, Q., Wang, B. and Yuan, F. (2021) Roles of phytohormones and their signaling pathways in leaf development and stress responses. *Journal of Agricultural and Food Chemistry* 69(12), 3566–3584. DOI: 10.1021/acs.jafc.0c07908.

Zhu, L., Li, S., Ma, Q., Yan, K., Ren, J. *et al.* (2023) An Acer palmatum R2R3-MYB gene, *ApMYB77*, confers freezing and drought tolerance in *Arabidopsis thaliana*. *Journal of Plant Growth Regulation* 42(2), 1017–1030. DOI: 10.1007/s00344-022-10611-7.

Zou, M., Zhang, D., Liu, Y., Chen, Z., Xu, T. *et al.* (2024) Integrative proteome and metabolome unveil the central role of IAA alteration in axillary bud development following topping in tobacco. *Scientific Reports* 14(1), 15309. DOI: 10.1038/s41598-024-66136-4.

Zulfiqar, S., Farooq, M.A., Zhao, T., Wang, P., Tabusam, J. *et al.* (2023) Virus-induced gene silencing (VIGS): A powerful tool for crop improvement and its advancement towards epigenetics. *International Journal of Molecular Sciences* 24(6), 5608. DOI: 10.3390/ijms24065608.

15 Strigolactone-mediated Defense in Plants: Mechanisms and Applications for Fungal Disease Control

Mashal Nadeem[1], Aimen Razzaq Khan[1], Saba Mazhar[1], Sajjad Hyder[1*], Javed Iqbal[2], Banzeer Ahsan Abbasi[3] and Mudassir Iqbal[4*]

[1]Department of Botany, Faculty of Natural Sciences, Government College Women University, Sialkot 51310, Pakistan; [2]Department of Botany, Bacha Khan University, Charsadda, Khyber Pakhtunkhwa, Pakistan; [3]Department of Botany, Rawalpindi Women University, Rawalpindi, Pakistan; [4]Department of Plant Protection Biology, Swedish University of Agricultural Sciences, Lomma, Sweden

Abstract

Strigolactones (SLs), a class of carotenoid-derived phytohormones, have gained significant attention in recent years for their multi-faceted roles in plant development and stress adaptation, particularly in modulating interactions with soil microbes and pathogens. This chapter explores the emerging evidence on SL-mediated defense mechanisms against fungal diseases in plants, presenting an integrative view of their biosynthesis, signaling, and regulatory pathways. Fungal pathogens account for the majority of plant disease-related yield losses globally, and there is an increasing demand for sustainable, environmentally friendly alternatives to chemical fungicides. SLs offer a promising avenue in this context, as they regulate plant architecture, promote arbuscular mycorrhizal symbiosis, and trigger systemic defense responses. This chapter delves into the molecular underpinnings of SLs biosynthesis and signaling, highlighting key genes such as β-carotene isomerase (*DWARF27*), *carotenoid cleavage dioxygenase 7* (*CCD7*), *carotenoid cleavage dioxygenase 8* (*CCD8*), and *MORE AXILLARY GROWTH 1* (*MAX1*), and describes how environmental and hormonal factors—including phosphate availability, auxins, and salicylic acid—influence SL production. Furthermore, it discusses how SLs suppress fungal growth and sporulation, induce expression of immunity-related genes, and modulate cross-talk with other phytohormones such as jasmonates, abscisic acid, and gibberellins. Case studies are presented demonstrating the efficacy of SLs against key pathogens such as *Botrytis cinerea*, *Magnaporthe oryzae*, and root-knot nematodes. In addition to elucidating their natural functions, the chapter reviews applied strategies including exogenous SL applications, genetic engineering for enhanced SL biosynthesis, and integration with other disease-management techniques. The potential of SLs as eco-friendly biocontrol agents is emphasized, while also addressing current challenges and future research directions. Ultimately, this chapter positions SLs as pivotal components of integrated plant disease-management strategies, aligning with global goals for sustainable agriculture.

Keywords: biological disease management, fungal disease resistance, hormonal signaling, plant immunity, plant–microbe interactions, strigolactones

*Corresponding authors: sajjad.hyder@gcwus.edu.pk and mudassir.iqbal@slu.se

DOI: 10.1079/9781836990598.0015

15.1 Introduction

Plant diseases continue to be a major contributor to global food insecurity and agriculture output losses, presenting a chronic and significant challenge to sustainable crop production. Fungi and fungus-like organisms make up around 70% of all plant diseases, making them the main culprits behind major agricultural obstacles among the wide variety of plant pathogens (Chelladurai *et al.*, 2023; Ekwomadu and Mwanza, 2023). Globally, fungal diseases alone are thought to be responsible for yield losses of up to 20% (McLaughlin *et al.*, 2023).

Traditionally, the primary method for controlling fungal diseases in crops has involved the use of chemical fungicides. However, scientists from all over the world are investigating a variety of novel and environmentally friendly methods of fungal control due to the significant rise in the development of fungal resistance to these fungicides, as well as the detrimental effects that synthetic fungicides have on the environment, human health, and animal health (El-Baky and Amara, 2021). In recent decades, several disease-management techniques have been developed to address plant fungal diseases as an alternative to traditional chemical fungicides. These include fungal cell deactivation and evacuation employing ghost methods (El-Baky and Amara, 2021), RNA interference, plant defense elicitors, biological control or biochemical fungicides (Iqbal *et al.*, 2022, 2023, 2025; McLaughlin *et al.*, 2023), agronanotechnology, and botanical fungicides (Atiq *et al.*, 2020; Mazhar *et al.*, 2025; Ramzan *et al.*, 2025).

To support sustainable agricultural production, strigolactones (SLs), a newly identified plant hormone, are crucial, as they increase productivity, reduce chemical inputs, and effectively control plant defense mechanisms against fungi and weeds (Mansoor *et al.*, 2023). SLs are natural carotenoid derivatives (Soliman *et al.*, 2022) and have several functions in plant growth and development as well as in response to environmental stressors (Li *et al.*, 2023; Wani *et al.*, 2023; Clark and Bennett, 2024). Use of SLs will reduce the negative environmental effects of agriculture in addition to lowering production costs (Mansoor *et al.*, 2023).

15.2 Biosynthesis and Signaling of Strigolactones

15.2.1 SL-biosynthesis pathways

Together with a group of other enzymes called carotenoid cleavage dioxygenases (CCDs), β-carotene isomerase (DWARF27 (D27)) is responsible for the biosynthesis of SL molecules and SL-like substances. CCD7 and CCD8 are two of the primary dioxygenase groups (Harrison *et al.*, 2015; Soliman *et al.*, 2022). By transforming all-*trans*-β-carotene into 9-*cis*-β-carotene and vice versa, D27 has two ways of dissociating. In the plastid, CCD7 and CCD8 cleave 9-*cis*-β-apo-10′-carotene into 9-*cis*-β-apo-10′-carotenal from which is generated carlactone (Lopez-Obando *et al.*, 2015; Wang *et al.*, 2019). For non-canonical SLs, the cytochrome P450 MORE AXILLARY GROWTH 1 (MAX1/CYP711A1) in *Arabidopsis* converts carlactone to carlactonic acid (CLA) in the cytoplasm, which is subsequently changed to methyl carlactonoate by an unidentified methyltransferase (Mashiguchi *et al.*, 2021, 2022). Later, methyl carlactonoate is converted by the lateral branching oxidoreductase enzyme into hydroxymethyl carlactonoate (Yoneyama *et al.*, 2020a).

Unlike *Arabidopsis*, rice contains five homologs of MAX1, and Os900 (CYP711A2) and Os1400 (CYP711A3), which like *Arabidopsis* MAX1, convert CL to CLA (Yoneyama *et al.*, 2018). Os900 can also transform CLA into 4-deoxyorobanchol, a significant rice SL (Zhang *et al.*, 2014; Yoneyama and Brewer, 2021). Os1400 then catalyzes the conversion of 4-deoxyorobanchol to orobanchol via hydroxylation. Note that CLA is converted to 5-deoxystrigol (5DS) by cotton CYP722C (Wakabayashi *et al.*, 2020). Subsequently, CYP728B35 from sorghum, another P450 from a different lineage, transforms 5-deoxystrigol into sorgomol (Wakabayashi *et al.*, 2021). In SL biosynthesis, the development of structurally varied SLs may depend on both the upstream and downstream enzymes of CL. Through the intermediary zeaxanthin, carotenoid isomerase, CCD7, and CCD8 are involved in the first conversion of all-*trans*-β-carotene to carlactone and its hydroxylated derivative, 3-hydroxy-carlactone (Baz *et al.*, 2018; Yoneyama and Brewer, 2021).

The biosynthetic pathway of SL biosynthesis is shown in Fig. 15.1.

15.2.2 Regulation of SL production

SL production is regulated by an intricate interaction of various factors:

- *Nutrient availability*. The production of SLs is regulated by the availability of nutrients, especially phosphate. To encourage root growth and symbiotic interaction with microorganisms, SL production is elevated in phosphate-deficient environments (Yoneyama, 2019; Yoneyama and Brewer, 2021).
- *Plant hormones*. SLs biosynthesis is positively regulated by auxin (Yoneyama and Brewer, 2021), while cytokinins and gibberellins are negative regulators (Ito *et al*., 2017; Yoneyama *et al*., 2020b), and

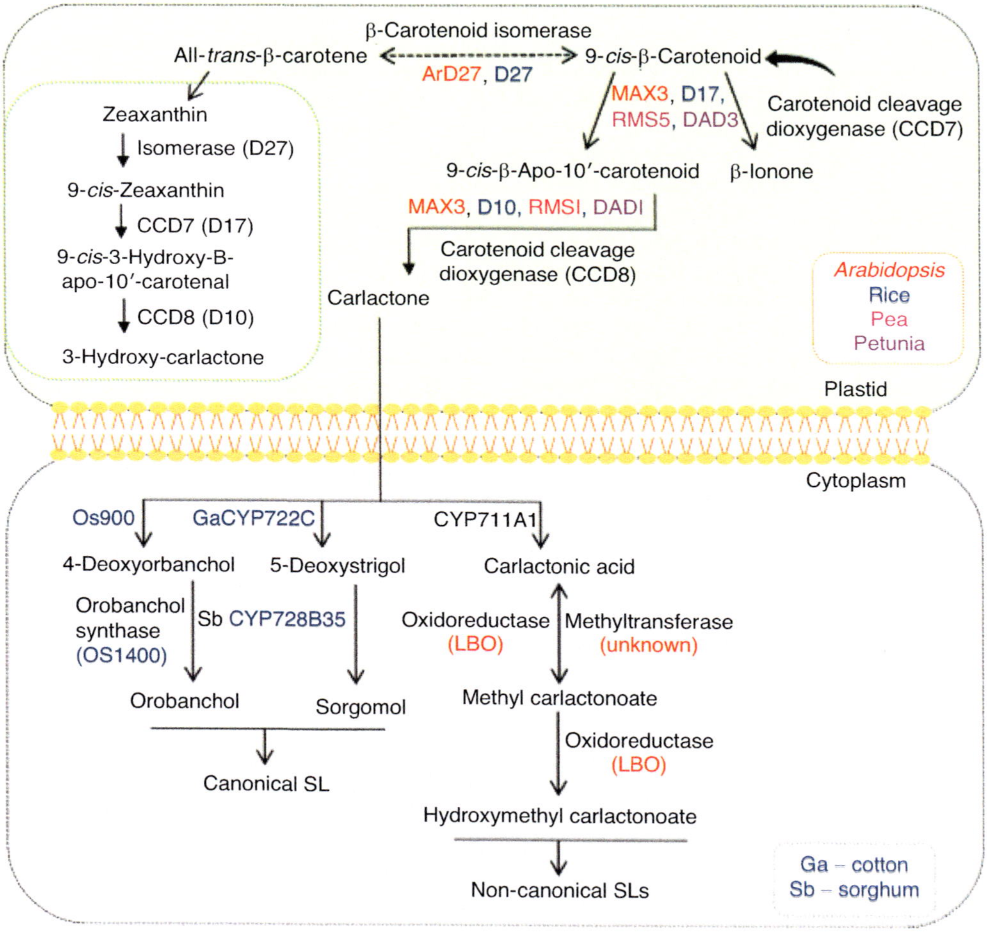

Fig. 15.1. The SL-biosynthesis pathway. All-*trans*-β-carotene undergoes many enzymatic interactions with carlactone in plastids. In the cytoplasm, carlactone is converted into carlactonic acid, which creates different kinds of SLs. Both canonical and non-canonical SLs are depicted in the figure. In *Arabidopsis* (red), peas (pink), petunias (purple), and rice (dark blue), the genes involved in SL biosynthesis have been identified. Novel genes and new pathways have been identified in recent studies employing transcriptomics and reverse genetic techniques, which are indicated in blue color.

mineral nutrients also affect their biosynthesis pathways (Vega *et al.*, 2019).

- *Feedback regulation.* SLs can be kept within ideal ranges by using a feedback system that allows SLs to control their own production (Waters *et al.*, 2017; Mashiguchi *et al.*, 2021).
- *Temperature.* SL synthesis is influenced by temperature, and different plant species have different ideal temperatures (Chi *et al.*, 2021; Zhang *et al.*, 2024). For example, SLs lower abscisic acid (ABA) levels, saving the development of dormant *Phelipanche ramosa* seed during warm stratification (Lechat *et al.*, 2015).
- *Light.* One of the main regulators of SL production is light. Genes involved in SL biosynthesis are regulated by cryptochromes and phytochromes (Ge *et al.*, 2022).
- *Developmental stages.* SL biosynthesis is controlled at many stages of development, including flowering, seed germination, and seed establishment (Soliman *et al.*, 2022).
- *Transcriptional regulation.* The expression of genes involved in SL biosynthesis is regulated by transcriptional factors, including MAX2 and SMXL6, −7, and −8 (Wang *et al.*, 2020).

15.2.3 Signaling pathways involved in SL regulation

SL activity is regulated by signal transduction mechanisms that are fundamentally similar to those of other phytohormones, especially in their dependence on the ubiquitin–proteasome system for the specific degradation of repressor proteins (Korek and Marzec, 2023). Key proteins involved in SL production and signaling were identified through studies conducted on mutant plants of *Oryza sativa* and *Arabidopsis thaliana* (Marzec, 2016a, b). The SL signal is transduced by the breakdown of repressor protein, as with the majority of phytohormones. The receptor (AtD14/OsD14), a member of the α/β-hydrolase protein family, recognizes and binds the SL molecules as the initial stage in cascade perception (Marzec *et al.*, 2016). The conformation modifications because of this contact are essential for the receptor's interaction with other signaling complex components (Marzec and Brewer, 2019). The SL repressor (AtSMXL6, -7,-8/OsD53) and the F-box protein (AtMAX2/OsD3) from the SKP1–CULLIN–F-BOX complex (SCF) can be bound by the receptor with a changed conformation (Chevalier *et al.*, 2014). SL-related transcription factors are then activated as a result of the SL repressor's degradation (Zhou *et al.*, 2013). Figure 15.2 summarizes this signaling pathway involved in SL regulation, showing the series of molecular processes from transcriptional activation to SL perception.

15.2.4 Role of SLs in plant–fungus interactions

SLs are secreted from the roots and work as contact chemicals with seeds, root parasites, and symbiotic arbuscular mycorrhizal fungi (AMF) (Mitra *et al.*, 2021). In the interactions between parasitic plants such as broomrapes (*Orobanche* and *Phelipanche* spp.) and witchweeds (*Striga* spp.), SLs are essential mediators. SLs generated by the roots of the host plant can be recognized by and respond to these parasitic plants (Mansoor *et al.*, 2023). To promote symbiotic relationships between plants and AMF, SLs are essential. Under biotic stress conditions, this symbiosis can enhance absorption of nutrients by promoting hyphal branching (Soliman *et al.*, 2022). For instance, in members of the Orobanchaceae, which comprises the genera *Phelipanche*, *Striga*, and *Orobanche*, the mycorrhizae reduced the predominance of parasite plant roots (López-Ráez *et al.*, 2011). Recent studies also suggest that SLs may have a role in plant defense mechanisms against weeds and fungus (Mansoor *et al.*, 2023).

SLs support plant defense processes such as the stimulation of systemic defense systems against pathogens and the induction of secondary metabolites in plants (Poveda *et al.*, 2020). According to Kalliola *et al.* (2020), SLs help plants defend themselves against pests and pathogens. They can prepare plants for greater resistance by promoting the synthesis of defense-related genes (Kalliola *et al.*, 2020) and phytoalexin (Chaki *et al.*, 2020). SLs were discovered to positively regulate tomato defense against root-knot nematodes (Xu *et al.*, 2019) and in rice against *Magnaporthe oryzae* (Nasir

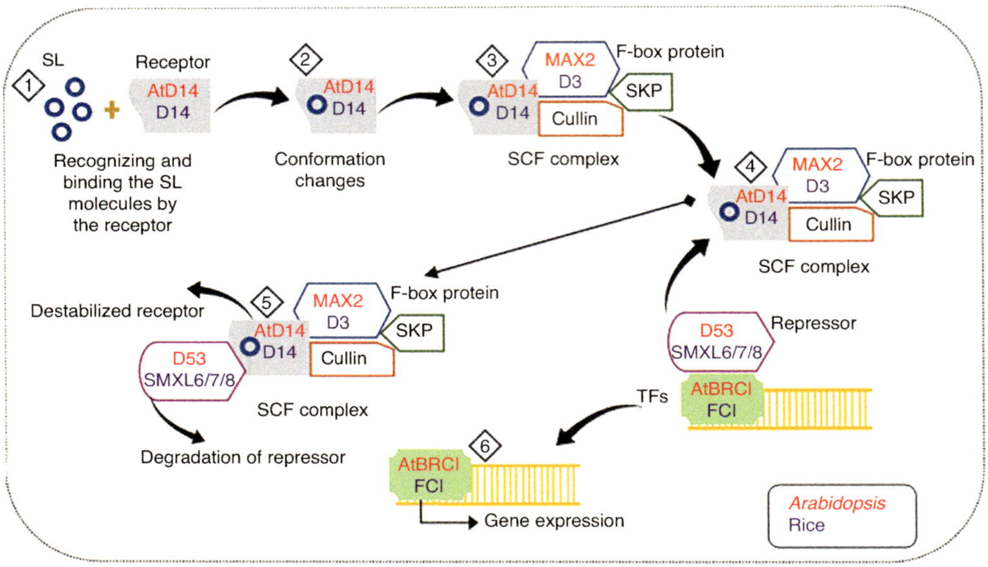

Fig. 15.2. Schematic diagram illustrating the strigolactone (SL) signaling pathway. (1) SL perception starts when the receptor recognizes and binds SL molecules. (2) As a result of structural changes in its structure, the receptor hydrolyses SL molecules. (3) This changed conformation allows the receptor to bind with an F-box protein, which then joins SKP, before cullin. (4) The SL repressor binds with the SCF complex. (5) Due to its structural alteration, the receptor is destabilized and the repressor is degraded. (6) The repressor's degradation allows the activation of SL-responsive transcription factors (TFs). Components of the SL signaling pathway are color coded by species origin: red denotes elements identified in *Arabidopsis thaliana*, while purple represents those characterized in *Oryza sativa* (rice).

et al., 2019). SLs have been shown to affect the production and signaling of auxin, cytokinins and gibberellins, among other plant hormones (Korek and Marzec, 2023). SLs may affect the immune responses of plants or the composition of the root microbiome. Thus, these effects might indirectly affect the ability of the plant to show resistance against pathogens (Rubiales *et al.*, 2018; Escudero Martinez *et al.*, 2019).

15.3 Strigolactones and Fungal Disease Mitigation

SLs are considered an important phytohormone group that not only help in plant formation processes but also initiate the synergistic association with some fungal species (Aliche *et al.*, 2020). SLs play a major role in the reduction of fungal growth and have the ability to induce plant defense mechanisms to resist harmful fungal infections (Belmondo *et al.*, 2017). SLs play a major role in maintenance of the relationship between host and parasitic plants (*Orobanche* and *Striga*) (Mansoor *et al.*, 2023). As promoters of hyphal branching and by improving the uptake of nutrients, SLs are also critical in enhancing the synergistic interaction between plants and AMF (Soliman *et al.*, 2022; Boyno *et al.*, 2023). SLs play a critical role in the suppression of various destructive and harmful fungal diseases including *Botrytis cinerea*, *Meloidogyne incognita*, and *Alternaria alternata* (Lahari *et al.*, 2019). SLs have ability to trigger induced systemic resistance in plants against phytopathogens by turning on the signaling of salicylic acid (SA) and jasmonic acid (JA) (Xu *et al.*, 2019; Wang *et al.*, 2020; Boyno *et al.*, 2023).

15.3.1 Suppression of fungal growth and sporulation

The artificial SL analog GR24 profoundly inhibits radial growth of fungus and massively induces hyphal branching of *Colletotrichum acutatum* (Dor *et al.*, 2011). Marzec (2016a, b) noted that GR24 significantly suppressed the fungal radial growth and enhanced hyphal branching in *Sclerotinia sclerotiorum*. The action of SLs toward fungal disease resistance has been illustrated in various plant species (Table 15.1).

15.3.2 Induction of plant defense responses

SLs play a key role in controlling plants' resistance to pathogens. Intriguingly, transcription factors associated with plant responses to infections caused by pathogens recognize sequences found in the promoters of genes involved in SL production (Marzec, 2016a). Furthermore, hormones such as SA, ethylene, and JA involved in plant defense modulate the expression of SL-biosynthesis genes in *O. sativa* and *A. thaliana* (Marzec and Muszynska, 2015). Multiple plant hormones, including auxin, gibberellins, ABA, ethylene, JA, SA, and SLs, are now understood to be pivotal parts of the signaling pathways of plant defense responses to phytopathogens (Robert-Seilaniantz *et al.*, 2011). Increased biotic stress tolerance in plants results from suppression of invasive microbe growth, stomatal closure, upregulation of hydrogen peroxide, and crosstalk with other defense hormones (Francis *et al.*, 2024). Table 15.2 presents mechanisms through which disease resistance is exerted by SLs in different plant species.

15.3.3 Modulation of plant–fungus symbiosis

The region where the roots interact with AMF is known as the rhizosphere. According to Faizan *et al.* (2020),the roots of the plant liberate SL phytohormones into the rhizosphere, which are necessary for a strong association of AMF. AMF symbiosis is known as a powerful interaction, which is basically a synergistic association between fungus and plants. According to Bonfante and Genre (2010), SLs play an important role in this mutualistic interaction. The utilization of artificial strigolactones (e.g. GR24) has been shown to enhance synergistic interaction in peas and petunias (Rozpądek *et al.*, 2018; Breuillin *et al.*, 2010). They also observed that treatments of *Mucor* sp. helped in the strong development of *Arabidopsis arenosa* and *Arabidopsis thaliana* (Rozpądek *et al.*, 2018).

15.4 Strigolactone-mediated Disease-resistance Mechanisms

SLs have a major role in the induction of plant defense mechanisms by enhancing resistance to destructive phytopathogens. Due to these properties, SLs are garnering attention in the field of plant pathology. Plant defense mechanisms include the production of antimicrobial compounds, regulation of plant hormone signaling pathways, and activation of defense-related genes in plants.

Table 15.1. Effect of strigolactones (SLs) on fungal disease resistance in different plant species.

Plant	Response	References
Tomato (*Solanum lycopersicum*)	SL helps tomato defense against foliar necrotrophic fungus	Blake *et al.* (2016)
Pea (*Pisum sativum*)	Increase resistance against the pathogenic fungus *Fusarium oxysporum*	Foo *et al.* (2016)
Thale cress (*Arabidopsis thaliana*)	Inhibit the growth of the fungus *Mucor* sp.	Rozpądek *et al.* (2018)
Brachypodium (*Brachypodium*)	Provide resistance against *Fusarium* head blight	Changenet (2018)
Rice (*Oryza sativa*)	Modulate defense against *Magnaporthe oryzae*	Nasir *et al.* (2019)

Table 15.2. Role of strigolactones (SLs) in disease-resistance development.

Crop	Disease	Pathogen	Role of SLs	Reference
Tomato (*Solanum lycopersicum*)	Root-knot	Root-knot nematodes	By suppression of MYC2, SLs provide resistance against root-knot disease in tomato	Xu *et al*. (2019)
Cotton (*Gossypium hirsutum*)	*Verticillium* wilt	*Verticillium dahliae*	By causing ROS accumulation and interacting with the ABA and JA signaling pathways, SLs enhance cotton's resistance to *Verticillium* wilt	Yi *et al*. (2023)
Arabidopsis thaliana	Gray mold	*Botrytis cinerea*	By boosting plant immunity, SL signaling improves disease resistance. The artificial SL analog *rac*-GR24 boosts resistance in *B. cinerea* by increasing the aggregation of camalexin	Fujita *et al*. (2024)
Rice (*Oryza sativa*)	Rice blast	*Pyricularia oryzae* (syn. *Magnaporthe oryzae*)	SLs provide resistance against the *P. oryzae* by triggering the JA and sugar signaling pathways	Lahari *et al*. (2024)
Tomato (*Solanum lycopersicum*)	Foliar blight	*Botrytis cinerea* and *Alternaria alternata*	SLs play a part in defense responses against *B. cinerea* and *A. alternata*	Torres-Vera *et al*. (2016)

ABA, abscisic acid; JA, jasmonate; ROS, reactive oxygen species.

15.4.1 Activation of plant immunity genes

SLs have the ability to withstand the transcriptional stressors released from phytopathogens due to the presence of PROTEINASE INHIBITOR II (PIN II) (Torres-Vera *et al.*, 2016). According to Fujita *et al.* (2024), *rac*-GR24 boosts resistance against the pathogenic fungus *B. cinerea* by activation of the ethylene and camalexin signaling pathways. Immunopriming boosted the expression of genes associated with ethylene, the gene that generates camalexin, and there was accumulation of camalexin after pathogen infection in plants treated with *rac*-GR24, suggesting that SL-mediated immune priming is an effective way to address different resistance mechanisms in plant defense systems.

15.4.2 Production of antimicrobial compounds

SLs are strong allelochemicals that are known to stimulate hyphal branching in AMF and promote parasitic plant germination. However, they also help the plant defend itself by inducing secondary metabolites that are antimicrobial. SLs are among the best examples of allelochemicals (Soto-Cruz *et al.*, 2021). Lahari *et al.* (2024) demonstrated that rice seedlings downregulate the SL-signaling pathway in response to *P. oryzae* infection, which results in a higher accumulation of flavonoids, phytoalexins, soluble sugars, and JA.

15.4.3 Regulation of hormone signaling pathways

Kusajima *et al.* (2022) conducted a study showing that SLs did not induce systemic acquired resistance, but they did favorably affect SA-mediated signaling during basal defense responses. Based on these results, it appears that SLs prime *A. thaliana* by promoting SA-mediated disease resistance. Tomato (*Solanum lycopersicum*) resistance to the fungal pathogens *A. alternata* and *B. cinerea* is strengthened by SL signaling. Different amounts of SA, JA and ABA are linked to this impact (Xu *et al.*, 2019). Thus, ABA, JA, and SA are important components of plant defensive responses (Robert-Seilaniantz *et al.*, 2011).

15.5 Strigolactone-interacting Fungi

SLs are considered a versatile phytohormonal group that not only help in plant body regulation and maintenance but also act as a signaling molecule to indicate the presence of microorganisms such as bacteria and fungi, and initiate the interaction between plants and these organisms (bacteria and fungi). Mostly, they interact with fungi such as AMF, root-associated fungi (RAF) and various pathogenic fungi (e.g. *Fusarium*, *Phytophthora*, *Rhizoctonia*). This mutualistic interaction of SL and fungi enhances plant productivity, changes the rhizosphere structure, promotes shoot and root growth, inhibits pathogenic fungal attack, and stimulates the growth of beneficial fungi in the rhizosphere (Begum *et al.*, 2023; Kee *et al.*, 2023; Francis *et al.*, 2025). However, the effects of each of these fungal groups, AMF, RAF, and pathogenic fungi, are different, as outlined below.

15.5.1 AMF

The mutualistic association of SLs and AMF is assumed to have been an effective primitive evolutionary adaptation for over 400 million years, similar to the association between plants and AMF (Liu and Chen, 2024; Liu *et al.*, 2025; Mishra *et al.*, 2017). In this mutualistic association, SLs are secreted by plants into the rhizosphere in response to biotic and abiotic stress. In the rhizosphere, SLs chemotactically attract AMF nearby the plant and stimulate AMF colonization by increasing the spore germination rate and hyphal branching system (Boyno *et al.*, 2023). Once the AMF have successfully colonized, they begin to penetrate cortical cells of the roots where they form arbuscles, the main nutrient-exchange sites. These arbuscles are the major symbiotic structures and boost essential nutrient levels (e.g. nitrogen, phosphorus, zinc, potassium, and various other micronutrients) in plants and improve the environmental stress tolerance ability of the plant (Bhantana *et al.*, 2021; Wahab *et al.*, 2023). In return, the plant provides the AMF with carbohydrates and lipids at the site of association. This synergistic association also enhances the soil fertility (Wahab *et al.*, 2023). Mitra *et al.* (2021) elaborated the mutualistic association between SLs and AMF as shown in Fig. 15.3.

There are many natural SLs, such as strigol and orobanchol and other canonical and non-canonical SLs that induce positive changes in AMF association; Akiyama *et al.* (2005) highlighted the effects of natural SLs such as 5-deoxystrigol in the AMF *Gigaspora margarita* that induce extensive hyphal branching. Synthetic SL analogs such as GR24, GR5, GR7, and others have also been shown to be associated with AMF and perform various activities; for example, Besserer *et al.* (2008) showed that GR24 boosted mitosis and growth of AMF by increasing mitochondrial activity. Similarly, there is much literature available on natural and synthetic SL and AMF association responses; some of these studies are given in Table 15.3. After successful AMF colonization, AMF regulate the activity of SL biosynthesis through negative feedback mechanisms (Fadiji *et al.*, 2023).

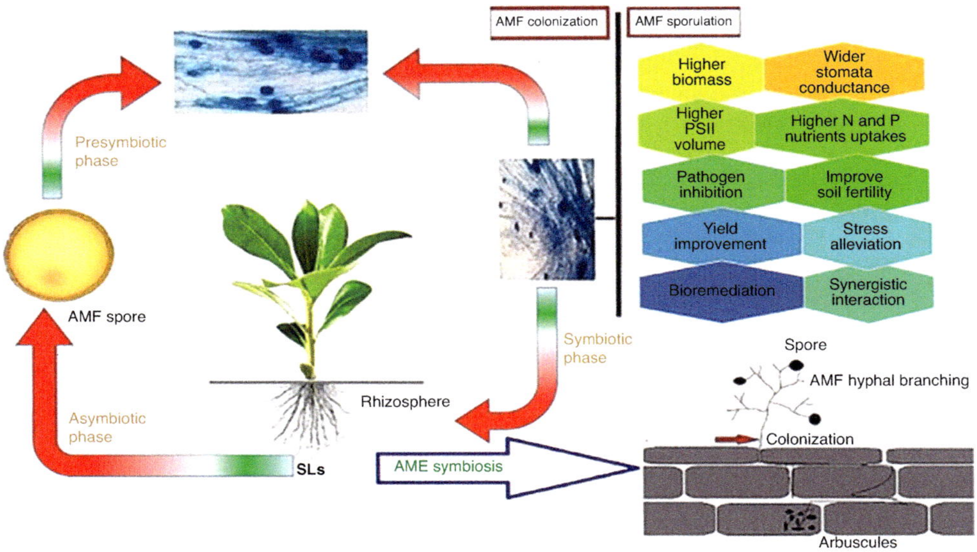

Fig. 15.3. Schematic diagram of the mutualistic association between AMF and strigolactones (Mitra *et al.*, 2021). AMF, arbuscular mycorrhizal fungi; N, nitrogen; P, phosphate; PSII, photosystem II. Figure used under license CC-BY-NC-ND.

Table 15.3. Natural and synthetic strigolactones (SLs) and arbuscular mycorrhizal fungi (AMF) association responses.

Type of SL	AMF species	Effect of SL interaction	Reference
Medicaol	*Gigaspora margarita*	Induces short tertiary hyphae	Tokunaga *et al.* (2015)
GR24 and 5-deoxystrigol	*Rhizophagus irregularis*	Enhances colonization, hyphal length, and branching	Klein *et al.* (2024)
GR24	*Gigaspora margarita*	Stimulates spore germination and host-root colonization	Kountche *et al.* (2018)
Apocarotenoid strigolactone	*Rhizophagus irregularis*	Boosts root colonization	Servanté *et al.* (2024)

15.5.2 RAF

As well as AMF, SLs are also involved in significant symbiotic relationships with RAF such as commensalism, parasitism, and amensalism. RAF mainly fall in the category of endophytes. This interaction RAF and SLs regulates the activity of host-root colonization and maintains equilibrium in rhizo-microbiome concentration (Omoarelojie and van Staden, 2020; Mattoo and Nonzom, 2021). Additionally in this association, SLs play a pivotal role as an intermediator or signaling molecule between the host and RAF.

This association also helps in mitigating the level of biotic or abiotic stress. Boorboori and Zhang (2022) reported the activity of *Serendipita indica*, an endophytic fungus, in reducing the effects of abiotic stresses such as waterlogging, salinity, and drought. Additionally, it also helps in stimulating plant growth by enhancing primary metabolite activity. For instance, SLs act as a regulatory molecule in the association between *A. thaliana* and *Mucor* sp., enhancing plant growth and development and mitigating heavy-metal stress (Rozpądek *et al.*, 2018). Despite these results, in-depth knowledge about non-mycorrhizal RAF and SLs is still required.

15.5.3 Pathogenic fungi

In case of pathogenic fungi such as *Fusarium*, *Phytophthora*, and *Rhizoctonia*, SLs perform dual activity, either escalating the beneficial effects or decreasing the harmful effects on host due to the infectious fungus association. Varying according to the circumstances, SL and pathogenic fungi interaction is a complicated type of interaction in which SLs act as a protective as well as a hazardous agent. There is no doubt that SLs encourage a significant host–fungi mutualistic association, but in some cases SLs unintentionally promote pathogen growth in the rhizosphere, which may damage the host plant from the roots to the shoots and make the plant tender (Pandey *et al.*, 2016; Mapuranga *et al.*, 2022).

Despite this, SLs can also stimulate the defense response in the plant body against pathogens. AbuQamar *et al.* (2017) and Pusztahelyi *et al.* (2017) highlighted the activity of SLs in case of *B. cinerea* attack: SLs inhibit the pathogen growth through various defense signaling mechanisms including immune signaling modulation and oxidation–reduction reactions. Most importantly, SLs inhibit pathogenic attack by *Fusarium*, *Rhizoctonia* spp., and *Phytophthora* spp. through immune signaling modulation pathways, which boost the plant's immunity and restrict growth and invasion of the pathogen (Torres-Vera, 2017; Wu *et al.*, 2022).

SLs restrict pathogen attack and their colonization in the rhizosphere by changing the root structure of the host plant. They promote the dense growth of root hairs and lateral roots in the vicinity of the rhizosphere, blocking the access of pathogens to the host plant. Some pathogens have adopted various innovative ways to overcome this effect and enter the plant cell and attack the host plant body. Despite this, SLs play an important role in sustainable agriculture, and various integrated methods are in practice to enhance the activity of targeted genes of SL signaling and biosynthesis. Such practices support the development of various disease-resistant varieties and various advanced sustainable agriculture strategies.

15.6 Application of Strigolactones in Disease Management

SLs are now well-known phytohormones that help in various developmental processes such as plant–microbe association, plant architectural modifications, and activation of various other necessary phytohormones (auxin, cytokinins, gibberellins). SLs play pivotal roles in the development of plants, as shown in Fig. 15.4.

SLs also help in mitigating the effects of biotic or abiotic stressors (Kapoor *et al.*, 2024). Most importantly, SLs support various disease-management strategies in multiple ways such as the establishment of beneficial mutualistic associations and inhibition of pathogenic associations (Aliche *et al.*, 2020). In recent research, these properties of SLs have been utilized for sustainable agriculture including crop productivity, disease management, and the development of resistant varieties though various means, as discussed below.

15.6.1 Exogenous application of SLs

Exogenous application of SLs is an innovative technique to promote plant productivity and defense responses against pathogenic attack. In the case of pathogen attack, SLs stimulate various hormone-based signaling pathways including the JA and SA pathways, which prompt their respective systemic immune responses in plants against pathogens (Kusajima *et al.*, 2022; Selwal *et al.*, 2023). Similarly, exogenous SLs chemotactically stimulate the nearby beneficial microbial community to form mutualistic associations with the host plant, while also restricting the interaction and growth of pathogenic microbes (Bukhat *et al.*, 2020). Additionally, SL application improves the root architecture, enriches nutrients uptake, and hinders the interactions of root-associated pathogens (Andreo-Jimenez *et al.*,

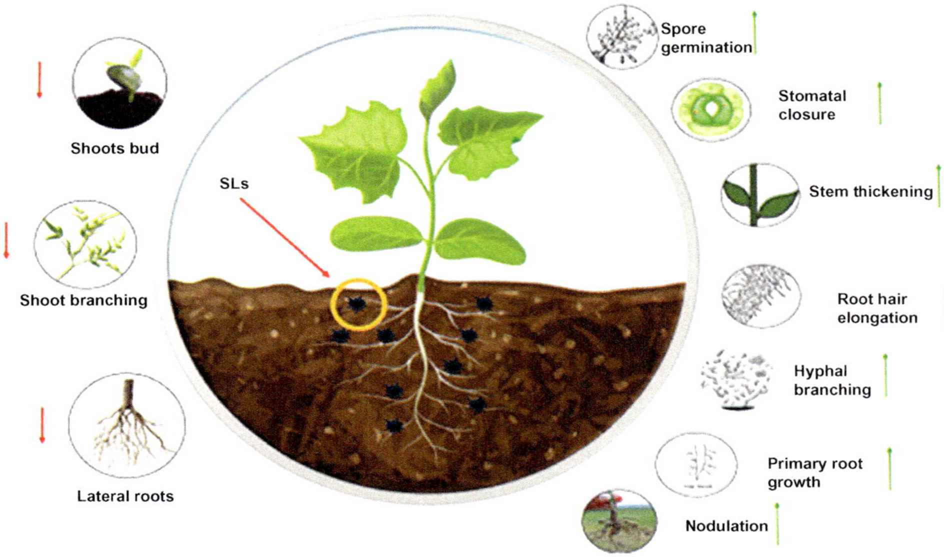

Fig. 15.4. The pivotal roles of strigolactones (SLs) in the development of plants.

2015). As Dor *et al.* (2011) noted, application of the synthetic SL GR24 enhanced the activity of defensive enzymes to significantly diminish fungal proliferation. Foliar applications of SLs also change the plant structure and increase resistance against pests, herbivores, and pathogens (Guo *et al.*, 2023; Mehrabi *et al.*, 2024).

SLs also play a pivotal role in mitigating the effect of abiotic stresses including heavy metals, drought, and salinity. In response to abiotic stress, SLs control stomatal activity (stomatal closure, stomatal conductivity, and density of stomata), regulate chlorophyll content, increase cell number, elongate roots and shoots, increase antioxidant content, and enhance the spore germination of beneficial microbes (Alvi *et al.*, 2022; Sattar *et al.*, 2022; Shu *et al.*, 2024). SLs also stimulate a strong AMF association, which increases the nutrient and moisture content and produces secondary metabolites to mitigate the pathogen invasion (Pusztahelyi *et al.*, 2017). SLs also act synergistically with auxin and cytokinins to maintain shoot and root development, thereby protecting the plant from pathogen attack (Mehmood *et al.*, 2025).

15.6.2 Genetic engineering of SL production

Advances in genetic engineering allow manipulation and integration of the genetic machinery of an organism. These techniques are being used to try and boost SL synthesis levels in plants by modifying SL-biosynthesis genes, which in return increases the plant's immunity level. This can help the plant to overcome biotic and abiotic stresses (Chesterfield *et al.*, 2020; van Esse *et al.*, 2020). Through this kind of modification, the plant root architecture may also be changed, which may induce a strong plant–microbe interaction to help tackle the biotic and abiotic pressures (Begum *et al.*, 2023). For instance, Xu *et al.* (2019) and Lahari *et al.* (2019) altered SL-related genes in rice and tomato plants to increase the SL levels in these transgenic plants, which in return showed higher resistance to root-knot nematodes and *Fusarium* wilt. However, such modifications require accurate genome-editing technologies such as clustered regularly interspaced short palindromic repeats/CRISPR-associated protein 9 (CRISPR-Cas9), transcription activator-like effector nucleases (TALENs), and zinc-finger nucleases to target specific SL genes to improve the plant–microbe interaction (Hao *et al.*, 2023). Most importantly, SL biosynthesis is regulated by both positive and negative feedback mechanisms. Positive feedback helps to upregulate the production of SLs in the case of biotic or abiotic stress, while negative feedback downregulates the production of SLs to reduce the attraction of pathogens to the host plant. Hence, both feedback mechanisms are specifically needed for crop protection (Srivastava *et al.*, 2022; Jain *et al.*, 2025).

15.6.3 Integration with other disease-management strategies

There are many diseases-management strategies are in practice, designed to prevent, eradicate, and control various diseases. These include cultural, biological, and chemical methods (Haq and Ijaz, 2020). Various pathogens have adapted to make them resistant to a specific disease-management strategy, and therefore another approach may be needed using integrated disease-management strategies, which combine two or more of the above strategies, making them more viable (Razdan and Sabitha, 2009; Pandey *et al.*, 2016).

There are many integrated disease-management strategies applicable to SLs to enhance their activity. Integrating SL disease-management strategies with traditional or advanced strategies enhances the protection of the host plant and may aid sustainable agriculture. For instance, the synergistic application of SLs with pesticides or fungicides amplifies their ability to hinder the pathogen attack by disturbing the pathogen's life cycle and structure, while also reducing the use of chemicals in the soil for sustainable agriculture (Shahid *et al.*, 2024). Correspondingly, the co-application of microbial inoculants with SLs enriches the soil with beneficial nutrients that interact with the microbial community and suppress disease invasion over the long term (Tsuzuki *et al.*, 2016; Rostami and Azhdarpoor, 2019; Chattaraj *et al.*, 2025). Likewise, integrated application of SLs and use of disease-resistant varieties provides a versatile defense mechanism that increases the plant's resistance (Han *et al.*, 2025). SLs also changes the root architecture, enrich essential nutrient uptake levels, and support the activity of microbes via secreted phytohormones, thereby contributing to soil health and overall plant immunity (Boyno *et al.*, 2023). Use of SLs together with integrated pest management strategies is thus a modern approach that promotes beneficial plant–microbe interactions and reduces the ecological impact on crops as a crop-protection strategy.

15.7 Conclusion and Future Directions

SLs are essential signaling phytohormones that help in the regulation of plant–pathogenic fungi interactions. They play a pivotal role in boosting plant immunity by activating various primary and secondary host defense mechanisms. These defense mechanisms promote the growth of beneficial fungi (AMF and RAF), as well as inhibit the sporulation, growth, and development of various pathogenic fungi such as *Fusarium*, *Rhizoctonia*, and *Phytophthora* spp. Generally, SLs are a complex of signals that harmonize various disease-resistance processes (in response to biotic and abiotic factors), transcriptional regulators, and many other primary phytohormones pathways (e.g. auxin, cytokinins).

Various analyses have shown that SLs protect plants through a number of defensive mechanisms such as activating immunity genes, modifying hormonal signals, and initiating favorable associations such as with AMF. SLs are also strongly associated with a broad range of fungal species including symbiotic, root-associated, and various pathogenic fungi.

Most importantly, SLs can be used to significantly decrease the spread of plant diseases in the plant by adopting various strategies, including extraneous application of SLs, genetic modification of SL-biosynthesis genes, and various integration methods for sustainable agriculture.

Despite these promising strategies, the use of SLs faces various challenges, and a deeper mechanistic understanding of SL-mediated disease resistance is necessary. Future research should aim to identify novel SL-biosynthetic and SL-signaling genes, characterize their functions across plant species, and explore their interactions with a wider range of pathogens. Additionally, the development of scalable, economically viable SL-based technologies—through selective breeding, gene editing, or microbial formulations—is essential for practical agricultural deployment.

Addressing these challenges will enable the wider adoption of SL-based approaches within integrated disease-management systems, thereby promoting crop resilience, reducing chemical inputs, and advancing the sustainability of global agriculture.

References

AbuQamar, S., Moustafa, K. and Tran, L.S. (2017) Mechanisms and strategies of plant defense against *Botrytis cinerea*. *Critical Reviews in Biotechnology* 37(2), 262–274. DOI: 10.1080/07388551.2016.1271767.

Akiyama, K., Matsuzaki, K. and Hayashi, H. (2005) Plant sesquiterpenes induce hyphal branching in arbuscular mycorrhizal fungi. *Nature* 435(7043), 824–827. DOI: 10.1038/nature03608.

Aliche, E.B., Screpanti, C., de Mesmaeker, A., Munnik, T. and Bouwmeester, H.J. (2020) Science and application of strigolactones. *New Phytologist* 227(4), 1001–1011. DOI: 10.1111/nph.16489.

Alvi, A.F., Sehar, Z., Fatma, M., Masood, A. and Khan, N.A. (2022) Strigolactone: An emerging growth regulator for developing resilience in plants. *Plants* 11(19), 2604. DOI: 10.3390/plants11192604.

Andreo-Jimenez, B., Ruyter-Spira, C., Bouwmeester, H.J. and Lopez-Raez, J.A. (2015) Ecological relevance of strigolactones in nutrient uptake and other abiotic stresses, and in plant–microbe interactions below-ground. *Plant and Soil* 394(1–2), 1–19. DOI: 10.1007/s11104-015-2544-z.

Atiq, M., Naeem, I., Sahi, S.T., Rajput, N.A., Haider, E. *et al.* (2020) Nanoparticles: A safe way towards fungal diseases. *Archives of Phytopathology and Plant Protection* 53(17–18), 781–792. DOI: 10.1080/03235408.2020.1792599.

Baz, L., Mori, N., Mi, J., Jamil, M., Kountche, B.A. *et al.* (2018) 3-hydroxycarlactone, a novel product of the strigolactone biosynthesis core pathway. *Molecular Plant* 11(10), 1312–1314. DOI: 10.1016/j.molp.2018.06.008.

Begum, N., Zhao, T., Antwi-Boasiako, A., Athar, H.U.R., Zhang, L. *et al.* (2023) Role of strigolactones in rhizobiology: Plant–microbe interactions. In: Ozturk, M., Bhat, R.A., Ashraf, M., Tonelli, F.M.P., Unal, B.T. *et al.* (eds) *Phytohormones and Stress Responsive Secondary Metabolites*. Academic Press, Cambridge, Massachusetts, pp. 13–25.

Belmondo, S., Marschall, R., Tudzynski, P., López Ráez, J.A., Artuso, E. *et al.* (2017) Identification of genes involved in fungal responses to strigolactones using mutants from fungal pathogens. *Current Genetics* 63(2), 201–213. DOI: 10.1007/s00294-016-0626-y.

Besserer, A., Bécard, G., Jauneau, A., Roux, C. and Séjalon-Delmas, N. (2008) GR24, a synthetic analog of strigolactones, stimulates the mitosis and growth of the arbuscular mycorrhizal fungus *Gigaspora rosea* by boosting its energy metabolism. *Plant Physiology* 148(1), 402–413. DOI: 10.1104/pp.108.121400.

Bhantana, P., Rana, M.S., Sun, X., Moussa, M.G., Saleem, M.H. *et al.* (2021) Arbuscular mycorrhizal fungi and its major role in plant growth, zinc nutrition, phosphorous regulation and phytoremediation. *Symbiosis* 84(1), 19–37. DOI: 10.1007/s13199-021-00756-6.

Blake, S.N., Barry, K.M., Gill, W.M., Reid, J.B. and Foo, E. (2016) The role of strigolactones and ethylene in disease caused by *Pythium irregulare*. *Molecular Plant Pathology* 17(5), 680–690. DOI: 10.1111/mpp.12320.

Bonfante, P. and Genre, A. (2010) Mechanisms underlying beneficial plant–fungus interactions in mycorrhizal symbiosis. *Nature Communications* 1(1), 48. DOI: 10.1038/ncomms1046.

Boorboori, M.R. and Zhang, H.Y. (2022) The role of *Serendipita indica* (*Piriformospora indica*) in improving plant resistance to drought and salinity stresses. *Biology* 11(7), 952. DOI: 10.3390/biology11070952.

Boyno, G., Rezaee Danesh, Y., Demir, S., Teniz, N., Mulet, J.M. *et al.* (2023) The complex interplay between arbuscular mycorrhizal fungi and strigolactone: Mechanisms, sinergies, applications and future directions. *International Journal of Molecular Sciences* 24(23), 16774. DOI: 10.3390/ijms242316774.

Breuillin, F., Schramm, J., Hajirezaei, M., Ahkami, A., Favre, P. *et al.* (2010) Phosphate systemically inhibits development of arbuscular mycorrhiza in *Petunia hybrida* and represses genes involved in mycorrhizal functioning. *Plant Journal* 64(6), 1002–1017. DOI: 10.1111/j.1365-313X.2010.04385.x.

Bukhat, S., Imran, A., Javaid, S., Shahid, M., Majeed, A. *et al.* (2020) Communication of plants with microbial world: Exploring the regulatory networks for PGPR mediated defense signaling. *Microbiological Research* 238, 126486. DOI: 10.1016/j.micres.2020.126486.

Chaki, M., Begara-Morales, J.C. and Barroso, J.B. (2020) Oxidative stress in plants. *Antioxidants* 9(6), 481. DOI: 10.3390/antiox9060481.

Changenet, V. (2018) Towards new roles for cytochrome p450s and strigolactones in *Fusarium* head blight of *Brachypodium distachyon*. PhD thesis, Université Paris Saclay (COmUE), Gif-sur-Yvette, France.

Chattaraj, S., Mitra, D., Ganguly, A., Das Mohapatra, P.K. and Thatoi, H. (2025) Harnessing microbe-based soil inoculums, strigolactones, and nanotechnology for sustainable agriculture: Mechanisms, innovations, and challenges. *Pedosphere* (in Press). DOI: 10.1016/j.pedsph.2025.01.002.

Chelladurai, K., Sujatha, N. and Valli, L.N. (2023) Fungal infection in plant leaves—a review. *E3S Web of Conferences* 399, 02010. DOI: 10.1051/e3sconf/202339902010.

Chesterfield, R.J., Vickers, C.E. and Beveridge, C.A. (2020) Translation of strigolactones from plant hormone to agriculture: Achievements, future perspectives, and challenges. *Trends in Plant Science* 25(11), 1087–1106. DOI: 10.1016/j.tplants.2020.06.005.

Chevalier, F., Nieminen, K., Sánchez-Ferrero, J.C., Rodríguez, M.L., Chagoyen, M. *et al.* (2014) Strigolactone promotes degradation of DWARF14, an α/β hydrolase essential for strigolactone signaling in *Arabidopsis*. *Plant Cell* 26(3), 1134–1150. DOI: 10.1105/tpc.114.122903.

Chi, C., Xu, X., Wang, M., Zhang, H., Fang, P. *et al.* (2021) Strigolactones positively regulate abscisic acid-dependent heat and cold tolerance in tomato. *Horticulture Research* 8(1), 237. DOI: 10.1038/s41438-021-00668-y.

Clark, J. and Bennett, T. (2024) Cracking the enigma: Understanding strigolactone signalling in the rhizosphere. *Journal of Experimental Botany* 75(4), 1159–1173. DOI: 10.1093/jxb/erad335.

Dor, E., Joel, D.M., Kapulnik, Y., Koltai, H. and Hershenhorn, J. (2011) The synthetic strigolactone GR24 influences the growth pattern of phytopathogenic fungi. *Planta* 234(2), 419–427. DOI: 10.1007/s00425-011-1452-6.

Ekwomadu, T.I. and Mwanza, M. (2023) *Fusarium* fungi pathogens, identification, adverse effects, disease management, and global food security: A review of the latest research. *Agriculture* 13(9), 1810. DOI: 10.3390/agriculture13091810.

El-Baky, N.A. and Amara, A.A.A.F. (2021) Recent approaches towards control of fungal diseases in plants: An updated review. *Journal of Fungi* 7(11), 900. DOI: 10.3390/jof7110900.

Escudero Martinez, C.M., Guarneri, N., Overmars, H., van Schaik, C., Bouwmeester, H. *et al.* (2019) Distinct roles for strigolactones in cyst nematode parasitism of *Arabidopsis* roots. *European Journal of Plant Pathology* 154(1), 129–140. DOI: 10.1007/s10658-019-01691-5.

Fadiji, A.E., Yadav, A.N., Santoyo, G. and Babalola, O.O. (2023) Understanding the plant–microbe interactions in environments exposed to abiotic stresses: An overview. *Microbiological Research* 271, 127368. DOI: 10.1016/j.micres.2023.127368.

Faizan, M., Faraz, A., Sami, F., Siddiqui, H., Yusuf, M. *et al.* (2020) Role of strigolactones: Signalling and crosstalk with other phytohormones. *Open Life Sciences* 15(1), 217–228. DOI: 10.1515/biol-2020-0022.

Foo, E., Blake, S.N., Fisher, B.J., Smith, J.A. and Reid, J.B. (2016) The role of strigolactones during plant interactions with the pathogenic fungus *Fusarium oxysporum*. *Planta* 243(6), 1387–1396. DOI: 10.1007/s00425-015-2449-3.

Francis, B., Aravindakumar, C.T. and Simon, S. (2024) Regulatory role of strigolactones in biotic stress tolerance. In: Bashri, G., Hayat, S. and Bajguz, A. (eds) *Strigolactones: Synthesis, Application and Role in Plants*. Academic Press, London, pp. 189–200. DOI: 10.1016/B978-0-443-13521-7.00010-5.

Francis, B., Aravindakumar, C.T. and Simon, S. (2025) Strigolactones and plant defense: Protection against pests and pathogens. In: Aftab, T. and Wani, K.I. (eds) *Strigolactones: Emerging Plant Hormones*. Scrivener Publishing, Beverly, Massachusetts, pp. 281–300. DOI: 10.1002/9781394302826.ch13.

Fujita, M., Tanaka, T., Kusajima, M., Inoshima, K., Narita, F. *et al.* (2024) Enhanced disease resistance against *Botrytis cinerea* by strigolactone-mediated immune priming in *Arabidopsis thaliana*. *Journal of Pesticide Science* 49(3), 186–194. DOI: 10.1584/jpestics.D24-019.

Ge, S., He, L., Jin, L., Xia, X., Li, L. *et al.* (2022) Light-dependent activation of HY5 promotes mycorrhizal symbiosis in tomato by systemically regulating strigolactone biosynthesis. *New Phytologist* 233(4), 1900–1914. DOI: 10.1111/nph.17883.

Guo, S., Wei, X., Ma, B., Ma, Y., Yu, Z. *et al.* (2023) Foliar application of strigolactones improves the desiccation tolerance, grain yield and water use efficiency in dryland wheat through modulation of non-hydraulic root signals and antioxidant defense. *Stress Biology* 3(1), 54. DOI: 10.1007/s44154-023-00127-9.

Han, Y., Sun, Y., Wang, H., Li, H., Jiang, M. *et al.* (2025) Biosynthesis and signaling of strigolactones act synergistically with that of ABA and JA to enhance *Verticillium dahliae* resistance in cotton (*Gossypium hirsutum* L.). *Plant, Cell & Environment* 48(1), 571–586. DOI: 10.1111/pce.15148.

Hao, J., Yang, Y., Futrell, S., Kelly, E.A., Lorts, C.M. *et al.* (2023) CRISPR/Cas9-mediated mutagenesis of *Carotenoid Cleavage Dioxygenase (CCD)* genes in sorghum alters strigolactone biosynthesis and plant biotic interactions. *Phytobiomes Journal* 7(3), 339–351. DOI: 10.1094/PBIOMES-08-22-0053-R.

Haq, I.U. and Ijaz, S. (eds) (2020) *Plant Disease Management Strategies for Sustainable Agriculture Through Traditional and Modern Approaches*. Springer Nature, Cham, Switzerland.

Harrison, P.J., Newgas, S.A., Descombes, F., Shepherd, S.A., Thompson, A.J. *et al.* (2015) Biochemical characterization and selective inhibition of β-carotene *cis-trans* isomerase D27 and carotenoid cleavage dioxygenase CCD8 on the strigolactone biosynthetic pathway. *FEBS Journal* 282(20), 3986–4000. DOI: 10.1111/febs.13400.

Iqbal, M., Jützeler, M., França, S.C., Wäckers, F., Andreasson, E. *et al.* (2022) Bee-vectored *Aureobasidium pullulans* for biological control of gray mold in strawberry. *Phytopathology* 112(2), 232–237. DOI: 10.1094/PHYTO-05-21-0205-R.

Iqbal, M., Andreasson, E. and Stenberg, J.A. (2023) Biological control of strawberry diseases by *Aureobasidium pullulans* and sugar beet extract under field conditions. *Journal of Plant Pathology* 105(3), 933–941. DOI: 10.1007/s42161-023-01408-y.

Iqbal, M., Flöhr, A., Andreasson, E. and Stenberg, J.A. (2025) Breeding for integrated pest management (B-IPM): A new concept simultaneously optimising plant resistance and biocontrol. *Frontiers in Plant Science* 16, 1659069. DOI: 10.3389/fpls.2025.1659069.

Ito, S., Yamagami, D., Umehara, M., Hanada, A., Yoshida, S. *et al.* (2017) Regulation of strigolactone biosynthesis by gibberellin signaling. *Plant Physiology* 174(2), 1250–1259. DOI: 10.1104/pp.17.00301.

Jain, S., Kumar, R., Gunsola, D., Živković, S., Vasić, T. *et al.* (2025) Strigolactones and control of parasitic weeds. In: Aftab, T. and Wani, K.I. (eds) *Strigolactones: Emerging Plant Hormones*. Scrivener Publishing, Beverly, Massachusetts, pp. 261–279.

Kalliola, M., Jakobson, L., Davidsson, P., Pennanen, V., Waszczak, C. *et al.* (2020) Differential role of MAX2 and strigolactones in pathogen, ozone, and stomatal responses. *Plant Direct* 4(2), e00206. DOI: 10.1002/pld3.206.

Kapoor, R.T., Alam, P., Chen, Y. and Ahmad, P. (2024) Strigolactones in plants: From development to abiotic stress management. *Journal of Plant Growth Regulation* 43(3), 903–919. DOI: 10.1007/s00344-023-11148-z.

Kee, Y.J., Ogawa, S., Ichihashi, Y., Shirasu, K. and Yoshida, S. (2023) Strigolactones in rhizosphere communication: Multiple molecules with diverse functions. *Plant And Cell Physiology* 64(9), 955–966. DOI: 10.1093/pcp/pcad055.

Klein, M., Bisot, C., Oyarte Gálvez, L., Kokkoris, V., Shimizu, T.S. *et al.* (2024) The potential of strigolactones to shift competitive dynamics among two *Rhizophagus irregularis* strains. *Frontiers in Microbiology* 15, 1470469. DOI: 10.3389/fmicb.2024.1470469.

Korek, M. and Marzec, M. (2023) Strigolactones and abscisic acid interactions affect plant development and response to abiotic stresses. *BMC Plant Biology* 23(1), 314. DOI: 10.1186/s12870-023-04332-6.

Kountche, B.A., Novero, M., Jamil, M., Asami, T., Bonfante, P. *et al.* (2018) Effect of the strigolactone analogs methyl phenlactonoates on spore germination and root colonization of arbuscular mycorrhizal fungi. *Heliyon* 4(11), e00936. DOI: 10.1016/j.heliyon.2018.e00936.

Kusajima, M., Fujita, M., Soudthedlath, K., Nakamura, H., Yoneyama, K. *et al.* (2022) Strigolactones modulate salicylic acid-mediated disease resistance in *Arabidopsis thaliana*. *International Journal of Molecular Sciences* 23(9), 5246. DOI: 10.3390/ijms23095246.

Lahari, Z., Ullah, C., Kyndt, T., Gershenzon, J. and Gheysen, G. (2019) Strigolactones enhance root-knot nematode (*Meloidogyne graminicola*) infection in rice by antagonizing the jasmonate pathway. *New Phytologist* 224(1), 454–465. DOI: 10.1111/nph.15953.

Lahari, Z., van Boerdonk, S., Omoboye, O.O., Reichelt, M., Höfte, M. *et al.* (2024) Strigolactone deficiency induces jasmonate, sugar and flavonoid phytoalexin accumulation enhancing rice defense against the blast fungus *Pyricularia oryzae*. *New Phytologist* 241(2), 827–844. DOI: 10.1111/nph.19354.

Lechat, M.M., Brun, G., Montiel, G., Véronési, C., Simier, P. *et al.* (2015) Seed response to strigolactone is controlled by abscisic acid-independent DNA methylation in the obligate root parasitic plant,

Phelipanche ramosa L. Pomel. *Journal of Experimental Botany* 66(11), 3129–3140. DOI: 10.1093/jxb/erv119.

Li, C., Lu, X., Liu, Y., Xu, J. and Yu, W. (2023) Strigolactone alleviates the adverse effects of salt stress on seed germination in cucumber by enhancing antioxidant capacity. *Antioxidants* 12(5), 1043. DOI: 10.3390/antiox12051043.

Liu, R. and Chen, Y. (2024) Mycorrhizal symbiosis: Evolution, opportunities, challenges, and prospects. In: Parihar, M., Rakshit, A., Adholeya, A. and Chen, Y. (eds) *Arbuscular Mycorrhizal Fungi in Sustainable Agriculture: Inoculum Production and Application*. Springer Nature, Singapore, pp. 1–35.

Liu, Z., Hu, B., Flemetakis, E., Haensch, R., Franken, P. *et al.* (2025) Convergent evolution and adaptive diversification of root symbioses. *Biological Reviews of the Cambridge Philosophical Society* 101(1), 147–162. DOI: 10.1111/brv.70077.

Lopez-Obando, M., Ligerot, Y., Bonhomme, S., Boyer, F.D. and Rameau, C. (2015) Strigolactone biosynthesis and signaling in plant development. *Development* 142(21), 3615–3619. DOI: 10.1242/dev.120006.

López-Ráez, J.A., Pozo, M.J. and García-Garrido, J.M. (2011) Strigolactones: A cry for help in the rhizosphere. *Botany* 89(8), 513–522. DOI: 10.1139/b11-046.

Mansoor, S., Mir, M.A., Karunathilake, E.M.B.M., Rasool, A., Ştefănescu, D.M. *et al.* (2023) Strigolactones as promising biomolecule for oxidative stress management: A comprehensive review. *Plant Physiology and Biochemistry* 206, 108282. DOI: 10.1016/j.plaphy.2023.108282.

Mapuranga, J., Chang, J., Zhang, L., Zhang, N. and Yang, W. (2022) Fungal secondary metabolites and small RNAs enhance pathogenicity during plant–fungal pathogen interactions. *Journal of Fungi* 9(1), 4. DOI: 10.3390/jof9010004.

Marzec, M. (2016a) Perception and signaling of strigolactones. *Frontiers in Plant Science* 7, 1260. DOI: 10.3389/fpls.2016.01260.

Marzec, M. (2016b) Strigolactones as part of the plant defence system. *Trends in Plant Science* 21(11), 900–903. DOI: 10.1016/j.tplants.2016.08.010.

Marzec, M. and Brewer, P. (2019) Binding or hydrolysis? how does the strigolactone receptor work? *Trends in Plant Science* 24(7), 571–574. DOI: 10.1016/j.tplants.2019.05.001.

Marzec, M. and Muszynska, A. (2015) *In silico* analysis of the genes encoding proteins that are involved in the biosynthesis of the RMS/MAX/D pathway revealed new roles of strigolactones in plants. *International Journal of Molecular Sciences* 16(4), 6757–6782. DOI: 10.3390/ijms16046757.

Marzec, M., Gruszka, D., Tylec, P. and Szarejko, I. (2016) Identification and functional analysis of the *HvD14* gene involved in strigolactone signaling in *Hordeum vulgare*. *Physiologia Plantarum* 158(3), 341–355. DOI: 10.1111/ppl.12460.

Mashiguchi, K., Seto, Y. and Yamaguchi, S. (2021) Strigolactone biosynthesis, transport and perception. *Plant Journal* 105(2), 335–350. DOI: 10.1111/tpj.15059.

Mashiguchi, K., Seto, Y., Onozuka, Y., Suzuki, S., Takemoto, K. *et al.* (2022) A carlactonoic acid methyltransferase that contributes to the inhibition of shoot branching in *Arabidopsis*. *Proceedings of the National Academy of Sciences USA* 119(14), e2111565119. DOI: 10.1073/pnas.2111565119.

Mattoo, A.J. and Nonzom, S. (2021) Endophytic fungi: Understanding complex cross-talks. *Symbiosis* 83(3), 237–264. DOI: 10.1007/s13199-020-00744-2.

Mazhar, S., Hyder, S., Khan, B.S., Gondal, A.S., Ahmed, R. *et al.* (2025) Green synthesis of silver nanoparticles using guava leaves: An effective strategy to control chilli fruit rot disease. *BMC Plant Biology* 25(1), 499. DOI: 10.1186/s12870-025-06528-4.

McLaughlin, M.S., Roy, M., Abbasi, P.A., Carisse, O., Yurgel, S.N. *et al.* (2023) Why do we need alternative methods for fungal disease management in plants? *Plants* 12(22), 3822. DOI: 10.3390/plants12223822.

Mehmood, I., Wani, K.I. and Aftab, T. (2025) Strigolactone interplay with other phytohormones under stressed and normal conditions. In: Aftab, T. and Wani, K.I. (eds) *Strigolactones: Emerging Plant Hormones*. Scrivener Publishing, Beverly, Massachusetts, pp. 161–191.

Mehrabi, S.S., Sabokdast, M., Bihamta, M.R. and Dedičová, B. (2024) The coupling effects of PGPR inoculation and foliar spraying of strigolactone in mitigating the negative effect of salt stress in wheat plants: Insights from phytochemical, growth, and yield attributes. *Agriculture* 14(5), 732. DOI: 10.3390/agriculture14050732.

Mishra, S., Upadhyay, S. and Shukla, R.K. (2017) The role of strigolactones and their potential cross-talk under hostile ecological conditions in plants. *Frontiers in Physiology* 7, 691. DOI: 10.3389/fphys.2016.00691.

Mitra, D., Rad, K.V., Chaudhary, P., Ruparelia, J., Sagarika, M.S. *et al.* (2021) Involvement of strigolactone hormone in root development, influence and interaction with mycorrhizal fungi in plant: Mini-review. *Current Research in Microbial Sciences* 2, 100026. DOI: 10.1016/j.crmicr.2021.100026.

Nasır, F., Tian, L., Shi, S., Chang, C., Ma, L. *et al.* (2019) Strigolactones positively regulate defense against *Magnaporthe oryzae* in rice (*Oryza sativa*). *Plant Physiology and Biochemistry* 142, 106–116. DOI: 10.1016/j.plaphy.2019.06.028.

Omoarelojie, L.O. and van Staden, J. (2020) Plant–endophytic fungi interactions: A strigolactone perspective. *South African Journal of Botany* 134, 280–284. DOI: 10.1016/j.sajb.2020.02.009.

Pandey, A., Sharma, M. and Pandey, G.K. (2016) Emerging roles of strigolactones in plant responses to stress and development. *Frontiers in Plant Science* 7, 434. DOI: 10.3389/fpls.2016.00434.

Pandey, A.K., Sain, S.K. and Singh, P. (2016) A perspective on integrated disease management in agriculture. *Bio Bulletin* 2(2), 13–29.

Poveda, J., Abril-Urias, P. and Escobar, C. (2020) Biological control of plant-parasitic nematodes by filamentous fungi inducers of resistance: Trichoderma, mycorrhizal and endophytic fungi. *Frontiers in Microbiology* 11, 992. DOI: 10.3389/fmicb.2020.00992.

Pusztahelyi, T., Holb, I.J. and Pócsi, I. (2017) Plant–fungal interactions: Special secondary metabolites of the biotrophic, necrotrophic, and other specific interactions. In: Mérillon, JM. and Ramawat, K. (eds) *Fungal Metabolites*. Springer, Cham, Switzerland, pp. 1–58. DOI: 10.1007/978-3-319-19456-1_39-1.

Ramzan, I., Khan, B.S., Hayat, S.S., Hyder, S., Asmat, N. *et al.* (2025) SDS stabilized CuO nanoparticles as dual functional agents for antifungal and photocatalytic applications. *Discover Sustainability* 6(1), 703. DOI: 10.1007/s43621-025-01654-3.

Razdan, V.K. and Sabitha, M. (2009) Integrated disease management: Concepts and practices. In: *Integrated Pest Management: Innovation-Development Process*, Vol. 1. pp. 369–389. DOI: 10.1007/978-1-4020-8992-3_15.

Robert-Seilaniantz, A., Grant, M. and Jones, J.D.G. (2011) Hormone crosstalk in plant disease and defense: More than just jasmonate–salicylate antagonism. *Annual Review of Phytopathology* 49(1), 317–343. DOI: 10.1146/annurev-phyto-073009-114447.

Rostami, S. and Azhdarpoor, A. (2019) The application of plant growth regulators to improve phytoremediation of contaminated soils: A review. *Chemosphere* 220, 818–827. DOI: 10.1016/j.chemosphere.2018.12.203.

Rozpądek, P., Domka, A.M., Nosek, M., Ważny, R., Jędrzejczyk, R.J. *et al.* (2018) The role of strigolactone in the cross-talk between *Arabidopsis thaliana* and the endophytic fungus *Mucor* sp. *Frontiers in Microbiology* 9, 441. DOI: 10.3389/fmicb.2018.00441.

Rubiales, D., Fernández-Aparicio, M., Vurro, M. and Eizenberg, H. (2018) Advances in parasitic weed research. *Frontiers in Plant Science* 9, 236. DOI: 10.3389/fpls.2018.00236.

Sattar, A., Ul-Allah, S., Ijaz, M., Sher, A., Butt, M. *et al.* (2022) Exogenous application of strigolactone alleviates drought stress in maize seedlings by regulating the physiological and antioxidants defense mechanisms. *Cereal Research Communications* 50(2), 263–272. DOI: 10.1007/s42976-021-00171-z.

Selwal, N., Wani, A.K., Akhtar, N., Kaur, M. and Jassal, P.S. (2023) Molecular insights of strigolactone biosynthesis, signalling pathways, regulatory roles, and hormonal crosstalks in plant systems. *South African Journal of Botany* 160, 9–22. DOI: 10.1016/j.sajb.2023.06.046.

Servanté, E.K., Halitschke, R., Rocha, C., Baldwin, I.T. and Paszkowski, U. (2024) Independent regulation of strigolactones and blumenols during arbuscular mycorrhizal symbiosis in rice. *Plant Journal* 119(3), 1289–1298. DOI: 10.1111/tpj.16848.

Shahid, M., Shafi, Z., Ilyas, T., Singh, U.B. and Pichtel, J. (2024) Crosstalk between phytohormones and pesticides: Insights into unravelling the crucial roles of plant growth regulators in improving crop resilience to pesticide stress. *Scientia Horticulturae* 338, 113663. DOI: 10.1016/j.scienta.2024.113663.

Shu, H., Xu, K., Li, X., Liu, J., Altaf, M.A. *et al.* (2024) Exogenous strigolactone enhanced the drought tolerance of pepper (*Capsicum chinense*) by mitigating oxidative damage and altering the antioxidant mechanism. *Plant Cell Reports* 43(4), 106. DOI: 10.1007/s00299-024-03196-w.

Soliman, S., Wang, Y., Han, Z., Pervaiz, T. and El-kereamy, A. (2022) Strigolactones in plants and their interaction with the ecological microbiome in response to abiotic stress. *Plants* 11(24), 3499. DOI: 10.3390/plants11243499.

Soto-Cruz, F.J., Zorrilla, J.G., Rial, C., Varela, R.M., Molinillo, J.M.G. *et al.* (2021) Allelopathic activity of strigolactones on the germination of parasitic plants and arbuscular mycorrhizal fungi growth. *Agronomy* 11(11), 2174. DOI: 10.3390/agronomy11112174.

Srivastava, R., Bajpai, R., Khan, Z., Singh, S.P., Mehrotra, R. *et al.* (2022) Insight into strigolactone hormone functions in plant parasitic weeds: A regulatory perspective. *Indian Journal of Experimental Biology* 60(9), 659–666. DOI: 10.56042/ijeb.v60i09.65154.

Tokunaga, T., Hayashi, H. and Akiyama, K. (2015) Medicaol, a strigolactone identified as a putative didehydro-orobanchol isomer, from *Medicago truncatula*. *Phytochemistry* 111, 91–97. DOI: 10.1016/j.phytochem.2014.12.024.

Torres-Vera, R. (2017) Role of strigolactones in plant defense: Hormonal cross-talk and implication in arbuscular mycorrhizal symbiosis. PhD thesis, Universidad de Granada, Granada, Spain.

Torres-Vera, R., García, J.M., Pozo, M.J. and López-Ráez, J.A. (2016) Expression of molecular markers associated to defense signaling pathways and strigolactone biosynthesis during the early interaction tomato–*Phelipanche ramosa*. *Physiological and Molecular Plant Pathology* 94, 100–107. DOI: 10.1016/j.pmpp.2016.05.007.

Tsuzuki, S., Handa, Y., Takeda, N. and Kawaguchi, M. (2016) Strigolactone-induced putative secreted protein 1 is required for the establishment of symbiosis by the arbuscular mycorrhizal fungus *Rhizophagus irregularis*. *Molecular Plant–Microbe Interactions* 29(4), 277–286. DOI: 10.1094/MPMI-10-15-0234-R.

van Esse, H.P., Reuber, T.L. and van der Does, D. (2020) Genetic modification to improve disease resistance in crops. *New Phytologist* 225(1), 70–86. DOI: 10.1111/nph.15967.

Vega, A., O'Brien, J.A. and Gutiérrez, R.A. (2019) Nitrate and hormonal signaling crosstalk for plant growth and development. *Current Opinion in Plant Biology* 52, 155–163. DOI: 10.1016/j.pbi.2019.10.001.

Wahab, A., Muhammad, M., Munir, A., Abdi, G., Zaman, W. *et al.* (2023) Role of arbuscular mycorrhizal fungi in regulating growth, enhancing productivity, and potentially influencing ecosystems under abiotic and biotic stresses. *Plants* 12(17), 3102. DOI: 10.3390/plants12173102.

Wakabayashi, T., Shida, K., Kitano, Y., Takikawa, H., Mizutani, M. *et al.* (2020) CYP722C from *Gossypium arboreum* catalyzes the conversion of carlactonoic acid to 5-deoxystrigol. *Planta* 251(5), 97. DOI: 10.1007/s00425-020-03390-6.

Wakabayashi, T., Ishiwa, S., Shida, K., Motonami, N., Suzuki, H. *et al.* (2021) Identification and characterization of sorgomol synthase in sorghum strigolactone biosynthesis. *Plant Physiology* 185(3), 902–913. DOI: 10.1093/plphys/kiaa113.

Wang, J.Y., Haider, I., Jamil, M., Fiorilli, V., Saito, Y. *et al.* (2019) The apocarotenoid metabolite zaxinone regulates growth and strigolactone biosynthesis in rice. *Nature Communications* 10(1), 810. DOI: 10.1038/s41467-019-08461-1.

Wang, L., Wang, B., Yu, H., Guo, H., Lin, T. *et al.* (2020) Transcriptional regulation of strigolactone signalling in *Arabidopsis*. *Nature* 583(7815), 277–281. DOI: 10.1038/s41586-020-2382-x.

Wani, K.I., Zehra, A., Choudhary, S., Naeem, M., Khan, M.M.A. *et al.* (2023) Exogenous strigolactone (GR24) positively regulates growth, photosynthesis, and improves glandular trichome attributes for enhanced artemisinin production in *Artemisia annua*. *Journal of Plant Growth Regulation* 42(8), 4606–4615. DOI: 10.1007/s00344-022-10654-w.

Waters, M.T., Gutjahr, C., Bennett, T. and Nelson, D.C. (2017) Strigolactone signaling and evolution. *Annual Review of Plant Biology* 68(1), 291–322. DOI: 10.1146/annurev-arplant-042916-040925.

Wu, F., Gao, Y., Yang, W., Sui, N. and Zhu, J. (2022) Biological functions of strigolactones and their crosstalk with other phytohormones. *Frontiers in Plant Science* 13, 821563. DOI: 10.3389/fpls.2022.821563.

Xu, X., Fang, P., Zhang, H., Chi, C., Song, L. *et al.* (2019) Strigolactones positively regulate defense against root-knot nematodes in tomato. *Journal of Experimental Botany* 70(4), 1325–1337. DOI: 10.1093/jxb/ery439.

Yi, F., Song, A., Cheng, K., Liu, J., Wang, C. *et al.* (2023) Strigolactones positively regulate *Verticillium* wilt resistance in cotton via crosstalk with other hormones. *Plant Physiology* 192(2), 945–966. DOI: 10.1093/plphys/kiad053.

Yoneyama, K. (2019) How do strigolactones ameliorate nutrient deficiencies in plants? *Cold Spring Harbor Perspectives in Biology* 11(8), a034686. DOI: 10.1101/cshperspect.a034686.

Yoneyama, K. and Brewer, P.B. (2021) Strigolactones, how are they synthesized to regulate plant growth and development? *Current Opinion in Plant Biology* 63, 102072. DOI: 10.1016/j.pbi.2021.102072.

Yoneyama, K., Mori, N., Sato, T., Yoda, A., Xie, X. *et al.* (2018) Conversion of carlactone to carlactonoic acid is a conserved function of MAX1 homologs in strigolactone biosynthesis. *New Phytologist* 218(4), 1522–1533. DOI: 10.1111/nph.15055.

Yoneyama, K., Akiyama, K., Brewer, P.B., Mori, N., Kawano-Kawada, M. *et al.* (2020a) Hydroxyl carlactone derivatives are predominant strigolactones in *Arabidopsis*. *Plant Direct* 4(5), e00219. DOI: 10.1002/pld3.219.

Yoneyama, K., Xie, X., Nomura, T. and Yoneyama, K. (2020b) Do phosphate and cytokinin interact to regulate strigolactone biosynthesis or act independently? *Frontiers in Plant Science* 11, 438. DOI: 10.3389/fpls.2020.00438.

Zhang, J., Tang, C., Xie, J., Li, J., Zhang, X. *et al.* (2024) Exogenous strigolactones alleviate low-temperature stress in peppers seedlings by reducing the degree of photoinhibition. *BMC Plant Biology* 24(1), 907. DOI: 10.1186/s12870-024-05622-3.

Zhang, Y., van Dijk, A.D.J., Scaffidi, A., Flematti, G.R., Hofmann, M. *et al.* (2014) Rice cytochrome P450 MAX1 homologs catalyze distinct steps in strigolactone biosynthesis. *Nature Chemical Biology* 10(12), 1028–1033. DOI: 10.1038/nchembio.1660.

Zhou, F., Lin, Q., Zhu, L., Ren, Y., Zhou, K. *et al.* (2013) D14–SCF^{D3}-dependent degradation of D53 regulates strigolactone signalling. *Nature* 504(7480), 406–410. DOI: 10.1038/nature12878.

16 Strigolactone-mediated Signaling and Regulation of Viral and Bacterial Diseases

Salma Noureen[1†], Javed Iqbal[1*†], Banzeer Ahsan Abbasi[2], Tabassum Yaseen[1], Muhammad Anas[3], Sadia Zafar[4], Kamran Akbar[1], Aneesa Gul[1], Sobia[1] and Farishta Zarshan[1]

[1]Department of Botany, Bacha Khan University, Charsadda, Khyber Pakhtunkhwa, Pakistan; [2]Department of Botany, Rawalpindi Women University, Satellite Town, Rawalpindi, Pakistan; [3]Department of Soil Science & Plant Nutrition, University of Selcuk, Konya, Turkey; [4]Department of Botany, Division of Science and Technology, University of Education, Lahore, Punjab, Pakistan

Abstract

Strigolactones (SLs), categorized as plant hormones and derived from carotenoids, perform a variety of physiological functions in plant development. These molecules were initially detected as strong seed germination enhancers of the host-dependent weed *Striga lutea* in root exudates, and are produced primarily in the root tissue of a diverse range of monocot and dicot flora, indicating their significance in nature. At least 20 SLs have been identified in plants, and are classified into two types based on their chemical structure: canonical and non-canonical. carotenoid cleavage dioxygenase 7 (CCD7), CCD8, and DWARF27 (D27) are among the enzymes that mediate the production of carlactone, the common precursor for all SLs. Synthetic SLs have been produced, such as GR24, with a role in both internal and external plant processes. In the rhizosphere, they function as signaling molecules controlling a variety of developmental processes and environmental reactions. They control the symbiotic and parasitic association with roots of the host plant and arbuscular mycorrhizal fungi, and regulate the formation of adventitious roots, reproduction, and secondary growth. New functions are constantly being discovered. There have been several advances in our knowledge of SL synthesis, hormone activity, and signaling molecules. However, there has been less discovered regarding the molecular pathways engaged in bacterial and viral diseases. The role of SLs could be revealed via modern imaging methods.

Keywords: bacterial diseases, growth regulation, molecular signaling, phytohormone, strigolactones, viral diseases

16.1 Introduction

Climate change poses a major threat worldwide and is expected to intensify in the near future. The effects of global climate variation include rising heat levels, unpredictable precipitation trends, flooding, and drought (Kumar and Mohanasundari, 2025). People tend to be

*Corresponding author: javed89qau@gmail.com

† These authors contributed equally.

DOI: 10.1079/9781836990598.0016

peripatetic, making them more likely to be able to adjust to the changing surrounding conditions via evasion. Unlike humans, plants are sessile and depend heavily on defensive mechanisms rather than evasion (Demis, 2024). Farming remains the key important area with severe significant consequences of environmental inequities. Scientists are becoming increasingly deliberate about justifying declines in yield and soil fertility (Steidl *et al.*, 2025). Protecting the environment and feeding the increasing population are the main aims of sustainable development (Emerenciano *et al.*, 2025). To lessen the damage and contribute to sustainable development, researchers will need to find novel resources and hidden pathways. Currently, plant researchers are focusing on reinforcement of plant immunity against environmental changes (Scarpin, 2024), with efforts concentrated on investigating gene manipulation, plant hormone signaling routes, and protein-based mechanisms (Fig. 16.1) (Yetgin *et al.*, 2025).

Strigolactones (SLs) are emerging plant hormones with significant potential in enhancing plant resilience by regulating diverse developmental processes and environmental responses (Das *et al.*, 2025). They were first discovered in 1966 as stimulants for germination of the parasitic weed *Striga lutea* using root-derived substances from cotton in a crystalline form (Wakabayashi *et al.*, 2025) and were commonly regarded as detrimental secondary metabolites produced by plants, due to the damage caused by parasitic weeds worldwide (Bendejacq-Seychelles *et al.*, 2024). They are carotenoid-derived hormones, named by Butler (1995), and are based on strigol and structurally related compounds that trigger the emergence of seeds, mostly of obligate parasitic flora (Thiombiano *et al.*, 2023). Additional studies

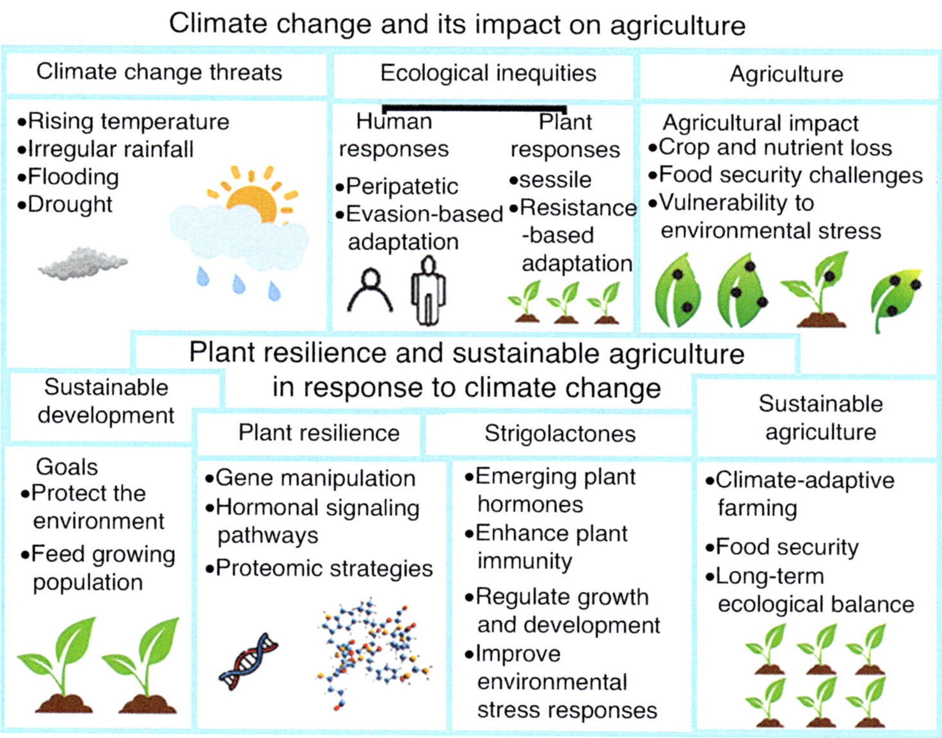

Fig. 16.1. Impact of climatic change on agriculture, sources of sustainable development, and their hidden pathways.

showed that they stimulated hyphal growth of arbuscular mycorrhizae fungi, thereby enhancing this symbiotic relationship. They also act as internal signaling molecules that move upward within the plant to suppress shoot branching (Naseer *et al.*, 2024). Since the discovery of SLs as phytohormones, the field has advanced swiftly, providing us with a deep understanding of their biosynthesis, transport, and perception (Aftab and Wani, 2025).

Recent studies have shown that the germination rate of *Orobanche minor* (a parasitic plant) seeds can be significantly increased (80%) by naturally occurring stimulants such as jasmonates and SLs (Brun *et al.*, 2025). However, SLs are now increasingly sought after for their ability to enhance plant growth and development. They affect the phenotypic behavior of PIN-FORMED (PIN) auxin transporters by preventing the formation of auxin-conducting channels, particularly after wounding or in response to external auxin sources (Alvi *et al.*, 2022). SLs are involved in regulating the plant's response to nutrient deficiencies, especially nitrogen and phosphorus (Marro *et al.*, 2022). Leaf senescence is a complex process influenced by various factors, including aging, flowering, exposure to darkness, nutrient scarcity, and different stressors. During this process, nutrients transferred from older to younger tissues. SLs have been shown to hasten the onset of leaf senescence (Sharma *et al.*, 2024).

Research has shown that SLs increase the development of primary roots and root hairs. This enables the root hairs to absorb more water and nutrients from the soil, facilitating the formation of a symbiotic relationship between *Rhizobia* and leguminous plants (Hajam and Bashri, 2025). They are also recognized as root-derived signaling compounds in the rhizosphere that control both parasitic associations with host plant roots and symbiotic relationships with arbuscular mycorrhizae fungi (Mehra *et al.*, 2025). SLs regulate a variety of important agronomic characteristics in plants, such as root growth, branching initiation, stress tolerance induction, and the regulation of secondary growth, among the developmental processes (Das *et al.*, 2025). These functions make SLs unique and have encouraged researchers to explore their full mechanism of action in plants. They also interact antagonistically with cytokinins (CKs), inhibiting shoot growth, axillary bud branching, tillering, and the formation of lateral roots (Helmich, 2025).

16.2 Strigolactone Chemistry

At least 20 SLs have been identified in plants. The chemistry of SLs has been reviewed extensively elsewhere (Helmich, 2025). Based on their chemical structure, they are classified into two categories: canonical and non-canonical (Fig. 16.2) (Nomura *et al.*, 2024). Canonical SLs featured a structure, consisting of three rings (A, B, and C), further linked to a butenolide group with the help of an enol-ether bridge. The butenolide group and bridge are essential for their biological activities, particularly in plants. Based on the stereochemistry of the C-ring (α- and β-oriented), canonical forms are categorized into orobanchol and strigol types (Păun *et al.*, 2024).

Non-canonical forms contain both the butenolide group and a bridge but lack the characteristic ABC ring system (Jain *et al.*, 2025). There are many other natural and synthetic compounds that lack the core ABC structure but show SL-like activity (Kapoor *et al.*, 2024), such as avenaol, debranones, heliolactone, methyl carlactonoate (MeCLA), GC242 (Guillaume Clave series), and Yoshimulactone Green (Yoshimura *et al.*, 2019). For all SLs found in plants, the butenolide group exhibits a 2′R configuration. This peculiarity is important for the role of SLs in regulating plant growth (Helmich, 2025). GR24, a synthetic form of SL, has proven valuable for studying SL responses and is widely used in research (Moosavi *et al.*, 2025).

16.3 Strigolactones Biosynthesis

In order to understand SL biosynthesis researchers used a series of mutants of many plants. These included: *max4* and *max3* of *Arabidopsis*, *rms1* and *rms5* of pea (*Pisum sativum*) and *d17* and *d10* of rice (*Oryza sativa*) (Valivand, 2023). These mutants, which were defective in carotenoid cleavage dioxygenase 7 (CCD7) and CCD8, were found to be deficient in SLs. Alder *et al.* (2012) performed an *in*

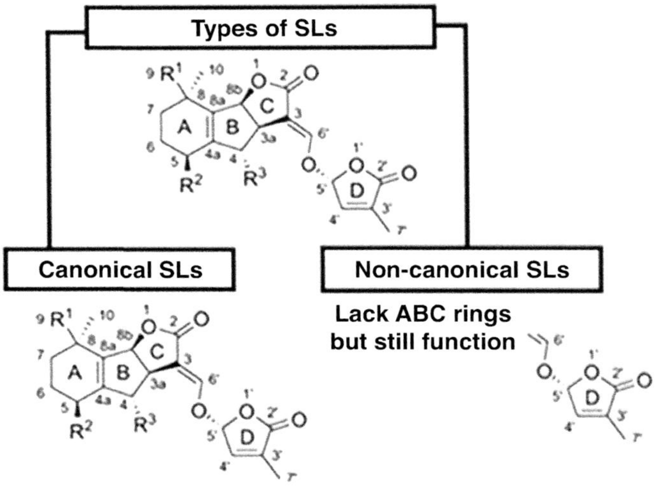

Fig. 16.2. Types of strigolactones (SLs).

vitro analysis using the recombinant proteins DWARF27 (D27; *Arabidopsis*), CCD7 (pea), and CCD8 (rice), and confirmed that these three types of enzyme catalyzed sequential reactions, resulting in the production of a non-canonical SL (carlactone) from all *trans*-β-carotene via 9-*cis*-β-carotene and 9-*cis*-β-apo-10′-carotenal (Bajguz and Piotrowska-Niczyporuk, 2023). Subsequently, liquid chromatography/tandem mass spectrometry (LC-MS/MS) analysis recognized carlactone as an endogenous compound in *Arabidopsis thaliana* and *O. sativa*. Carlactone serves as the common precursor for all SLs (Kuno *et al.*, 2025). In *Arabidopsis* and *Oryza*, biosynthesis is facilitated by various enzymes: 9-*cis*-β-carotene is formed from *trans*-β-carotene through the action of carotenoid isomerase, while CCD7 and CCD8 catalyze the formation of caprolactone, a key intermediate in SL biosynthesis (Fig. 16.3) (Jain *et al.*, 2025).

In *Arabidopsis*, MORE AXILLARY GROWTH 1 (MAX1) comprises a group of genes that encodes cytochrome P450 enzymes and convert carlactone into carlactonoic acid (Zaidi *et al.*, 2024). Carlactonoic acid is then methylated by an enzyme (unidentified) to form MeCLA. MeCLA is transformed into an unknown compound by LATERAL BRANCHING OXIDOREDUCTASE (LBO) (Zhou *et al.*, 2025). It is argued that *Arabidopsis* makes only MeCLA and derived products rather than canonical SLs. Similarly, *Oryza* has five types of MAX1 with various functions (Nomura *et al.*, 2024). For example, Os01g0700900 forms the precursor for orobanchol-type SLs by converting carlactone to 4-deoxy-orobanchol, which is further converted to orobanchol by Os01g0701400 (Mukherjee and Aftab, 2023).

16.4 Transportation and Perception

A major breakthrough in understanding SL transport was the discovery of an ABC-type (AT-binding cassette) transporter, PLEIOTROPIC DRUG RESISTANCE 1 (PDR1), responsible for their exudation from roots in *Petunia hybrida* (Demurtas *et al.*, 2023). Although similar genes have been suggested to play a role in SL movement, the specific pathways governing their secretion from roots and their transformation within the plant are still not clear (Mehmood *et al.*, 2025). It has established that SLs are released from roots into the soil, particularly under conditions of low nitrogen and phosphorus availability. Additionally, they also function as long-distance signals, moving unidirectionally from the roots to the shoots (Helmich, 2025). In 2008, SLs were identified as plant hormones inhibiting shoot branching (Jain *et al.*, 2025).

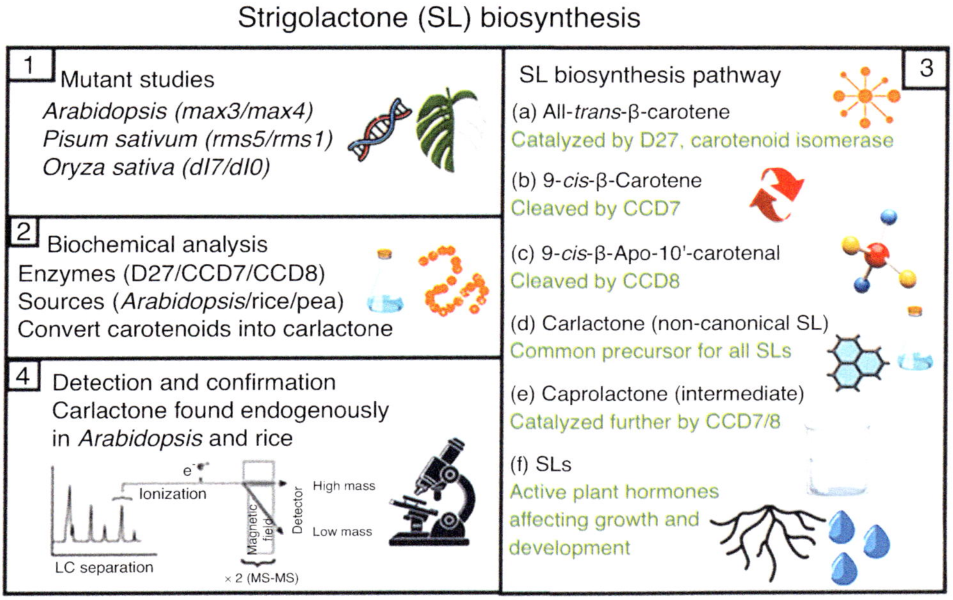

Fig. 16.3. Biosynthesis of strigolactones.

A series of grafting experiments conducted using the *DECREASED APICAL DOMINANCE* (*DAD*) and *RMS* (*Ramosus*) branching mutants of *MAX* mutants of *Petunia*, *Arabidopsis* and pea suggested that a small section of wild-type shoot had adequate amounts of SLs to inhibit axillary bud development: when wild-type scion was grafted on to the rootstock of a mutant (SL deficient), it produced the usual phenotype (Liu *et al.*, 2024; Verma *et al.*, 2024). It was concluded that SLs were synthesized in roots and then moved acropetally to shoots (Shivran *et al.*, 2022). Confirmation for the one-way transport of SLs was also provided by the examination of nodulation in pea mutants (SL deficient), which showed fewer nodules (Roy *et al.*, 2024). The pea scions had no impact on nodulation in SL-deficient mutants (Sharma *et al.*, 2024).

LC-MS/MS analysis of xylem sap has offered insights into the movement of SLs. Tomato and *Arabidopsis* xylem sap contain numerous SLs, such as orobanchol (Zhang *et al.*, 2020). Xie *et al.* (2015) collected xylem fluid from various plant species, including tomato and *Arabidopsis*, and concluded that SL biosynthetic intermediates (carlactone, carlactonoic acid and MeCLA) were absent (Mashiguchi *et al.*, 2021). LC-MS/MS and grafting analyses suggested that multiple molecular forms of SLs and their biosynthetic intermediates may serve as mobile signals (Seto, 2024). Although some evidence is available about transport from the root to the shoot, graft studies have suggested synthesis of SLs in shoots as well (Torres *et al.*, 2021). However, they are generally present in shoots at lower levels (30–1000 times), or are below the limit of detection (Perdomo *et al.*, 2024).

D14 (α/β-hydrolase enzyme family protein) plays a key role in SL perception in angiosperms. Unlike other plant receptors, D14 is unique in that it acts as both a receptor and an enzyme (Zhang *et al.*, 2024). It possesses a highly conserved Ser-His-Asp sequence catalytic triad, necessary for both signaling functions and SL hydrolysis (Wang *et al.*, 2025). During SL hydrolysis, D14 can undergo conformational changes resulting in signal transduction. The D-ring is cleaved and forms an intermediate product, covalently bonded to the His residue of the catalytic triad (Deng *et al.*, 2025). This results a closed confirmation in which the cover structure collapsed from four to three helices and the volume is reduced (Lazou *et al.*, 2024). This significant advancement was made through the structural characterization of *Arabidopsis* SKP1-LIKE1 (ASKI) in associated with D3 and AtD14 (*A. thaliana* D14) (Liu *et al.*, 2023). The complex maintains the closed state and reduces hydrolysis of SLs analogs through D14. Without D3–ASK1, SL hydrolysis is enzymatically slow (Palayam *et al.*, 2024). Additional evidence supporting D14 as a SL receptor comes from the SL-insensitive D14-5 protein, which shows a significantly low interface with D3, exhibiting a higher rate of hydrolysis compared with the wild-type protein (Struk, 2018).

16.5 Strigolactone Signaling and Biotic Stress

Biotic stress is a condition in which detrimental biological factors interact with plants and disrupt their normal growth (Francis *et al.*, 2025). These biological factors include both macro- and microorganisms including viruses, bacteria, fungi, nematodes, insects, and weeds. They take nutrients from their host plants, affecting optimal growth and finally causing death to the host plants (Adigun *et al.*, 2024). Biotic stress is a major problem, causing massive loss in agriculture by disrupting crop productivity. The main morphological symptoms of biotic stress are stunting, rotting, chlorosis, and formation of lesions (Nagaraja *et al.*, 2021). In order to combat the damage caused by biotic stresses, plants have established wide-ranging defense mechanisms (Varadharajan *et al.*, 2025). These include biochemical, morphological, and molecular defensive mechanisms, triggered by several phytohormone signaling pathways. Phytohormones are vital for plant growth and development and are only needed in extremely small amounts (10^{-5} to 10^{-6} mol l^{-1}) (Saha *et al.*, 2025).

As a new emerging phytohormone, SLs play a key role in mediating plant defense responses to different biotic stresses. They are involved in defending the host plants against attacks by various pathogens, thus providing tolerance to stress in plants (Choubey *et al.*, 2025). They protect plants from pathogenic attacks by encouraging stomatal closure, amassing the production of hydrogen peroxide, and preventing their growth (Yang *et al.*, 2025). The communication between plant host and parasite depends on signaling molecules such as SLs released from host roots into the rhizosphere.

Studies on the synthesis and signaling of SLs in *A. thaliana* have revealed their role in plant resistance against specific bacteria: *Rhodococcus fascians*, *Pseudomonas syringae*, and *Pectobacterium carotovorum* (Nicolle, 2023). Infection by *R. fascians* causes leafy gall syndrome and involves signaling (MAX2) and SL biosynthesis (MAX1, MAX3, and MAX4) (Verma *et al.*, 2024). Studies have shown that when D2 (an inhibitor of SL biosynthesis) and GR24 (a synthetic analog of SLs) were added to the medium of wild-type plants, they exhibited increased susceptibility to *R. fascians* (Khatun *et al.*, 2024). Furthermore, the infection resulted in regulation of genes involved in SL production (MAX1, MAX3, and MAX4). Knockout of *Arabidopsis MAX2* reduced resistance to *Pseudomonas syringae* (Zhao *et al.*, 2024). Silencing the SL biosynthetic genes *CCD7* and *CCD8* in tomatoes increased the plant's response to root-knot nematodes. Additionally, the key biosynthesis genes *GbCCD7* and *GbCCD8b* are involved in regulating resistance to *Verticillium* wilt in *Gossypium hirsutum* (Huang *et al.*, 2024).

16.6 Strigolactone Signaling and Bacterial Diseases

SLs are involved in regulating plant responses to bacterial pathogens such as *R. fascians*, *Pseudomonas syringae*, and *Pectobacterium carotovorum*.

Leafy gall syndrome is a contagious disease, affecting various herbaceous dicots (Francis *et al.*, 2025), caused by *R. fascians* (a Gram-positive bacterium), and characterized by the formation of multiple shoots (Fig. 16.4). The formation of these multiple shoots occurs through stimulation of dormant axillary meristems, leading to altered plant architecture (Grich *et al.*, 2025). In *A. thaliana*, *R. fascians* infection causes a delay in senescence, triggers the activation of axillary meristems, disrupts apical dominance, and promotes the formation of inflorescences from compact rosettes, ultimately yielding a

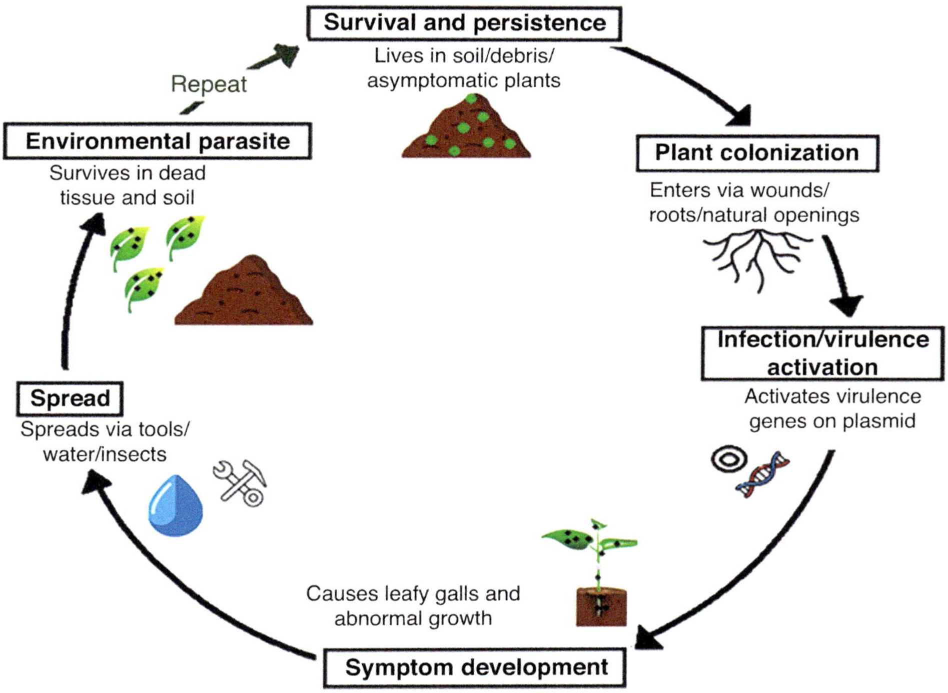

Fig. 16.4. Life cycle of *Rhodococcus fascians*.

highly branched morphology (Dhaouadi and Ali Rhouma, 2020). Interestingly, leafy gall syndrome shares similarities with mutants exhibiting impaired SL sensing and/or production. Thus, the syndrome serves as a tool to evaluate the importance of SLs in pathogen-induced changes (Mishra *et al.*, 2024). Researchers compared the phenotypes of four *max* and *brc1* mutants with the effects of *R. fascians* (strain D188) on wild-type *Arabidopsis* plants. The synthetic SL combination of GR24 and CCD inhibitor D2 were then added to evaluate the impact of endogenous SL levels on symptom onset (Claassens, 2024). Using a variety of methods, SL levels in infected tissues have been measured. One study examined the impact of CKs on transcription (Nuzhnaya *et al.*, 2024), by analyzing the expression patterns of *MAX* and *BRANCHED1* (*BRC1*) genes in infected tissues of wild-type Columbia-0 (Col-0) plants and various mutants. The findings led to the development of thr current model explaining the molecular basis of leafy gall disease (Xu *et al.*, 2024).

In conclusion, it is becoming increasingly obvious that balanced reactions are essential for the production of leafy gall syndrome. By influencing phytohormone levels, *Arabidopsis* defends itself against *R. fascians* and lessens the effects of bacterial-based CKs (Mishra *et al.*, 2024). The plant initially attempts to overcome the pressure exerted by *R. fascians* by activating its CK homeostasis systems. This response is triggered by the recognition of the bacterial CKs by AHK3/AHK4, histidine kinase homologs that function as CK receptors (Clark and Bennett, 2024). In response, the bacteria cause the infected plant tissues to produce auxin and putrescine, which serve as auxiliary signals that promote the leafy galls (Oyedele *et al.*, 2024). Finally, the plant overproduces SLs, which are strong inhibitors of shoot branching, further counteracting the activity of *R. fascians*. The freshly stimulated shoot outgrowth is inhibited by *BRC1* genes (Mishra *et al.*, 2024). It is impossible to exclude other direct effects of SLs on symptom generation because *BRC1* is just one component of the

SL signal transduction pathway. Although the molecular response triggered by this pathogen has not been fully explored in *Arabidopsis* and *R. fascians* pathosystem (Helmich, 2025), it would be interesting to know whether the plant's immunity exerts some degree of control over the pathogen (Ali *et al.*, 2024).

Arabidopsis mutants (*d14*, *max2*, *max3*, *max4*) exhibit increased susceptibility to the biotrophic and virulent bacterium *Pseudomonas syringae* pv. *tomato* DC3000 (*Pst*) (Fig. 16.5). (Hu *et al.*, 2021). *Pst* serves as a model microorganism for studying host–pathogen interactions. It not only infects *Arabidopsis* but also causes speck disease in tomatoes. It is known for its type III secretion system (T3SS) and the virulence factors it uses to suppress plant immune responses (Jardim-Messeder *et al.*, 2025). The sequenced genome of *Pst* revealed a substantial number of virulence genes, including those encoding transporters for nutrient acquisition and T3SS effectors, which play a crucial role in the bacterium's pathogenicity (Jung *et al.*, 2025). Gene expression analyses and pathogen inoculation assays showed that GR24 is actively involved in regulation of disease resistance against *Pst* infection (O'Malley and Anderson, 2021). These findings are further reinforced by the observation that *in planta*, *Pst* growth is higher in *max* mutants compared with wild-type plants. However, the absence of SLs during germination in *max* mutants showed more effects on the plant's stress response systems (Fujita *et al.*, 2024).

To verify the involvement of endogenous SLs in resistance against *Pst*, the effects of TIS108, a SL biosynthesis inhibitor, were examined on disease resistance, metabolite accumulation, and gene expression (Kusajima *et al.*, 2022). The expression levels of a jasmonate-related gene (*PDF1.2*), ethylene-related gene (ERF1) and *SAR* (systemic acquired resistance)-related

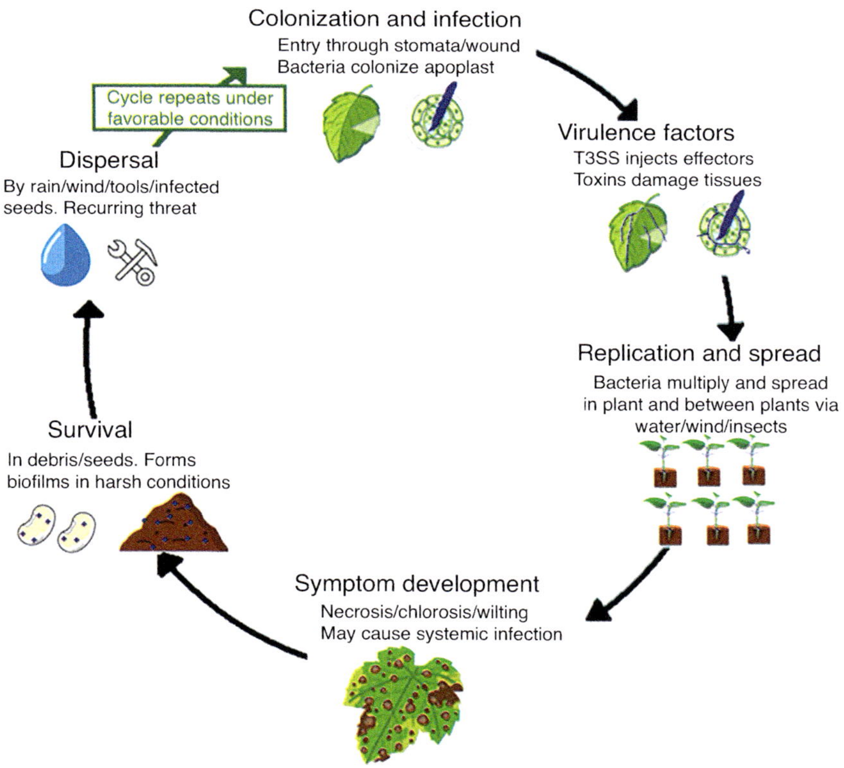

Fig. 16.5. Life cycle of *Pseudomonas syringae*. T3SS, type III secretion system.

genes (*PR1*, *PR2*, *PR5*) remined unaffected by TIS108 treatment (Zhong *et al.*, 2024). Treatment with a low concentration of TIS108 (1 μM) significantly reduced disease resistance in wild-type plants compared with untreated plants, indicating that endogenous SLs play a role in conferring resistance to *Pst*.

Studies have shown that plants such as *A. thaliana max2* mutant lacking SL signaling are more susceptible to bacterial pathogens such as *Pectobacterium carotovorum* (Kusajima *et al.*, 2022; Pontiggia *et al.*, 2023). *Pectobacterium carotovorum* is a Gram-negative, host-specific and necrotrophic bacterium, causing soft rot disease in various plants (Fig. 16.6) (Zhao *et al.*, 2024). The increased susceptibility of the *max2* mutant to *Pectobacterium carotovorum* is probably due to modifications in abscisic acid signaling (Temmerman *et al.*, 2022). It colonizes the intercellular spaces of host tissues and carries potent effector molecules (avirulence) via T3SS (Verma *et al.*, 2024).

16.7 Strigolactone Signaling and Viral Diseases

SLs are involved in regulating plant response to various viruses, including tobacco mosaic virus (TMV) (López-Pagán *et al.*, 2025). TMV has a single strand of RNA and infects plant species in the families Brassicaceae and Solanaceae (Huang *et al.*, 2024). The virus persists for a long time, resulting symptoms such as curling, mottling, stunted growth, and leaf deformation, resulting in reduced yields and productivity (Fig. 16.7) (Wang *et al.*, 2025).

SLs play a significance role in the resistance of *Nicotiana benthamiana* (tobacco plants) to TMV (Fig. 16.8). The exogenous application of SLs decreased resistance, whereas treatment with Tis108, a SL biosynthesis inhibitor, increased resistance to TMV. Knockdown of SL genes (*NtCCD7* and *NtCCD8*) led to enhanced resistance against TMV (Guha, 2025). Further analysis

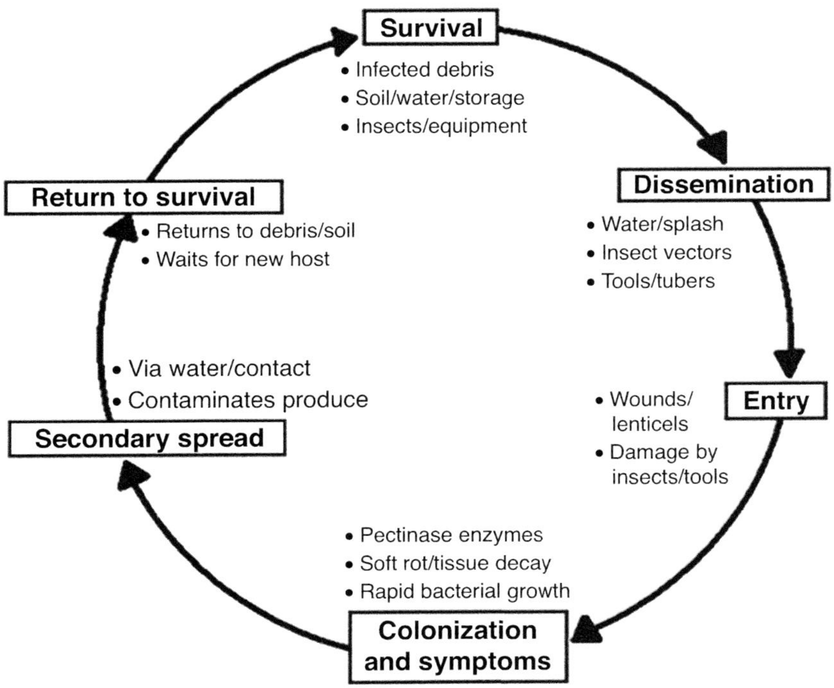

Fig. 16.6. Life cycle of *Pectobacterium carotovorum*.

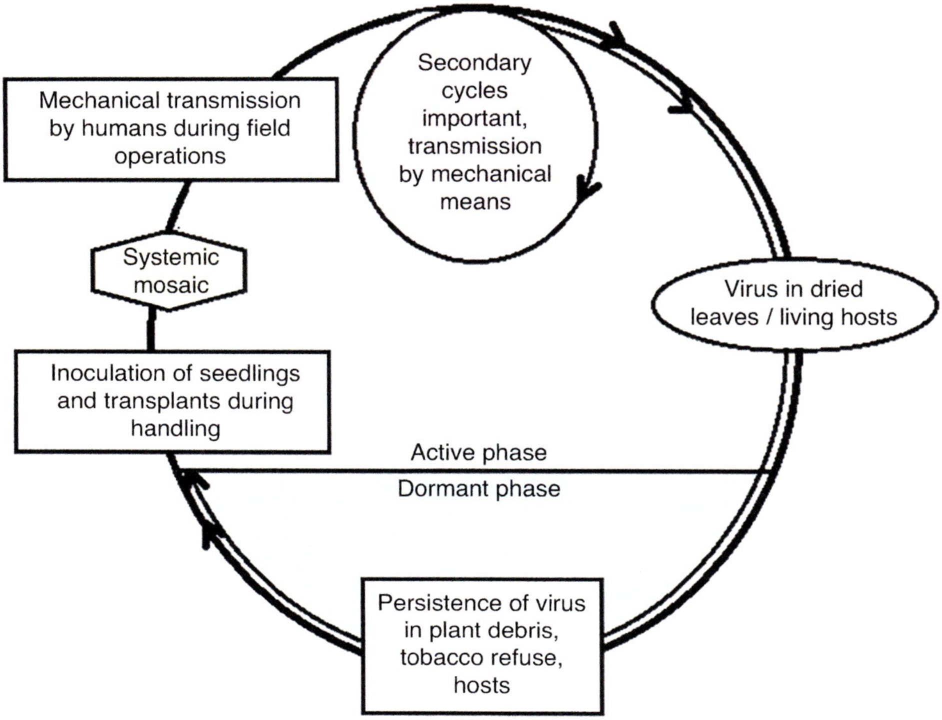

Fig. 16.7. Disease cycle of tobacco mosaic virus.

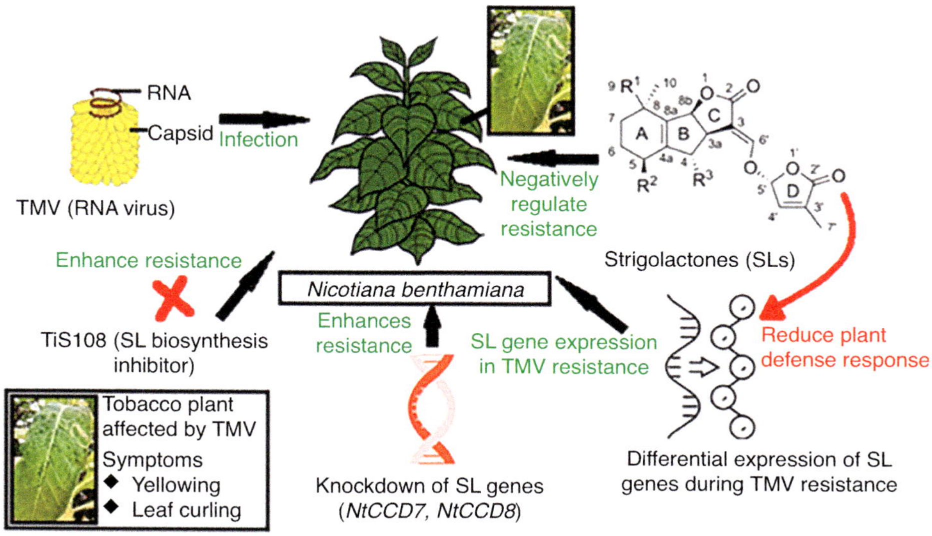

Fig. 16.8. Role of strigolactones in resistance of *Nicotiana benthamiana* to tobacco mosaic virus (TMV).

showed that genes involved in the synthesis and regulation of SLs are differentially expressed during the process of effective resistance against TMV. It can be concluded that SLs negatively regulate *N. benthamiana* defense against TMV.

Tobacco plants are significantly affected when they are exposed to cucumber mosaic virus (CMV). CMV comprises isometric particles 30 nm in diameter (Huang *et al.*, 2024). They are transmitted in a non-persistent manner via aphids and the virus a wide host range of at least 800 plant species (Ganchozo-Mendoza *et al.*, 2024). The genome of the virus is organized into RNAs (1, 2, and 3) and subgenomic RNAs (4A and 4). RNA1 encodes components of the replicase complex only, while RNA2 codes for both replicase complex components and the 4A protein, which suppresses gene silencing. RNA3 encodes the 3A protein, vital for virus movement (Ma and Lin, 2025). SL biosynthesis can be inhibited by CMV infection, which leads to reduced β-carotene content in tobacco plants (Brilli *et al.*, 2024), with a significant reduction in concentration of up to 50% compared with uninfected plants. Carotenoid levels have also been found to decline drastically (19–52%) in *Amaranthus* and *Chenopodium* plants infected with CMV (Ibdah *et al.*, 2014). To study transcription changes in β-carotene biosynthesis, researchers investigated the effect of CMV on accumulation of phytoene desaturase (PDS) transcripts, an enzyme involved in β-carotene biosynthesis in tobacco plants (Paudel *et al.*, 2022). The *PDS* gene was found to be crucial in controlling carotenoid levels and gene expression in infected tissue (Li *et al.*, 2022; Mehboob *et al.*, 2023).

16.8 Strigolactones and their Antiviral Activity in Humans

SLs have been studied extensively for their role in plant biology, but their potential applications in the medical field, particularly their antiviral activity, remain relatively unexplored (Janeeshma *et al.*, 2024). Synthetic analogs of SLs induce apoptosis and G_2/M cell-cycle arrest in various cancerous cells, such as human fibroblasts, normal primary prostate cells, and mammary epithelial cells (Sharmin *et al.*, 2021). Cancerous cells are more sensitive to the inhibitory effects of SL analogs than heterogeneous cell populations. SLs induce DNA double-strand breaks and impair cellular DNA repair (Zahoor *et al.*, 2025). Research on human cytomegalovirus (HCMV) suggests that SLs may be a promising alternative to nucleoside analogs for treating for HCMV infections (Chen *et al.*, 2022). HCMV is a prevalent pathogen and leading cause of congenital malformations in developed countries (Modrzejewska *et al.*, 2024). While nucleoside analogs have been effective against HCMV, their use is limited by severe side effects (Chen *et al.*, 2022). Researchers found that EDOT-EGO (ethylene doxytriazole ego) and TH-EGO (triazole hydroxy ego) significantly inhibited HCMV replication by impairing viral protein expression (Biolatti *et al.*, 2020).

16.9 Spatial Regulation of Strigolactones

Following the development of new imaging technologies, fluorescent SL analogs may be considered as molecular tools to study the distribution of SLs (Chen *et al.*, 2024). Yoshimulactone Green and GC242 are fluorescent when cleaved by SL receptors (Wang *et al.*, 2022). Therefore, these probes could reveal the sites of recognition of SLs. The StrigoQuant biosensor system has been developed, utilizing a genetically encoded *SMXL6* gene fused with the firefly luciferase and *Renilla* luciferase (REN) coding sequences, driven by the 35S promoter (Sun *et al.*, 2021). Firefly luciferase is directly fused to *SMXL6*, whereas *Renilla* luciferase is linked via a 2A peptide coding sequence, which is cleaved during translation resulting in two separate peptides (Thomas, 2022). The construct generates two chimeric proteins, SMXL6–FF and REN, from a single transcript. SL signals are detected through luminescence resulting from SL-dependent degradation of SMXL6-FF (Kameoka and Kyozuka, 2018).

16.10 Conclusion

SLs play a dual role: they function as both plant hormones regulating growth and development, and as signaling molecules in the soil influencing interactions with beneficial and pathogenic microbes. Although SLs play a role in plant defense, their function is not all-encompassing; some research indicates that SLs are also involved in resistance to numerous fungal pathogens and root-knot nematodes. In *Arabidopsis*, SLs are involved in the attraction of plant-cyst nematodes to the host, potentially through a connection with the karrikin (phytohormone) pathway. Interactions with other hormones, such as abscisic acid, salicylic acid, and jasmonic acid, probably have an impact on how SLs affect plant defense. A better understanding of SLs in plant defense may lead to new crop-protection techniques, such SL-based agrochemicals that increase plant resistance. To completely comprehend the ways in which SLs affect plant defense and identify the precise plant–pathogen combinations, more investigation is required.

References

Adigun, B.A., Nahar, S.B.M., Oyebamiji, Y.O., Daba, T.M., Sinumvayo, J.P. *et al.* (2024) Opportunistic fungi, plant, and nematode interactions in agricultural crops. In: Akhtar, M.S. (ed.) *Opportunistic Fungi, Nematode and Plant Interactions: Interplay and Mechanisms*. Springer Nature Singapore, Singapore, pp. 11–32.

Aftab, T. and Wani, K.I. (eds) (2025) *Strigolactones: Emerging Plant Hormones*. Scrivener Publishing, Beverly, Massachusetts. DOI: 10.1002/9781394302826.

Alder, A., Jamil, M., Marzorati, M., Bruno, M., Vermathen, M. *et al.* (2012) The path from β-carotene to carlactone, a strigolactone-like plant hormone. *Science* 335(6074), 1348–1351. DOI: 10.1126/science.1218094.

Ali, S., Tyagi, A. and Mir, Z.A. (2024) Plant immunity: At the crossroads of pathogen perception and defense response. *Plants* 13(11), 1434. DOI: 10.3390/plants13111434.

Alvi, A.F., Sehar, Z., Fatma, M., Masood, A. and Khan, N.A. (2022) Strigolactone: An emerging growth regulator for developing resilience in plants. *Plants* 11(19), 2604. DOI: 10.3390/plants11192604.

Bajguz, A. and Piotrowska-Niczyporuk, A. (2023) Biosynthetic pathways of hormones in plants. *Metabolites* 13(8), 884. DOI: 10.3390/metabo13080884.

Bendejacq-Seychelles, A., Gibot-Leclerc, S., Guillemin, J.-P., Mouille, G. and Steinberg, C. (2024) Phytotoxic fungal secondary metabolites as herbicides. *Pest Management Science* 80(1), 92–102. DOI: 10.1002/ps.7813.

Biolatti, M., Blangetti, M., d'Arrigo, G., Spyrakis, F., Cappello, P. *et al.* (2020) Strigolactone analogs are promising antiviral agents for the treatment of human cytomegalovirus infection. *Microorganisms* 8(5), 703. DOI: 10.3390/microorganisms8050703.

Brilli, F., Gallo, F. and Balestreri, C. (2024) Making use of biogenic volatile organic compounds in sustainable agriculture and from aquatic ecosystems. In: Brilli, F. and Decesari, S. (eds) *Biogenic Volatile Organic Compounds and Climate Change*. Elsevier, Amsterdam, pp. 117–163.

Brun, G., Billard, E., Hýlová, A., Montiel, G., Plačková, L. *et al.* (2025) Divergent control of seed germination by cytokinins in weedy broomrapes and witchweeds. *Plant And Cell Physiology (in Press)*. DOI: 10.1093/pcp/pcaf032.

Butler, L. (1995) Chemical communication between the parasitic weed striga and its crop host. In: Inderjit, A., Dakshini, K. and Einhellig, F. (eds) *Allelopathy*. American Chemical Society, Washington DC, pp. 158–168.

Chen, S.J., Wang, S.C. and Chen, Y.C. (2022) Challenges, recent advances and perspectives in the treatment of human cytomegalovirus infections. *Tropical Medicine and Infectious Disease* 7(12), 439. DOI: 10.3390/tropicalmed7120439.

Chen, Y., He, B., Hu, M., Bao, J., Yan, W. *et al.* (2024) Fluorescent probes for imaging and detection of plant hormones and their receptors. *Advanced Agrochem* 3(1), 83–98. DOI: 10.1016/j.aac.2023.09.002.

Choubey, K., Dey, S.K., Dey, G. and Boruah, T. (2025) Understanding the role of phytohormones in mitigation of biotic and abiotic stress: Current status and future prospects. *Proceedings of the*

National Academy of Sciences, India Section B: Biological Sciences 95(1), 9–14. DOI: 10.1007/s40011-024-01566-8.

Claassens, A.P. (2024) *Role of Strigolactones in Adventitious Rooting of Arabidopsis thaliana callus in vitro*. PhD thesis, Stellenbosch University, Stellenbosch, South Africa.

Clark, J. and Bennett, T. (2024) Cracking the enigma: Understanding strigolactone signalling in the rhizosphere. *Journal of Experimental Botany* 75(4), 1159–1173. DOI: 10.1093/jxb/erad335.

Das, S., Shil, S., Rime, J., Alice, A.K., Yumkhaibam, T. *et al*. (2025) Phytohormonal signaling in plant resilience: Advances and strategies for enhancing abiotic stress tolerance. *Plant Growth Regulation* 105(2), 329–360. DOI: 10.1007/s10725-025-01279-6.

Demis, E. (2024) Mechanism of plant resistance to insects, weeds and pathogens. *Middle East Research Journal of Agriculture and Food Science* 4(2), 76–85. DOI: 10.36348/merjafs.2024.v04i02.005.

Demurtas, O.C., Nicolia, A. and Diretto, G. (2023) Terpenoid transport in plants: How far from the final picture? *Plants* 12(3), 634. DOI: 10.3390/plants12030634.

Deng, Q., Wang, H., Qiu, Y., Wang, D., Xia, Y. *et al*. (2025) The multifaceted impact of karrikin signaling in plants. *International Journal of Molecular Sciences* 26(6), 2775. DOI: 10.3390/ijms26062775.

Dhaouadi, S. and Ali Rhouma, M.A.H. (2020) The plant pathogen *Rhodococcus fascians*: History, disease symptomatology, host range, pathogenesis and plant–pathogen interaction. *Annals of Applied Biology* 177(1), 4–15. DOI: 10.1111/aab.12600.

Emerenciano, M.G.C., Khanjani, M.H., Sharifinia, M. and Miranda-Baeza, A. (2025) Could biofloc technology (BFT) pave the way toward a more sustainable aquaculture in line with the circular economy? *Aquaculture Research* 2025(1), 1020045. DOI: 10.1155/are/1020045.

Francis, B., Aravindakumar, C.T. and Simon, S. (2025) Strigolactones and plant defense: Protection against pests and pathogen. In: Aftab, T./ and Wani, K.I. (eds) *Strigolactones: Emerging Plant Hormones*. Scrivener Publishing, Beverly, Massachusetts, pp. 281–300. DOI: 10.1002/9781394302826.ch13.

Fujita, M., Tanaka, T., Kusajima, M., Inoshima, K., Narita, F. *et al*. (2024) Enhanced disease resistance against botrytis cinerea by strigolactone-mediated immune priming in *Arabidopsis thaliana*. *Journal of Pesticide Science* 49(3), 186–194. DOI: 10.1584/jpestics.D24-019.

Ganchozo-Mendoza, E., Quadros, A.F.F., Zerbini, F.M., Flores, F.J. and Garcés-Fiallos, F.R. (2024) Characterization and genome analysis of cucumber mosaic virus on commercial tobacco plants in Ecuador. *Tropical Plant Pathology* 49(4), 480–488. DOI: 10.1007/s40858-024-00643-7.

Grich, N., Huynh, T., Kisiala, A., Palberg, D. and Emery, R.J.N. (2025) The biosynthesis and impacts of cytokinins on growth of the oyster mushroom, *Pleurotus ostreatus*. *Mycologia* 117(1), 76–94. DOI: 10.1080/00275514.2024.2401320.

Guha, A. (2025) *Common Diseases of Field Crops and Their Management*. Educohack Press, Delhi, India.

Hajam, A.H. and Bashri, G. (2025) Strigolactones: Key phytohormones in plant–microbe interactions and development. In: Aftab, T. and Wani, K.I. (eds) *Strigolactones: Emerging Plant Hormones*. Scrivener Publishing, Beverly, Massachusetts, pp. 235–259. DOI: 10.1002/9781394302826.

Helmich, R. (2025) Unveiling the strigolactone signaling pathway: From receptors to responses. In: Aftab, T. and Wani, K.I. (eds) *Strigolactones: Emerging Plant Hormones*. Scrivener Publishing, Beverly, Massachusetts, pp. 35–67. DOI: 10.1002/9781394302826.

Hu, Z., Ma, Q., Foyer, C.H., Lei, C., Choi, H.W. *et al*. (2021) High CO2- and pathogen-driven expression of the carbonic anhydrase βCA3 confers basal immunity in tomato. *New Phytologist* 229(5), 2827–2843. DOI: 10.1111/nph.17087.

Huang, R., Bie, S., Li, S., Yuan, B., Zhang, L. *et al*. (2024) Strigolactones negatively regulate tobacco mosaic virus resistance in *Nicotiana benthamiana*. *International Journal of Molecular Sciences* 25(15), 8518. DOI: 10.3390/ijms25158518.

Ibdah, M., Dubey, N.K., Eizenberg, H., Dabour, Z., Abu-Nassar, J. *et al*. (2014) Cucumber mosaic virus as a carotenoid inhibitor reducing *Phelipanche aegyptiaca* infection in tobacco plants. *Plant Signaling & Behavior* 9(10), e972146. DOI: 10.4161/psb.32096.

Jain, S., Kumar, R., Gunsola, D., Živković, S., Vasić, T. *et al*. (2025) Strigolactones and control of parasitic weeds. In: Aftab, T. and Wani, K.I. (eds) *Strigolactones: Emerging Plant Hormones*. Scrivener Publishing, Beverly, Massachusetts, pp. 261–279. DOI: 10.1002/9781394302826.

Janeeshma, E., Habeeb, H., Shackira, A.M., Sinisha, A.K., Mirshad, P.P. *et al*. (2024) Strigolactone and analogues: A new generation of plant hormones with multifactorial benefits in environmental sustainability. *Environmental and Experimental Botany* 223, 105775. DOI: 10.1016/j.envexpbot.2024.105775.

Jardim-Messeder, D., Souza-Vieira, Y. and Sachetto-Martins, G. (2025) Dressed up to the nines: The interplay of phytohormones signaling and redox metabolism during plant response to drought. *Plants* 14(2), 208. DOI: 10.3390/plants14020208.

Jung, H., Han, G., Lee, D., Jung, H.-K., Kim, Y.-S. *et al.* (2025) Understanding the impact of salt stress on plant pathogens through phenotypic and transcriptomic analysis. *Plants* 14(1), 97. DOI: 10.3390/plants14010097.

Kameoka, H. and Kyozuka, J. (2018) Spatial regulation of strigolactone function. *Journal of Experimental Botany* 69(9), 2255–2264. DOI: 10.1093/jxb/erx434.

Kapoor, R.T., Alam, P., Chen, Y. and Ahmad, P. (2024) Strigolactones in plants: From development to abiotic stress management. *Journal of Plant Growth Regulation* 43(3), 903–919. DOI: 10.1007/s00344-023-11148-z.

Khatun, M., Nasir, N.A.N.M., Zakarya, I.A. and Islam, A.A. (2024) Role of strigolactone in the alleviation of biotic stress in plants. In: Abd-Elsalam, K.A. and Mohamed, H.I. (eds) *Plant Growth Regulators to Manage Biotic and Abiotic Stress in Agroecosystems*. CRC Press, Boca Raton, Florida, pp. 99–124.

Kumar, A. and Mohanasundari, T. (2025) Assessing climate change risk and vulnerability among Bhil and Bhilala tribal communities in Madhya Pradesh, India: A multidimensional approach. *Scientific Reports* 15(1), 7096. DOI: 10.1038/s41598-025-90390-9.

Kuno, M., Miyamoto, A., Takano, H., Homma, M., Shiotani, N. *et al.* (2025) CYP722A1-mediated 16-hydroxylation of carlactonoic acid regulates the floral transition in *Arabidopsis*. *Plant & Cell Physiology* 66(4), 645–657. DOI: 10.1093/pcp/pcaf022.

Kusajima, M., Fujita, M., Soudthedlath, K., Nakamura, H., Yoneyama, K. *et al.* (2022) Strigolactones modulate salicylic acid-mediated disease resistance in *Arabidopsis thaliana*. *International Journal of Molecular Sciences* 23(9), 5246. DOI: 10.3390/ijms23095246.

Lazou, M., Khan, O., Nguyen, T., Padhorny, D., Kozakov, D. *et al.* (2024) Predicting multiple conformations of ligand binding sites in proteins suggests that AlphaFold2 may remember too much. *Proceedings of the National Academy of Sciences USA* 121(48), e2412719121. DOI: 10.1073/pnas.2412719121.

Li, X., Fu, W., Meng, F. and Guan, C. (2022) Carotenoid-mediated regulation of photosynthetic performance and antioxidant defense confer tolerance to perfluorooctanoic acid stress in *Nicotiana tabacum*. *Plant Growth Regulation* 98(2), 307–319. DOI: 10.1007/s10725-022-00856-3.

Liu, S., Wang, J., Song, B., Gong, X., Liu, H. *et al.* (2023) Conformational dynamics of the D53–D3–D14 complex in strigolactone signaling. *Plant & Cell Physiology* 64(9), 1046–1056. DOI: 10.1093/pcp/pcad067.

Liu, Y., Chen, S., Pal, S., Yu, J., Zhou, Y. *et al.* (2024) The hormonal, metabolic, and environmental regulation of plant shoot branching. *New Crops* 1, 100028. DOI: 10.1016/j.ncrops.2024.100028.

López-Pagán, N., Rufián, J.S., Luneau, J., Sánchez-Romero, M.-A., Aussel, L. *et al.* (2025) *Pseudomonas syringae* subpopulations cooperate by coordinating flagellar and type III secretion spatiotemporal dynamics to facilitate plant infection. *Nature Microbiology* 10(4), 958–972. DOI: 10.1038/s41564-025-01966-0.

Ma, L. and Lin, Y. (2025) Orthogonal RNA replication enables directed evolution and darwinian adaptation in mammalian cells. *Nature Chemical Biology* 21(3), 451–463. DOI: 10.1038/s41589-024-01783-2.

Marro, N., Lidoy, J., Chico, M.Á., Rial, C., García, J. *et al.* (2022) Strigolactones: New players in the nitrogen–phosphorus signalling interplay. *Plant, Cell & Environment* 45(2), 512–527. DOI: 10.1111/pce.14212.

Mashiguchi, K., Seto, Y. and Yamaguchi, S. (2021) Strigolactone biosynthesis, transport and perception. *Plant Journal* 105(2), 335–350. DOI: 10.1111/tpj.15059.

Mehboob, I., Mughees, M., Baig, A., Ali, S., Sajjad, Y. *et al.* (2023) An efficient virus-induced gene silencing of PDS gene in *Solanum lycopersicum* (cv. Rio Grande) and its functional analysis. *Brazilian Journal of Botany* 46(4), 881–892. DOI: 10.1007/s40415-023-00941-5.

Mehmood, I., Wani, K.I. and Aftab, T. (2025) Nature, structural diversity, biosynthetic pathway, and strigolactone transport in plants. In: Aftab, T. and Wani, K.I. (eds) *Strigolactones: Emerging Plant Hormones*. Scrivener Publishing, Beverly, Massachusetts, pp. 13–34. DOI: 10.1002/9781394302826.

Mehra, P., Banda, J., Ogorek, L.L.P., Fusi, R., Castrillo, G. *et al.* (2025) Root growth and development in "real life": Advances and challenges in studying root–environment interactions. *Annual Review of Plant Biology* 76(1), 467–492. DOI: 10.1146/annurev-arplant-083123-074506.

Mishra, P., Saini, P. and Patni, V. (2024) Biochemical dynamics during development of insect-induced plant galls: A review. *Journal of Plant Diseases and Protection* 131(6), 1803–1818. DOI: 10.1007/s41348-024-00931-z.

Modrzejewska, M., Połubiński, P. and Zdanowska, O. (2024) Ophthalmic complications, diagnosis, and treatment of congenital human cytomegalovirus infection. *Journal of Clinical Medicine* 13(12), 3379. DOI: 10.3390/jcm13123379.

Moosavi, M., Khorassani, R. and Tavakkol Afshari, R. (2025) Strigolactone (GR24) and abscisic acid induced drought tolerance in wheat by ameliorating nutrient uptake. *Journal of Plant Growth Regulation* 44(2), 972–987. DOI: 10.1007/s00344-024-11508-3.

Mukherjee, S. and Aftab, T. (eds) (2023) *Strigolactones, Alkamides and Karrikins in Plants: Recent Updates and Future Prospects*. CRC Press, Boca Raton, Florida.

Nagaraja, A., Chethana, B.S. and Jain, A.K. (2021) Biotic stresses and their management. In: Singh, M. and Sood, S. (eds) *Millets and Pseudo Cereals: Genetic Resources and Breeding Advancements*. Woodhead Publishing, Cambridge, UK, pp. 119–142.

Naseer, M.A., Zhang, Z.Q., Mukhtar, A., Asad, M.S., Wu, H.Y. *et al.* (2024) Strigolactones: A promising tool for nutrient acquisition through arbuscular mycorrhizal fungi symbiosis and abiotic stress tolerance. *Plant Physiology and Biochemistry* 215, 109057. DOI: 10.1016/j.plaphy.2024.109057.

Nicolle, C. (2023) Molecular mechanisms of the interaction of a rhizospheric *Streptomyces* sp. strain with plant roots. PhD thesis, Université Paul Sabatier-Toulouse III, Toulouse, France.

Nomura, T., Seto, Y. and Kyozuka, J. (2024) Unveiling the complexity of strigolactones: Exploring structural diversity, biosynthesis pathways, and signaling mechanisms. *Journal of Experimental Botany* 75(4), 1134–1147. DOI: 10.1093/jxb/erad412.

Nuzhnaya, T.V., Sorokan, A.V., Burkhanova, G.F., Maksimov, I.V. and Veselova, S.V. (2024) The role of cytokinins and abscisic acid in the growth, development and virulence of the pathogenic fungus *Stagonospora nodorum* (Berk.). *Biomolecules* 14(5), 517. DOI: 10.3390/biom14050517.

O'Malley, M.R. and Anderson, J.C. (2021) Regulation of the pseudomonas syringae type III secretion system by host environment signals. *Microorganisms* 9(6), 1227. DOI: 10.3390/microorganisms9061227.

Oyedele, A.O., Ezaka, E. and Taiwo, L.B. (2024) Microbial biosynthesis of the classical phytohormones by plant growth-promoting microorganisms in plants. In: Chauhan, P.S., Bisht, N. and Agarwal, R. (eds) *Microbial Biostimulants for Plant Growth and Abiotic Stress Amelioration*. Elsevier, Amsterdam, pp. 345–366. DOI: 10.1016/B978-0-443-13318-3.00004-5.

Palayam, M., Yan, L., Nagalakshmi, U., Gilio, A.K., Cornu, D. *et al.* (2024) Structural insights into strigolactone catabolism by carboxylesterases reveal a conserved conformational regulation. *Nature Communications* 15(1), 6500. DOI: 10.1038/s41467-024-50928-3.

Paudel, L., Kerr, S., Prentis, P., Tanurdžić, M., Papanicolaou, A. *et al.* (2022) Horticultural innovation by viral-induced gene regulation of carotenogenesis. *Horticulture Research* 9, uhab008. DOI: 10.1093/hr/uhab008.

Păun, A.M., Mernea, M. and Avram, S. (2024) Bioinformatical and experimental approaches of strigolactones receptors. *Romanian Journal of Biophysics* 34(3), 103–116. DOI: 10.59277/RJB.2024.3.01.

Perdomo, S.A., Valencia, D.P., Velez, G.E. and Jaramillo-Botero, A. (2024) Advancing abiotic stress monitoring in plants with a wearable non-destructive real-time salicylic acid laser-induced-graphene sensor. *Biosensors & Bioelectronics* 255, 116261. DOI: 10.1016/j.bios.2024.116261.

Pontiggia, D., Giulietti, S., Gramegna, G., Lionetti, V., Lorrai, R. *et al.* (2023) The ancient battle between plants and pathogens: Resilience of the plant cell wall and damage-associated molecular patterns (DAMPs) drive plant immunity. In: Geitmann, A. (ed.) *Plant Cell Walls: Research Milestones and Conceptual Insights*. CRC Press, Boca Raton, Florida, pp. 393–411. DOI: 10.1201/9781003178309-20.

Roy, S., Kang, Y., Zhang, S., Torres-Jerez, I., Jain, D. *et al.* (2024) A multitrait genome-wide association study reveals a requirement for the strigolactone receptor MtDWARF14 in optimal GOLVEN signaling. *bioRxiv (Preprint)*. DOI: 10.1101/2024.06.24.599968.

Saha, P., Debnath, B., Sharma, N., Dutta, S., Barman, A. *et al.* (2025) Phytohormones: Crucial factors for plant stress tolerance. In: Chen, J.-T. (ed.) *Plant Stress Tolerance: Molecular Mechanisms and Breeding Strategies*, Vol. 2. CRC Press, Boca Raton, Florida, pp. 113–133.

Scarpin, D. (2024) Innovative technologies enabling plants to increase resilience as a pivotal feature in the conservation of agro-ecosystems. PhD thesis. Università degli Studi di Udine, Udine, Italy. Available at: https://air.uniud.it/handle/11390/1276924 (accessed 1 February 2026).

Seto, Y. (2024) Latest knowledge on strigolactone biosynthesis and perception. *Bioscience, Biotechnology, and Biochemistry* 88(1), 1–7. DOI: 10.1093/bbb/zbad150.

Sharma, P., Jha, A.B. and Dubey, R.S. (2024) Strigolactones: Coordination with other phytohormones and enhancement of abiotic stress responses. *Environmental and Experimental Botany* 223, 105782. DOI: 10.1016/j.envexpbot.2024.105782.

Sharmin, S., Rahaman, M.M., Martorell, M., Sastre-Serra, J., Sharifi-Rad, J. *et al.* (2021) Cytotoxicity of synthetic derivatives against breast cancer and multi-drug resistant breast cancer cell lines: A literature-based perspective study. *Cancer Cell International* 21(1), 612. DOI: 10.1186/s12935-021-02309-9.

Shivran, M., Sharma, N., Dubey, A.K., Singh, S.K., Sharma, N. *et al.* (2022) Scion–rootstock relationship: Molecular mechanism and quality fruit production. *Agriculture* 12(12), 2036. DOI: 10.3390/agriculture12122036.

Steidl, P., Hackett, P.M. and Gordley-Smith, A. (2025) *Environmental and Social Justice Issues: A Declarative Mapping Literature-based Approach for Achieving Pro-ecological and Social Change*. Routledge, London. DOI: 10.4324/9781003477167.

Struk, S.M. (2018) Elucidation of the protein–protein interaction network at play during strigolactone and karrikin signaling. PhD thesis, Ghent University, Ghent, Belgium.

Sun, Y., Zhao, J., Patil, S.B., Fang, J., Liu, J. *et al.* (2021) Improved dual luciferase reporter (DLR) assay to determine the protein stability. *Analytical Biochemistry* 612, 114021. DOI: 10.1016/j.ab.2020.114021.

Temmerman, A., Guillory, A., Bonhomme, S., Goormachtig, S. and Struk, S. (2022) Masks start to drop: suppressor of MAX2 1-like proteins reveal their many faces. *Frontiers in Plant Science* 13, 887232. DOI: 10.3389/fpls.2022.887232.

Thiombiano, B., Floková, K., Walmsley, A. and Bouwmeester, H.J. (2023) Striga germination stimulant analysis. In: Ghanim, A.M.A., Sivasankar, S. and Rich, P.J. (eds) *Mutation Breeding and Efficiency Enhancing Technologies for Resistance to Striga in Cereals*. Springer, Berlin/Heidelberg, pp. 115–141. DOI: 10.1007/978-3-662-68181-7_8.

Thomas, R. (2022) A quantitative RT-PCR assay to monitor luciferase reporter mRNA levels. PhD thesis, UC, Santa Cruz, Santa Cruz, California.

Torres, L.F., de Andrade, S.A.L. and Mazzafera, P. (2021) Split-root, grafting and girdling as experimental tools to study root-to shoot-to root signaling. *Environmental and Experimental Botany* 191, 104631. DOI: 10.1016/j.envexpbot.2021.104631.

Valivand, M. Mukherjee, S. and Aftab, T. (eds). (2023) *Biosynthetic Regulation of Strigolactones in Plants*. CRC Press, Boca Raton, Florida, pp. 7–18.

Varadharajan, V., Rajendran, R., Muthuramalingam, P., Runthala, A., Madhesh, V. *et al.* (2025) Multi-omics approaches against abiotic and biotic stress—A review. *Plants* 14(6), 865. DOI: 10.3390/plants14060865.

Verma, R., Sharma, B. and Rather, G.A. (2024) Strigolactones: Biosynthesis, regulation, signaling, roles, and response to stress. In: Bashri, G., Hayat, S. and Bajguz, A. (eds) *Strigolactones: Synthesis, Application and Role in Plants*. Academic Press, London, pp. 147–188. DOI: 10.1016/B978-0-443-13521-7.00009-9.

Wakabayashi, T., Nakayama, M., Kitano, Y., Homma, M., Miura, K. *et al.* (2025) Discovery of strigol synthase from cotton (*Gossypium hirsutum*): The enzyme behind the first identified germination stimulant for striga. *Plants, People, Planet* 7(2), 354–359. DOI: 10.1002/ppp3.10420.

Wang, D., Pang, Z., Yu, H., Thiombiano, B., Walmsley, A. *et al.* (2022) Probing strigolactone perception mechanisms with rationally designed small-molecule agonists stimulating germination of root parasitic weeds. *Nature Communications* 13(1), 3987. DOI: 10.1038/s41467-022-31710-9.

Wang, M., Yuan, Y., Zhao, Y., Hu, Z., Zhang, S. *et al.* (2025) PhWRKY30 activates salicylic acid biosynthesis to positively regulate antiviral defense response in petunia. *Horticulture Research* 12(5). DOI: 10.1093/hr/uhaf013.

Wang, Q., Ye, Y., Wang, L., Guan, Y., Wang, S. *et al.* (2025) Independent horizontal transfer of genes encoding α/β-hydrolases with strigolactone binding and hydrolytic activities from bacteria to fungi and plants. *Molecular Plant* 18(11), 1949–1961. DOI: 10.1016/j.molp.2025.09.021.

Xie, X., Yoneyama, K., Kisugi, T., Nomura, T., Akiyama, K. *et al.* (2015) Strigolactones are transported from roots to shoots, although not through the xylem. *Journal of Pesticide Science* 40(4), 214–216. DOI: 10.1584/jpestics.D15-045.

Xu, Y., Lv, Z., Manzoor, M.A., Song, L., Wang, M. *et al.* (2024) VvD14c-VvMAX2-VvLOB/VvLBD19 module is involved in the strigolactone-mediated regulation of grapevine root architecture. *Molecular Horticulture* 4(1), 40. DOI: 10.1186/s43897-024-00117-z.

Yang, L., Fang, S., Liu, L., Zhao, L., Chen, W. *et al.* (2025) WRKY transcription factors: Hubs for regulating plant growth and stress responses. *Journal of Integrative Plant Biology* 67(3), 488–509. DOI: 10.1111/jipb.13828.

Yetgin, A., Srivastava, R.K. and Mandal, N. (2025) Insights into plant hormone signaling networks for environmental responses. In: Srivastava, R.K. and Chakraborty, A. (eds) *Mitigation and Adaptation Strategies Against Climate Change in Natural Systems*. Springer Nature, Cham, Switzerland, pp. 505–523.

Yoshimura, M., Fonné-Pfister, R., Screpanti, C., Hermann, K., Rendine, S. *et al.* (2019) Total synthesis and biological evaluation of heliolactone. *Helvetica Chimica Acta* 102(11), e1900211. DOI: 10.1002/hlca.201900211.

Zahoor, A., Khazer, R., Mehraj, I., Gani, U., Fayaz, F. *et al.* (2025) Aberrant DNA methylation as a key modulator of cell death pathways: Insights into cancer progression and other diseases. *Functional & Integrative Genomics* 25(1), 50. DOI: 10.1007/s10142-025-01552-x.

Zaidi, S., Arif, Y., Imtiaz, H., Shiraz, M. and Hayat, S. (2024) Structural chemistry, biosynthesis, and signaling of multifaceted plant growth regulator: Strigolactone. *Journal of Plant Growth Regulation* 43(8), 2489–2502. DOI: 10.1007/s00344-024-11285-z.

Zhang, C., Wang, F., Jiao, P., Liu, J., Zhang, H. *et al.* (2024) The overexpression of Zea mays strigolactone receptor gene D14 enhances drought resistance in *Arabidopsis thaliana* L. *International Journal of Molecular Sciences* 25(2), 1327. DOI: 10.3390/ijms25021327.

Zhang, X., Wang, G., Xue, H., Zhang, J., Wang, Q. *et al.* (2020) Metabolite profile of xylem sap in cotton seedlings is changed by K deficiency. *Frontiers in Plant Science* 11, 592591. DOI: 10.3389/fpls.2020.592591.

Zhao, X., Liu, Q. and Tan, L. (2024) Callose and salicylic acid are key determinants of strigolactone-mediated disease resistance in *Arabidopsis*. *Plants* 13(19), 2766. DOI: 10.3390/plants13192766.

Zhong, L., Yang, C., Chen, Y., Guo, L., Liu, D. *et al.* (2024) Reduced strigolactone synthesis weakens drought resistance in tall fescue via root development inhibition. *Agronomy* 14(4), 725. DOI: 10.3390/agronomy14040725.

Zhou, A., Kane, A., Wu, S., Wang, K., Santiago, M. *et al.* (2025) Evolution of interorganismal strigolactone biosynthesis in seed plants. *Science* 387(6731), eadp0779. DOI: 10.1126/science.adp0779.

17 Recent Developments in Strigolactone–Heavy-metal Research in Plants

Aribah[1], Nadia Sharif[2], Muhammad Faheem Siddiqui[1], Hina Jabeen[2], Sanjay Gairola[3] and Zainul Abideen[3,4*]

[1]Department of Botany, University of Karachi, Karachi, Pakistan; [2]Department of Biotechnology, Women University Mardan, Mardan, Khyber Pakhtunkhwa, Pakistan; [3]Sharjah Seed Bank and Herbarium, Environment and Protected Areas Authority, Sharjah, United Arab Emirates; [4]Dr. Muhammad Ajmal Khan Institute of Sustainable Halophyte Utilization, University of Karachi, Karachi, Pakistan

Abstract

Strigolactone (SL) levels can be critical for plant performance to enhance biomass under higher heavy-metal-affected soils. SL application can modify nutrients status by enhancing the supply of basic nutrients and altering the fate of heavy metals by upgrading microbial interactions. This chapter defines the role of SLs in soil fertility, plant growth, photosynthesis, and promotion of plant tolerance to heavy-metal toxicity. Results have demonstrated that exogenous application of SLs, particularly the synthetic analog GR24, is also applicable in the regeneration of plant biomass and chlorophyll concentration, as well as the root systems of plants in cadmium- and chromium-contaminated soils. Photosynthetic performance is improved and thus SLs maintain energy balance despite oxidation damage. They participate in the homeostasis of reducing oxidation by activating antioxidant defense through increased superoxide dismutase, catalase, and peroxidase activities. In contrast, mutants deficient in SLs have poor resilience, indicating the necessity of endogenous SLs in regulation of the stress response.

Keywords: arbuscular mycorrhizal fungi (AMF), biogenic strigolactones, heavy-metal stress, phytohormone, reactive oxygen species (ROS), solidification/stabilization

17.1 Introduction

The living conditions of people are greatly influenced by their surroundings. The biosphere is generally recognized as a vital component of the natural world, primarily due to its essential role in providing a habitat for a diverse array of living species. Environmental pollution refers to alterations in the atmosphere, hydrosphere, and lithosphere of physical, chemical, and biological features, which can pose a threat to the existence of a wide range of species, including both plants and animals (Masindi and Muedi, 2018). Soil is the basis of agricultural production and has a wide variety of mineral components, including hazardous and necessary trace elements, such as zinc. Optimum amounts of copper and other heavy metals are necessary for healthy human,

*Corresponding author: zuabideen@uok.edu.pk

DOI: 10.1079/9781836990598.0017

animal, and plant functions, although too much of them can be dangerous, while cadmium, mercury, arsenic, and other elements are non-essential and harmful to living things, even in modest quantities (Tiecher *et al.*, 2016).

Heavy metals can be released into the environment via industrial activities such as mining, as well as natural causes such as volcanic eruptions. As populations and industries grow, the need for clean water arises. Many regions of the world have limited drinking water, making water contamination a worldwide issue. A variety of contaminants released from various human sources have harmed the quality of water supplies. More specifically, heavy-metal discharge is a primary contributor to water pollution (Jagaba *et al.*, 2024). The presence of heavy metals in wastewater discharged from laboratories and industries is a significant source of environmental worry, as they are durable environmental contaminants that cannot be completely biodegraded and must be transformed into harmless forms before the water can be ingested (Zainon *et al.*, 2025). Factories and other industrial facilities often discharge hazardous compounds into aquatic systems through their effluents. The direct dumping of dangerous metals into rivers, lakes, and seas, as well as mining, metal smelting, and other industrial activities, contribute to pollution of the seas. Heavy metals may build up in soil as a result of the application of fertilizers, pesticides, and herbicides that include them. These metals may pollute soil and water and degrade over time, dissolving into groundwater or being carried into nearby water sources during rainfall (Maftouh *et al.*, 2024).

Heavy metals have a detrimental impact on the distribution, functional activity, and diversity of soil microbes, as well as the production of enzymes that cycle carbon and nutrients (Xu *et al.*, 2019). Plant hormones are small compounds produced through various essential metabolic routes. They play essential roles in all developmental stages of plants, from early embryogenesis to maturity. Plant hormone research began at the turn of the 20th century and resulted in the discovery of auxin, ethylene, cytokinins, and gibberellins. Some of the plant hormones that control plant growth and development include abscisic acid (ABA), brassinosteroids, jasmonic acid, salicylic acid, and the newly discovered strigolactones (SLs). With a few minor exceptions, such as the omission of particular processes in SL synthesis, the manufacturing pathways of plant hormones are largely understood. In general, plant hormones affect plants. Local hormones are found at or around the point of production, whereas circulating hormones travel throughout the plant. The mechanisms by which hormones communicate differ, and include hormone-signaling pathways whereby hormones interact with gene expression. A common cross-talk mechanism involves controlling specific key elements of other hormone-signaling pathways (Santner and Estelle, 2010; Bajguz and Piotrowska-Niczyporuk, 2023).

SLs play an important role in regulating plant growth and developmental stages and have generated much scientific interest. SLs were initially discovered by Cook *et al.* (1966), who reported the discovery of (+)-strigol from cotton root exudates. It took almost two decades to fully define its structure. More than 35 natural SLs have now been identified from a variety of plant species, with the majority found in root exudates or roots (Guercio *et al.*, 2023; Daignan-Fornier *et al.*, 2024). Environmental challenges, such as drought and high salinity, impair plant development and productivity, and various phytohormones such as ABA, brassinosteroids, and cytokinins play critical roles in the response of the plant to these types of stress, working together to govern the plant's adaptive responses. The effect of SLs on plant stress responses are currently under investigation. In one study, a number of SL-deficient and SL-signaling mutants were functionally examined to see whether SLs plays a role in modulating drought- and salt-stress responses (Liao *et al.*, 2025). According to the findings, SLs act as a positive regulator of stress signaling networks and a variety of ABA signaling pathways by regulating the expression of several stress-related genes and/or ABA-responsive genes involved in plant development and the abiotic stress response.

In recent years, water-related challenges have intensified dramatically, including drought and salinity, severely affecting plant growth and development worldwide. Drought is a chronic abiotic stress, intensified by global warming, that accounts for around 70% of total global crop losses. According to projections, drought will cause productivity losses of 30% by the year 2050. Drought occurs when plant transpiration exceeds water uptake by the roots due to inadequate rainfall or a decrease in groundwater level. To satisfy the population food requirements, research on enhanced drought-tolerant crops, as well as the discovery of inexpensive and sustainable alternatives, is necessary (Muhammad Fasih *et al.*, 2024). Photosynthesis is critical for sustaining plant development, with photosynthetic pigments playing an important role. Exogenous SL treatment considerably boosts photosynthetic pigment concentration under drought stress by allowing plants to increase chlorophyll concentration in the leaves, which promotes photosynthesis. According to research, the exogenous administration of SLs has a beneficial regulatory effect on minimizing abiotic stress in plants (Yan *et al.*, 2024).

SLs modify root architecture to increase water absorption during drought. The synthetic SL analog GR24 has been shown to promote root growth, boost water absorption, and improve drought resilience in a variety of plants, such as crab apple (Xu *et al.*, 2023). SLs respond to a variety of stressors at the transcriptional level. Abiotic stress causes SL biosynthesis in plants, as seen by increased expression. SLs promote plant stress tolerance by regulating the expression of relevant genes such as *carotenoid cleavage dioxygenase 7* (*CCD7*), *carotenoid cleavage dioxygenase 8* (*CCD8*), and *DWARF27* (*D27*) (Zhang *et al.*, 2020; Yang *et al.*, 2023). Global climate change is a serious problem. Since 1850, the average temperature of land and sea has risen by an average of around 0.08°C per decade, although the rate of warming has been more than three times faster since 1982, at approximately 0.20°C per decade (Dunn *et al.*, 2024). Over the previous century, significant greenhouse-gas emissions, rapid population growth, and fossil fuel burning have been recognized as the major contributors to global warming. Between 1850 and 2019, the global surface temperature rose by 0.8–1.3°C (Calvin *et al.*, 2023). This large temperature rise directly affects biodiversity, frequently with noxious effects. Increased temperatures have a range of impacts on climate, for example increased South-east Asian monsoon rains and markedly rising sea levels, and mega-droughts and agricultural collapses across parts of Africa and the southern USA. It can result in the disappearance of ice on mountains, and the falling of river and lake flows. Cutbacks in water supply also affect irrigation and the supply of drinking water (Muluneh, 2021). Climate change has far-reaching consequences regarding biodiversity loss, including species extirpation, population loss, alterations in species abundance, and changes in species' geographical range (Habibullah *et al.*, 2022).

17.2 Soil Pollution by Heavy Metals

Soil, as one of the primary components of the ecosystem, is composed of particles of many different dimensions. This valuable resource is under threat due to the negative impacts of pollution (Nakai, 2018). The most significant cause of soil pollution is usually human activities and interventions (Cachada *et al.*, 2018), while natural factors such as forest fires and volcano eruptions can also add poisonous chemicals to the environment. These can have major effects on the ecosystem, which depends heavily on soil formation and the presence of diverse chemicals and heavy metals. The natural variation in trace metals is related to the parent rocks from which the soil is formed, and affects the extent to which the amount of heavy metals in soil is significantly different between regions (Shetty *et al.*, 2023).

By causing increased precipitation, climate change has severely interfered with soil and its ability to sustain the most significant microbial life. Thus, excessive rainfall leads to soil erosion, which has been one of the largest contributors to land degradation and reduction of fertile top soil containing large quantities of organic matter. This leaves the soil with a dramatically reduced ability to retain water, hence impairing its functionality (Oishy *et al.*, 2025). The physical and chemical properties of the environment are affected by pollutants of the soil, and significantly influence a wide range of changes including changes in soil temperature, profiles, and texture, rainfall, and absorption of nutrients, among others. Most of these pollutants are the causes of soil degradation, thus affecting the health of vegetation and reducing the quality of water in the soil (Biswas *et al.*, 2018). The industrialization process has led to extreme manipulation of the environment due to the excessive exploitation of natural and mineral resources and the unsustainability of human activity. Some of the numerous effects include contamination of the environment and the subsoil, usually as an end result of dumping of urban and industrial wastes directly in dug pits in the earth (Isele *et al.*, 2025).

Heavy metals represent a category of pollutants that pose significant risks to human health. Exposure to these hazardous substances occurs through various environmental pathways. Naturally occurring in the earth's ecosystem, these metal compounds are resistant to decomposition and tend to accumulate in soil, water, and sediment layers within lakes and rivers. The impact of climate change exacerbates this issue by reducing soil moisture and increasing evapotranspiration, which in turn increases the resuspension of soil dust particles, further contributing to environmental contamination. Heavy metals in the atmosphere appear to be present at relatively low concentrations; however, they contribute significantly to deposition and accumulation in soils. These metals are known for their persistence in the environment and their tendency to bioaccumulate within food chains. Human activities have been major contributors to the introduction of heavy metals in the environment, which has been aggravated by naturally occurring environmental processes. Their transport across the atmosphere as well as water entails dispersion and deposition processes, with their eventual accumulation in soil and water sediments. These sediments act as reservoirs, and the release of heavy metals into the atmosphere and other environmental systems is secondary and semi-natural. This release and spatial distribution are also related to the prevailing climatic conditions. As an example, an increase in temperature leads to decreased soil water content in addition to higher levels of evapotranspiration, which increases resuspension of soil dust particles and has additional influence on the redistribution of heavy metals among the different media (Jeričević *et al.*, 2011).

17.3 Methods for the Removal of Heavy Metals From Soil

Two commonly used methods for heavy metals removal from soil are solidification/stabilization (S/S) technology and soil washing (Fig. 17.1).

The standard approaches of mediating the heavy metal in soil usually involve removal of the contaminated soil, followed by the use of S/S technology to fix the contaminant. The materials are disposed either on site after treatment or transported to a certified landfill (Wang *et al.*, 2025). However, the application of S/S remedial technology is becoming unfeasible when it comes to solving stubborn environmental issues because of a number of constraints. For example, heavy metals are not effectively isolated in the contaminated medium during the S/S process, and thus there must be continuous on-site monitoring of the levels of heavy metals. Moreover, the sustainability of S/S materials in the long run is not certain, leading to concerns over the reliability of such materials in the long term. Furthermore, S/S-treated contamination is treated by materials that need regular maintenance such as soil caps to suppress soil erosion, which further complicates the landfill management procedures. Thus, there is a pressing need to seek new technologies for soil remediation to curb the challenges of heavy-metal-contaminated soils (Petrillo *et al.*, 2025).

Soil washing is one promising approach that can be used to remediate polluted soils in the long term. This technique involves the use of physical or chemical means of washing to eliminate heavy metals in polluted soils. The method tries to identify soil matrices that have been affected as a result of industrial and military activities, such as recycling sites and shooting ranges among others. However, it is not used for treating soils contaminated with radioactive metals or organic pollutants. The two major techniques of soil washing are physical separation and chemical separation. Physical separation uses differences in physical properties such as magnetism, density, particle

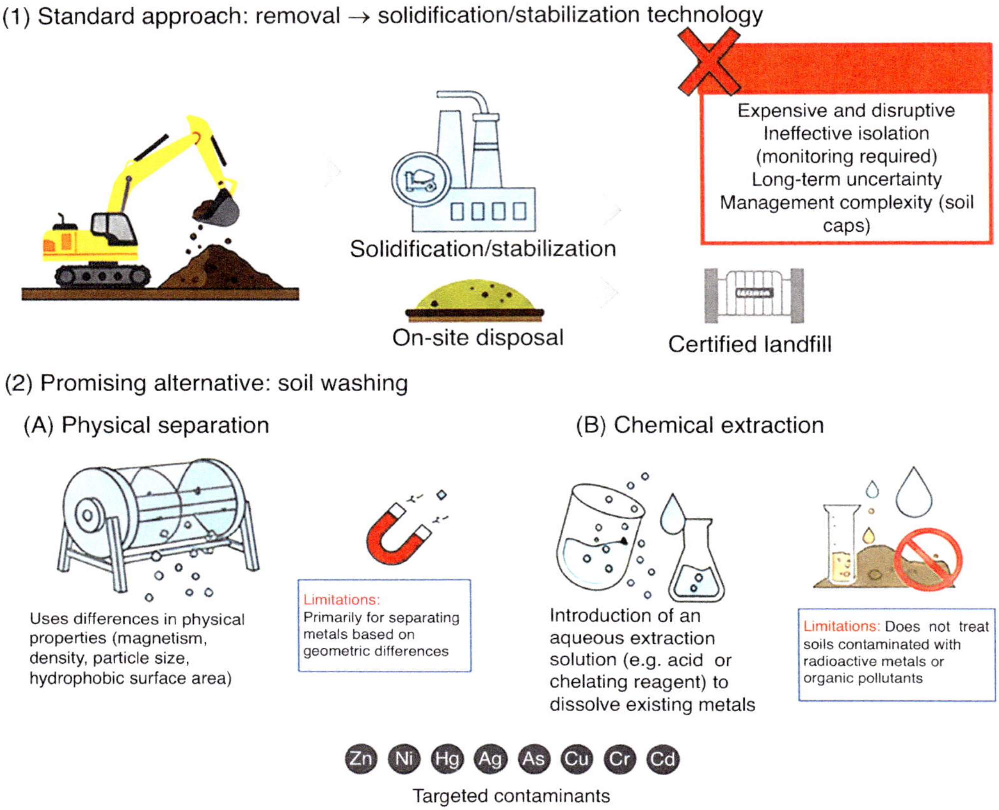

Fig. 17.1. Standard and alternate methods for the removal of heavy metals from soil. Ag, silver; As, arsenic; Cd, cadmium; Cr, chromium; Cu, copper; Hg, mercury; Ni, nickel; Zn, zinc.

size, and hydrophobic surface area in order to remove small amounts of metal contaminants from the soil. Any of these geometrical differences make the metal-carrying particles and the surrounding soil easy to separate using specific mechanisms. Chemical extraction involves the addition of agents such as chelators or acids to dissolve the existing metals in the contaminated soils and form an extractable aqueous solution (Zheng *et al.*, 2022).

These strategies coupled with others will help in the development of effective remediation mechanisms that will be able to correct a number of contamination states. Methods of soil washing will require field testing and practice tests, which are carried out on sites polluted with zinc, lead, nickel, mercury, arsenic, copper, chromium, and cadmium, to determine whether they can be adopted in the real world and utilized on a large scale to clean the environment and ensure its sustainability, including soil management (Dermont *et al.*, 2008).

17.3.1 SLs inhibit accumulation of heavy metals in plants

SLs play a vital role in chromium reduction in plants. Recent research has confirmed their importance in the growth and resistance of plants to varied stresses using numerous mechanisms. SLs are active stimulators of physiological process with a significant role in the adaptation of plants. SLs can also significantly increase root length, an important component for increased uptake of water and nutrients. This is complemented by a significant increase in plant weight, chlorophyll content, and protein levels, which is essential to ensure vigorous plant growth and production on a large scale (Younis *et al.*, 2024). In addition to stimulating plant growth, SLs increases the self-defense mechanisms of the plant against heavy-metal toxicity by promoting the production of stress-related proteins and other compounds, and thus the plant can withstand the enhanced photo-oxidative stress of the plants caused by to chromium toxicity (Rai *et al.*, 2021; Raja *et al.*, 2023). The protective effect is observed in the quantified reduction of reactive oxygen species (ROS) such as hydrogen peroxide (H_2O_2) and malondialdehyde (MDA), and superoxide dismutase (SOD) activity, which is reduced due to chromium stress (Jomova *et al.*, 2023). It has been noted that there is increased activity of SLs in association with de-ashed biochar (Younis *et al.*, 2024). Such a synergistic relationship enhances the effects of SLs even further, leading to even greater increases in root length, plant dry weight, and protein content. The joint influence not only enhances the growth and hardiness of plants but also offers a sustainable and efficient remedial method for chromium-contaminated soils (Younis *et al.*, 2024).

17.3.2 Effects of SLs on the properties of soil

SLs have been increasingly identified to be of importance in affecting soil properties, especially through their interactions with the soil microorganisms. One of the first studies to show that SLs are host-derived signals that arbuscular mycorrhizal fungi (AMF) respond to was conducted by Akiyama *et al.* (2005) and showed that SLs stimulated the germination of spores and branching of hyphae. This observation associated SLs with recruitment of symbiotic fungi, thereby enhancing soil nutrient dynamics. Specifically, plants enhance (indirectly) colonization by different fungi, thus altering nutrient solubility in the soil such as that of phosphorus, affecting its assimilation. These microorganisms thus alter the nutrient status of the rhizosphere, which is a significant soil characteristic in stress conditions.

Later studies looked at the ability of SLs to change the composition of microbial communities in soil. For example, Besserer *et al.* (2006) demonstrated that SL exudation had the potential to enhance AMF activity in soil microcosms by a significant margin. This increased microbial activity changes soil enzymatic activities and leads to enhanced mineral cycling and organic matter decomposition. Such changes represent direct improvements in soil functionality, creating conditions more favorable for plant growth under stress. More recently, research has expanded to consider the effect of SLs on heavy-metal-contaminated soils (Shah and Daverey, 2021). SL analogs applied to rice under cadmium stress influenced root exudation patterns, which in turn altered cadmium availability in the soil solution. By affecting the chemical form of metals and their mobility in soils, SLs indirectly help detoxify the rhizosphere. This highlights the important regulatory role of SLs in soil chemistry under conditions of environmental stress (Mitra *et al.*, 2021). Beyond fungal associations, SLs also shape broader rhizosphere communities. Xie *et al.* (2010) described SLs as multi-functional signals that can impact rhizobacterial populations. By favoring beneficial microbes, SLs indirectly influence soil pH stability, enzymatic activity, and nutrient turnover. Collectively, these microbial shifts contribute to improved soil structure and long-term fertility.

17.4 Mechanisms of Strigolactone Action

SLs are capable of reducing heavy-metal toxicity by multiple mechanisms, as described in Table 17.1. SLs mitigate the damage to photosynthesis, increase osmolyte production, and improve the antioxidant defense system of the plant. Additionally, SLs are able to maintain plant water and nutrient uptake, and are also able to regulate stomatal conductance.

17.4.1 Impact of biogenic SLs on plant growth under heavy-metal stress

Numerous experimental studies support the role of SLs in mitigating growth inhibition under

Table 17.1. Impact of strigolactone (SL) application on plant performance under heavy-metal stress.

Mechanism of action	Response	Metal/plant	Reference
Photosynthesis protection	SLs help to preserve photosynthetic pigments and maintain gas exchange, mitigating heavy-metal-induced damage to the photosynthetic machinery	Cadmium	Mehmood *et al.* (2025)
Antioxidant defense	SLs stimulate the synthesis and activity of antioxidant enzymes, which helps to scavenge reactive oxygen species and reduce the oxidative stress caused by heavy metals	Cadmium, lead/*Artemisia annua*, *Arabidopsis thaliana*	Wani *et al.* (2023)
Root architecture	SLs can reshape root morphology, promote deeper root growth, and increase the efficiency of water and nutrient uptake, which is beneficial under stress conditions	Arsenic/*Arabidopsis*, *Pisum sativum*, *Oryza sativa*	Pandey *et al.* (2016)
Stomatal regulation	SLs regulate stomatal apertures, helping to control water loss while ensuring sufficient gas exchange for photosynthesis	Wheat	Guo *et al.* (2023)
Osmolyte accumulation	SLs induce the production of osmolytes, such as proline, which protect against cellular damage and maintain water balance under stress	*Gossypium arboreum*/*O. sativa*/*A. thaliana*	Dong *et al.* (2025)
Heavy-metal detoxification/ sequestration	SLs enhance phytochelatin/ metallothionein synthesis and vacuolar sequestration	Arsenic/*A. thaliana*	Geng *et al.* (2023)
Interaction with other hormones	SLs interact with other plant hormones such as auxin, jasmonic acid, and abscisic acid to coordinate the plant's overall stress response	Cadmium, arsenic/ *Gossypium hirsutum*, *Sorghum bicolor*, *Zea mays*, *Panicum miliaceum*	Bhoi *et al.* (2021)

heavy-metal stress. Ha *et al.* (2014), demonstrated that the *Arabidopsis* SL mutant *max4* exhibited greater sensitivity to cadmium stress compared with wild-type plants. The mutants had reduced root and shoot biomass, whereas exogenous application of a synthetic SL analog (GR24) alleviated these symptoms, suggesting that endogenous SLs are critical in sustaining growth under toxicity. In rice, it has been reported that SL analog treatment improved plant height and biomass under cadmium exposure. Treated plants showed more extensive root systems and less chlorosis compared with untreated controls. These improvements were attributed to changes in root exudation and nutrient uptake, highlighting the dual role of SLs in root growth and soil interaction (Cai *et al.*, 2020; Balliu *et al.*, 2021). Studies also suggest that SLs interact with other hormones to regulate growth under stress. Kapulnik *et al.* (2011) found that SLs modulate auxin transport, thereby shaping root architecture in ways that enhance plant tolerance to unfavorable conditions. This regulatory effect becomes especially beneficial under heavy-metal stress, where nutrient imbalance and toxicity can restrict growth.

17.4.2 Impact of biogenic SLs on photosynthesis under heavy-metal stress

Evidence from multiple studies confirms that SLs influence photosynthetic resilience under heavy-metal stress (Mehmood *et al.*, 2025). It has been reported that application of GR24 to cadmium-stressed wheat plants preserved chlorophyll content and photosystem II efficiency compared with untreated controls. This suggests that SLs directly or indirectly stabilize photosynthetic pigments and machinery under toxic conditions (Zulfiqar and Ashraf, 2021). In rice, exogenous SL analog application improved chlorophyll fluorescence parameters and increased net photosynthetic rates under cadmium exposure. The treated plants demonstrated better gas exchange and reduced stomatal closure, reflecting a key role of SLs in regulating stomatal function during stress. These findings align with earlier results indicating hormonal cross-talk between SLs and ABA in stomatal regulation (Wani *et al.*, 2023). Further evidence was provided by Ha *et al.* (2014), who found that *Arabidopsis* SL-deficient mutants exhibited greater photoinhibition and chlorophyll loss under stress. The mutants also accumulated more ROS in chloroplasts, while GR24-treated plants displayed improved protection of photosystems. This indicates a significant role of SLs in safeguarding the photosynthetic apparatus against oxidative damage induced by heavy metals.

17.4.3 Impact of biogenic SLs on mineral nutrition under heavy-metal stress

Research findings provide strong evidence that SLs help sustain mineral nutrition under heavy-metal stress. Akiyama *et al.* (2005) first showed that SLs act as host-derived signals for AMF, initiating hyphal branching and symbiotic colonization. This discovery linked SL secretion with improved nutrient acquisition, particularly phosphorus uptake, in soils where availability is limited by metal contamination. Subsequent studies confirmed that enhanced AMF colonization improves overall plant nutrient balance. Yoneyama *et al.* (2012) demonstrated that phosphorus deficiency strongly stimulates SL biosynthesis and exudation. Under heavy-metal stress, phosphorus availability is typically restricted due to metal-induced precipitation, making SL-mediated mycorrhizal signaling crucial for nutrient acquisition. This adaptive increase in SLs helps plants recruit fungal partners that bypass the limitations of metal-affected soils (Mostofa *et al.*, 2021).

Synthetic SL analogs have an impact on cadmium-stressed rice plants in terms of improvement of nutrient status, particularly for phosphorus and potassium. This improvement is attributed to modified root exudation patterns and enhanced nutrient transport, indicating that SLs directly or indirectly promote nutrient uptake efficiency under toxic conditions (Mostofa *et al.*, 2021; Liao *et al.*, 2024). Additional evidence from Kapulnik *et al.* (2011) suggested that SLs regulate root-hair elongation, further enhancing nutrient absorption capacity. Together, these studies show that SLs protect mineral nutrition by adjusting root morphology, stimulating symbiotic associations, and maintaining nutrient homeostasis. This role is central to plant resilience, as balanced nutrition is fundamental for sustaining growth and physiological processes in contaminated soils.

17.4.4 Impact of biogenic SLs on antioxidant enzymes under heavy-metal stress

Experimental studies have shown clear links between SL signaling and antioxidant enzyme regulation under heavy-metal stress. Cadmium-stressed rice plants treated with the SL analog GR24 reported significantly higher activities of SOD, catalase (CAT), and ascorbate peroxidase (APX) compared with untreated plants. These enhancements were associated with reduced oxidative damage, improved photosynthetic performance, and greater tolerance to cadmium toxicity (Wani *et al.*, 2023). GR24 application improves antioxidant defense systems under cadmium stress. Treated plants had elevated levels of both enzymatic (SOD, CAT, APX) and non-enzymatic (glutathione, ascorbate) antioxidants. This comprehensive protection reduced lipid peroxidation and maintained chlorophyll stability, directly linking SL signaling to oxidative stress mitigation (Duan *et al.*, 2025; Mehmood *et al.*, 2025). Additional evidence comes from *Arabidopsis* studies. Ha *et al.* (2014) observed that SL-deficient mutants (*max4*) accumulated higher levels of H_2O_2 and MDA under stress, markers of oxidative damage. In contrast, wild-type plants or mutants treated with GR24 showed improved antioxidant capacity, indicating that endogenous SLs are essential for maintaining ROS homeostasis.

17.4.5 Impact of biogenic SLs on gene regulation under heavy-metal stress

Research has provided evidence that SLs regulate gene expression under stress, including heavy-metal toxicity. Ueda and Kusaba (2015) reviewed the broad regulatory role of SLs and noted their involvement in transcriptional reprogramming of stress responses. This includes the activation of stress-responsive transcription factors and genes controlling shoot branching and root architecture, both of which are vital for adapting to toxic soils. Ha *et al.* (2014) used SL-deficient and SL-response (*max*) mutants in *Arabidopsis* and found that gene expression of drought- and salt-responsive genes varied. The interactions of SLs with heavy metals was not investigated in their study, but their findings show that there must exist an active route through which SLs can act as transcriptional regulators of the stress response. The similar way in which the responses occur to different abiotic stresses indicate that SLs activate the same defense genes that are relevant in tolerance response to heavy metals. For example, Liu *et al.* (2013) showed that SLs are able to respond to ABA signaling to control gene expression during abiotic stress conditions. These hormones have cross-sensitivity to the gene related to stomatal control and ROS scavenging and root development. As heavy-metal stress also activates these pathways, it can be assumed that SLs also provide transcriptional regulation under heavy-metal stress, enhancing the resilience of plants (Khalid *et al.*, 2024).

17.5 Conclusion

SLs have been identified as playing a key role in mitigating the effects of heavy-metal stress and enhancement of soil–plant interactions over the last 20 years, and this has become one of the major themes of plant biology. Study results indicate a dual role of SLs in facilitating plant growth while also being utilized by the environment to regulate the interactions between soil and microbes. Such duality renders them unavoidable in the context of stability of a given ecosystem, particularly with rising industrial pollution and global climate change. Nevertheless, the use of SLs in soil management and agriculture is still in its infancy, although it holds special promise. Transplantation of laboratory results into viable field-based policies for sustainable agricultural practices is an area for future research. SLs are considered one of the new directions for plant–soil biology. Not only do they protect plants against heavy-metal poisoning, but they also enhance the fertility of the soil in the long-term and increase microbial diversity. Future utilization of the potential restructure resilience of crops in polluted or degraded environments will depend on interdisciplinary research in the areas of soil science, molecular biology, biotechnology, and ecological engineering.

References

Akiyama, K., Matsuzaki, K. -i. and Hayashi, H. (2005) Plant sesquiterpenes induce hyphal branching in arbuscular mycorrhizal fungi. *Nature* 435(7043), 824–827. DOI: 10.1038/nature03608.

Bajguz, A. and Piotrowska-Niczyporuk, A. (2023) Biosynthetic pathways of hormones in plants. *Metabolites* 13(8), 884. DOI: 10.3390/metabo13080884.

Balliu, A., Zheng, Y., Sallaku, G., Fernández, J.A., Gruda, N.S. *et al.* (2021) Environmental and cultivation factors affect the morphology, architecture and performance of root systems in soilless grown plants. *Horticulturae* 7(8), 243. DOI: 10.3390/horticulturae7080243.

Besserer, A., Puech-Pagès, V., Kiefer, P., Gomez-Roldan, V., Jauneau, A. *et al.* (2006) Strigolactones stimulate arbuscular mycorrhizal fungi by activating mitochondria. *PLOS Biology* 4(7), e226. DOI: 10.1371/journal.pbio.0040226.

Bhoi, A., Yadu, B., Chandra, J. and Keshavkant, S. (2021) Contribution of strigolactone in plant physiology, hormonal interaction and abiotic stresses. *Planta* 254(2), 28. DOI: 10.1007/s00425-021-03678-1.

Biswas, B., Qi, F., Biswas, J.K., Wijayawardena, A., Khan, M.A.I. *et al.* (2018) The fate of chemical pollutants with soil properties and processes in the climate change paradigm—a review. *Soil Systems* 2(3), 51. DOI: 10.3390/soilsystems2030051.

Cachada, A., Rocha-Santos, T. and Duarte, A.C. (2018) Soil and pollution: An introduction to the main issues. In: Duarte, A.C., Cachada, A. and Rocha-Santos, T. (eds) *Soil Pollution: From Monitoring to Remediation*. Academic Press, London, pp. 1–28. DOI: 10.1016/B978-0-12-849873-6.00001-7.

Cai, Y., Zhang, S., Cai, K., Huang, F., Pan, B. *et al.* (2020) Cd accumulation, biomass and yield of rice are varied with silicon application at different growth phases under high concentration cadmium-contaminated soil. *Chemosphere* 242, 125128. DOI: 10.1016/j.chemosphere.2019.125128.

Calvin, K., Dasgupta, D., Krinner, G., Mukherji, A., Thorne, P.W. *et al.* (2023) *IPCC, 2023: Climate Change 2023: Synthesis Report. Contribution of Working Groups I, II and III to the Sixth Assessment Report of the Intergovernmental Panel on Climate Change*. IPCC, Geneva, Switzerland.

Cook, C.E., Whichard, L.P., Turner, B., Wall, M.E. and Egley, G.H. (1966) Germination of witchweed (*Striga lutea* Lour.): Isolation and properties of a potent stimulant. *Science* 154(3753), 1189–1190. DOI: 10.1126/science.154.3753.1189.

Daignan-Fornier, S., Keita, A. and Boyer, F.-D. (2024) Chemistry of strigolactones, key players in plant communication. *Chembiochem* 25(12), e202400133. DOI: 10.1002/cbic.202400133.

Dermont, G., Bergeron, M., Mercier, G. and Richer-Laflèche, M. (2008) Soil washing for metal removal: A review of physical/chemical technologies and field applications. *Journal of Hazardous Materials* 152(1), 1–31. DOI: 10.1016/j.jhazmat.2007.10.043.

Dong, J., Fu, H., Wang, Z., Zhang, L., Liu, Z. *et al.* (2025) Mechanisms of strigolactone-regulated abiotic stress responses in plants. *Plants* 14(16), 2582. DOI: 10.3390/plants14162582.

Duan, S., Zhao, L., Chen, W., Zhang, Q., Ya, J. *et al.* (2025) Sowing methods and strigolactones alleviate damage to the photosynthetic system of rice seedlings under salt stress by enhancing antioxidant capacity. *Antioxidants* 14(8), 1020. DOI: 10.3390/antiox14081020.

Dunn, R., Blannin, J., Gobron, N., Miller, J., Willett, K. *et al.* (2024) Global climate in 2023. *Bulletin of the American Meteorological Society* 105(8), S12–S155. DOI: 10.1175/BAMS-D-24-0116.1.

Geng, A., Lian, W., Wang, X. and Chen, G. (2023) Regulatory mechanisms underlying arsenic uptake, transport, and detoxification in rice. *International Journal of Molecular Sciences* 24(13), 11031. DOI: 10.3390/ijms241311031.

Guercio, A.M., Palayam, M. and Shabek, N. (2023) Strigolactones: Diversity, perception, and hydrolysis. *Phytochemistry Reviews* 22(2), 339–360. DOI: 10.1007/s11101-023-09853-4.

Guo, S., Wei, X., Ma, B., Ma, Y., Yu, Z. *et al.* (2023) Foliar application of strigolactones improves the desiccation tolerance, grain yield and water use efficiency in dryland wheat through modulation of non-hydraulic root signals and antioxidant defense. *Stress Biology* 3(1), 54. DOI: 10.1007/s44154-023-00127-9.

Ha, C.V., Leyva-González, M.A., Osakabe, Y., Tran, U.T., Nishiyama, R. *et al.* (2014) Positive regulatory role of strigolactone in plant responses to drought and salt stress. *Proceedings of the National Academy of Sciences USA* 111(2), 851–856. DOI: 10.1073/pnas.1322135111.

Habibullah, M.S., Din, B.H., Tan, S.-H. and Zahid, H. (2022) Impact of climate change on biodiversity loss: Global evidence. *Environmental Science and Pollution Research International* 29(1), 1073–1086. DOI: 10.1007/s11356-021-15702-8.

Isele, T.O., Alabi, B.E. and Aderogba, K.A. (2025) Industrial revolutions, industrialization, and impacts on human environment. In: Agbai, E., Esmail, A., Okoro, H.M., Udeorah, S. and Hamann, B. (eds) *Twenty-First Century Issues: Perspective on Climate Change, Pandemic, and Digital Transformation*. Bloomsbury Publishing, London, pp. 61–92. DOI: 10.5040/9780761880493.

Jagaba, A.H., Lawal, I.M., Birniwa, A.H., Affam, A.C., Usman, A.K. *et al.* (2024) Sources of water contamination by heavy metals. In: Jaafar, J., Zaidi, A.A. and Naseer, M.N. (eds) *Membrane*

Technologies for Heavy Metal Removal from Water. CRC Press, Boca Raton, Florida, pp. 3–27. DOI: 10.1201/9781003326281-2.

Jeričević, A., Ilyin, I. and Vidič, S. (2011) Modelling of heavy metals: Study of impacts due to climate change. In: Fernando, H.J.S., Klaić, Z. and McCulley, J.L. (eds) *National Security and Human Health Implications of Climate Change*. Springer, Dordrecht, Netherlands, pp. 175–189. DOI: 10.1007/978-94-007-2430-3_15.

Jomova, K., Raptova, R., Alomar, S.Y., Alwasel, S.H., Nepovimova, E. *et al.* (2023) Reactive oxygen species, toxicity, oxidative stress, and antioxidants: Chronic diseases and aging. *Archives of Toxicology* 97(10), 2499–2574. DOI: 10.1007/s00204-023-03562-9.

Kapulnik, Y., Delaux, P.-M., Resnick, N., Mayzlish-Gati, E., Wininger, S. *et al.* (2011) Strigolactones affect lateral root formation and root-hair elongation in *Arabidopsis*. *Planta* 233(1), 209–216. DOI: 10.1007/s00425-010-1310-y.

Khalid, M.F., Shafqat, W., Khan, R.I., Jawaid, M.Z., Hussain, S. *et al.* (2024) Unveiling the resilience mechanism: Strigolactones as master regulators of plant responses to abiotic stresses. *Plant Stress* 12, 100490. DOI: 10.1016/j.stress.2024.100490.

Liao, Y., Huang, S., Hareem, M., Hussain, M.B., Alarfaj, A.A. *et al.* (2024) Addressing cadmium stress in rice with potassium-enriched biochar and *Bacillus altitudinis* rhizobacteria. *BMC Plant Biology* 24(1), 1084. DOI: 10.1186/s12870-024-05793-z.

Liao, Z., Chen, B., Boubakri, H., Farooq, M., Mur, L.A.J. *et al.* (2025) The regulatory role of phytohormones in plant drought tolerance. *Planta* 261(5), 98. DOI: 10.1007/s00425-025-04671-8.

Liu, J., Novero, M., Charnikhova, T., Ferrandino, A., Schubert, A. *et al.* (2013) CAROTENOID CLEAVAGE DIOXYGENASE 7 modulates plant growth, reproduction, senescence, and determinate nodulation in the model legume *Lotus japonicus*. *Journal of Experimental Botany* 64(7), 1967–1981. DOI: 10.1093/jxb/ert056.

Maftouh, A., El Fatni, O., Ben Moussa, A., Boukir, F. and Subha, W. (2024) Heavy metals in the ecosystem; sources and their effects. In: Kumar, N. (ed.) *Heavy Metal Remediation: Sustainable Nexus Approach*. Springer, Cham, Switzerland, pp. 27–44. DOI: 10.1007/978-3-031-53688-5_2.

Masindi, V. and Muedi, K.L. (2018) Environmental contamination by heavy metals. *Heavy Metals* 10(4), 115–133. DOI: 10.5772/intechopen.76082.

Mehmood, I., Wani, K.I. and Aftab, T. (2025) Role of strigolactones in heavy metal tolerance: A case study on cadmium. In: Aftab, T. and Wani, K.I. (eds) *Strigolactones: Emerging Plant Hormones*. Scrivener Publishing, Beverly, Massachusetts, pp. 137–159. DOI: 10.1002/9781394302826.

Mitra, D., B.E., G.S., Khoshru, B., De Los Santos Villalobos, S., Belz, C. *et al.* (2021) Impacts of arbuscular mycorrhizal fungi on rice growth, development, and stress management with a particular emphasis on strigolactone effects on root development. *Communications in Soil Science and Plant Analysis* 52(14), 1591–1621. DOI: 10.1080/00103624.2021.1892728.

Mostofa, M.G., Ha, C.V., Rahman, M.M., Nguyen, K.H., Keya, S.S. *et al.* (2021) Strigolactones modulate cellular antioxidant defense mechanisms to mitigate arsenate toxicity in rice shoots. *Antioxidants* 10(11), 11. DOI: 10.3390/antiox10111815.

Muhammad Fasih, K., Shafqat, W., Khan, R.I., Jawaid, M.Z., Hussain, S. *et al.* (2024) Unveiling the resilience mechanism: Strigolactones as master regulators of plant responses to abiotic stresses. *Plant Stress* 12, 100490. DOI: 10.1016/j.stress.2024.100490.

Muluneh, M.G. (2021) Impact of climate change on biodiversity and food security: A global perspective—a review article. *Agriculture & Food Security* 10(1), 36. DOI: 10.1186/s40066-021-00318-5.

Nakai, J. (2018) Food and agriculture organization of the United Nations and the sustainable development goals. *Sustainable Development* 22. Available at: www.okayama-u.ac.jp/user/kouhou/ebulletin/contribution/vol22/contribution_001.html (accessed 5 February 2026).

Oishy, M.N., Shemonty, N.A., Fatema, S.I., Mahbub, S., Mim, E.L. *et al.* (2025) Unravelling the effects of climate change on the soil–plant–atmosphere interactions: A critical review. *Soil & Environmental Health* 3(1), 100130. DOI: 10.1016/j.seh.2025.100130.

Pandey, A., Sharma, M. and Pandey, G.K. (2016) Emerging roles of strigolactones in plant responses to stress and development. *Frontiers in Plant Science* 7, 434. DOI: 10.3389/fpls.2016.00434.

Petrillo, A., Fraternali, F., Acampora, A., Di Chiara, G., Colangelo, F. *et al.* (2025) Innovative solidification and stabilization techniques using industrial by-products for soil remediation. *Applied Sciences* 15(7), 4002. DOI: 10.3390/app15074002.

Rai, K.K., Pandey, N., Meena, R.P. and Rai, S.P. (2021) Biotechnological strategies for enhancing heavy metal tolerance in neglected and underutilized legume crops: A comprehensive review. *Ecotoxicology and Environmental Safety* 208, 111750. DOI: 10.1016/j.ecoenv.2020.111750.

Raja, V., Qadir, S.U., Kumar, N., Alsahli, A.A., Rinklebe, J. *et al.* (2023) Melatonin and strigolactone mitigate chromium toxicity through modulation of ascorbate–glutathione pathway and gene expression in tomato. *Plant Physiology and Biochemistry* 201, 107872. DOI: 10.1016/j.plaphy.2023.107872.

Santner, A. and Estelle, M. (2010) The ubiquitin–proteasome system regulates plant hormone signaling. *Plant Journal* 61(6), 1029–1040. DOI: 10.1111/j.1365-313X.2010.04112.x.

Shah, V. and Daverey, A. (2021) Effects of sophorolipids augmentation on the plant growth and phytoremediation of heavy metal contaminated soil. *Journal of Cleaner Production* 280, 124406. DOI: 10.1016/j.jclepro.2020.124406.

Shetty, S.S., Deepthi, D., Harshitha, S., Sonkusare, S., Naik, P.B. *et al.* (2023) Environmental pollutants and their effects on human health. *Heliyon* 9(9), e19496. DOI: 10.1016/j.heliyon.2023.e19496.

Tiecher, T.L., Ceretta, C.A., Tiecher, T., Ferreira, P.A.A., Nicoloso, F.T. *et al.* (2016) Effects of zinc addition to a copper-contaminated vineyard soil on sorption of Zn by soil and plant physiological responses. *Ecotoxicology and Environmental Safety* 129, 109–119. DOI: 10.1016/j.ecoenv.2016.03.016.

Ueda, H. and Kusaba, M. (2015) Strigolactone regulates leaf senescence in concert with ethylene in *Arabidopsis*. *Plant Physiology* 169(1), 138–147. DOI: 10.1104/pp.15.00325.

Wang, M.-H.S., Wang, L.K. and Shammas, N.K. (2025) Advances in the remediation of metal finishing brownfield sites. In: Wang, L.K., Wang, M.-H.S., Hung, Y.-T. and Chen, J.P. (eds) *Control of Heavy Metals in the Environment*, Vol. 3. CRC Press, Boca Raton, Florida, pp. 517–576. DOI: 10.1201/9781003541615-13.

Wani, K.I., Naeem, M., Khan, M.M.A. and Aftab, T. (2023) Insights into strigolactone (GR24) mediated regulation of cadmium-induced changes and ROS metabolism in *Artemisia annua*. *Journal of Hazardous Materials* 448, 130899. DOI: 10.1016/j.jhazmat.2023.130899.

Xie, X., Yoneyama, K. and Yoneyama, K. (2010) The strigolactone story. *Annual Review of Phytopathology* 48(1), 93–117. DOI: 10.1146/annurev-phyto-073009-114453.

Yan, W., Lu, Y., Guo, L., Liu, Y., Li, M. *et al.* (2024) Effects of drought stress on photosynthesis and chlorophyll fluorescence in blue honeysuckle. *Plants* 13(15), 2115. DOI: 10.3390/plants13152115.

Xu, Y., Seshadri, B., Bolan, N., Sarkar, B., Ok, Y.S. *et al.* (2019) Microbial functional diversity and carbon use feedback in soils as affected by heavy metals. *Environment International* 125, 478–488. DOI: 10.1016/j.envint.2019.01.071.

Xu, J., Li, L., Liu, Y., Yu, Y., Li, H. *et al.* (2023) Molecular and physiological mechanisms of strigolactones-mediated drought stress in crab apple (*Malus hupehensis* Rehd.) seedlings. *Scientia Horticulturae* 311, 111800. DOI: 10.1016/j.scienta.2022.111800.

Yang, Y., Gu, M., Chen, J., Zhang, R., Liu, Z. *et al.* (2023) Comparative transcriptomes reveal the mitigation effect of GR24 in alfalfa under drought stress. *Journal of Plant Growth Regulation* 42(5), 3150–3161. DOI: 10.1007/s00344-022-10779-y.

Yoneyama, K., Xie, X., Kim, H.I., Kisugi, T., Nomura, T. *et al.* (2012) How do nitrogen and phosphorus deficiencies affect strigolactone production and exudation? *Planta* 235(6), 1197–1207. DOI: 10.1007/s00425-011-1568-8.

Younis, U., Danish, S., Datta, R., Alahmadi, T.A. and Ansari, M.J. (2024) Sustainable remediation of chromium-contaminated soils: Boosting radish growth with deashed biochar and strigolactone. *BMC Plant Biology* 24(1), 115. DOI: 10.1186/s12870-024-04791-5.

Zainon, M.R., Aziz, M.Y., Ishak, A.R., Abdullah, M.A., Tualeka, A.R. *et al.* (2025) Heavy metals pollution in drinking water sources: A case study from Kulim Hi-Tech Park, Malaysia. *Jurnal Kesehatan Lingkungan* 17(3), 256–267. DOI: 10.20473/jkl.v17i3.2025.256-267.

Zhang, X., Zhang, L., Sun, Y., Zheng, S., Wang, J. *et al.* (2020) Hydrogen peroxide is involved in strigolactone induced low temperature stress tolerance in rape seedlings (*Brassica rapa* L.). *Plant Physiology and Biochemistry* 157, 402–415. DOI: 10.1016/j.plaphy.2020.11.006.

Zheng, X.-J., Li, Q., Peng, H., Zhang, J.-X., Chen, W.-J. *et al.* (2022) Remediation of heavy metal-contaminated soils with soil washing: A review. *Sustainability* 14(20), 13058. DOI: 10.3390/su14201305.

Zulfiqar, F. and Ashraf, M. (2021) Bioregulators: Unlocking their potential role in regulation of the plant oxidative defense system. *Plant Molecular Biology* 105(1–2), 11–41. DOI: 10.1007/s11103-020-01077-w.

Index

Locators in *italics* and **bold** indicate figures and table, respectively.

Printed and bound by CPI Group (UK) Ltd, Croydon, CR0 4YY

18/08/2026

14944364-0002